VERIFICATION AND VALIDATION OF 3D FREE-SURFACE FLOW MODELS

SPONSORED BY
Task Committee on 3D Free-Surface Flow Model Verification and Validation

Environmental and Water Resources Institute (EWRI)
of the American Society of Civil Engineers

EDITED BY
Sam S. Y. Wang, Ph.D., P.E.
Patrick J. Roche, Ph.D.
Richard A. Schmalz, Jr., Ph.D.
Yafei Jia, Ph.D.
Peter E. Smith, Ph.D., P.E.

ASCE

EWRI

Published by the American Society of Civil Engineers

Library of Congress Cataloging-in-Publication Data

Verification and validation of 3D free-surface flow models / sponsored by Task Committee on 3D Free-Surface Flow Model Verification and Validation, Environmental and Water Resources Institute (EWRI) of the American Society of Civil Engineers ; edited by Sam S. Y. Wang … [et al.].

p. cm.

Includes bibliographical references and index.

ISBN 978-0-7844-0957-2

1. Fluid dynamics--Mathematical models. 2. Numerical calculations--Verification. I. Wang, S. Y. II. Environmental and Water Resources Institute (U.S.). Task Committee on 3D Free-Surface Flow Model Verification and Validation.

TA357.V47 2008
620.1'064015118--dc22 2008036309

American Society of Civil Engineers
1801 Alexander Bell Drive
Reston, Virginia, 20191-4400

www.pubs.asce.org

ISBN 978-0-7844-0957-2
Manufactured in the United States of America.

Preface

In recent years, more and more computational models for free surface flow simulations are needed in engineering analysis, design, and in making policy, planning and management decisions. Many computational models have been quickly developed and released to the clients without adequate verification and validation. As projects being planned and designed today have to include the considerations of multi-disciplinary interactions; their local, regional and even global effects; and short-/long-term impacts, the capability, quality and reliability of these computational models used as research, planning and design tools, are of great importance. Responding to these concerns model developers have applied various ways to verify the mathematical correctness and physical validity of the models which they have developed. Due to the lack of an established rigorous and systematic verification and validation process, some of the verification and validation tests found in the open literatures are from the Test of Symmetry and Mass/Volume Conservation to Tests of Translation/rotation of a Gussian Cone and Vortex Shedding, Tests of Analytic Verification and Grid Convergence, and the Validation by field data alone, just to list a few.

A method used most often has been the comparison of model results directly to the field measurements. Once agreement is obtained, either the model developer or the user would claim that the model is validated. Since it is so easy to do, this method has been widely adopted. Recently, more and more professionals have found out that the method was often irresponsibly abused and thus raised doubts about its adequacy and dependability. After careful examination, it was found that in many cases the agreement obtained was by "fine tuning" the model parameters exclusively. The fine-tuning is acceptable only during the model parameter calibration process, provided that it is conducted properly. After the calibration is completed, the values of the model parameters should not be changed to perform a validation test. The field data used during this validation test should not be the ones, which have been used already to do the calibration. Some modelers and users justified the use of the same set of field data for both calibration and validation by saying that because of the amount of field data measured is insufficient, so they have to accept the calibration as also validation. This is not acceptable. Furthermore, instances have also been found that the fine-tuning was used for the sole purpose of matching the modeled value to the field measurements. For example, some skillful fine-tuning expert have changed the flowfield, say stream or pathlines, by adjusting the Manning's n-values or bed friction factors of the depth-averaged flow models, just for the sake of obtaining a flowfield of his/her liking. This is wrong.

The ASCE/EWRI Task Committee is actually a defacto international committee with members representing academic, governmental and private research institution from six countries. After having spent nearly 10 years, a comprehensive, rigorous and systematic three dimensional free surface flow model verification and validation procedure has been developed, which is presented in this report. It consists of three steps, namely the Mathematical (Code) Verification, Physical Process Validation, and Application site Validation. In the Mathematical Verification Step, the numerical model results are compared with a known analytic or prescribed or manufactured solution of the same boundary value problem. Since all physical, mathematical and geometric parameters are identical, neither calibration nor fine-tuning is needed. During the verification, if the discrepancies are found, they must be due to the calculation errors and/or model (code) mistakes. In addition, the order of convergence and error can be determined quantitatively. Therefore, this step is very important. In the Physical Process Validation, the model's capability of representing all the essential and basic physical processes of the problem is determined. The validation is confirmed, if the simulated results are in good agreement of the physical process with the measurements of the same process in laboratories. After the model is proven to be able to simulate all the essential and basic physical processes, it is then ready to go through the Application Site Validation. This last step requires two sub-steps: The first sub-step is the model parameter calibration, which is to insure that the unique and site-specific characteristics of a natural system are taken into account in the model. Without this calibration, the governing equations can not be assured to be realistic when applied to specific site applications. Then, the calibrated model is to be tested by comparing the simulated results with the rest of the collected data, which have not been used in performing calibration. A reasonable agreement is needed to confirm the success of the Application Site Validation.

It is important to call reader's attention to the two key points affecting the success of this validation step: (1) the field data collected must be sufficient in the amount and high quality in accuracy to meet the rigorous calibration and validation needs, and (2) the calibration and validation are required on a case to case basis at each particular site and from time to time, especially when the lapse of time from one case to the next is of significance and/or during which major hydrologic event(s) has happened. During each step of the tests, the tester is recommended to perform Calculation Verification to estimate the error accumulated during the process of calculations.

It is the Committee's belief, that the Verification and Validation Procedure for 3D Free Surface Flow Model developed by this committee has been proven to be the most comprehensive in the open literature. Therefore, the Committee recommends its adoption by professionals in the field to carry out a free surface flow model verification and validation before applying the model to the investigation of real life problems.

The Committee fully realizes, however, that the comprehensive and systematic verification and validation procedure as presented in this report is far from being perfected. Therefore, the researchers in the field are strongly encouraged to further advance the state of the art. Some of the areas need further advancements include (but not limited to): the method for analytic verification of non-linear models, such as the methods using prescribed solution forcing, manufactured solutions, and other new approaches; the systematic method for conducting calculation verification; additional test cases for physical process validation; more complete sets of sufficient quantity and high quality field data for application site validation, etc. Most of these advancements can be quite involved and require considerable effort and time to accomplish. Therefore, it is hoped that the publication of this report is just a step forward at the beginning of a long term task, and a stimulant to the fellow researchers to devote their wisdom, knowledge, expertise and energy to continue the march to a better system for more rigorous and comprehensive verification and validation of the three-dimensional free surface flow models in particular, and all computational models in general.

Sam S.Y. Wang
Committee Chair
and Editor

Contributing Authors

Carlos V. Alonso, Ph.D., M.ASCE
National Sedimentation Laboratory
USDA Agricultural Research Service, USA

Yan Ding, Ph.D., D.E., M.ASCE
National Center for Computational Hydroscience and Engineering
The University of Mississippi, USA

Wenrui Huang, Ph.D., P.E., M.ASCE
Department of Civil & Environmental Engineering
Florida State University, Tallahassee, FL, USA

Klaus-Peter Holz, Dr.Ing., M.IAHR
Institute fur Bauinformatik
Brandenburg Technical University Cottbus, Cottbus, Germany

Bernard B. Hsieh, Ph.D., M.ASCE
Coastal and Hydraulics Laboratory
U.S. Army Engineer Research and Development Center, Vicksburg, MS, USA

Roger A. Kuhnle, Ph.D., M.ASCE
National Sedimentation Laboratory
USDA Agricultural Research Service, USA

Gerald S. Janowitz, Ph.D.
Department of Marine, Earth, and Atmospheric Sciences
North Carolina State University, NC, USA

Yafei Jia, Ph.D., M.ASCE, M.IAHR
National Center for Computational Hydroscience and Engineering
The University of Mississippi, USA

Billy H. Johnson, Ph.D., P.E., M.ASCE, D.WRE
Computational Hydraulics and Transport LLC
Edwards, MS, USA

William K. Jones
PBS&J
Tallahassee, FL, USA

Mutsuto Kawahara, Ph.D., M.JSCE
Department of Civil Engineering
CHUO University, Tokyo, Japan

Toshio Kodama, Ph.D., M.JSCE
Numerical Analysis Department
INA Corporation, Tokyo, Japan

Peter Mewis, Dr.Ing., M.IAHR
Institute of Hydraulic Engineering
University of Technology, Darmstadt, Germany

Patrick J. Roache, Ph.D., M.ASCE
Consultant
Socorro, New Mexico, USA

Francisco J. Rueda, Ph.D.
Centro Andaluz del Medio Ambiente CEAMA
Universidad de Granada, Granada, Spain

Roberto Mayerle, Ph.D., M.ASCE
Coastal Research Laboratory
University of Kiel, Kiel, Germany

Steven C. McCutcheon, Ph.D., F.ASCE
U. S. Environmental Protection Agency
Athens, GA, USA

H.P. Morvan, Ph.D.
School of Civil Engineering
University of Nottingham, Nottingham, UK

Gilmar Rodriguez
Northwest Florida Water Management District
Havana, FL, USA

Richard A. Schmalz, Jr., Ph.D., P.E., F.ASCE
National Oceanic and Atmospheric Administration
National Ocean Service, Silver Spring, MD, USA

Mac Sisson, Ph.D.
Virginia Institute of Marine Sciences
Gloucester Point, VA, USA

Peter E. Smith, Ph.D., P.E., M.ASCE
U.S. Geological Survey
Sacramento, CA, USA

Malcolm L. Spaulding, Ph.D., P.E., M.ASCE
Department of Ocean Engineering
University of Rhode Island, Kingston, RI, USA

Harry V. Wang, Ph.D., M.ASCE
Virginia Institute of Marine Sciences
Gloucester Point, VA, USA

Sam S.Y. Wang, Ph.D., P.E., F.ASCE, D. WRE, M.IAHR
National Center for Computational Hydroscience and Engineering
The University of Mississippi, USA

Christian Winter, Ph.D.
Coastal Research Laboratory
University of Kiel, Kiel, Germany

Nigel G. Wright, Ph.D., M.IAHR
Environmental Fluid Mechanics, School of Civil Engineering
University of Nottingham, Nottingham, UK

Tien-Shuenn Wu, Ph.D., M.ASCE
Florida Department of Environmental Protection
Tallahassee, FL, USA

Weiming Wu, Ph.D., M.ASCE, M.IAHR
National Center for Computational Hydroscience and Engineering
The University of Mississippi, USA

Task Committee Members[1]

Sam S.Y. Wang, Ph.D., P.E., F.ASCE, D. WRE, M.IAHR, USA

Billy H. Johnson, Ph.D., P.E., M.ASCE, D.WRE, USA

John D. Wang, Ph.D., P.E., M.ASCE, USA

Malcolm L. Spaulding, Ph.D., P.E., M.ASCE, USA

Peter E. Smith, Ph.D., P.E., M.ASCE, USA

Joe V. Letter, Jr., M.ASCE, USA

Mutsuto Kawahara, Ph.D., M.JSCE, Japan

Klaus-Peter Holz, Dr. Ing. M. IAHR, Germany

Alan F. Blumberg, Ph.D., F.ASCE, USA

Wilhelm Bechteler, Dr. Ing. M. IAHR, Germany

Daniel R. Lynch, Ph.D., M.ASCE, USA

Dick P. Dee, Ph.D., M.ASCE, The Netherlands

George Mellor, Ph.D., M.ASCE, USA

Toshio Kodama, Ph.D., M.ASCE, M.JSCE, Japan

Tien-Shuenn Wu, Ph.D., M.ASCE, USA

Rainer Lehfeldt, Dr. Ing., M. IAHR, Germany

Richard A. Schmalz, Jr., Ph.D., P.E., F.ASCE, USA

Warren W. Bell, Ph.D., M.ASCE, Canada

Alan Davies, Ph.D. M. IAHR, UK

---

[1] The order is according to seniority.

Acknowledgments

The task committee on Three-dimensional Free Surface Flow Model Verification and Validation wishes to express their appreciation and admiration to the leadership of ASCE and EWRI. More than 10 years ago, these organizations had the vision to initiate this important project when the three-dimensional free surface flow modeling methodology was still at its infancy. Also, the task committee recognizes the strong support and infinite patience in making exceptions for extending the life of the task committee, allowing it to produce a quality work. Acknowledgment is also due to several chairs of the Computational Hydraulics Committee, the Hydraulics and Waterways Council, the Executive Committees of EWRI, and the Hydraulic Division of ASCE before EWRI was founded. Without their continuous support, understanding, patience, and encouragement this task and the resulting publication of the report would not have been possible.

In addition, both the long-time (grandfather) and newer members deserve credit because of their firm belief and commitment to completing this task of lasting value. These members especially merit recognition on account of their perseverance and dedicated work. As the chair of the task committee, I sincerely hope that they can finally be proud of, and satisfied by, their contributions to the profession.

A large number of nationally and internationally recognized experts in the field are also greatly responsible for the finished product, with their generous contributions of guidance, reviews, criticism, expertise, knowledge, and encouragement. Their support is highly appreciated.

The editor is deeply grateful to his Colleagues at the National Center for Computational Hydroscience and Engineering (NCCHE) at the University of Mississippi. The NCCHE staff contributed much painstaking effort to complete the detailed work, without which this report could not have been completed at such a high standard. Despite efforts to keep these acknowledgements relatively ambiguous (so as not to forget any single contributor), there are two individuals who truly deserve individual recognition. With this in mind the NCCHE Administrative Coordinator, Miss Janice Crow, and the Research Scientist and Technical Assistant to the Director, Dr. Tingting Zhu, merit special thanks for their dedicated contributions throughout this task and especially at the final stage of report completion.

Several other groups should be recognized for their vital assistance, including those who developed and contributed verification and validation test cases, the Chapter Editors and Co-editors, and the Members of the Editorial Sub-committee. Their names are given individually with each Chapter and Test Case. Obviously, the most important technical contributions are credited to them.

Myself and the other editors of this report offer a sincere apology for the length of time taken to finalize this report. We attempted to minimize errors in the editorial process but the editors assume full responsibility for any mistakes or imperfections that may have been overlooked. We would be very appreciative if readers could share any criticisms and advice for future improvements. In the future, the contributors and organizers of the project would be very willing to work with experts in the field to further advance the state of the art methodologies for verification and validation of three-dimensional free surface flow models, as well as computational models in general.

Sam S.Y. Wang
Chair of the Task Committee

Contents

CHAPTER 1

VERIFICATION AND VALIDATION OF FREE SURFACE FLOW MODELS

Sam S.Y. Wang

1.1 INTRODUCTION

With rapid advancement of computer's capacity and speed, and significant breakthrough in numerical solution methodology of problems governed by nonlinear partial differential equations in highly irregular solution domain in the past decades, more and more disciplines have been applying numerical models to scientific research, engineering analyses, project designs, impact assessments, planning, management and policy decisions. Because of their cost-effectiveness, numerical models are often used in multi-objective optimization and/or parametric trade-off analysis for decision making, project design of large and complex systems and predictions of the outcomes of "what if" cases of scenarios. This development has increased the attention to the validity and accuracy of numerical models. To meet this need, many numerical modelers have adopted a convenient validation approach, which is essentially based on the comparison of numerical model solutions to a limited number of field or laboratory data available at the time. Once a reasonable or good agreement is obtained, the model is claimed to have been validated, and the findings are, then, submitted to a professional journal and/or a professional conference for publication. If the paper reporting the findings of a numerical model with the so-called "validation" is accepted for publication either in a conference proceedings or a professional journal, the author of the numerical model usually claims that it has been fully accepted by the peers in the field.

Due to the fact that the good or reasonable agreement between the numerical solutions and the physical measurements in either laboratory or field could be usually obtained by adjusting (or tuning) the model parameters, professionals and their institutions have begun to express concern or doubt about the value of the so-called validation practice described above. More and more scientific researchers and model users have realized that such tuning should have been called calibration of numerical models' parameters, rather than validation. Some users, especially those without the model development experience, were so anxious to obtain results to meet the deadlines. They used the good agreements between simulated results and measured field data as the justification to openly declare that there is no need of validation, because they have the model calibrated. They can apply this calibrated model to any problem with full confidence. As more modeling studies continued to adopt this calibration only approach, its defects have begun to appear. Therefore, a better validation methodology is needed.

As an example, during a presentation at a conference, when the presenter was asked why his simulated particle traces of the flow around a river bendway were somewhat uniformly distributed rather than having most of those traces shifted to the outer (concaved) bank as he had observed in both field and laboratory. The presenter quickly apologized for not having time to adjust the Manning's *n*-value, a model parameter, to achieve more accurate results and added that if he had enough time to adjust the *n*-value, he could match his simulated result as close to the physical observation as he wanted. In fact, he should have explained that his results were obtained from a Depth-Averaged 2D Free Surface Flow Model and the particle traces plotted were based on the simulated depth-averaged velocities rather than the velocities on the free surface, which was observed by the person who raised the question. Therefore, the disagreements were justifiable, because in the particular case, the solution and the observation were not the same and should not be compared. Had he used a 3D Free Surface Flow Model, the simulated particle traces on the free surface should have been in agreement with those observed. This example is a reminder that the modeler should not arbitrarily tune the model parameters just for matching certain physical measurements, which was not intended to simulate by the model. More importantly, this example can serve as a warning to the model users that if the model matches the physical measurements for the wrong reasons, the model has not been validated, and it should not be adopted for the project.

As another example, in a review to select one model for simulating a river mouth flowfield, all five competing models claimed to have been validated because they presented simulated results in excellent agreement with the same set of field data provided by the prospect user. After careful examination, from additional information from the model developers, it was found that the Manning's *n*-values adopted by them were all different. The highest value was more than 10 times larger than the lowest. During question and answer session, some modelers admitted that in order to match their results well to the measured field data, they sometimes needed to adjust the "n-value" locally and repeatedly near the data points. From this example, it further demonstrated that if you are skillful and patient enough, you may be able to almost always "validate" your model by "calibrations" of model parameters, provided that you have the measured field data. This experience again revealed that one should not accept a model just because its simulations match field data well by parametric calibration only. All numerical models should go through an additional validation step after the parametric calibration.

Most modelers would not like to admit that their model could have any error, because they have checked it many many times. More often than not, the carefully "checked" and calibrated models, supported by the near perfect match between the simulations and the field observations, have been found to have errors or imperfections months or even years later. The errors or mistakes could have happened in the long process of mathematical derivations or manipulations, implementations of numerical solution schemes, applications of special features for speeding-up/stabilizing, calculation algorithms, coding, etc. Some of these errors or imperfections are hard to find, even by spending hours and days. It has become a common practice adopted by more and more modelers to test the results' symmetry, mass or volume conservation, rotation and translation of a Gaussian cone, etc. to debug or to verify the models, and then to calibrate or tune the model parameters to validate the model. It has been found years later that the models passed this kind of verification and validation approach still have errors and/or bugs, which have been concealed by the parametric tuning.

In order to lessen the amount of effort required for the modelers to fine tune the model parameters, especially in the case when a large number of model parameters have to be tuned one by one by trial and error scheme, a mathematical method has been applied. It is called "Parametric Identification." It is based on the principle of minimization of the discrepancies between the measured data and model solutions. By applying an optimization analysis the over-all or the sum of weighted average error between numerical model results and field data at all or selected data points can be minimized to obtain the values of the parameters. This method can indeed save time and effort in performing parametric tuning. But the model developers and users are recommended to carefully examine the values of those tuned parameters to make sure that they are physically reasonable, and only to use this powerful tool for parametric calibration. A rigorous model validation should be followed.

For insuring the quality and integrity of numerical modeling research and the aforementioned reasons, several professional journals have implemented changes in policy on the acceptance of the submitted papers reporting results of numerical modeling of physical systems. For example, the Journal of Fluids Engineering, of the American Society of Mechanical Engineers, officially published the following Editorial Statement:

"The Journal of Fluids Engineering will not accept for publication any paper reporting the numerical solution of a fluids engineering problem that fails to address the task of systematic truncation, error testing, and the accuracy estimation." (Roache et al., 1986; Freitas, 1993, 1995). This means that papers reporting modeling results of field problems supported by calibration only are not accepted for publications.

Even though the statement emphasizes numerical accuracy, the required tests can insure the numerical model's consistency and convergence as well as determine quantitatively the order of convergence. The tests covering these aspects are called the Mathematical (or Code) Verification by this Task Committee. It has clearly shown that the editorial board has emphasized the importance of mathematical (or code) verification of numerical models when they announced the new publication policy in 1986. Subsequently, more rigorous policies have been adopted by other leading professional journals as well. Two examples are: AIAA Journal, 1994, and ASME Heat Transfer Editorial Board, 1994.

It has been generally agreed that a rigorous mathematical test of a numerical model or mathematical (code) verification is an important and necessary first step, even though it is not sufficient. After a numerical code has been proven to be mathematically correct and without coding errors, one needs to find out whether it is capable of predicting the physics of fluid flow phenomena in laboratory and nature. Therefore, the validation of a numerical model's capability to simulate the correct physics and realistic phenomena of real-life problems has also gained attention in scientific research and engineering applications. One of such example is the MetOcean Modeling Program (MOMOP) sponsored by the OKN group, a consortium of 11 oil companies administered by the MOMOP Committee (Roed and Hackett, 1989). The purpose of MOMOP is to assess all available 3D physical oceanographic numerical models' capabilities and accuracy for application to the studies of ocean currents in the oil exploration field off-shore of Norway. More specifically, they wanted to insure that the numerical model to be selected for this study would be able to predict the flow-field in reasonable agreement with the measurements and allow interpolation and extrapolation continuously in space and time. In addition, they wanted the model to be capable of predicting the outcomes of alternative management and evaluation scenarios. Six modeling groups responded to the request for participation in this contest and submitted their results for five pre-designed test cases to the MOMOP Committee. After a careful evaluation, one of the six models, which produced the most satisfactory results, was accepted to carry out the studies.

The examples presented above are only two of numerous efforts conducted in the past two decades trying to gain confidence in numerical models' capability and quality before using it to study real-life problems. A comprehensive literature survey is not in the scope of this Task Committee's work.

In the late 1980's, several task groups were organized to develop means for verifying and validating three-dimensional free surface flow models. One such example was a task group organized by several hydraulic research institutes from several countries with Dr. Richard Dee as one of the major contributors. Their primary recommendation was to request the model developers or model developing institutions to conduct their own validation tests and publish a validation document for each model at the time of its release. One of the products of this task group was published by IAHR in June 1994, which was entitled "Guidelines for documenting the validity of computational modeling software" (Dee, 1994) with a preface written by Dr. Torkild Carstens, the President of IAHR (International Association of Hydraulic Research) at that time.

In 1989, a Task Committee on Data Collection for 3D Free Surface Hydrodynamic Model Verification was established under the Computational Hydraulics Committee (CHC) of the ASCE Hydraulics Division. It was planned as the Phase I of a more comprehensive, long-term task on 3D Free Surface Flow Model Verification. An interim report was completed entitled, "Data and Information Base for 3D Free Surface Flow Model Verifications" (Wang, 1993). Due to budgetary problems, the task committee's continuation was postponed for about two years. In 1995, it was reactivated as a subcommittee of the ASCE CHC without financial support. Several die-hard members kept the committee work going at a slower pace on their own initiatives. The effort was reinvigorated after the establishment of the Environment and Water Resources Institute (EWRI), as a part of the ASCE reorganization in 2001. Dr. Sam Wang was then invited to chair a new task committee under a new title, "The Task Committee on 3D Free Surface Flow Model Verification and Validation Monograph" with the primary task of the publication of a monograph. Since then, test cases based on analytic solutions, laboratory experiments and field data having been developed by the preceding committees were peer reviewed and revised. Additional test cases were developed, reviewed, revised and added to the suite of verification and validation test cases. All of the findings are included in this report. A brief outline of the background information of the task committee is given in this section. Additional information on the modeling of physical systems, the development of free surface flow models, the basic needs and means of verification and validation of free surface flow models, and the application of a systematic procedure for verification and validation of numerical models are briefly summarized in the following sections of this chapter.

1.2 MODELING OF PHYSICAL SYSTEMS

The continuing pursuit of better understanding of natural phenomena and more effective utilization of natural resources (while upgrading the ecological and environmental quality), has driven mankind to advance the forefront of sciences as well as to sharpen the tools of research and engineering throughout the history of human civilization. By careful observations of natural phenomena quantitatively with instrumentations and describing them by mathematics, many hypothesized physical laws and principles guiding the variation trends of natural phenomena have been developed quantitatively by laboratory experimentations and validated, at least qualitatively, by field observations. Even though these laws have been found and proven under highly idealized assumptions and simplifications, they have nevertheless enhanced our knowledge of the behavior of the nature.

In another significant advancement of basic and applied sciences, mathematical modeling has been developed and applied to better understand how to do these idealized and simplified natural or man-made systems respond to external forcings. The greatness of early philosophers, mathematicians and scientists was in their ability to idealize the hopelessly complex and seemingly random natural phenomena of real-world systems into simplified mathematical models, which can be solved analytically. The analytic solutions of highly simplified mathematical models have provided us with the predictive capability of systems' responses to various external forcings, at least under the idealized assumptions. This development further enhanced our understanding of nature, and our ability to utilize natural resources to benefit mankind. One beauty of nature is that the simplest illustrations often reveal the most fundamental characteristics of natural systems.

To better predict the response of systems under known external forcing, physical modeling methodology has been conducted in the laboratories by well-designed experiments utilizing the latest and sophisticated instrumentations. It has produced more forcing-response relations, commonly called empirical laws, which have also provided us with additional predictive capabilities. Because not all idealization, linearization, and simplifying assumptions needed by analytic solutions are required to develop the empirical laws by physical modeling, the physical modeling approach has been considered as more realistic. Consequently, they have been adopted by engineers to solve real-life problems more often than the highly idealized mathematical models. For many decades, engineering designs have relied on the scaled physical modeling to confirm the effectiveness of conceptual designs of experienced engineers. In fact, even at the present, a few less involved problems are still being studied or designed by this physical or scaled modeling methodology.

The need to enhance the mathematical model's capability to simulate more realistic physical systems has accelerated the development of numerical modeling methodologies. Several new families of numerical techniques, such as Finite Difference, Finite Element, Finite Volume and other similar methods have been developed, which are capable of converting the nonlinear partial differential equations into algebraic equivalents by discretizing the continuous domain. Rapid solution of a huge number of coupled algebraic systems is amendable to computer implementations. With the aid of highly efficient modern digital computers, more and more realistic physical systems can be simulated quickly. Details of the numerical modeling are to be presented in the next section.

As multiple objective optimization in system analysis and design has become more prevalent, numerical models have been widely adopted as the tool of preference not only for engineering analysis and design, but also for decision making in planning and management. For example, in modern engineering design, one needs to conduct a parametric trade-off analysis in order to identify a set of optimal design parameters, which is the "best" compromised choice to achieve the pre-selected multiple objectives while satisfying all the required constraints. Some of these objectives and constraints are cost-effectiveness, safety of the community, protection of the ecological and environmental quality, and adherence to the laws, interests of stakeholders of the society, etc. If physical modeling is used alone to conduct such an investigation, numerous tests of different combinations of a large set of design parameters must be studied sequentially to determine their short- and long-term effects. Such a study would often be too costly in terms of funds and time. A more cost-effective approach is a well-planned integration of physical modeling, numerical modeling, and field measurements. In this approach, numerical models are used as much as possible in order to reduce the time and costs of the investigation. Of course, only those models that have been systematically verified and validated for modeling the physical system in question should be used. Physical modeling and field measurement are used primarily to validate the numerical models' capability to represent all essential physical processes and predict the correct responses of real-life systems under various forcings. In the case of more important projects, a scaled physical model experiment may be used to reconfirm the effectiveness and reliability of the design or decision obtained by integrated modeling methodology based on physical and numerical models and field studies.

1.3 FREE SURFACE FLOW NUMERICAL MODEL DEVELOPMENT

From the open literature, one finds a general pattern that most of the numerical models for free surface flow simulations have been developed by faculty members with assistance from students at academic institutions. Generally, a model is developed by applying either a new or an improved numerical technique, such as Finite Difference, Finite Element, or Finite Volume, because of their demonstrated capability in solving systems of nonlinear partial differential equations in highly irregular domains. After the completion of the code development, the numerical model was usually "validated" by comparing its solutions to a set of laboratory data, and/or field measurements. After good agreement(s) is achieved, usually by adjusting or tuning physical and numerical model parameters, a paper reporting the development and "validation" is submitted for publication. The acceptance for publication in a professional journal or a conference proceedings was based on a claim that the model was both validated and accepted by the peers in the field.

Some of these models were later generalized and wrapped in a highly sophisticated and user-friendly Graphic User Interface (GUI), and released to the market. Due to the competition in the marketplace, some of the model developers might have been induced to exaggerate the capabilities of their product and neglect to detail their limitations. If one is not careful enough, he/she might be led to believe that these models can accurately solve a large number of different real-life problems in highly complicated natural or artificial systems. Some of the model developers even claimed that their models could be applied by users without much basic technical knowledge of the problem; instead they only need to simply answer a series of questions, then the computer will do the rest to produce desired simulations of complex geophysical systems. Complaints and dissatisfactions were heard frequently from the professionals in the field.

Numerous studies on CFD models' error estimations, quality assurance, grid convergence, verification, quantification of uncertainty, certification, calibration, and validation, have been reported. (Bobitt, 1988; Bredehoeft and Konikow, 1992; Celik, et. al. 1993; Ferziger and Peric, 1996; Gresho and Taylor, 1994; Johnson and Hughes, 1995; Melnik, et. al., 1995; Oberkampf, et. al., 1995; Oreskes, et. al., 1994; Roache, 1989,1997, 1998, 1999; Wang, 1991, 1992, 1994; Dee, 1991a, 1991b; Toro and Wang, 1993). Even though their findings and opinions varied widely, from "the numerical model of natural systems can't be validated," to "verification of a numerical solution's consistency and convergence is a must" to "one can always use calibrations to achieve good agreements, so there is no need to validate" to " a systematic verification and validation procedure is desirable." By and large, the majority opinion is for conducting both verification and validation by utilizing the best techniques and data available. Without this important step, the confidence and validity of numerical model results can not be established. Efforts of several groups trying to establish the quality standards of free surface flow numerical models have been mentioned in the Section 1.1. Here, the work of the ASCE Task Committee on 3D Free-Surface Flow Model Verification and Validation is to be briefly outlined.

Since its formation, the Committee membership has varied from 12 to 45 from year to year representing 6 countries leading in hydraulic research. The Committee has been meeting only once a year over approximately 9 active years, the years without ASCE supported meetings being omitted. The task committee seems to have been existed for a long time, but without the hard work of the faithful members using their own spare time, the findings compiled in this report could not have been possible. The discussions and deliberations during the early years of this Task Committee were diverse but interesting and stimulating. The diverse views and opinions on whether the Committee should hold a competition of existing models, offer certifications by putting a stamp of approval, or provide a test methodology, among others were discussed thoroughly. After having all the advantages and disadvantages evaluated in detail, the Committee finally decided to develop a methodology or procedure for both the model developers and model users to conduct a pragmatic verification and validation study of numerical models, so that they can confirm a model can be used in their application with confidence. The modelers can follow the methodology to verify the model's correctness as well as to validate its capability to represent the correct physics during the model development, and use the positive or negative outcomes to confirm, improve or refine the model. On the other hand, the model users can apply the same methodology to determine the model's capability and validity in simulating a real-life problem realistically with acceptable accuracy before making their decision to adopt it to do the job. The Committee decided further on how the required model Verification and Validation should be conducted. This is discussed in the next section.

1.4 VERIFICATION AND VALIDATION

The ASCE-EWRI Task Committee on 3D Free Surface Flow Model Verification and Validation Monograph has extensively discussed numerous ways for conducting free surface flow model verification and validation over a period of 9 active years excluding those years without support and without meetings. Its decision was to develop and publish a systematic procedure for conducting an effective model verification and validation, for reasons detailed in the previous section.

The procedure recommended by the Task Committee is to include three steps, namely, the Mathematical (code) Verification, the Physical Processes Validation and the Application Site Validations. Several test cases developed by the committee members are provided to the model developers and users to conduct a series of tests at each step. Each test case is based on a given analytic, prescribed or manufactured solution, laboratory experimental measurements, or collected field data. The analytic results or physical measurements provided can be used as a measure to decide a model's correctness in mathematics, its capability in representing basic physical processes and in predicting with reasonable accuracy, and the trend of variation of physical properties of the real-life flow systems to be investigated. Each test case included is to have specific and relevant objectives of mathematical and/or physical significance and easy to understand instructions. All cases have been designed to minimize the effort required to prepare the test runs and to be computationally efficient to run. The test cases presented are intended to be only exemplary rather than comprehensive, however. Users are encouraged to design their own test cases, which may be more appropriate for evaluating a model to conduct their specific application.

The three major verification and validation steps are briefly described below.

1.4.1 Mathematical (Code) Verification

The purpose of this step is to insure that the mathematical model is correctly derived and converted into its corresponding numerical model, that the numerical solution technique is consistent and convergent to the model equations, with the solutions having the designed order of accuracy (or error) quantitatively, and that there are no program coding errors. This Verification step is conducted by comparing the results of a numerical code to the values of the selected analytic, prescribed or manufactured solution of the same set of differential equations and boundary/initial conditions. This is the only step where both the numerical solution and analytic solution are obtained from the same set of partial differential equations, same modeling domain, same boundary and initial conditions, and same geometrical and physical parameters (no tunings are needed). If there is any discrepancy found between the numerical and analytic solutions, it must be due to the mistakes committed in mathematics, numerics, or coding during the numerical model's development, solution, and computing processes. Due to the fact that there is no tuning or calibration of model parameters, the discrepancies cannot be concealed by adjusting those parameters to achieve a false agreement between the two results. If the numerical solutions agree well with the analytic solution, it is said that the numerical model (code) has been verified, which implies that it basically does not have mathematical formulation mistakes, discretization mistakes, numerical solution errors, computational coding errors, etc. However, the truncation and round-off errors or inaccuracies in the discretization, approximation and computation cannot be avoided.

A frequently asked question is "can the mathematical and code verification be carried out for those systems governed by nonlinear partial differential equations?" It is true that the state of the art in the mathematical solution of partial differential equations has not been advanced to the level that allows the researchers to obtain close-form analytic solutions of a system of nonlinear equations. However, it has been demonstrated by many applied mathematicians and scientists that an inverse method can be used to conduct mathematical verification of numerical solutions to nonlinear systems. Details are to be covered in Chapter 4 of this report.

Furthermore, through a convergence test of grid refinement, as a part of this mathematical code verification, one can determine whether the numerical solution is consistent with the differential equation and convergent to the true solution. This verification test can also determine quantitatively the order of convergence error. More details are given in Chapters 2 to 4.

1.4.2 Physical Process Validation and Calculation Verification

The Physical Process Validation involves testing a numerical model's capability in reproducing the basic physics that are fundamental to a selected application. This step is conducted by comparing the results of a numerical model to the measurements of a selected laboratory experiment of the physical processes essential to the physical problem to be studied. Due to the fact that the laboratory experiments are performed in a relatively controllable and repeatable environment, a variety of unimportant or irrelevant complications can be eliminated or controlled. For this reason, the characteristics of a simplified physical system tested in a laboratory are better represented by the idealized mathematical system and solved by a numerical model. And thus, the comparison of results between the laboratory model and numerical model is more meaningful in evaluating the capability to simulate basic physical phenomena than a similar comparison between the measurements of the real-life system and the solutions of the numerical model. This is because that in laboratory tests, many non-essential complication and uncertainties of real-life system can be controlled. The examples of those complications are effects of local heterogeneity, non-uniformity, boundary irregularities (bed form, roughness, vegetations, etc.) and wind gusts over the free surface, etc. If a good agreement between numerical model results and laboratory measurement can be obtained, it cannot be denied that the numerical model is capable to reproducing physics or basic physical processes of the system tested.

If a model fails the Physical Process Validations, one may want to determine whether it was caused by not solving the right equations or having unsound physical assumptions and/or simplifications. For examples, the chosen simplified turbulence closure, the hydrostatic pressure assumptions, etc. may be inappropriate or inadequate for reproducing a physical process. If so, necessary corrections must be implemented in the governing differential equations, numerical model, solution techniques, and coding. Therefore, this test step is very important to both model developers as well as users.

The Physical Process Validation described above can be used for another purpose, which is to test the numerical model simulation against the measured flowfield of a **scaled physical model** in a laboratory. In addition to testing the capability of reproducing basic physical processes, it can also be used to "confirm" the effectiveness of the engineering designs, obtained from numerical modeling. For highly important engineering designs, which cost a great deal of funds to build and have long-term effects to the society, a scaled model physical validation or confirmation is recommended.

Test cases for this purpose are included in Chapter 5. There is a general desire to determine the overall error bands (or error-bar) of a numerical model's prediction of the value of flowfield property. Discretization errors may be estimated and banded by a Grid Convergence Test (GCT) or simply by determining a Grid Convergence Index (GCI). More information is given in Chapter 2. Even though it is agreed that to conduct a comprehensive GCT of a large and highly complex real-life problem could be difficult or at least tedious; in order to have an estimate of error band resulting from calculations, one is highly recommended to perform GCT or Calculation Verification. More detailed Physical Process Validation and test cases are given in Chapter 5.

1.4.3 Application Site Validation and Calculation Verification

It would be highly desirable if a numerical model could be comprehensively validated by field data of all possible geohydrological conditions with spatial and temporal variations of a natural free surface flow system. Unfortunately, the state of the technology and the collected data available do not allow us to achieve this ultimate goal within the constraint of the available funds and time. Therefore, the only practical approach to validate numerical model's capability in reproducing reasonably accurate and realistic physical behavior of a natural or real-life system at the present is on a case by case basis for each application problem at a specific site and time.

Before applying a numerical model to investigate a particular geophysical system in nature, it is necessary to have access to a sufficient amount of physical data measured at well-selected data collection locations with appropriate temporal distribution for each Application Site Validation. Sometimes, it is advisable to perform a few selected preliminary simulations to identify critically important spatial and temporal data requirements, which include not only how much and accurate of what data to collect; but also where and when to collect. One is reminded that at least an extra set of data is needed for the purpose of calibration of the model parameters to represent the site unique characteristics of the system. Calibration is a very important and necessary process or a sub-step, which is just like calibration of instruments before making accurate physical measurements. Once the model parameters are calibrated, they should not be changed without justification. The calibrated numerical model is then tested by comparing its results with the remaining field data (unused for calibration) for the purpose of validation. After the numerical model is validated, it is ready to be applied to carry out the investigation intended at the designated site over a reasonable duration of time.

In our present society, the model developers and users are almost always frustrated by the scarcity and quality of available data, especially the field data. Consequently, both model developers and users have no choice but to use all data available to calibrate the parameters without conducting validation. As a result, there is no way to know how accurate the prediction of such a model is in predicting a future response of the flow system to an event of external forcing. This fact has defeated one of the usefulnesses of the model as a predictive tool. Therefore, if the model user needs the numerical model to serve as a predictive tool for engineering designs, environmental and ecological impact assessments, management, planning and policy decisions, he/she must demand to have sufficient amount of data accurately collected at appropriate location and time. This shall allow the Application Site Validation to be properly conducted, so that the validated model can serve as a reliable predictive tool.

Again, a Calculation Verification is highly recommended so that the model user can make an estimate of the values of calculation errors (or inaccuracies). This can be done by a grid refinement convergence analysis or by the evaluation of the Grid Convergence Index.

It is important for both model developers and model users to note that the Application Site Validation should be repeated for a different site of study because every natural system has its unique characteristics at each study site and there is no site which has exactly the same characteristics as those at any other site. Rigorously speaking, the model having passed the Application Site Validation at a particular time period may have to be revalidated for a different period of time even at the same site, because the system characteristics may be changed from time to time especially after major hydrological events. Of course, when a model is applied to the study of numerous cases at a particular site of a field system during a relatively steady period of time, only one Application Site Validation may be adequate. Test cases suggested for use in this Application Case & Site-Specific Validation are given in Chapter 6.

1.5 OUTLINE OF A SYSTEMATIC MODEL VERIFICATION AND VALIDATION PROCEDURE

There has been a trend that numerical models have become an increasingly important tool in scientific research, engineering design, planning, management, and policy decision making. And, more numerical models are applied to solve problems in large and highly complex water and environmental systems with long-term effects and costing large sums of monies. The users are demanding more rigorous model verification and validation before a model is used for applications. For this reason, the Task Committee recommends the following systematic verification and validation procedure for use by both the model developers and users.

After a numerical model has been developed, it should be mathematically verified first by using at least one of the tests based on analytic solutions, or prescribed or manufactured solution for linear or nonlinear case. Once the model is proven correct mathematically, it should be validated by one or more test cases based on laboratory experimental results to determine whether the model is capable of reproducing the basic physical processes relevant to the physical problems to be studied. The level of accuracy required will depend upon the specific objectives of the study. These two basic steps may be needed once for each model. Once a numerical model satisfies these two basic test steps, it is confirmed to be free of mathematical mistakes, coding errors, and capable of reproducing basic physical processes. One should note, however, that these two fundamental steps have to be repeated, whenever the model has been upgraded with major changes in either the physical representation or the numerical implementation, especially when the model is to be applied to the study a new problem requiring additional capabilities to reproduce additional basic physical processes.

Before applying a numerical model to an investigation of a real-world problem, one more step, the Application Site Validation, is required. To carry out this step, one needs to use an appropriate portion of field data collected at the study site to calibrate the site-specific values of the model parameters first. The calibrated model is then, used to predict field characteristics and processes of real-life problems, and compare with the predicted results with those data measured at the same site under the same forcing conditions. If a reasonable agreement between the model simulations and the field measurements is achieved, the numerical model is validated, but only for the application case at the study site intended. It is not recommended to apply this model to the study of a similar problem at a different site, nor a different case at the same site, if there are significant changes in site characteristics during the elapse of time, e.g., a few months or years, especially after the occurrence of a major hydrological event. To conduct an Application Site Validation successfully, one must have a sufficient amount of high quality data collected at well-designed locations with proper spatial and temporal distribution. The users are reminded that it is more important to calibrate the values of model parameters to achieve a reasonable over-all accuracy of most of the measured data in the entire study domain, than to achieve highly accurate agreements at a small number of selected data points. One should keep in mind that the numerical model represents an idealized and simplified system of the real-life problem, which is highly likely different from the complex real-life system in nature. Spending a tremendous amount of effort to achieve a near perfect agreement between the results of the two somewhat different systems by tuning the model parameters is not necessary. For the same reason, the prediction of correct trends of field properties' variations in space and time are more important than the accurate magnitudes of the field variables themselves. This is especially true, when a numerical model is used to compare the effectiveness of several conceptual designs and/or to identify the optimal values of all design parameters during a parametric trade-off analysis, etc. Due to the fact that the numerical model and the real-life system obey the same set of physical principles, the trend of variation of a system property predicted by the numerical model should be approximately correct. Therefore, the validated numerical model is more useful for comparative studies to predict the outcomes of "what if" cases of scenarios, parametric optimization analyses, etc. than for determining the precise values of design parameters.

Should one need to determine the accurate values of design parameters, a Calculation Verification is needed to estimate the error bounds of the numerical solutions. The detailed discussions of the systematic Verification and Validation are given in Chapter 7.

1.6 OTHER ISSUES OF IMPORTANCE

Numerous issues have been brought up by the members of the Task Committee throughout the years. Due to their relevance and importance to model validation in general, considerable time has been spent in discussions on these issues. Written contributions have been submitted by members. In the final stage of the Committee work, it was found that there was not enough time to fully integrate all of these issues for inclusion into this written document without further delaying the completion date. Therefore, it was decided to include only a few selected issues in a chapter, for the benefit of some readers, who may be interested in these related issues; although they are not directly tied to the Validation and Verification procedure proposed in the report. Nevertheless, they are related in a broader sense.

The issues included in Chapter 8 concern the followings:

(1) How to plan the optimal data collection locations so that a sufficient amount of data can be collected at well-distributed and crucially important locations, at least cost, which is obviously important to the success of model validation;

(2) How to estimate the missing data by analytic methods, in case one finds that the data set has been collected is not quite adequate for conducting both calibration and validation of a numerical model and there are no additional funds and time available for collecting the missing field data; and

(3) How to reduce the uncertainty, when the solutions of deterministic numerical model are compared to the statistical mean of both laboratory and field measurements of field properties of somewhat stochastic in nature.

Detailed discussions on these and other related issues as well as techniques to resolve some of these problems based on the system analysis are given in Chapter 8.

Also included in Chapter 8 is the development of a nowcast/forecast system by the National Ocean Service (NOS), of the National Oceanic and Atmospheric Administration (NOAA), for supplementing its Physical Oceanographic Real Time Systems (PORTS). Although this development was not intended originally for model validation, it can be regarded to as a continuous model calibration and prediction process. During the nowcasting, the numerical model's site-specific parameters can be calibrated by the field measurements. Then, the field state properties can be predicted at the next time step or a short time (say some hours) later, which may be considered as forecasting, prediction or extrapolation by the numerical model. At the new time step, additional field measurements are available and they can be used to validate the numerical predictions. If the discrepancies between the predictions and measurements exceed the allowed accuracy, the model needs to be corrected or refined. If, on the other hand, the discrepancies are allowable, the model is validated.

If one wants to enhance the accuracy of the next prediction, one may re-calibrate the parameter by using the latest measurement before making the next prediction. Therefore, through the combination of a hindcast/nowcast (between $t\text{-}\Delta t$ and t), and a nowcast/forecast (between t and $t + \Delta t$), one can repeat this continuous calibration, validation and prediction cycle as a real-time field predictor system of a free surface flowfield. Further research is needed to advance this methodology to improve the capability of longer time-step extrapolations or forecasts. Again, even the information presented in Chapter 8 is not directly applied to the Verification and Validation procedure developed by the Committee, but they are certainly useful to the Verification and Validation in general. In fact, one may refer to the hindcasting/nowcasting and newcasting/forecasting as a Dynamic Validation process, which serves a special and useful purpose.

1.7 FINDINGS, CONCLUSIONS, RECOMMENDATIONS, AND INPUT DATA

In Chapter 9, findings, conclusions and recommendations of importance are summarized.

Finally, the data files of test cases are described and included in Appendix A at the end of the report. The Fortran Codes for calculating the numerical values of source terms at all nodal points in the computational domain needed by the Method of Manufactured Solution (MMS) are provided in Appendix B given at the end of this report. Some Task Committee Members, the contributors of test cases, have agreed to maintain and upgrade the data files on their web-sites for users to download and use. Detailed and additional information on methods and data sets shall be announced on a Verification and Validation User's Website: http://www.ncche.olemiss.edu/publishing/.

1.8 REQUEST FOR FEEDBACK

The editors request feedback from users of the verification and validation procedure presented in this report. We are very interested in any applications of these techniques, and any use of the 22 example problems in any verification and validation study, including journal publications, meetings papers, and any in-house or other model documentation. Please give a short description of your application, with any references, at website: http://www.ncche.olemiss.edu/publishing/. Your feedback shall be considered for further improvement of the procedure in the near future. We appreciate your contributions.

1.9 REFERENCES

AIAA (1994), "Editorial Policy Statement on Numerical Accuracy and Experimental Uncertainty," *AIAA Journal,* Vol. 32, No. 1, January 1994, p.3.

ASME (1994), Editorial Board, "Journal of Heat Transfer Editorial Policy Statement on Numerical Accuracy," *ASME Journal of Heat Transfer*, Vol. 116, November 1994, pp. 797-798.

Bobbit, P.J. (1988), "The Pros and Cons of Code Validation," AIAA Paper 88-2535. Also, NASA TM 100657, July 1988.

Bredehoeft, J.D., and L.F. Konikow, (1992), Ground-water models cannot be validated, *Advances in Water Resources*, Vol. 15, pp.75-83.

Celik, I., Chen, C.J., Roache, P.J., and Scheurer, G. (1993), *Quantification of Uncertainty in Computational Fluid Dynamics*, ASME FED Vol. 158. ASME Fluids Engineering Division Summer Meeting, Washington, D.C., June 20-24, 1993.

Dee, D.P. (1991a). "Prescribed solution forcing method for model verification in hydraulic engineering." *Proc. 1991 National Conf. of Hydr. Engrg., ASCE*, Nashville, July 29-August 2, 1991.

Dee, D.P. (1991b). "Prescribed solution forcing for TRISULA: waves in a uniform channel, Part I: design and implementation." *Delft Hydraulics Report X85*, Delft, The Netherlands.

Dee. D.P., Toro, F.M., and Wang, S.S.Y. (1992). "Numerical model verification by prescribed solution forcing—a test case." *Proc. of the Hydr. Engrg. Sessions at the Water Forum 92, ASCE,* Baltimore, August 1992, 416-421.

Dee, D.P. (1994), "Guidelines for Documenting the Validity of Computational Modeling Software," IAHR, June, 1994.

Ferziger, J.H. and Peric, M. (1996), "Further Discussion on Numerical Errors in CFD," *International Journal for Numerical Methods in Fluids*, Vol. 23, pp. 1263-1274.

Freitas, C.J. (1993), "Editorial Policy Statement on the Control of Numerical Accuracy," *ASME Journal of Fluids Engineering,* Vol. 115, No.3, September 1993, pp. 339-340.

Freitas, C.J. (1995), "The End of Justification, the Beginning of Implementation," *ASME Journal of Fluids Engineering*, Vol. 117, No. 1, March 1995, p.9.

Gresho, P.M. and Taylor, C. (1994), "Editorial," *International Journal for Numerical Methods in Fluids*, Vol. 19, 1994, p. iii.

Johnson, R.W. and Hughes, E.D. (1995), *Quantification of Uncertainty in Computational Fluid Dynamics*—1995, ASME FED Vol. 213. ASME/JSME Fluids Engineering and Laser Anemometry Conference, Hilton Head, South Carolina, August 13-18, 1995.

Melnik, R.E., Siclari, M.J., Marconi, F., Barber, T., and Verhoff, A. (1995), "An Overview of a Recent Industry Effort at CFD Code Certification, " AIAA Paper 95-2229, AIAA 26th Fluid Dynamics Conference, San Diego, California, 19-22 June 1995.

Oberkampf, W.L., Blottner, F.G., and Aeschliman, D.P. (1995), "Methodology for Computational Fluid Dynamics Code Verification/Validation," AIAA Paper 95-2226, 26th AIAA Fluid Dynamics Conference, San Diego, California, 19-22 June 1995.

Oreskes, N,. K. Shrader-Frechette, K. Belitz, (1994), "Verification, Validation, and Confirmation of Numerical Models in the Earth Sciences", *Science*, Vol. 263, pp. 641-646.

Roach, P.J., Ghia, K., White, F., (1986), "Editoral Policy Statement on the Control of Numerical Accuracy," *ASME Journal of Fluids Engineering*, Vol. 108, No. 1, March 1986, p. 2.

Roache, P.J., (1989), Need for Control of Numerical Accuracy, *J. Spacecraft*, Vol. 27(2), pp. 98-102.

Roache, P.J.(1997), "Quantification of Uncertainty in Computational Fluid Dynamics," *Annual Review of Fluid Mechanics*, Vol. 29, 1997, pp. 123-160.

Roache, P.J., (1998), Verification and Validation in Computational Science and Engineering, Hermosa Publishers, 1998.

Roache, P.J., (1999), Numerical Model Verification Techniques for Hydraulic Engineers, *Proceedings, ASCE International Conference on Water Resources Engineering* 1999.

Roed, L.P., and B. Hackett, (1989), *Final Report: Results of Model Intercomparison, Veritas Offshore Technology and Services A/S.*

Toro, F.M. and Wang, S.S. Y. (1993). "Validation of a 3-D channel flow model by the prescribed solution forcing technique." *Advances in Hydro-Science & Engineering*, Edited by Sam S.Y. Wang, Vol. 1, part B, 2109-2114.

Wang, S.Y. (1991). "Task Committee data collection for verification of 3-D free surface flow hydrodynamic models." *Proc. 1991 National Conf. on Hydr. Engrg., ASCE,* Nashville, July 29-August 2, 1991.

Wang, S.Y., (1992), On Verification of Computational Hydro-System Models, *Proceedings, International Conference on Interaction of Computational Methods and Measurements in Hydraulics and Hydrology* 1992, pp. 163-169.

Wang, S.Y., (1993), "Data and Information Base for 3D Free Surface Flow Model Verification," *Interim Report (for Task Committee member circulation only), ASCE Task Committee on Data Collection for 3D Free Surface Flow Model Verification.*

Wang, S.Y., (1994), 3D Free Surface Flow Model Verification—A Panel Discussion on Test Cases, *Hydraulic Engineering*, pp. 910-912.

CHAPTER 2

TERMINOLOGY AND BASIC METHODOLOGY

Patrick J. Roache

2.1 INTRODUCTION [1]

Background discussion, definitions and descriptions will be given for some terms related to Verification, Validation, and related confidence-building exercises for 3D free surface flow models and computational hydraulics in general. The critical distinction will be made between Verification vs. Validation. Distinctions will be made between numerical errors vs. conceptual modeling errors; adequate and inadequate error taxonomies; Confirmation, Calibration, Tuning, and Certification; Verification of numerical accuracy of codes vs. Verification of individual calculations; and Physical Process Validation vs. Site Validation.

Next, the basic methodology for Calculation Verification will be presented by way of specific examples on easily replicated problems. The chapter closes with an examination of the incremental costs and benefits of Calculation Verification.

2.2 SEMANTICS

The definitions and distinctions of this field have developed over years, with less standardization in the earlier years. The definitions did not flow inexorably from either the etymology or the common use of terms but rather developed as the need arose to make the distinctions. None of the terms has enjoyed a pristine use in the literature, i.e., all have been used in ways inconsistent with the present definitions, and some continue to be so. Nevertheless, the definitions and usage herein are becoming standardized, enabling communications to become more efficient and precise. Also, the ideas behind the words are important. The growing recognition of the importance of the subject is attested to by the policy statements given by several professional societies and journals (Roache 1998a).

---

[1] This introductory chapter on terminology and basic methodology is developed from Chapters 2 and 5 of Roache (1998a); see also Chapters 18 and 19 of Roache (1998b).

Semantic distinctions can have major consequences to projects. "Semantics" is "the study of meaning, especially in language." The negative connotation of "mere semantics" arises when people argue uselessly about words without looking beyond, to the ideas behind the words. For example, the choice of *Verification* or *Validation* was originally arbitrary, and is now recommended solely because of widespread developing technical use. In a common English thesaurus, *verify*, *validate*, and *confirm* are all synonyms, but the words are used herein, and generally in code Quality Assurance (QA), as *technical terms* with more context-specific meaning. Such technical terms are preferably related to common use (regrettable exceptions occur) but the term's *technical meaning* is defined independent of common use, and in a *specific technical context*. The same word can have different technical meanings in different technical contexts. Even the word "error", which a non-technical person would never suspect of ambiguity, has multiple technical meanings depending on context. Programming errors are *mistakes*, while discretization errors are not. In common terminology, "error" and "mistake" are virtually synonymous. It is important to recognize that one cannot determine the meaning of a technical term simply by inquiring about its common meaning or less, its etymology. Failure to recognize this has led to serious misunderstanding and pernicious claims, e.g. that it is impossible to validate any codes. These controversies are not just the stuff of academic debates, but have serious consequences, especially on public policy. For examples and discussion, see Chapter 2 of (Roache 1998a).

We are interested herein in normal speech and practical definitions, applied in the context of engineering and science accuracy. The definitions and descriptions given here are technical terms defined in a technical context, not just common language.

2.3 CODE VERIFICATION AND VALIDATION: NUMERICAL VS. CONCEPTUAL MODELING

First and foremost, we make the essential distinction between Code Verification and Validation. We adopt the succinct description of "Verification" as "solving the equations right" and of "Validation" as "solving the right equations."

"Verification" ~ "solving the equations right"
"Validation" ~ "solving the right equations"

The code author defines precisely what continuum partial differential equations and continuum boundary conditions are being solved, and convincingly demonstrates that they are solved correctly, i.e. usually with some order of accuracy, and always consistently, so that as some measure of discretization (e.g., the mesh increments) $\Delta \rightarrow 0$, the code produces *a* solution to the continuum equations; this is Verification. Whether or not those equations and that solution bear any relation to a physical problem of interest to the code user is the subject of Validation. In a meaningful though perhaps overly scrupulous sense, a "code" (or better, the model embodied in the code) cannot be Validated, but only a calculation (or range of calculations with a code, for a specific class of problems) can be Validated.

Another way to make the distinction (i.e., to get to the idea behind the words, beyond "mere" semantics) is to speak of numerical errors vs. conceptual modeling errors. An example of conceptual modeling vs. numerical modeling is the assumption of constant density. This is clearly a conceptual modeling assumption. It is not the code builder's fault, or any criticism of the code itself, if the user incorrectly applies it. For example, the constant density assumption will be acceptable for most lake and river calculations, but may be inaccurate in a brackish estuary. It is no criticism of the code if it fails Validation because it was applied to variable density flow, although we may have some sympathy for the user. (But no one would have sympathy for a user who applied an incompressible flow code to a space shuttle at Mach 10.) In this case, and in many practical cases, the lack of agreement with experiment is not a *code* problem, but a *modeling* problem.

The word *model* includes more than the code; it also includes conceptual modeling assumptions (e.g., constant density vs. stratified flow, choice of frictional law, turbulence formulation, dimensionality, etc.). The word *model* can also be used in a weak or strong sense; in the strong sense, *model* includes parameter values. There are corresponding strong and weak senses of Validation, based on whether or not *model* includes parameters or is only a conceptual model. These distinctions can lead to failure of validation of a *model*, with possibly no criticism of the *code*. Also, in geophysical modeling (including free surface flow site models) in which a particular mesh will see long term use (e.g., a discretization of San Francisco Bay which will be used for years) the word "model" can include the particular mesh. In free-surface flow simulation practice, the word *model* is often used synonymously with *code*. In other fluid dynamics disciplines, one would not speak of "verifying a model" because the model (conceptual model, including turbulence formulations, etc.) is to be validated (physics) whereas the numerical algorithms and coding are to be verified (mathematics). It would be impossible to revise or ignore this existing practice in the free surface simulation community, so the context will have to guide the reader.

Another way to make the distinction between Verification and Validation is to follow the classical distinction between mathematics and science. Mathematics is a tool of science, often the predominant language of science. But mathematics exists by itself. It would be "true" regardless of any correspondence to the natural world. Verification is seen to be essentially and strictly an activity in mathematics, the mathematics of numerical analysis. Validation is essentially and strictly an activity in science: physics, chemistry, fluid dynamics, etc.

The typical Computer Science view of "Code Verification" often includes all aspects of code QA. This is not the definition adopted herein. For example, it does not include the important concerns of version control, archiving of input data, documentation, etc. (See Chapter 2 of Roache (1998a) for further discussion.)

It is important to state that certain activities, though desirable, do *not* constitute Verification or even partial Verification. These include reading the source code, journal publication, wide code distribution, or extensive use of a code. Especially in free-surface codes applied to field data where the physical parameters and initial conditions are poorly known, extensive experience with "real world" calculations (even hundreds of cases calculated by many different users) would be an inefficient and, more importantly, *inconclusive* way to approach Verification. Extensive code use and journal publication certainly have value. They add to confidence building, and are worthwhile for general code QA, and are helpful to improve code documentation, gather suggestions for input/output improvements, obtain data on performance measures for a wide range of problem parameters, demonstrate robustness and portability, etc. They are just not part of Verification as technically defined herein.

A code failure (i.e., a divide by zero, or instability) is not a failure of Code Verification in the present context. As long as the code does not lie to the user, Code Verification has not been breached. Code robustness is certainly desirable, and is properly part of QA and overall code evaluation. For real-time applications, robustness would be required for Verification. But in the present use of the terms, robustness is not an issue in Verification.

Validations, unlike Verifications, are based on comparisons with experiments. For complex systems, it is useful to distinguish two levels of physical Validation. For free-surface flow, these are described as Physical Process Validation vs. Site Validation. (In other fields of engineering, the terms Unit Process Validation and System-Level Validation are used; the latter can be expanded into a hierarchy.) The distinguishing characteristics are that the Physical Process Validation involves controlled laboratory-scale experiments with relatively high-resolution precise measurements and a focus on one or a few distinct physical processes, whereas Site Validation involves parameter variation by nature (e.g., rainfall, winds, shifting bathymetry) which are not only uncontrolled by the experimenter but also imprecisely measured at sampled space-time locations.

Since Validation involves comparisons of simulations with experiment, there are necessarily error tolerances involved. There are three distinct sets of error bars (uncertainty estimates) involved in a Validation: one associated with the experiments, one with the calculations, and one that defines agreement between the two. What levels are acceptable depends on the use intended. By definition then, one can never claim to Validate and Certify a computer model without specifying qualifying phrases on range of parameters, error metrics, and error tolerances, for both the calculations and the experiments.

Calculation Verification (by grid convergence studies of single-grid error estimators) should be performed for both Physical Process Validation and Application Site Validation. Grid convergence studies are easier for Physical Process Validation because the geometries are more regular. Formal, quantitative reporting of numerical uncertainty is preferable to informal and un-reported assessment of resolution adequacy (or worse, arbitrary grid selection.)

Variance exists in the use of the word "Validation" in regard to whether or not an acceptable tolerance for the agreement between experiment and simulation is specified, i.e. a pass/fail evaluation. Full Validation of a model can be considered in two steps: comparison of model predictions with experiments, and determination of pass/fail for a particular application. In some usage, a model whose results have been compared to experiments is labeled "Validated" regardless of the agreement achieved. "Validation" then is not a quality of the code/model *per se*, but just refers to the process. Carried to an extreme, this viewpoint gives the designation "Validated" even to very poor models. The other extreme makes Validation project-specific by specifying the error tolerance *a priori*. This neglects the fact that agreement may be acceptable for one application and not for another. Not all comparisons should result in a code being given the value-laden designation of "Validated" because some minimal agreement should be required. The general (and necessarily vague) level of acceptable agreement must be determined by common practice in the discipline. (Certainly incorrect trend prediction should be enough to categorically reject a model, i.e. to fail Validation.) The simulation results are compared to experiments, and if "reasonable" agreement as determined by the state-of-the-art standards is achieved, then the code/model can be termed "Validated." This does not necessarily mean that the model will be adequate for all applications. Such project-specific pass/fail tolerance should be relegated to Certification, which is within the purview of Engineering Management rather than Engineering Science. The value of this pass/fail tolerance tends to vary over time with management decisions, even though the objective results of the Validation comparison itself have more permanent value.

Note also that the design of Validation *metrics*, i.e. the direct (e.g. surface height h(x,y,t)) or derived (e.g. total discharge) quantities to be compared between experiment and simulation, is a research area in its own right, and is an essential part of Validation in any definition (Oberkampf and Trucano 2002).

The question of *solution uniqueness* always arises with nonlinear partial differential equations (PDE's) and its position in the "Verification" or "Validation" distinction is important. If, as stated carefully above, we take "Verification" to mean simply that "*a* solution" to the continuum PDE's is obtained, then the problem of PDE non-uniqueness is avoided. In most cases, physically inadmissible solutions should be eliminated by the code, e.g., a hydraulic jump in the wrong direction, but there do exist in nature physically non-unique solutions in fluid dynamics. When nature cannot decide which solution to produce, a user cannot expect more of a code. Some non-unique solutions may not be stable or observable in nature, but such Verified simulations are still mathematical solutions. Sorting out these solutions is not an aspect of Verification; it is an aspect of Validation, but it must be approached with great respect for the subtlety of nature.

Code Verification can and *should* be completed without appeal to physical experiments. The first part of this statement claims that Code Verification should be *completed*, i.e., Code Verification is *not* an ongoing exercise. Verification is an exercise in mathematics, not science. When one proves a theorem, the work is completed. Proving the formula for solution of a quadratic equation is not ongoing work. This is not to say that one could not have made an error in the proof of a theorem, nor that Confirmation exercises (see below) are not valuable in confidence - building. It is to say that Code Verification is a mathematical activity that in principle comes to a conclusion, e.g. a code is, or is not, 2nd order accurate. The second part of the statement claims that Verification can and should be achieved without using physical experiments. That it *can* is unarguable - very general methods are available (see the Method of Manufactured Solutions in Chapter 3 of Roache (1998a)). That it *should* is the clear opinion of most people who have been immersed in such work. Not only are physical experiments not needed, they may just be confusing, because one now has to sort out two sources of error, the rational approximation error and the physical experimental error.

Validation has highest priority to engineers and scientists because "nature" is the final jury. But any experience with laboratory experiments or field data will quickly discredit the absolute quality of experimental data. It is asking too much of a code to agree with test data when these data do not agree with other test data. Thus, it is recommended that complete Verification of a code (and a calculation - see below) should *precede* any comparisons with experimental data: Verification first, then Validation.

Another contrast between Verification and Validation is the following. Verification is *completed* (at least in principle, first for the code, then for a particular calculation) whereas Validation is *ongoing* (as experiments are improved and/or parameter ranges are extended).

Complaints are sometimes voiced that a code needs additional data that experimenters typically do not measure, e.g. upstream turbulence properties. This neglects a fundamental fact. *Codes require no more information than the physics.* In fact, a code may be expected to require *less* information than the full physics, due to simplifying approximations in the turbulence theory. If experimenters have not measured these quantities, then they have run an uncontrolled experiment, regardless of whether codes will be used. The question, of course, is whether these unmeasured quantities are important to the physics of interest.

When we say that Verification is completed, rather than ongoing, we are addressing only a completed code, or code version, and perhaps only a limited set of option combinations. Large scientific codes develop over years, and have multitudinous combinations of options. In this sense, we can say that Code Verification is an ongoing process, but only because the word "Code" as used is amorphous, really referring to many codes (all with the same name, but hopefully different version numbers), each of which in turn must be Verified.

2.4 CODE CONFIRMATION

Some Computer Science or QA people would have Code Verification necessarily performed by someone other than the code developer, sort of an "arm's length transaction" philosophy. In our view, it is ridiculous to expect code builders to not Verify their codes! Verification is a necessary part of code development. Code authors would be remiss in their duty if they released a code without Verification. We would usually trust a code builder's Verification results if presented in full; fraud is not usually the issue. But if it is, and if further tests are required (or repeats of the original Verifications to check for fraud or oversight) then we use the term "Confirmation" for calculations independently run by someone other than the code builder.

Also, the suite of problems can be used for re-Verification (e.g. after porting to a new computer or a new compiler, or to be run after addition of new options) and for user training. Code use independent of the code builder is probably more of a genuine issue in Validation than in Verification, especially when judicious "tweaking" or Tuning of code parameters can be involved.

We recognize five distinct regimes where errors can be made in Verifying a code, even without considering the Validation question of whether the right equations are being solved for the target problem:

- **(1)** in code generation (either by hand or using computer Symbolic Manipulation);
- **(2)** in code instructions (e.g., in a user manual or comment lines);
- **(3)** in problem set-up;
- **(4)** in defining and coding a test case (analytical solutions are often more difficult to code than numerical solutions); and
- **(5)** in the interpretation of code results.

The first two are errors of the code author. The last three are errors of the code user, although ambiguous or scant code documentation can put some of the responsibility back onto the code author. "Verification" of a code removes (1) and, if done thoroughly, (2), but (3-5) still contain the potential for errors in any new application. Therefore, there will be a continuing need for users to construct and exercise Confirmation test cases *even when using Verified codes*.

Obviously, it is good common sense to build more confidence with more problems, even if the code is Verified in the sense of a theorem. Such an exercise is pragmatic and builds confidence, not only in the Verification but in one's understanding of the code. Thus, although Code Verification should be completed and therefore *is not* on ongoing activity, Code Confirmation *is* naturally an ongoing activity (as is Validation). Also, as a practical political consideration, not everyone can appreciate the generality of a mathematical proof or a thoroughly executed Code Verification. This is particularly true when a physically unrealistic solution is used to Verify a code. Although mathematically sound, the exercise will not inspire confidence in the mathematically unsophisticated, including the general public (e.g., "stakeholders" in environmental studies) and many engineers. For such people, an extensive suite of realistic-*looking* Confirmation problems, even if they alone fail to constitute a rigorous Code Verification, may build more user confidence than a single rigorous but unreal-appearing Code Verification problem. Indeed, this skeptical attitude is not only forgivable but often wise, prudent, and pragmatic. Still, the skeptic should keep in mind the possibility, in principle and often in practice, of rigorous Code Verification in the sense of a mathematical theorem.

2.5 BENCHMARKS AND INTER-CODE COMPARISONS

The term "benchmarking" of codes is a less specific term than Verification or Validation. Some authors prefer to reserve the term "benchmarking" for code-to-code comparisons; this has the advantage of clearly distinguishing it from Verification and Validation, i.e. comparisons with closed form mathematical solutions and with experiments, respectively. Unfortunately, the term is often used more inclusively. It commonly refers to comparisons of results of the code in question to some other standard or "benchmark" which might be an analytical solution, a physical experiment (usually on simple configurations involving a single dominant physical phenomenon), a rigorously performed numerical solution, or just another code solution at comparable grid resolution, which has been termed "inter-code comparison."

Analytical solutions obtained by perturbation methods can be useful but require some care. Since they are exact only in the unattainable limit of some physical parameter (e.g. Froude number $\rightarrow$ 0, wave height/depth $\rightarrow$ 0) there will remain an error in the limit of grid convergence. Grid convergence testing applied mechanically and thoroughly will produce a false negative Code Verification; the perturbation solution error will be indistinguishable from a coding error.

Often, the term "benchmarking" has been used in meeting papers wherein rigorous grid convergence testing has not been performed, and comparisons have not been precise; such benchmarking exercises could be classified with Confirmation exercises. The more rigorous benchmarking exercises can be classified with Verification or Validation exercises, depending on whether the benchmark is a rigorous numerical solution or a physical experiment.

Unfortunately, in practice, inter-code comparisons usually have been rather loose. Although agreement of a new code with an old and widely used code can legitimately build some confidence, it is not usually the same as a rigorous Verification. Note there is nothing to be gained by demonstrating agreement with an inaccurate code.

2.6 CODE CERTIFICATION, QUALITY ASSURANCE, AND ACCREDITATION

The term "Code Certification" goes beyond Code Verification and Validation, and is used with less uniformity. It can include all the elements of Verification, Validation, Confirmation and Calibration; it can include pass/fail error tolerances, documentation, QA procedures including version control, programming issues such as logic checks and programming style, model options, robustness, portability, and risk assessment. Although occasionally used for Verification type activities, more often, Certification includes project-oriented Validation activities as well as Verification, so that Code Certification is practically indistinguishable from Code QA. As contrasted to Code Verification and Validation, Code Certification or QA is usually not associated with mathematics or science so much as engineering management. Certification is more of a *programmatic* concept than a scientific concept. "Code Accreditation" is simply the process of some authority (perhaps legal or regulatory) officially declaring a code to be useable for a specific project; no general guidelines are discernible in the literature.

2.7 VERIFICATION OF CALCULATIONS

Just as it is critical to distinguish Code Verification (correctness of the mathematics) from Code Validation (correctness of the science), so we must distinguish between Verification of Codes vs. Verification of (individual) Calculations.

A code may be rigorously Verified to be (say) 2nd order accurate, but when applied to a new problem, this fact provides no *a priori* estimate of accuracy or confidence interval. It is still necessary to band the numerical error for the individual calculation, usually (and most reliably) by performing grid convergence tests. It would be preferable to have different words for these two "Verification" activities, but no clarifying term has been offered. The very important point, independent of the semantics, is that *the use of a Verified Code is not enough*.

The methodology for Verification of Calculations is a major theme of Roache (1998a). Emphasis is given to rigorous Code Verification via systematic grid convergence, and to a simple method for uniform reporting of grid convergence studies using the Grid Convergence Index (GCI). Detailed examples on easily replicated simple problems will be given below in Section 2.17.

2.8 ESTIMATION OF UNCERTAINTY

The term "Quantification of Uncertainty" can refer both to Verification of Codes and to Verification of (individual) Calculations as described above. The term also allows for inclusion of both error estimation and the more conservative error banding, which includes a factor of safety as in the Grid Convergence Index (Section 2.17) and related methods. Quantification of Uncertainty is distinguished from the more amorphous term "Confidence Building" by the key word "Quantification," and is less project-oriented than Code Certification or Quality Assurance. Etymologically, it could include Validation, but whether or not this is intended must be inferred by its context.

2.9 GRID CONVERGENCE VS. ITERATIVE CONVERGENCE

The literature commonly uses the term "convergence" in two completely different ways. Readers probably will know the distinction between iterative convergence vs. grid convergence (or residual "accuracy" vs. discretization accuracy). Usually, the meaning is clear from the context, but sometimes confusion occurs, e.g. when some new variant of an iteration algorithm is presented as being "more accurate." The "accuracy" claimed here often is residual accuracy, i.e. what is better called iterative convergence accuracy or iterative speed, and has nothing to do with the order of accuracy of the discretization.

For the present subject, we note that iterative convergence can muddy the distinction between Code Verification and Calculation Verification, since iterative tuning parameters (e.g., multigrid cycles, relaxation factors, etc.) can be problem dependent.

2.10 ERROR TAXONOMIES

Several taxonomies of errors given in the literature are inadequate and misleading. Not all lists are taxonomies. For example, "grid generation errors" or "grid resolution / distortion" errors or "factorization errors" are not separate from discretization errors for purposes of error estimation. For the Verification of a code or a calculation, there are no such things as separate grid generation errors nor are there separate errors associated with coordinate transformations. Although bad grids do add to discretization error *size*, they do not add new terms. This does not mean that one grid is as good as another, or that a really bad grid cannot magnify errors, but only that these so-called grid generation errors do not have to be considered separately from other discretization errors in a grid convergence test. If a grid convergence test is performed, and the errors are shown to reduce as $O(\Delta^2)$, for example, then all discretization errors are verified. One does not need to separately estimate or band the grid generation errors.

Likewise, numerical errors from *boundary conditions, computational domain size, temporal errors,* and *spatial errors* are not always independent. The category *boundary errors* can include ordered terms that will tend to zero as the discretization improves, so that a *boundary error* need not be considered separately from the grid convergence study. The category *boundary errors* would also include far-field boundary conditions, which are not ordered in Δ, i.e. the error persists even in the limit of $\Delta \to 0$. This does not mean that all far-field boundary conditions are equally bad; some are better than others, but *none* vanish as $\Delta \to 0$. On the other hand, *all* vanish as distance to the far-field boundary $L_b \to \infty$. This *boundary error* will improve as the *computational domain size* increases. The taxonomy already includes *temporal errors* and *spatial errors* (which are evaluated in the grid convergence study) and *computational domain size errors*, so that both the ordered boundary errors and the non-ordered (outflow, or "free") boundary errors are already counted elsewhere in the taxonomy. (Note, however, that free outflow boundary errors may prove to be ordered not in Δ but in $1/L_b$ where L_b is the distance from the region of interest to the outflow boundary. See Chapter 6 of Roache (1998a) for an example.)

The following taxonomy of errors is appropriate for purposes of Verification of Codes and Calculations.

Table 2.10 An Error Taxonomy Appropriate for Verification of Codes and Calculations.

- Errors ordered in discretization measures Δ ; these errors can be evaluated by grid convergence studies.
- Errors ordered in some numerical (rather than physical) parameter not associated with discretization (like L_b); these errors can be tested numerically in the code being Verified.
- Errors ordered in some physical parameter.
- Non-ordered approximations (like $\partial\rho/\partial n = 0$ or $\partial P/\partial n = 0$) that are conceptual modeling errors (approximations).
- Programming Errors (unlike the other errors, these errors are *mistakes*); these can be detected by grid convergence studies for a problem with an exact solution.
- Computer Round-Off Errors; these errors can be identified by grid convergence studies or *ad hoc* approaches, but often they are simply demonstrated to not be significant.

For elaboration on this error taxonomy, see Chapter 2 of Roache (1998a).

2.11 TRUNCATION ERROR VS. DISCRETIZATION ERROR

"Truncation error" is an unfortunate term. Strictly speaking, it refers to the truncation at some finite value of a series, which could be analytical or more commonly in the present context, the Taylor-series expansion of the solution. It is a worthwhile concept because it allows one to define the order of the finite difference (or finite element, finite volume, etc.) method. Unfortunately, it is often used loosely in the sense of "discretization error", i.e. the error that is caused by the fact that we can only use a finite number of elements or grid points (or another measure of discretization). In a FDM, one cannot take the limit of infinite order (i.e., limit of zero truncation error) without also taking the limit of infinite number of grid points, since high order methods require higher support. This terminology makes the limit process somewhat misleading. Also, it confuses the issue of solution smoothness with discretization error, since the Taylor series expansion depends on smoothness.

In the context of grid convergence tests, it is preferable to not speak of evaluating the "truncation error" of a numerical approximate solution, but rather the "discretization error" that arises because of the finite discretization of the problem. This terminology applies to every consistent methodology: FDM, FVM, FEM, spectral, pseudo-spectral, vortex-in-cell, etc., regardless of solution smoothness. (By "consistent" we mean that the continuum equations are recovered in the limit of infinite discretization.) The term "truncation error" is then reserved just for the examination of the order of convergence rate of the discretization. Note again the point of a taxonomy; these two errors are not independent. For any finite grid calculation, we do not have a truncation error (arising from the use of finite-ordered FVM) that we *add* to the discretization error (arising from the use of a finite number of grid points). And it is not possible to approach the limit of zero truncation error by arbitrarily increasing the order of the FDM or FEM without increasing the discretization. (Note in FEM we could fix a finite number of elements but we would still have to increase the discretization, i.e. the support within the elements.) However, the alternate is true: we can in fact approach the limit of eliminating all the discretization error by arbitrarily increasing the number of grid points without changing the order of the method. Thus, "discretization error" is the preferable term for speaking of the numerical error of a calculation, and truncation error is not separate in the taxonomy of an error estimate for a calculation. (However, the order of the truncation error is still verified in the Code Verification.)

Truncation error has also been defined as the residual resulting from substitution of the continuum (exact) solution values into the discrete equations on some grid. This definition can be useful for analysis of discrete methods. But again, it is not distinguishable from discretization error in an accuracy estimate of a real calculation.

2.12 CALIBRATION AND TUNING

The term "Calibration" is used with more latitude than Verification, Validation, or Confirmation. "Code Calibration" means the adjustment of parameters needed to fit experimental data, e.g. the 6 closure coefficient values necessary for two-equation turbulence models, especially to accommodate applications for geometries and conditions outside the envelope of their original Validation. Experiments specifically designed for this can be referred to as Calibration experiments. Often, Calibration is simply equated with the adjustment of parameters that is called "Tuning," and perhaps without scientific justification. The faintly pejorative association of Tuning is deserved if every new data set requires re-tuning, or if physically unrealistic values are used (e.g. unrealistic bottom-friction coefficients) but not so if reasonable universality is obtained. Tuning can also corrupt Validation. Validation should be predictive, so Calibration should be performed on a set of experiments separate from the Validation experiments to prevent the Validation from becoming merely a self-fulfilling prophecy.

It is important to recognize that all of these activities (Verification of Calculations, Calibration, Validation, and especially Certification) have associated with them error tolerances that cannot be arbitrarily defined universally but must be defined with reference to their intended use. As noted above, Certification is a *programmatic* concept, rather than a scientific concept.

2.13 SOFTWARE AND CODES

Another distinction often made in the software engineering community is between "software" and "code." Briefly, code is defined as computer instructions and data definitions, whereas software is a more comprehensive term, including programs, procedures, rules, and any associated documentation pertaining to the operation of a computer system. This distinguishes system software from applications software (like PDE codes) but, as readers will know, physical modelers usually overlap the terms, to no practical detriment.

2.14 CODE OPTIONS

A serious concern in Code Verification is the number of user options in a code, especially for general-purpose commercial CFD codes. This is a genuine practical problem, but does not nullify claims of Verification; it just limits those claims. The exponentially expanding complexity of the option tree does not nullify the definition of Verification of Code; it simply qualifies the definition. "Code Verification" is restricted to that combination of options claimed to be Verified. There may be a gray area here, in the judgment of the independence of options. Some knowledge of algorithm and code structure may be necessary to infer the reasonableness of simplifications of option interactions (essentially, partitioning of the full matrix of option interactions). The overly scrupulous approach of considering no independence of options will often be impractical and unnecessary. A more intelligent and economical approach is possible, always bearing in mind that subtle and unanticipated option interactions have occurred, especially in codes built before the acceptance of structured coding and modularity.

In regard to "canned" or highly modular elliptic solvers, one can make a very strong case for independence of the options. The argument becomes unassailable if separate residual checks are made after the canned solver has obtained the solution. Essentially, the solver is treated as providing an initial guess for a simple point-iterative method. (See Roache (1998a) for examples.)

2.15 ZONAL VERIFICATION

A powerful simulation approach involves zonal modeling or splitting, in which different governing equations are applied in different physical zones. Even problem dimensionality can differ between zones, as when a 2-D top layer including external forcing is used with a 3-D lower layer. The calculations are performed with much smaller time-steps in the top layer, and the effects must be communicated through the interface. This approach poses special considerations for both Code Verification and Calculation Verification. The basic approach is the same, but the coding for each zone is verified independently, as is the patching. Only the discretization error can be estimated through grid convergence testing, not the patching error; the latter, like far-field approximations, is not an ordered approximation, and therefore does not vanish as $\Delta \to 0$.

2.16 ETYMOLOGY AND NEAR-SYNONYMS

As noted earlier, *validation, verification* and *confirmation* in common (non-technical) use are given as synonyms in a common English thesaurus. More complete definitions indicate extensive overlap of these words in some contexts, and with other related words such as *substantiation, sanction, approval, authoritative, cogent, convincing, truth, authenticity, accuracy, genuineness, definite assurance,* even *logical*. The Middle English / Latin root of *verification* is the equivalent of *truth*, whereas the Latin root of *validation* is the equivalent of *strong*. This etymology contributes, at best, a shade of meaning in favor of the distinction used herein, but mainly serves to support the position that these are technical terms defined specifically in a technical context.

2.17 LIMITATIONS OF SEMANTIC DISTINCTIONS

The semantic distinctions involved in the general area of confidence-building are important and worthwhile. While sometimes arbitrary, it is worthwhile to try to maintain uniformity of terminology, or at least to recognize the underlying conceptual distinctions and to define one's terms with appropriate precision. It is well to recognize the limitations of semantic distinction; although these efforts are worthwhile, it is clear that the scientific - mathematical - engineering community are *not* going to achieve completely uniform, non-overlapping terminology. The most important distinction, and fortunately the most generally agreed upon one, is the distinction between Verification (mathematics) and Validation (science).

We *Verify a Code* by convincingly demonstrating (if not "proving" as in a mathematical theorem) that the code *can* solve the equations right. When done properly, the exercise also Verifies the *order of convergence* of the code. Then, we *Verify a Calculation* by convincingly demonstrating that the code *has* solved the equations right, to a rationally estimated accuracy or error band. These two exercises are purely mathematical; neither appeals to experimental data for their justification. Only in *Validation* do we demonstrate that we have solved the right equations, with an understood context of engineering or scientific accuracy, by appealing to experimental data.

2.18 BASIC METHODOLOGY OF CALCULATION VERIFICATION

2.18.1 Informal Verification of Calculations

Verification of Calculations has often been done informally, especially for site-specific models. Most site-specific grids were not chosen completely arbitrarily. Rather, they evolved over much use (sometimes years) with experience-based intuition and from experimentation with varying grid density, but perhaps not from a systematic study. Although it would not be surprising that such grids indeed provide adequate numerical accuracy for the intended use, it is also true that the numerical accuracy estimate is un-quantified and the justification is often un-documented. This shortcoming of previous studies can be addressed *a posteriori*. For long-running parametric studies at one site, it may not be necessary to repeat the Calculation Verification for each problem. The fact that there may be hundreds of problems simulated makes it *easier*, not more difficult, to justify the expense of rigorous Calculation Verification for one or a few samples. However, note that it is likely that a grid that is marginally acceptable for typical conditions will be unacceptable for extreme events (e.g., floods).

Even for more formal Verification of Calculations, some studies are more thorough than others. The various examples presented in this publication cover a range of formality and thoroughness. The reader is asked to bear in mind that this publication project was originally conceived as primarily a compilation of experimental data bases for Validations. Over years, the scope has been extended to add value, but some of the calculations go back over a decade, well before the publication of Roache (1998a) and the widespread acceptance of terminology and the development of methodologies. Here, we present examples of a more thorough Calculation Verification that may be taken as a paradigm. The examples embody several levels of thoroughness or rigor (although the term "rigor" is not deserved in a strict mathematical sense).

2.18.2 Formal Verification of Calculations using the GCI

The following examples of Calculation Verification use the Grid Convergence Index (GCI) as defined in Roache (1998ab). The GCI provides a uniform method of reporting grid convergence studies through a simple formula based on (a) asymptotic theory that rationally normalizes the tolerances of various studies by including the effect of grid refinement ratio r and order of convergence p, and (b) an empirically based Factor of Safety Fs that correlates roughly with the goal of providing numerical error bars that are conservative in 95% of the cases. For development of the GCI and discussion on the numerical experiments, see Roache (1998ab); for later evaluations, see Roache (2003). The "numerical error bars" refer only to ordered discretization errors (due to finite mesh size), not to other modeling errors such as non-ordered boundary condition errors and position of the outflow boundaries, turbulence modeling, patching errors, etc. as discussed earlier in this Chapter.

What is often reported from a grid convergence study is the difference ε in the solutions f_1 obtained in the fine grid and f_2 in the coarse grid. The quantities represented by f can be local values of flow or transport variables, or integrated functionals of the solutions such as bottom drag, sediment load, etc. The latter functionals are usually more of engineering interest and usually produce better convergence properties, since they are only slightly affected by local problem areas of slow or non-monotone convergence. The f values used may be fractional, as in the examples here, or absolute. For fractional values based on a local value normalization, ε is defined as

$$\varepsilon = \frac{f_2 - f_1}{f_1} \tag{2.1}$$

(For a % indication, multiply by 100%.) The GCI for the fine grid solution using the very conservative Factor of Safety Fs is given by

$$\text{GCI}\left[\text{fine grid}\right] = F_s \frac{|\varepsilon|}{r^p - 1} \quad , \quad F_s = 3 \tag{2.2}$$

where r is the grid refinement ratio, $r = \Delta_2/\Delta_1$ and Δ is some representative size of the grid elements, and p is the rate of convergence, e.g. $p = 2$ for a second-order method.

Obviously, if the denominator of Eqn. 2.1 is small, ε should be normalized by (denominator replaced by) some other characteristic quantity for the calculation. Alternately it can be evaluated as an absolute (rather than relative) quantity (i.e., without the division by f_1 or any normalizing value), in which case Eqn. 2.2 produces an "absolute error" GCI.

Note that for a grid doubling ($r = 2$) with a 2nd-order method ($p = 2$), the denominator = 3, and we obtain GCI = $|\varepsilon|$.

It is well recognized that ostensibly 2nd-order algorithms may fail to attain 2nd-order performance in a particular calculation, due to subtleties in nonlinear problems, overly strong grid stretching, failure to attain the asymptotic range, etc. (Roache 1998ab). Unless the author has convincingly Verified that the code actually attains the theoretical order, at least on a "nearby" problem, the conservative value of $F_s = 3$ should be used in reporting the GCI in Eqn. 2.2. For more rigorous grid convergence studies using 3 or more grids to actually calculate rather than assume an observed order of convergence p (examples are given below) one may use the less conservative value $Fs = 1.25$ in Eqn. 2.2, giving a more optimistic GCI.

In some practical situations, we actually use the coarse grid solution. Consider a parametric study in which hundreds of variations are to be run. A scrupulous approach would require a grid convergence study for each case, but most engineers would be satisfied with one or a few good grid convergence tests on representative problems. For the bulk of the stack of calculations, we would be using the coarse grid solution, and we want a GCI for it. In this case, the coarse-grid GCI must be less optimistic.

$$GCI\ [coarse\ grid\ ] = F_s \left|\varepsilon\right| r^p / (r^p - 1),\ \ F_s = 3 \tag{2.3}$$

2.18.3 Examples of Formal Verification of Calculations using the GCI

The following simple examples of the calculation of the GCI are based on the easily reproduced case of a steady-state nonlinear Burgers equation, often used as a one-dimensional prototype of fluid dynamics problems.

$$-U\,U_x + U_{xx} / Re = 0,\ \ U(0) = 1,\ \ U(1) = 0 \tag{2.4}$$

For the solution value, we evaluate the one-dimensional "shear" $f = dU/dx$ at $x = 1$. The examples are given in numerical detail to facilitate confirmation by the reader.

For $Re = 1000$ solved with 2nd-order centered differences on a uniform grid, a very fine grid calculation with 2000 cells produces $f_1 = -529.41$.

Next, we *coarsen* the grid to 1600 cells ($r = 2000/1600 = 1.25$) which produces coarse grid solution $f_2 = -544.48$.

The quantity typically reported (from Eqn. 2.1) would be

$$\begin{aligned}|\varepsilon| &= 100\% \times |(f_2 - f_1) / f_1| \\ &= 100\ |(-544.48+529.41) / 529.41| = 2.85\%.\end{aligned} \tag{2.5}$$

In Eqn. 2.2 for the GCI, the factor

$$(r^p - 1) = (1.25^2 - 1) = 0.5625. \tag{2.6}$$

The GCI from Eqn. 2.2 with the conservative $F_s = 3$ is

$$\begin{aligned}\text{GCI [fine grid]} &= 3\ |\varepsilon|/(r^p - 1) \\ &= 3 \times 2.85\% / 0.5625 \\ &= 15.20\%.\end{aligned} \tag{2.7}$$

Comparison with the exact solution $f_{exact} = -500.00$ indicates that the exact magnitude of the fine-grid error A_1 is

$$\begin{aligned} A1 &= 100\% \times |(f_2 - f_{exact}) / f_{exact}| \\ &= 100\% \times |(-529.41 + 500.00)| / 500.00 = 5.88\%. \end{aligned} \quad (2.8)$$

As is typical, the ε is not conservative (2.85% < 5.88%) and the GCI with $F_s = 3$ is conservative (15.20% > 5.88%), in the spirit of a 95% error band.

If a 3-grid study is performed (on this problem or a nearby one) and it is determined that the actual observed rate of convergence is close to the theoretical $p = 2$, then the recommended value of $F_s = 1.25$ can be used. This produces

$$GCI = 6.34\% > 5.88\% \quad (2.9)$$

still conservative but more palatable than the 15.2% from using $Fs = 3$.

If the coarse grid solution of 1600 cells were to be used, the GCI with $F_s = 3$ would be increased from Eqn. (2.3) to

$$GCI\ [\text{coarse grid}] = 15.20\% + 3 \times 2.85\% = 23.75\%. \quad (2.10)$$

The actual magnitude of the coarse grid error is

$$\begin{aligned} A2 &= 100\% \times |(f_2 - f_{exact}) / f_{exact}| \\ &= 100\% \times |-544.48 + 500.00| / 500.00 \\ &= 8.90\%. \end{aligned} \quad (2.11)$$

Again, the GCI using $Fs = 3$ is very conservative (23.75% > 8.90%). Using $Fs =$ 1.25, the coarse-grid is

$$GCI = 9.90\% > 8.90\% \quad (2.12)$$

which is still adequately conservative.

Summary recommendations for the GCI:

(a) Use $F_s = 1.25$ for convergence studies with a minimum of three grids (on either the subject problem or a "nearby" one) to experimentally demonstrate the observed order of convergence p on the actual problem,
(b) Use $F_s = 3$ for two-grid convergence studies.

2.18.4 Examples of Calculation of Observed Order of Convergence

We illustrate the calculation of the observed order of convergence p again using the easily reproduced case of a steady-state nonlinear Burgers equation.

This requires a minimum of 3 grid solutions.

It is not necessary that the grid refinement ratio r be (approximately) constant over the 3 grids, but this avoids solving a transcendental algebraic equation for p. Nor is it necessary for these grid sizes to be integer multiples. The first example will be for a simple case of three grids with $r = 2$, and the second will be more general and realistic.

2.18.4.1 Extracting Observed p for Constant r

For the grid refinement factor r = constant over a 3 grid set, p can be solved from (corrected) Eqn. (5.10.6.4) of Roache (1998a).

$$p = ln\ \{\varepsilon 23/\varepsilon 12\}/\ \ln\{r\} \tag{2.13}$$

The numerical solutions of the one-dimensional "shear" $f = dU / dx$ at $x = 1$ given below were produced using formally second-order methods (theoretical $p = 2$) and Re = 10. (They correspond to Cases 26 and 28 of Table 5.9.1, Chapter 5 of Roache (1998a).)

Ncell	f	r
25	$f3$ =-5.1932	
		2
50	$f2$ = -5.0504	
		2
100	$f1$ = -5.0134	

$$\varepsilon 23 = (f3 - f2)/f2 = (-5.1932 + 5.0504)/(-5.0504) = 0.0282749 \text{ or } 2.83\%$$

$$\varepsilon 12 = (f2 - f1)/f1 = (-5.0504 + 5.0134)/(-5.0134) = 0.0073802 \text{ or } 0.74\%$$

Then Eqn. (2.13) gives observed p as

$$p = ln\ \{\varepsilon 23/\varepsilon 12\}/\ \ln\{r\}$$
$$= ln\ \{0.0282749/0.0073802\}/\ \ln\{2\} = 1.3414/0.69315$$

$$\text{observed } p = 1.935 \tag{2.14}$$

over the 3-grid triplet (100, 50, 25). Intuitively, this appears to be close enough to use the theoretical value of 2.0 in the GCI calculation. It is even somewhat better than it appears. The actual performance will be better as the grids are refined. Whereas this observed p = 1.935 is an average over grids of 25, 50, 100 cells, the 100 cell grid is further into the asymptotic regime. This is confirmed for this case where the exact solution is known. In this situation, the observed p may be calculated from only 2 grid solutions, 50 and 100, using a formula easily derived from Eqn. 5.4.1 of Roache (1998a) (but not included in Roache (1998a)).

$$d = (f1 - f2) / (f_{\text{exact}} - f1) \qquad (2.15a)$$

$$p = ln\ \{d + 1\} / ln\{r\} \qquad (2.15b)$$

(Obviously, this approach is not applicable to a practical calculation, but is used here to demonstrate that typically the observed p calculated over a widely separated grid triplet, as here with overall $r = 4$, is often pessimistic.) Applying Eqn. (2.15) with the $f_{\text{exact}} = -5.001$ gives

$$\textit{observed}\ p\ (25{,}50) = ln\{d{+}1\}/\ ln\{r\}$$
$$= ln(3.8907)/\ ln(2) = 1.3586/0.6931$$

$$\textit{observed}\ p = 1.9600 \qquad (2.16)$$

This is even closer to the theoretical $p = 2$ than the value 1.935 obtained from the 3-grid set. Also, note that we do not expect observed $p = 2$ exactly, but only asymptotically. These observed values are in the asymptotic range.

2.18.4.2 Extracting Observed p for Variable r

The more general procedure for extracting p, not restricted to constant r, is to solve the following equation Roache (1998a) for p.

$$\frac{\varepsilon_{23}}{r_{23}^{p} - 1} = r_{12}^{p}\left[\frac{\varepsilon_{12}}{r_{12}^{p} - 1}\right] \qquad (2.17)$$

(This equation differs from Eqn. 2.13 only in higher order terms and only for the local normalization definition of ε.)

Usual solution techniques can be applied to solve this transcendental equation for p, e.g., direct substitution iteration, Newton-Raphson, even graphical. When considering the stability of the iteration, one should allow for observed $p < 1$. This can happen even for simple problems at least locally and in some cases the observed p is < 0. (Unfortunately, behavior far away from asymptotic convergence can be non-monotone.) Also, $r \sim 2$ will be easier to solve than $r \sim 1$, and $r >> 2$ is probably not of much interest. For well behaved synthetic cases tested, direct substitution iteration with a relaxation factor $\omega \sim 0.5$ works well and is robust, but faster convergence can usually be obtained $\omega = 0$ (no relaxation). With ρ = previous iterate for p, the iteration equation is

$$p = \omega\rho + (1 - \omega)\frac{\ln(\beta)}{\ln(r_{12})} \qquad (2.18a)$$

$$\beta = \frac{(r_{12}^p - 1)\varepsilon_{23}}{(r_{23}^p - 1)\varepsilon_{12}} \tag{2.18b}$$

Note this form of the iteration gives the exact answer in one step for the case of r = constant and $\omega = 0$.

Again, the numerical solutions of the one-dimensional "shear" $f = dU / dx$ at $x = 1$ given below were produced using formally second-order methods (theoretical $p = 2$) and Re = 10. (They correspond to Cases 2 and 28 of Table 5.9.1, Chapter 5 of Roache (1998a).)

Ncell	f	r
40	$f3$ =-5.0778	
		1.25
50	$f2$ = -5.0504	
		2
100	$f1$ = -5.0134	

ε23 = (f3 - f2)/f2 = (-5.0778 + 5.0504)/(-5.0504)
= 0.0054253 or 0.54%

ε12 = (f2 - f1)/f1 = (-5.0504 + 5.0134)/(-5.0134)
= 0.0073802 or 0.74%

Using the expected theoretical value of $p = 2$ to start the iteration, and the relaxation factor $\omega = 0$, we obtain the following sequence of estimates for observed p.

$$\begin{aligned} p &= 2 \\ &= 1.9710798 \\ &= 1.9584008 \\ &= 1.9528545 \\ &= 1.9504307 \\ &\ldots \\ p &= 1.948551 \text{ (converged)} \end{aligned} \tag{2.19}$$

This observed p = 1.949 determined from the grid triplet (100, 50, 40) differs only slightly from p = 1.935 obtained from the triplet (100, 50, 25), again indicating that the grids are within the asymptotic range. Also, note that the value p = 1.960 obtained above from the grid doublet (100, 50) using the known exact solution is in better agreement with the result from (100, 50, 40). This is an important practical consideration, since there is a widespread erroneous belief that integer grid refinement is preferable; here, we have shown that the observed p extracted from solution on the non-integer triplet (100, 50, 40) is more accurate than that from the integer triplet (100, 50, 25). See also discussion and examples in Roache (1998a), Chapter 5.

Once p is known with some confidence, one may predict the next level of grid refinement r^* necessary to achieve a target accuracy. See Roache (1998a), Chapter 5 for details and for discussion of additional features of grid convergence studies.

2.18.4.3. Extracting Observed p for Noisy Convergence

In many practical problems, convergence is not monotonic even on the finest grid triplets used, especially for local variables. (A major cause of noise is lack of strict geometric similarity within the grid sets (Eça and Hoekstra 2002)) When observed p over 4 or more grids is not constant, the recommended approach is to use the least-squares GCI procedure developed by Eça and Hoekstra (2002).

This procedure requires a minimum of 4 grid solutions for determination of effective convergence rates, which provide improved error estimation for the difficult problems. For difficult realistic problems, more than the minimum 4 grids may be necessary; (Eça and Hoekstra 2002) obtains "fairly stable results using about 6 grids with total refinement ratio near 2." We recommend the Eça-Hoekstra procedure for noisy p problems, with the additional step of limiting the maximum p used in the GCI to theoretical p.

The least squares approach has been applied (Eça and Hoekstra 2002) to several models of convergence. The simplest method works as well as any, and is recommended as follows. The assumed one-term expansion of the discretization error is

$$f_i - f_\infty \cong \alpha \Delta_i^p \tag{2.20}$$

The least squares approach is based on minimizing the function

$$S(f_\infty, \alpha, p) = \sqrt{\sum_{i=1}^{Ng} \left[f_i - (f_\infty + \alpha \Delta_i^p) \right]^2} \tag{2.21}$$

where the notation f_∞ (different from that of Eça and Hoekstra (2002)) suggests the limit of fine resolution (in the absence of round-off error). Setting the derivatives of S with respect to f_∞, α, p equal to zero leads to

$$f_\infty = \frac{1}{Ng} \left\{ \sum_{i=1}^{Ng} f_i - \alpha \sum_{i=1}^{Ng} \Delta_i^p \right\} \tag{2.22a}$$

$$\alpha = \frac{Ng\sum_{i=1}^{Ng} f_i\Delta_i^p - \left(\sum_{i=1}^{Ng} f_i\right)\left(\sum_{i=1}^{Ng}\Delta_i^p\right)}{Ng\sum_{i=1}^{Ng}\Delta_i^{2p} - \left(\sum_{i=1}^{Ng}\Delta_i^p\right)\left(\sum_{i=1}^{Ng}\Delta_i^p\right)} \tag{2.22b}$$

$$\sum_{i=1}^{Ng} f_i\Delta_i^p \log(\Delta_i) - f_\infty\sum_{i=1}^{Ng}\Delta_i^p \log(\Delta_i) - \alpha\sum_{i=1}^{Ng}\Delta_i^{2p}\log(\Delta_i) = 0 \tag{2.22c}$$

The last equation is nonlinear and is solved iteratively by a false position method for observed p. (Eça and Hoekstra 2002)

To repeat, it clearly would be imprudent to calculate GCI with observed p > theoretical p. Although such superconvergence can occur, and would be appropriate to use if one were actually using the extrapolated solution (see Roache (1998a) for discussion), we recommend for uncertainty calculations that max p = theoretical p be used, e.g. maximum $p = 2$ for a nominally second-order method.

On the other hand, there seems to be no reason to categorically reject observed $p < 1$. If observed p is < 1, it probably means that the coarsest grid is somewhat outside the asymptotic range, and the resulting uncertainty estimate of the GCI will be overly conservative (Roache 2003). This is not an impediment to publication or reporting.

2.18.5 Grid Convergence with Reduced Dimensionality

Often, parts of a conceptual model have reduced dimensionality. For example, a small inlet stream flowing into a large bay might be conceptualized as a one-dimensional sub-model of the two- or three-dimensional main model. The 1-D sub-model has no boundary layers, etc. It is not necessary, nor even desirable, to change this dimensionality in a grid convergence study. The sub-model remains 1-D as the grid is refined, but the grid resolution along that one dimension increases by the same factor as the rest of the 2-D or 3-D model grid.

2.18.6 Incremental Costs and Benefits of Calculation Verification

While CFD practitioners may shrink from the idea of performing multiple-grid studies, the following observations on the incremental costs and benefits of thorough Calculation Verification are relevant.

First, unless your code contains some single-grid error indicator (Roache 1998a), at least two grids are absolutely necessary in order to obtain any *quantified* estimate of numerical uncertainty. Even if the code uses something like the Zhu-Zienkiewicz estimators (Pelletier and Roache, 2006), which is highly recommended (Roache 1998a), these do not provide error bands for the quantities of engineering interest, and by themselves (in a single grid calculation) can provide no indication that the actual rate of convergence is adequate.

Thus, two grids are a minimum for *quantification* of error bands. But these two grids need not represent a doubling in each direction; in fact, there are advantages besides economy to using less than integer coarsening/refining (Roache 1998a), e.g. ~10% coarsening. For example, we may want to verify the calculations on a grid of 100 cells in a direction, and obtain our error bars for this grid solution by obtaining another solution with 90 cells. Note this is cheaper than the original solution, so the computer cost is less than doubled. More importantly, the human cost is far from doubled. The incremental human cost to run the second grid solution should be amortized over the problem set-up time, including geometry specification, decisions on type of boundary conditions, etc.

Given the fact that two grids are a minimum, and that integer-grid coarsening is not required, consider the incremental costs and benefits of obtaining even more grid solutions. An obvious upper bound for the incremental work (computer and human) to obtain the third grid solution is 50% (3 solutions instead of 2) but this is not at all a tight upper bound. If the third grid uses 80 cells, the problem is 3-D time dependent, optimal efficient solvers are used, and the time step is correctly scaled along with the grid size, then the computation time varies (Roache 1998a,b) as Δ^4. If the base (100-cell) grid cost is 1, then the 90-cell and 80-cell solutions cost 0.66 and 0.41. The incremental computer cost for the third grid is 0.41/1.66 = 0.25. (The incremental cost is even less than 25% if non-optimal solvers are used. For typical point-iterative elliptic solvers, which dominate the costs, the variation is Δ^5 or more.) The incremental human cost is probably less than this 25%.

For this roughly 25% incremental cost, what is the incremental benefit? The gain in credibility and confidence is very significant indeed. Convergence rate is now *verified*, rather than assumed. The GCI factor of safety can be confidently reduced from 3 to 1.25, so that a previously reported error band of (say) 12% can be reduced to a more optimistic 5%. The confidence of all readers (or customers, regulators, stakeholders, etc.) is justifiably increased as *evidence* is provided to demonstrate that the modelers are serious about assessing accuracy.

Further work produces yet higher benefit-to-cost ratios, provided that the coarsest grids used are still in a reasonable range. With least-squares determination of observed order of convergence *p*, or perhaps other intuitive approaches to use multiple grid triplets (e.g. averaging *p* from each of the four triplets in a four-grid sequence, or weighting the *p* from the finest grid triplet) reader confidence turns to admiration. If the results are demonstrably well-behaved (e.g. monotonic and in reasonable agreement with theoretical *p*) then numerical accuracy is no longer a divisive issue.

Further, it is important to bear in mind that these verifications need not be performed on every calculation in a large study if many of the problems are "nearby" to one another, in the sense of parameter ranges and dominant physics. This will be the situation when large numbers of parametric runs are performed on the same site, and in design studies (e.g., dams and spillways) in which the various geometries are highly derivative of one another.

Even more favorable cost-benefits accrue when we consider the overall modeling costs including Validation. In Chapter 8 it is acknowledged that "The field data collection program in most cases consumes a major portion of the modeling budget." It is clear that the easiest gains in V&V quality can be obtained by Calculation Verifications. The accuracy of the agreement with experiments may or may not be improved, but the confidence in the *reasons* for the disagreement will be greatly improved. This part of the V&V problem can be solved convincingly, with straightforward, well established techniques. The prevalence of powerful computing resources removes any excuses for not doing this job well. Neither Code Verification nor Calculation Verification should contribute substantially to any doubts about the overall V&V process. In contrast, Validation and experimental difficulties will always be with us.

In summary, consideration of the incremental costs and benefits of thorough Calculation Verification makes a very good professional and economic case for increasing thoroughness in Verification of Calculations using the methodologies described herein.

2.19 REFERENCES

Eça, L. and Hoekstra, M., 2002. "An Evaluation of Verification Procedures for CFD Algorithms", *Proc. 24th Symposium on Naval Hydrodynamics*, Fukuoka, Japan, 8-13 July 2002.

Oberkampf, W. L. and Trucano, T. G., 2002. "Verification and Validation in Computational Fluid Dynamics," *Progress in Aerospace Sciences*, vol. 38, no. 3, 2002, pp. 209-272.

Pelletier , D. and Roache, P. J., 2006, "Verification and Validation of Computational Heat Transfer", Chapter 13 of *Handbook of Numerical Heat Transfer*, Second Edition, W. J. Minkowycz, E. M. Sparrow, and J. Y. Murthy, eds., Wiley, New York.

Roache, P. J., 1998a. *Verification and Validation in Computational Science and Engineering*, Hermosa Publishers, Albuquerque, August 1998. http://www.hermosa-pub.com/hermosa.

Roache, P. J., 1998b. *Fundamentals of Computational Fluid Dynamics*, Hermosa Publishers, Albuquerque, December 1998. *http://www.hermosa-pub.com/hermosa*.

Roache, P. J., 2003. "Error Bars for CFD," AIAA Paper 2003-0408, AIAA 41st Aerospace Sciences Meeting, January 2003, Reno, Nevada.

CHAPTER 3

ANALYTICAL SOLUTIONS FOR MATHEMATICAL VERIFICATION

Peter E. Smith

3.1 INTRODUCTION

Contributors: Peter E. Smith and Francisco J. Rueda

In this chapter we present a suite of analytical solutions and one perturbation solution that can be used in verifying 3-D hydrodynamic models of free-surface flows. The objective is to present the solutions in a concise and consistent form that can readily be used by a model developer or user to test a model computer code. Although we include here only a relatively small subset of the analytical solutions available in the published literature, the solutions were selected to be useful in testing many of the individual terms and groups of terms from the governing equations for 3-D shallow water flow. An emphasis has been placed on geophysical problems that are within the purview of hydraulic and coastal engineers. All solutions are for the hydrodynamic equations rather than the advection-diffusion (mass transport) equation, for which solutions are readily available elsewhere (e.g., Baptista et al., 1995).

The analytical solutions are included as part of a set of 8 test cases contributed by different authors. Each test case is presented in a similar format. For each test case the continuum partial differential equations (PDEs) and continuum boundary and initial conditions to be solved are presented. Because no known analytical solutions exist for the complete nonlinear forms of the continuum PDEs for 3-D shallow water flow, the equations have been simplified by removing certain terms and equation nonlinearities. The analytical solution for each test case is given first in general mathematical terms along with the assumptions that are needed in simplifying the continuum equations. Then the test case (or test problem) itself is defined by a set of physical parameters that specify the domain geometry and a set of boundary and initial conditions. The use of standard test cases is meant to encourage a comparison of solutions between modeling groups. A minimum of one high-resolution and one low-resolution numerical mesh of uniform square cells in the horizontal plane are suggested for each test case. Meshes of square cells are suggested because the simple structure of these meshes can be accommodated by most models. Square-celled meshes also are most appropriate for grid-convergence testing. The user is free, however, to test a model using a different choice of structured mesh or to use a fully unstructured mesh. The number of "degrees of freedom" is presented for each square-celled mesh so that other types of meshes can be used with an equivalent level of resolution. In each test case, at least one numerical solution is presented using an existing hydrodynamic code and is compared with the analytical solution. These applications are mostly not rigorous mathematical verifications of the code done in the manner recommended in this report, but they do serve to demonstrate a code application to the test case and to provide examples of metrics that can be applied in the evaluation of errors.

These test cases can be used in verifying a PDE code. In Chapter 2, code verification is defined in general terms as "solving the equations right." This means that a code must be verified (1) to be free of coding errors; (2) to be consistent, in the limit of infinite discretization, with the continuum equations being solved; and (3) to establish the order of convergence of the underlying numerical scheme. The order of convergence is expressed by the overall order of accuracy for the numerical scheme as determined, not by the theoretical Taylor series analysis of the discretized equations, but by the observed behavior of errors in the discrete solution obtained from systematic grid convergence tests. A methodology for verifying the accuracy of codes and examples of applications are provided by Roache (1998) in chapters 3 through 8.

The accuracy of a code should be verified before the code is applied to any laboratory or field problem such as the test cases presented in chapters 5 or 6 of this report. In theory, verification should be done by the code developers and should not be a concern of the model user. In practice, however, many of the codes in widespread use today, especially 3-D codes which are generally newer and have more complex features and options than 1-D or 2-D codes, have not been convincingly and rigorously verified in the manner recommended in this report. Model users should verify a code themselves if they are not firmly convinced that it has been done previously. In part, the failure of some developers to verify their codes fully has been a result of not having analytical solutions available in a convenient and clearly posed form. With this chapter we hope to remedy that problem at least partially. Chapter 4 is closely linked to this chapter and provides additional analytical test cases for code verification developed with the Method of Manufactured Solutions (MMS). The MMS provides a general procedure for deriving analytical solutions when the nonlinearities in the governing equations have not all been removed.

When running the test cases described herein, some modifications to the program instructions for nonlinear hydrodynamic codes may be necessary to "turn off" certain terms or remove equation nonlinearities. In some nonlinear codes used today, the code developers have included input options (or switches) that allow users to choose among standard forms of equation linearizations. This is a useful enhancement to a code. Not only can these options help the code developers verify their own code, but they also conveniently allow code users to do their own verification testing to confirm or add to the testing done by the developer and to assure their version of the compiled code is working correctly on standardized analytical solutions.

In writing this report we wish to highlight a need within the civil engineering profession to place greater emphasis on the importance of verifying the accuracy of computational hydraulics codes and calculations. Often results from computational hydraulics models are published that cannot be evaluated for accuracy because formal verification of the model code and a systematic grid convergence study has not been done or made available. The editors of this report believe that a formal written policy to establish acceptable standards for the publication of numerical results should be adopted by the editors of the ASCE Journal of Hydraulic Engineering (JHE). Such an editorial policy has already been adopted by several other leading fluid mechanics-related journals (see Roache, 1998, appendices A and B, for an informative discussion on these editorial policies and their implementation). Although adopting a publication policy on standards for numerical accuracy would likely be controversial, it would do much to improve the quality of published numerical work in the JHE.

3.1.1 Mathematical Problem: Assumptions, Equations and Boundary Conditions

Contributor: Francisco J. Rueda

The governing equations that form the basis of 3-D hydrodynamic free-surface models are the continuity (or mass conservation) equation for incompressible fluids, the Reynolds averaged form of the Navier-Stokes equations for momentum, the transport equation for active scalar fields, and an equation of state relating the scalar variables to the density field. By assuming that there are no point withdrawals or inflows of discharge, that the hydrostatic pressure and Boussinesq approximations are valid, and that there are no volumetric source or sink terms in the scalar transport equation, the governing equations may be written in Cartesian rectangular coordinates as

Mass conservation: continuity equation

$$\frac{\partial u}{\partial x}+\frac{\partial v}{\partial y}+\frac{\partial w}{\partial z}=0 \tag{3.1}$$

Momentum conservation: Reynolds-averaged Navier-Stokes equations

$$\underbrace{\frac{Du}{Dt}}_{\mathbf{A}}-\underbrace{fv}_{\mathbf{B}}=-\frac{1}{\rho_0}\frac{\partial p}{\partial x}+\frac{\partial}{\partial x}\left(K_H\frac{\partial u}{\partial x}\right)+\frac{\partial}{\partial y}\left(K_H\frac{\partial u}{\partial y}\right)+\underbrace{\frac{\partial}{\partial z}\left(K_V\frac{\partial u}{\partial z}\right)}_{\mathbf{C}} \tag{3.2}$$

$$\overbrace{\frac{Dv}{Dt}}+\overbrace{fu}=-\frac{1}{\rho_0}\frac{\partial p}{\partial y}+\frac{\partial}{\partial x}\left(K_H\frac{\partial v}{\partial x}\right)+\frac{\partial}{\partial y}\left(K_H\frac{\partial v}{\partial y}\right)+\overbrace{\frac{\partial}{\partial z}\left(K_V\frac{\partial v}{\partial z}\right)} \tag{3.3}$$

$$0 = -\frac{\rho}{\rho_0} g - \frac{1}{\rho_0}\frac{\partial p}{\partial z} \tag{3.4}$$

Scalar transport equation:

$$\frac{\partial S}{\partial t} + u\frac{\partial S}{\partial x} + v\frac{\partial S}{\partial y} + w\frac{\partial S}{\partial z} = \frac{\partial}{\partial x}\left(D_H \frac{\partial S}{\partial x}\right) + \frac{\partial}{\partial y}\left(D_H \frac{\partial S}{\partial y}\right) + \frac{\partial}{\partial z}\left(D_V \frac{\partial S}{\partial z}\right) \tag{3.5}$$

Equation of state:

$$\rho = \rho(S_i) \qquad i = 1, \ldots, n_S \tag{3.6}$$

A right-handed coordinate system fixed on the earth's surface with horizontal coordinates x(east) and y(north) and vertical coordinate z is adopted here. The bold letters **A**, **B**, and **C** are used as identifiers of terms. The total derivative operator $D(\ )/Dt$ used in the x- and y-momentum equations (terms **A**) includes both the unsteady and the advective terms ($= \partial(\ )/\partial t + u\partial(\ )/\partial x + v\partial(\ )/\partial y + w\partial(\ )/\partial z$). The velocity components in the x-, y-, and z-directions are given by u, v, and w; f is the Coriolis parameter (assumed zero in all the test cases in this chapter); g is the acceleration of gravity; ρ is the local water density; ρ_0 is a mean reference value of the water density; S_i is one of n_S active scalars (such as temperature, salt, suspended solids, etc.); and p is pressure. Owing to the Boussinesq approximation, the variability of the density was neglected everywhere in the governing equations except in the gravity term. Also, the well-known Boussinesq eddy viscosity concept was used in equations 3.2 and 3.3 to model the turbulent transport of momentum; the eddy diffusivity concept was used in equation 3.5 to model the turbulent transport of scalar variables. The coefficients K_H and K_V represent horizontal and vertical turbulent momentum transport coefficients (or eddy viscosities), and D_H and D_V stand for horizontal and vertical turbulent transport coefficients (or eddy diffusivities) for the active scalars S_i.

In invoking the hydrostatic pressure approximation, it is assumed that the weight of the fluid identically balances the fluid pressure. As a result of this assumption, the z-momentum equation reduces to the hydrostatic pressure equation (eq. 3.4). The non-hydrostatic terms can be safely neglected so long as the wave phenomena being represented have horizontal wavelengths that are much longer than their vertical wave heights, which is the case for basin-scale oscillations. Using the hydrostatic equation, the terms involving the gradients of pressure in the x- and y-momentum equations (eqs. 3.2 and 3.3) can be reformulated in terms of the position of the free surface and the density field as (Smith, 2006)

$$\frac{1}{\rho_0}\frac{\partial p}{\partial x} = \frac{1}{\rho_0}\frac{\partial p_a}{\partial x} + \underbrace{g\frac{\partial \zeta}{\partial x}}_{\mathbf{D}} + \underbrace{g\frac{1}{\rho_0}\int_z^{\zeta}\frac{\partial \rho}{\partial x}dz'}_{\mathbf{E}} \tag{3.7}$$

$$\frac{1}{\rho_0}\frac{\partial p}{\partial y} = \frac{1}{\rho_0}\frac{\partial p_a}{\partial y} + \overbrace{g\frac{\partial \zeta}{\partial y}} + \overbrace{g\frac{1}{\rho_0}\int_z^{\zeta}\frac{\partial \rho}{\partial y}dz'} \tag{3.8}$$

where p_a is the atmospheric pressure and ζ is the water surface elevation above an undisturbed level that can be taken as $z = 0$. The integral terms in equations 3.7 and 3.8 (terms **E**) are referred to as the baroclinic pressure terms and couple the hydrodynamic solution to the solution of the active scalar field. These terms are a function of the vertical coordinate z. The water surface slope terms (terms **D**), on the other hand, are referred to as the barotropic pressure terms and are not a function of depth.

Equations 3.7 and 3.8 allow the free-surface elevation ζ to replace the pressure as the unknown in the horizontal momentum equations 3.2 and 3.3. The terms in equations 3.7 and 3.8 involving the atmospheric pressure will be neglected for the test cases in this chapter. The baroclinic pressure terms are required only in test cases 6 and 7. The continuity equation, in terms of ζ, is obtained by integrating equation 3.1 over the water column and combining the result with the equations for the kinematic boundary conditions at the free surface and bottom boundaries (see below, eqs. 3.12 and 3.16). The integrated form of the continuity equation is essentially two-dimensional because the integrals of the 3-D velocity components over the depth of flow represent depth-integrated volumetric fluxes. That continuity equation is then

$$\frac{\partial \zeta}{\partial t} + \frac{\partial}{\partial x}\left[\int_{-H}^{\zeta} u dz\right] + \frac{\partial}{\partial y}\left[\int_{-H}^{\zeta} v dz\right] = 0 \qquad \mathbf{F} \tag{3.9}$$

where $z = -H(x, y)$ is the depth of the bottom boundary measured from the undisturbed free surface ($z = 0$). In 3-D calculations the vertical velocity component, w, typically is obtained by using the 3-D form (eq. 3.1) of the continuity equation.

The boundary conditions at the free surface, $z = \zeta(x, y, t)$, are

$$\rho_0 K_V\left(\frac{\partial u}{\partial z}, \frac{\partial v}{\partial z}\right)\bigg|_{z=\zeta} = \left(\tau_{sx}, \tau_{sy}\right) \qquad \mathbf{G} \tag{3.10}$$

$$\rho_0 D_V\left(\frac{\partial S}{\partial z}\right) = H_S \qquad \mathbf{H} \tag{3.11}$$

$$w\big|_{z=\zeta} = \frac{\partial \zeta}{\partial t} + u\frac{\partial \zeta}{\partial x} + v\frac{\partial \zeta}{\partial y} \qquad (3.12)$$

where (τ_{sx}, τ_{sy}) are the component directional wind stresses acting on the free surface, and H_S represents any flux of scalars across the free surface (scalar transport is simulated only in test case 6, and H_S is assumed to be zero there). Equation 3.12 represents the kinematic boundary condition, already incorporated in equation 3.9, which requires that a particle that starts on the free surface remains on the free surface. At the lower (bottom) boundary, $z = -H(x, y)$, the conditions are

$$\rho_0 K_V \left(\frac{\partial u}{\partial z}, \frac{\partial v}{\partial z} \right)\Bigg|_{z=-H} = \left(\tau_{bx}, \tau_{by}\right) \qquad \mathbf{I} \qquad (3.13)$$

or

$$(u, v)\big|_{z=-H} = (0, 0) \qquad \mathbf{J} \qquad (3.14)$$

and

$$\frac{\partial S}{\partial n}\bigg|_{z=-H} = 0 \qquad \mathbf{K} \qquad (3.15)$$

$$w\big|_{z=-H} = -u\frac{\partial H}{\partial x} - v\frac{\partial H}{\partial y} \qquad \mathbf{L} \qquad (3.16)$$

Here n is the direction normal to the bottom boundary, and (τ_{bx}, τ_{by}) are the component directional bottom frictional stresses exerted by the boundary on the fluid. Equations 3.13 and 3.14 impose two different boundary conditions for the hydrodynamic solution at the bottom boundary. The formal hydrodynamic condition at a wall boundary is zero velocity, as represented by equation 3.14. However, in the numerical modeling of flows in natural channels that have irregular bottom profiles and that use course resolution grids near the boundaries, this boundary condition, known as "no-slip," is often not appropriate. In this chapter the no-slip boundary condition is used only in test case 5. Equation 3.13, known as the dynamic boundary condition, equates the component frictional stresses exerted by the boundary on the fluid with the stresses in the fluid exerted on the boundary. In engineering practice the boundary frictional stresses are usually determined from the bottom velocity by a quadratic stress law derived from the well-known Law of the Wall. To avoid the quadratic nonlinearity in the boundary condition, most test cases in this chapter assume the bottom frictional stresses are zero ($\tau_{bx} = \tau_{by} = 0$), which is known as a "perfect-slip" condition. The perfect-slip condition allows water to move parallel to the bottom without any resistance. Test case 4 and the second part of test case 5 do not use the perfect-slip condition, but instead they each employ a linear stress model to replace the usual quadratic one.

Along the shoreline, in the horizontal direction, a perfect-slip boundary condition also is imposed. No flux of water or any scalar is allowed across the shoreline (**M**). At open boundaries, either the velocity or the water surface elevation must be defined (**N**). The test cases in this chapter do not require the simulation of any scalar variable at an open boundary.

3.1.2 Test Cases for the Verification of 3-D Numerical Models: the Linear Problem

As stated in the previous section, the mathematical problem for solving 3-D free-surface flows is nonlinear because the unknowns appear in combinations within several terms in the governing equations. Nonlinearities exist in the integral terms of the continuity equation (eq. 3.9) and the advective terms ($u\partial(\ )/\partial x + v\partial(\ )/\partial y + w\partial(\ )/\partial z$) within the total derivative operator ($D(\ )/Dt$) in the horizontal momentum equations (eqs. 3.2 and 3.3). The continuity equation is linear when reduced to the form

$$\frac{\partial \zeta}{\partial t} + \frac{\partial}{\partial x}\left[\int_{-H}^{0} u dz\right] + \frac{\partial}{\partial y}\left[\int_{-H}^{0} v dz\right] = 0 \tag{3.9.a}$$

For the special case when u and v are not functions of the vertical coordinate z and H is constant (a horizontal bottom boundary), equation 3.9a simplifies further to the form

$$\frac{\partial \zeta}{\partial t} + H\frac{\partial u}{\partial x} + H\frac{\partial v}{\partial y} = 0\,, \tag{3.9.b}$$

which is used in the first three test cases. The momentum equations are linearized by retaining only the unsteady term ($\partial(\ )/\partial t$) in the total derivative operator $D(\ )/Dt$.

Once the governing equations and boundary conditions are linear, one can test the numerical discretization and implementation of the computer algorithm using test problems that have analytical solutions. Here we present eight simple test problems having analytical solutions that are intended to test different features of a 3-D numerical model. Those features include (1) the discretization and implementation of individual terms in the governing equations, (2) the representation of arbitrary domains, and (3) the correct implementation of boundary conditions at the free-surface, bottom, and lateral (open and closed) boundaries. The suite of test cases presented here (table 3.1.1) is by no means a complete compilation of available problems having analytical solutions. There is, for example, no solution given for testing the numerical discretization used in a model for the Coriolis terms (terms **B**). The test cases, however, do provide a good starting point and a consistent framework for verifying 3-D numerical models that solve the equations describing shallow water, free-surface flows.

Table 3.1.1 Numerical test cases for model verification.
[Note: The capital letters identify the terms or boundary conditions in the governing equations (see previous section) in which their discretization/implementation is verified.]

TEST CASE	FEATURES VERIFIED
1. Free-surface seiching in a closed rectangular basin with a horizontal bottom	**A**, **D**, **F**, **M**
2. Free-surface seiching in a closed circular basin with a horizontal bottom	**A**, **D**, **F**, **M** and representation of curved lateral boundary
3. Tidal forcing in a rectangular basin with a horizontal bottom	**A**, **D**, **F**, **M**, **N**
4. Tidal forcing in a rectangular basin with a varying bottom slope and linear bottom friction	**A**, **C**, **D**, **F**, **M**, **N**, **I**, **L** and representation of varying bottom slope
5. Wind-driven flow in a closed rectangular basin with a horizontal bottom	**C**, **D**, **G**, **I** or **J**
6. Internal-wave seiching in a closed rectangular basin with a horizontal bottom	**A**, **E**, **F** and advective scalar transport
7. Density-driven flow in a rectangular basin with a horizontal bottom	**C**, **D**, **F**, **E**, **I**
8. A three step procedure to check nonlinear effects for tidal circulation in a square basin with a horizontal bottom	**D**, **M**, **N** and nonlinear momentum advection

3.1.3 Quantitative Evaluation of Errors

The differences between the analytical solution and its numerical representation were mostly quantified by using the set of error measures defined in Table 3.1.2. The Root Mean Square Error (RMSE) is not dimensionless, hence its value will depend on the units used in the numerical and analytical solutions. In most test cases, we have used centimeters per second (cms^{-1}) for the magnitude of velocity, centimeters (cm) for water surface elevation, and kilograms per cubic meter (kgm^{-3}) for density. Phase and dissipation errors were also evaluated in some test cases by comparing the results from the models with the analytical solutions. They are given here in the form of a celerity ratio Q and dissipation factor ε_A. Calculating Q and ε_A can be explained by considering a wave of angular frequency ω, whose real and imaginary components are Ω and ν respectively (i.e., $\omega = \Omega + j\,\nu$). A given numerical scheme will represent this wave as having an angular frequency ω_{num} ($= \Omega_{num} + j\,\nu_{num}$), which in general is different from ω. The ratio Ω_{num}/Ω defines the velocity or celerity ratio Q and characterizes the phase error introduced by the algorithm. The dissipation-amplification factor, ε_A ($= \exp\{-\nu_{num}\,\Delta t\}$), relates the amplitudes of the numerical representation of the wave at times $t + \Delta t$ and t (Cathers and O'Connor 1985) and determines the stability and dissipative properties of the method. A necessary condition for stability of a scheme is that $\varepsilon_A \leq 1 + O(\Delta t)$ for all wave numbers (Fletcher, 1991, Pinder and Gray, 1977). A method with $\varepsilon_A < 1$ is said to be dissipative, whereas the method is considered to be neutrally stable if $\varepsilon_A = 1$. Estimates of the frequency and the amplitude of the analytical and numerical waves are obtained with the non-linear Levenberg-Marquardt method (e.g., Press et al., 1992). The implementation of this method in the MATLAB® (The Mathworks, Inc.) function *lsqcurvefit* has been used in some of the problems presented in this monograph.

Table 3.1.2 Definition of error measures used to evaluate numerical results. A_i = analytical solution at time step (or node) i, N_i = numerical result at that same time (or node). Primes denote deviations from the mean analytical solution. The symbol i denotes the i-th data point in a time series (or grid), and i_{max} is the total number of data points in the time series (or grid).

Error Measures	**Definition**
Index of agreement (d)	$d = 1 - \dfrac{\sum_{i=1}^{i\max}(A_i - N_i)^2}{\sum_{i=1}^{i\max}(\lvert A_i'\rvert - \lvert N_i'\rvert)^2}$
ℓ_1 *error norm*	$l_1 = \dfrac{\sum_{i=1}^{i\max}\lvert N_i - A_i\rvert}{\sum_{i=1}^{i\max}\lvert A_i\rvert}$
ℓ_2 *error norm*	$l_2 = \dfrac{\left[\sum_{i=1}^{i\max}(N_i - A_i)^2\right]^{1/2}}{\left[\sum_{i=1}^{i\max}(A_i)^2\right]^{1/2}}$
ℓ_∞ *error norm*	$l_\infty = \dfrac{\max_{\forall i}\lvert N_i - A_i\rvert}{\max_{\forall i}\lvert A_i\rvert}$
Root Mean Square Error (RMSE)	$RMSE = \sqrt{\dfrac{\sum_{i=1}^{i\max}(N_i - A_i)^2}{i_{\max}}}$
Error (E) for a given node i in steady state solutions	$E = \lvert N_i - A_i\rvert$

3.1.4 Overview of Models

SI3D: The model SI3D uses a semi-implicit, three-level, leapfrog-trapezoidal finite difference scheme on a staggered Cartesian grid (Arakawa C-grid) to solve the layered-averaged form of the governing equations. The semi-implicit approach is based on treating the gravity-wave and vertical-diffusion terms implicitly to prevent limitations on the size of the model time step for these terms from affecting the stability of the method. All other terms, including advection, are treated explicitly.

The approach avoids using mode splitting to solve the problem posed by a system of equations that supports both fast (barotropic or external) and slow (baroclinic or internal) waves. The iterative leapfrog-trapezoidal algorithm for time stepping gives second order accuracy in both time and space. The objective of using trapezoidal iterations is to remove the well-known computational mode associated with the leapfrog discretization (Durran, 1998) and to increase the stability of the code. Further details of the algorithm were reported by Smith (2006).

SHULIN: The model SHULIN is a two-dimensional model that uses a fully implicit finite difference scheme on a staggered Cartesian grid (Arakawa C-grid) to determine the horizontal-averaged velocities and water surface elevations. The system of nonlinear equations is solved by the well-known Newton iteration method.

CCHE3D: The model CCHE3D was developed at the National Center for Computational Hydroscience and Engineering at the University of Mississippi. It was designed to simulate unsteady, free-surface, three-dimensional turbulent flows and sediment/pollutant transport and water quality. The model is based on the Efficient Element Method, a collocation version of the Galerkin approach. The basic element is a cube having twenty-seven nodes. Either the hydrostatic or the dynamic (non-hydrostatic) pressure of the flow can be simulated, depending on the nature of the problem. A staggered grid is used for the pressure solution. The velocity correction method is adopted to compute the dynamic pressure and enforce mass conservation. A solution of the kinematic equation for the free surface determines the location of the free surface. Five different methods for computing the eddy viscosity are available in the model (Jia et. al., 2001, 2005).

HUANG/SPAULDING MODEL: This model is described by Huang and Spaulding (1995). The model consists of components for hydrodynamics, pollutant transport, and turbulence closure. The conservation equations are solved by finite difference techniques. A semi-implicit algorithm is used to simulate the vertically averaged flow, and a vertically implicit procedure is used to determine the interior flow and also model the salinity and turbulent kinetic energy and dissipation. The model uses an algebraic transformation within the vertical sigma coordinate system. Vertical grid stretching is dependent on local water depth. The model has been tested using analytical solutions, laboratory experimental data, and field observations.

3.1.5 List of Symbols and Abbreviations

Note: Other symbols that are mostly minor or for mathematical substitution are defined in the text. The symbols for error measures are defined in table 3.1.2.

A_s	Amplitude of the Bessel function of order s in test case 2
a	Amplitude of oscillations in free-surface elevation for test cases 1, 3, 4, and 8
B	Width of rectangular basin (dimension in y-direction)
c_D	Non-dimensional drag coefficient used in test case 4

D/Dt	Total derivative ($\partial/\partial t + u\partial/\partial x + v\partial/\partial y + w\partial/\partial z$)
D_H, D_V	Horizontal and vertical eddy diffusivity coefficients
f	Coriolis parameter
g	Acceleration of gravity (9.81 m/s^2)
H	Constant depth of water below equilibrium position (datum) in all test cases but 4 and 5.
$H(x)$	Depth of water below equilibrium position (datum) in test case 4
H_0	Parameter used in defining the depth in test case 4
HR(n)	High resolution test case, n referring test node n
H_S	Flux of scalars through the free surface
h	Total depth of water ($H + \zeta$)
j	$= \sqrt{(-1)}$
J_s	Bessel function of the first kind and order s
J_s'	Derivative of Bessel function of first kind and order s
K	Non-dimensional slip coefficient used in test case 4
K_H, K_V	Horizontal and vertical eddy viscosity coefficients
k	Coefficient in linear friction law relating bottom velocity to bottom stress in test cases 4 and 5
L	Length of rectangular basin in x-direction
LR(n)	Low resolution test case, n referring to test node n
l	Horizontal wave number for the internal mode of oscillation in test case 4
m	Vertical wave number for internal mode of oscillation in test case 4
N	Constant Brunt-Väisäla frequency used in test case 6
n	Direction normal to the boundary
n_S	Number of active scalar constituents
p	Hydrostatic pressure
Q	Celerity ratio for definition of phase-lag error in numerical solution
R_0	Radius of circular basin in test case 2
Re	Real part of a complex number
r	Radial coordinate for circular basin in test case 2
S	Active scalar concentration (salinity or temperature)
$\hat{S}$	Mean background scalar concentration used to represent temperature in test case 6
s	Number of nodal diameters in free-surface oscillations (seiches) for a circular basin in test case 2
T	Tidal wave period in test case 8
T_s	Total simulation time in example test cases
t	Time
t' ($= t \times \omega_T$)	Non-dimensional time used in test case 8
U, V	Depth-averaged velocity components in x- and y-directions in test case 8
U_0	Depth-invariant, high-frequency limit for u in test case 4

U_w	Resultant wind speed in test case 5
u, v, w	Velocity components in the x-, y-, and z-directions
$u_{\max}$	Water velocity in the x-direction at the free surface in test case 5
u_w, v_w	Wind velocity components in the x- and y- directions in test case 5
ν	Imaginary part of the angular frequency ω of an analytical wave
ν_{num}	Numerical representation of ν
x, y, z	Spatial coordinates in a right hand system (z upwards)
x_1, x_2	Geometric parameters used in defining the basin geometry for test case 4
y' $(= y/L)$	Non-dimensional spatial coordinate used in test case 8
Z	Normalized depth, $Z = z/H(x)$, in test case 4
λ	Wavelength in test case 4
Δt	Time step in example test cases
Δx, Δy, Δz	Size of computational cells in x-, y- and z-directions
$\partial\Omega$	Lateral boundaries of the solution domain in test case 2
$\in (= \omega_T^2 L^2 / (gh))$	Perturbation parameter in the analytical solution for test case 8
ε_A	Dissipation factor (= 1 for neutrally stable algorithms)
ζ_0	Zero-order solution for perturbation method in test case 8
ζ_1	First-order solution for perturbation method in test case 8
ζ	Displacement of free surface from equilibrium position (datum)
θ	Azimuthal coordinate for circular basin in test case 2
κ	Parameter in solution of test case 2 (radial wave number)
μ_0	Constant (s^{-1}) used in definition of vertical eddy viscosity K_V in test case 4
ρ	Local water density
$\hat{\rho}$	Mean background density used in test case 6
ρ_0	Mean reference value for the water density
ρ_a	Density of air above free surface in test case 5
σ	Frequency of oscillations in test case 2
τ_{bx}, τ_{by}	Bottom frictional stresses in the x- and y-directions
τ_{sx}, τ_{sy}	Surface wind stresses in the x- and y-directions
Ω	Real part of the angular frequency ω of an analytical wave
Ω_{num}	Numerical representation of Ω
ω_T	Frequency of tidal oscillations in test cases 3, 4, and 8
ω	Frequency of baroclinic oscillations in test case 6

3.1.6 References

Baptista, A. M., Adams, E. E., and Gresho, P., 1995, Benchmarks for the transport equation—the convection-diffusion forum and beyond, *in* Lynch, D. R., and Davies, A. M., eds., Quantitative Skill Assessment for Coastal Ocean Models: American Geophysical Union, Coastal and Estuarine Studies, Series no. 47, p. 241-268.

Cathers, B. and O'Connor, B.A., 1985, The group velocity of some numerical schemes: International Journal for Numerical Methods in Fluids, v. 5, no. 3, p. 201-224.

Durran, D. R., 1998, Numerical Methods for Wave Equations in Geophysical Fluid Dynamics, Springer, New York, 465 p.

Fletcher, C. A. J., 1991, Computational Techniques for Fluid Dynamics (2nd Edition): Springer Series in Computational Physics, 401 p.

Huang, W. and Spaulding, M. L., 1995, 3D model of estuarine circulation and water quality induced by surface discharges: American Society of Civil Engineers, Journal of Hydraulic Engineering, v. 121, no. 4, p. 300-311.

Jia, Y., Kitamura, T., and Wang, S. S. Y., 2001, Simulation of scour process in a plunging pool of loose bed-material: American Society of Civil Engineers, Journal of Hydraulic Engineering, v. 127, no. 3, p. 219-229.

Jia, Y., Scott, S., Xu, Y. C., Huang, S. L., and Wang, S. S. Y., 2005, Three-dimensional numerical simulation and analysis of flows around a submerged weir in a channel bendway: American Society of Civil Engineers, Journal of Hydraulic Engineering, v. 131, no. 8, p. 682-693.

Pinder, G. F., and Gray, W. G., 1977, Finite Element Simulation in Surface and Subsurface Hydrology: Academic Press, New York, 307 p.

Press, W. H., Teukolsky, S. A., Vetterling, W. T., and Flannery, B. P., 1992, Numerical Recipes in Fortran—The Art of Scientific Computing (2nd edition): Cambridge University Press, 963 p.

Roache, P.J., 1998, Verification and Validation in Computational Science and Engineering, Hermosa Publishers, Albuquerque, New Mexico, 446 p.

Smith, P. E., 2006, A semi-implicit, three-dimensional model for estuarine circulation: U.S. Geological Survey Open-File Report 2006-1004, 176 p. (accessed August 10, 2007 at the URL http://pubs.usgs.gov/of/2006/1004/).

3.2 TEST CASE 1: FREE-SURFACE SEICHING IN A CLOSED RECTANGULAR BASIN WITH A HORIZONTAL BOTTOM

Contributors: Francisco J. Rueda and Peter E. Smith

3.2.1 Description and Objectives

The physical phenomenon of an oscillating free surface (seiche) in a natural basin is the result of an interaction between inertia and gravity. The analytical solution for this problem using the simplest geometry—a rectangular basin with horizontal bottom—is well-known and can be easily implemented even in spreadsheets. This solution can be used to test the basic numerical scheme in a model without the complications of open boundaries, varying bottom topography, or friction. In particular, this test problem checks the discretization and implementation of the temporal acceleration ($\partial u/\partial t$) and water surface slope terms (terms **D** from equations 3.7 and 3.8) in the simplified form of the horizontal momentum equations given in equations 3.19 and 3.20. It also tests the implementation of the continuity equation (in the form of equations 3.17 and 3.18), both for the solution of the free-surface position and for the calculation of the vertical velocities throughout the 3-D domain. Furthermore, phase, amplitude and mass conservation errors in the model can be examined as a function of discretization parameters, allowing one to test the convergence of a numerical scheme.

For this test case, the initial location of the free surface is defined as a cosine wave of small amplitude (eq. 3.25) in the longitudinal (x-coordinate) direction. The fluid is assumed to be frictionless (inviscid) so the fluid seiching continues undamped forever in the analytical solution. The lateral (y-direction) velocity component v and the variation in the water surface elevation in the lateral direction remain everywhere equal to zero. Although zero in magnitude, the terms $\partial v/\partial y$, $\partial v/\partial t$, and $\partial \zeta/\partial y$ in the governing equations are included in the mathematical statement of the problem. This is a reminder in testing a fully 3-D code that it is preferable that these terms not be hard-wired to zero. It should be checked in the numerical solution that these terms do indeed remain equal to zero (or negligibly small).

For notational simplicity in presenting the governing equations for this test case and others, subscripts are used to represent partial differentiation (e.g., $u_t = \partial u/\partial t$, $u_x = \partial u/\partial x$).

3.2.2 The Mathematical Problem

3.2.2.1 Governing Equations

Continuity

$$u_x + v_y + w_z = 0 \tag{3.17}$$

$$\zeta_t + u_x H + v_y H = 0 \tag{3.18}$$

Assumptions:

- u, v are not functions of the vertical coordinate z (note, however, that the vertical velocity w is a function of z)

- $\zeta << H$ (allows a linearization of the code
- H is constant

Momentum

$$u_t + g\zeta_x = 0 \tag{3.19}$$

$$v_t + g\zeta_y = 0 \tag{3.20}$$

Assumptions:

- Inviscid fluid (interfacial and bottom friction are neglected)
- Homogeneous fluid (the water density, ρ, is assumed to be constant)
- Nonrotating reference frame (the Coriolis parameter, f, is assigned to be zero)
- $uu_x, vu_y << u_t, g\zeta_x$ (the x-direction advection terms are neglected)
- $uv_x, vv_y << v_t, g\zeta_y$ (the y-direction advection terms are neglected)

3.2.2.2 Domain

The domain for test case 1 is shown in figure 3.2.1.

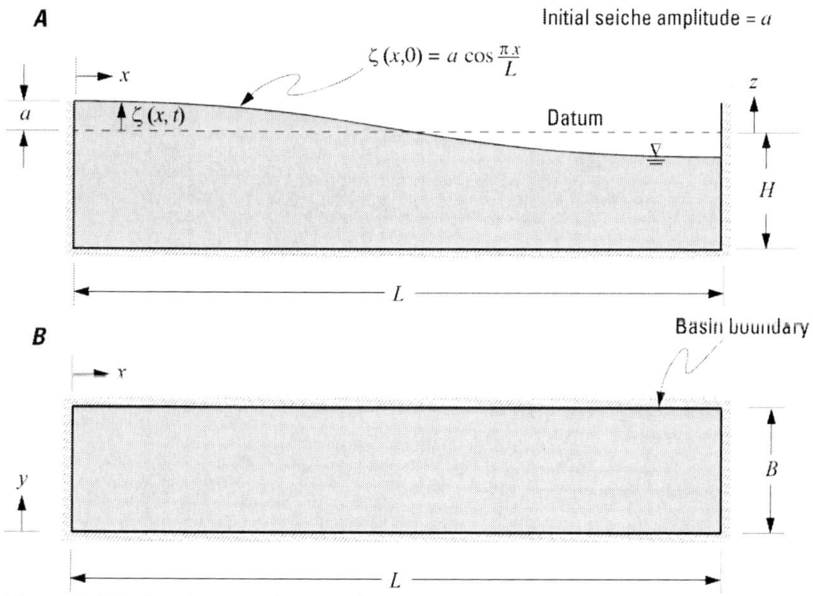

Figure 3.2.1. Test Case 1: Diagram of the domain for the test problem of free-surface seiching in a closed rectangular basin with a horizontal bottom.

(***A***) Cross section of rectangular basin showing initial position of the free surface, and (***B***) top view. The vertical scale in (***A***) is greatly exaggerated and distorted.

3.2.2.3 Boundary Conditions

$$\left.\frac{\partial u}{\partial y}\right|_{y=0} = \left.\frac{\partial u}{\partial y}\right|_{y=B} = u|_{x=0} = u|_{x=L} = 0 \tag{3.21}$$

$$\left.\frac{\partial v}{\partial x}\right|_{x=0} = \left.\frac{\partial v}{\partial x}\right|_{x=L} = v|_{y=0} = v|_{y=B} = 0 \tag{3.22}$$

$$\left.\frac{\partial u}{\partial z}\right|_{z=-H} = \left.\frac{\partial v}{\partial z}\right|_{z=-H} = \left.\frac{\partial u}{\partial z}\right|_{z=\zeta} = \left.\frac{\partial v}{\partial z}\right|_{z=\zeta} = 0 \tag{3.23}$$

$$w|_{z=-H} = 0 \tag{3.24}$$

3.2.2.4 Initial Conditions

$$\zeta|_{t=0} = a\cos\left(\frac{\pi}{L}x\right) \tag{3.25}$$

$$u, v, w|_{t=0} = 0 \tag{3.26}$$

3.2.3 Analytical Solution (Neuman and Pierson, 1966)

$$\zeta(x,t) = a\cos\left(\frac{\pi}{L}x\right)\cos\left(\frac{\pi\sqrt{gH}}{L}t\right) \tag{3.27}$$

$$u(x,t) = \frac{a\sqrt{gH}}{H}\sin\left(\frac{\pi}{L}x\right)\sin\left(\frac{\pi\sqrt{gH}}{L}t\right) \tag{3.28}$$

$$v(x,t) = 0 \tag{3.29}$$

$$w(x,z,t) = -\frac{a\pi\sqrt{gH}}{L}\frac{z+H}{H}\cos\left(\frac{\pi}{L}x\right)\sin\left(\frac{\pi\sqrt{gH}}{L}t\right) \tag{3.30}$$

3.2.4 Test Case

3.2.4.1 Physical Parameters

L = 40,000 m
B = 8,000 m
H = 12 m
a = 0.25 m

3.2.4.2 Numerical Parameters

High resolution

A regular mesh of 960 (20 x 4 x 12) computational cells, or an equivalent mesh of 2,960 degrees of freedom (3 degrees of freedom per computational cell plus 1 for each water column), can be used for a high-resolution test case. The calculated total number for the degrees of freedom includes all nodes on the basin boundary. The regular mesh cells in the horizontal plane are squares ($\Delta x = \Delta y$). It is assumed that a mesh of twelve vertical layers is used. Only 1/12 of the count for degrees of freedom should be used to discretize the rectangular basin horizontally. Of course, the user is not restricted to using regular meshes; irregular finite difference, finite element, or body-fitted meshes are encouraged, subject to the specifications on degrees of freedom.

For the regular mesh, the suggested values for the discretization parameters are

$$\Delta x = \Delta y = 2{,}000 \text{ m} \quad \text{implying} \quad \lambda/\Delta x = 40$$
$$\Delta z = 1 \text{ m}$$
$$T_s = 24 \text{ hours}$$

where $\lambda = 80{,}000$ m is the wavelength of the seiche and T_s is the duration of the simulation.

Modelers are asked to report the numerical treatment of the time domain in full reproducible detail. For time-stepping schemes, the following time step is proposed, which guarantees a stable solution for explicit methods:

$$\Delta t = 150 \text{ seconds} \quad \text{leading to} \quad \sqrt{gH}\,\Delta t/\Delta x \text{ (Courant Number)} = 0.81$$

Low resolution

A regular mesh of 240 (10 x 2 x 12) computational cells, or an equivalent mesh of 740 degrees of freedom (3 per computational cell plus 1 for each water column), can be used for a low-resolution test case. For the regular mesh, the suggested values for the discretization parameters are

$$\Delta x = \Delta y = 4{,}000 \text{ m} \quad \text{implying} \quad \lambda/\Delta x = 20$$
$$\Delta z = 1 \text{ m}$$
$$T_s = 24 \text{ hours}$$

For time stepping schemes a step size Δt of 300 seconds is proposed, leading to a Courant number ($\sqrt{gH}\,\Delta t/\Delta x$) of 0.81.

3.2.4.3 Reporting of Model Results

Modelers should report the $\ell 1$, $\ell 2$ and $\ell \infty$ error norms, RMSE, and the index of agreement d for the longitudinal velocity (u, in cm/s), vertical velocity (w, in cm/s), and the free-surface elevation (ζ, in cm) at the following nodal locations

Low resolution:
$x = 6{,}000$ m $y = 2{,}000$ m $z = -0.5$ m

High resolution:
$x = 7{,}000$ m $y = 1{,}000$ m $z = -0.5$ m

As an option, modelers are also invited to submit phase and dissipation errors in the form of celerity ratio Q and dissipation factor ε_A.

3.2.5 Example Verification: Low Resolution Test Case

A comparison of the analytical and numerical solutions for the low-resolution test case is given below in figure 3.2.2. The SI3D model was used to compute the numerical solution. The results for the high-resolution test case mimic the analytical solution very well. The computed error measures for the low-resolution numerical solution are given in table 3.2.1.

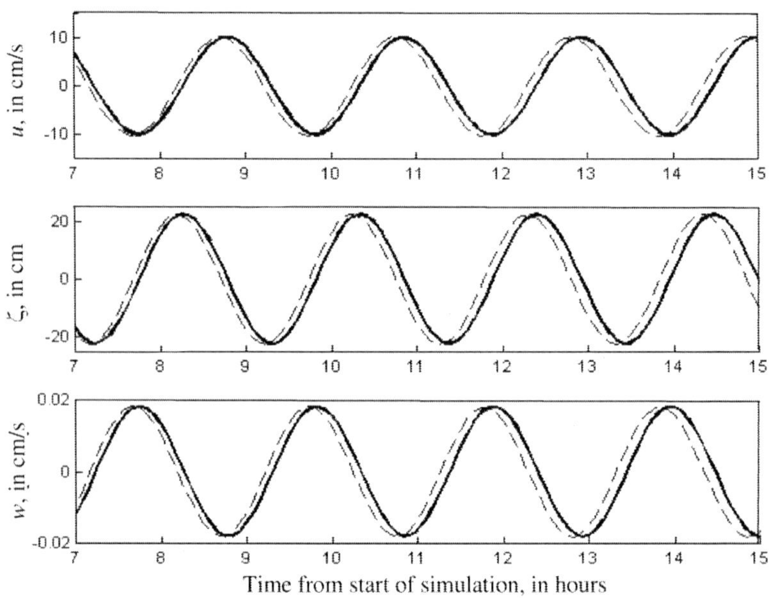

Figure 3.2.2 Test Case 1: Graphical display of analytical and numerical solutions for the low-resolution test case.

Analytical solution is dashed line; numerical solution is solid line. The results are reported for the location x = 6,000 m, y = 2,000 m, and z = – 0.5 m. The numerical calculations were done with the SI3D model using 1 leapfrog and 1 trapezoidal iteration for time stepping.

Table 3.2.1 Test Case 1: Error measures for the low-resolution solution.

	u	w	ζ
ε_A	—	—	—
Q	—	—	—
d	0.9612	0.9612	0.9634
$\ell 1$	0.3346	0.3357	0.3273
$\ell 2$	0.3855	0.3870	0.7751
$\ell \infty$	0.6467	0.6519	0.6379
RMSE	2.7867	4.9596×10^{-3}	5.9649

The results in figure 3.2.2 show that a lagging phase error exists in the numerical solution. The differences in the solutions becomes larger over time as the error accumulates. There is very little error in the computed amplitude of the seiche.

3.2.6 References

Neuman, G., and W. J. Pierson, 1966, Principles of Physical Oceanography: Prentice Hall, Englewood Cliffs, N.J.

3.3 TEST CASE 2: FREE-SURFACE SEICHING IN A CLOSED CIRCULAR BASIN WITH A HORIZONTAL BOTTOM

Contributors: Francisco J. Rueda and Peter E. Smith

3.3.1 Description and Objectives

Although the underlying physics in this problem consists, as in test case 1, of a simple interaction between inertia and gravity, the curved boundary introduces an additional element of complexity. A curved boundary cannot easily be represented with Cartesian grids, and hence this test provides a simple way to check the ability of 3-D models that are based on structured grids to represent complex plan-form geometries and their effects on hydrodynamic processes. Also, the imperfect representation of the boundaries can influence the phase, amplitude, and mass conservation errors of the model. The effect of discretization parameters on these aspects of a numerical solution can be checked with this test case.

3.3.2 The Mathematical Problem

3.3.2.1 Governing Equations

Continuity

$$u_x + v_y + w_z = 0 \tag{3.31}$$

$$\zeta_t + u_x H + v_y H = 0 \tag{3.32}$$

Assumptions:

- u, v are not functions of the vertical coordinate z (note, however, that the vertical velocity w is a function of z)
- $\zeta << H$ (allows a linearization of the code)
- H is constant

Momentum

$$u_t + g\zeta_x = 0 \tag{3.33}$$

$$v_t + g\zeta_y = 0 \tag{3.34}$$

Assumptions:

- Inviscid fluid (interfacial and bottom friction are neglected)
- Homogeneous fluid (ρ = constant)
- Nonrotating framework ($f = 0$)
- $uu_x, vu_y << u_t, g\zeta_x$ (the x-direction advection terms are neglected)
- $uv_x, vv_y << v_t, g\zeta_y$ (the y-direction advection terms are neglected)

3.3.2.2 Domain

The domain for test case 2 is shown in figure 3.3.1.

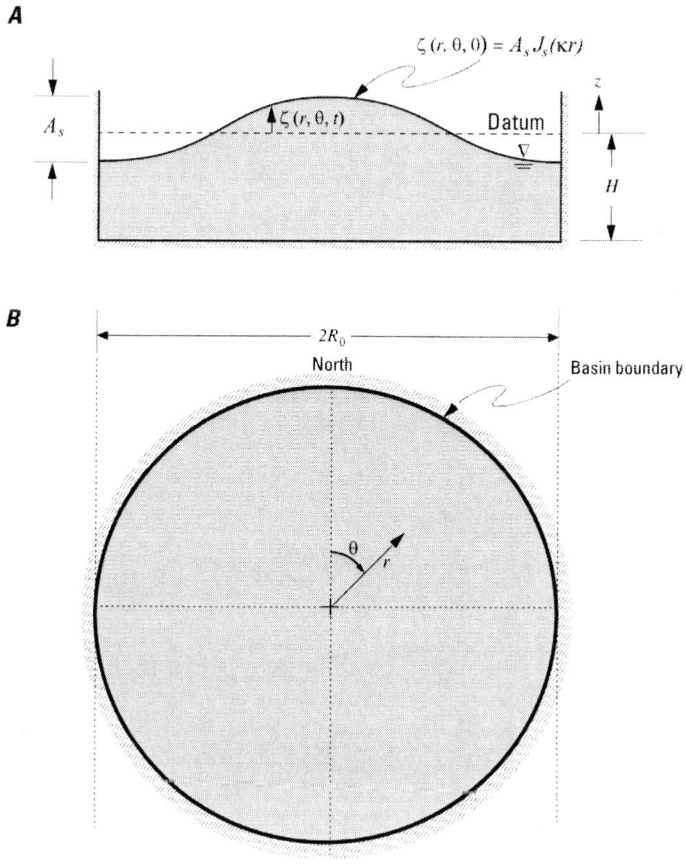

Figure 3.3.1 Test Case 2: Diagram of the domain for the test problem of free-surface seiching in a closed circular basin with a horizontal bottom.

(***A***) Cross section of circular basin showing initial position of the free surface, and (***B***) top view. For the specific test case, $s = 0$ (J_0 is a Bessel function of the first kind and zero order) and $\kappa = 3.83171/R_0$. The vertical scale in (***A***) is greatly exaggerated and distorted.

3.3.2.3 Boundary Conditions

$$\left.\frac{\partial \zeta}{\partial n}\right|_{\partial\Omega} = 0 \qquad (3.35)$$

$$(u, v) \cdot \vec{n} \,\big|_{\partial\Omega} = 0 \tag{3.36}$$

$$\left.\frac{\partial u}{\partial z}\right|_{z=-H} = \left.\frac{\partial v}{\partial z}\right|_{z=-H} = \left.\frac{\partial u}{\partial z}\right|_{z=\zeta} = \left.\frac{\partial v}{\partial z}\right|_{z=\zeta} = 0 \tag{3.37}$$

$$w\big|_{z=-H} = 0 \tag{3.38}$$

where $\partial\Omega$ refers to the lateral boundary of the circular basin

3.3.2.4 Initial Conditions

$$\zeta\big|_{t=0} = A_s J_s(\kappa r)\cos(s\theta) \tag{3.39}$$

$$(u, v, w)\big|_{t=0} = (0, 0, 0) \tag{3.40}$$

3.3.3 Analytical Solution (Derived from Lamb, 1945, p.285) for an Arbitrary Number of Nodal Circles and Diameters

$$\zeta(r, \theta, t) = \mathrm{Re}\left[A_s J_s(\kappa r)\cos(s\theta)\exp(j\sigma t)\right] \tag{3.41}$$

$$u(r, \theta, t) = \mathrm{Re}\left[\frac{jg}{\sigma}\left(\cos(\theta)\frac{\partial\zeta}{\partial r} - \frac{1}{r}\sin(\theta)\frac{\partial\zeta}{\partial\theta}\right)\right] \tag{3.42}$$

$$v(r, \theta, t) = \mathrm{Re}\left[\frac{jg}{\sigma}\left(\sin(\theta)\frac{\partial\zeta}{\partial r} + \frac{1}{r}\cos(\theta)\frac{\partial\zeta}{\partial\theta}\right)\right] \tag{3.43}$$

$$w(r, \theta, z, t) = \left(z + H\right) \cdot \mathrm{Re}\left[\frac{jg}{\sigma}\left(\frac{\partial^2\zeta}{\partial r^2} + \frac{1}{r^2}\frac{\partial^2\zeta}{\partial\theta^2} + \frac{1}{r}\frac{\partial\zeta}{\partial r}\right)\right] \tag{3.44}$$

in which

$\kappa =$ radial wave number defined by the condition that $\partial J_s/\partial r = 0$ on the boundary $r = R_0$, or

$$\left.\frac{\partial J_s(\kappa r)}{\partial r}\right|_{r=R_0} = 0 \tag{3.45}$$

$$\sigma = \sqrt{gH}\,\kappa \tag{3.46}$$

$$\frac{\partial \zeta}{\partial r} = A_s J_s'(\kappa\, r)\cos(s\,\theta)\,exp(j\sigma\, t) \tag{3.47}$$

$$\frac{\partial^2 \zeta}{\partial r^2} = A_s J_s''(\kappa\, r)\cos(s\,\theta)\,exp(j\sigma\, t) \tag{3.48}$$

$$\frac{\partial \zeta}{\partial \theta} = -s\; A_s J_s(\kappa\, r)\sin(s\,\theta)\exp(j\sigma\, t) \tag{3.49}$$

$$\frac{\partial^2 \zeta}{\partial \theta^2} = -s^2\; A_s J_s(\kappa\, r)\cos(s\,\theta)\exp(j\sigma\, t) \tag{3.50}$$

3.3.4 Test Case

3.3.4.1 Physical Parameters

R_0 = 16,000 m
H = 12 m
s = 0 (No nodal diameters)
A_s = 1 m
κ = 3.83171/ R_0
Only one zero of $J_0(r)$ occurs in the interval (0, R_0) (one unique nodal circle)

3.3.4.2 Numerical Parameters

High resolution

A regular mesh of 75,576 (12,596 x 6) computational cells, or an equivalent mesh of 239,324 degrees of freedom (assuming 3 degrees of freedom per computational cell plus 1 for each water column), can be used for a high-resolution test case. The calculated total number for the degrees of freedom includes all nodes on the basin boundary. The regular mesh cells in the horizontal plane are squares ($\Delta x = \Delta y$). It is assumed that a mesh of 6 vertical layers is used. Only 1/6 of the degrees of freedom should be used to discretize the circular basin horizontally. Of course, the user is not restricted to using regular meshes; irregular finite difference, finite element, or body-fitted meshes are encouraged, subject to the specifications on degrees of freedom.

For the regular mesh, the suggested values for the discretization parameters are

Δx = Δy = 250 m
Δz = 2 m
T_s = 10 hours

where T_s is the duration of the simulation.

Modelers are asked to report the numerical treatment of the time domain in full reproducible detail. For time-stepping schemes, the following time step is proposed, which guarantees a stable solution for explicit methods:

$$\Delta t \quad = 15 \text{ seconds} \qquad \text{leading to} \quad \sqrt{gH}\Delta t/\Delta x \text{ (Courant Number)} = 0.65$$

Low resolution

A regular mesh of 18,480 (3,080 x 6) computational cells, or an equivalent mesh of 58,520 degrees of freedom, can be used for a low-resolution test case. For a regular mesh, the suggested values for the discretization parameters are

$$\begin{aligned} \Delta x \quad &= \Delta y = 500 \text{ m} \\ \Delta z \quad &= 2 \text{ m} \\ T_s \quad &= 10 \text{ hours} \end{aligned}$$

For time stepping schemes a step size Δt of 30 seconds is proposed, leading to a Courant number ($\sqrt{gH}\Delta t/\Delta x$) of 0.65.

3.3.4.3 Reporting of Model Results

Modelers should report the $\ell 1$, $\ell 2$ and $\ell \infty$ error norms, RMSE, and the index of agreement d for the longitudinal (u, in cm/s) and vertical (w, in cm/s) velocity solutions and the free surface elevation (ζ, in cm) at the locations indicated below, measured from the center of the circular basin.

Low resolution:

LR(1) $x = -250$ m $y = -250$ m $z = -1$ m
LR(2) $x = 9{,}250$ m $y = -250$ m $z = -1$ m
LR(3) $x = 14{,}250$ m $y = -250$ m $z = -1$ m

High resolution:

HR(1) $x = -125$ m $y = -125$ m $z = -1$ m
HR(2) $x = 9{,}125$ m $y = -125$ m $z = -1$ m
HR(3) $x = 13{,}625$ m $y = -125$ m $z = -1$ m

As an option, modelers are also invited to submit phase and dissipation errors in the form of celerity ratio Q and dissipation factor ε_A.

3.3.4.4 Initial Conditions and Analytical Solution at Selected Points

To define the initial condition, the modeler only needs to calculate the radial location, r, of each computational node relative to the center of the domain and multiply by κ ($= 3.83171/ R_0$). The initial condition at the location r is then determined by the Bessel function of zero order J_0 (κr). Routines for calculating the Bessel function are given, for example, by Press *et al.* (1992, p. 225). Figure 3.3.2 is a graph of the initial free surface displacement for the circular basin with radius $R_0 = 16{,}000$ m.

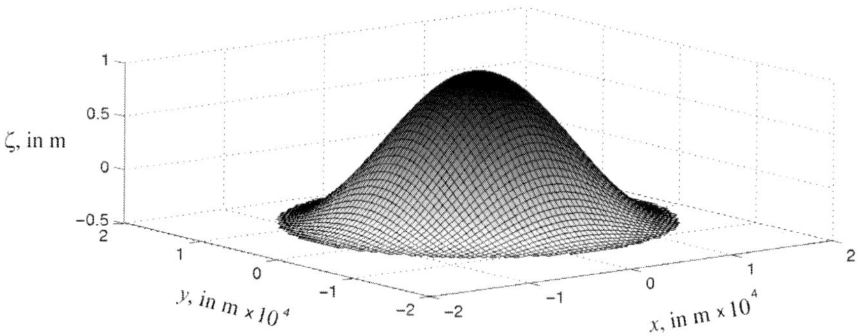

Figure 3.3.2 Test case 2: Initial conditions for the test case using the proposed parameters.

3.3.5 Example Verification: High Resolution Test Case

A comparison of the analytical and numerical solutions for node HR(3) in the high-resolution test case is given in figure 3.3.3. The computed error measures for all of the test nodes in the low- and high-resolution numerical solutions are given in table 3.3.1.

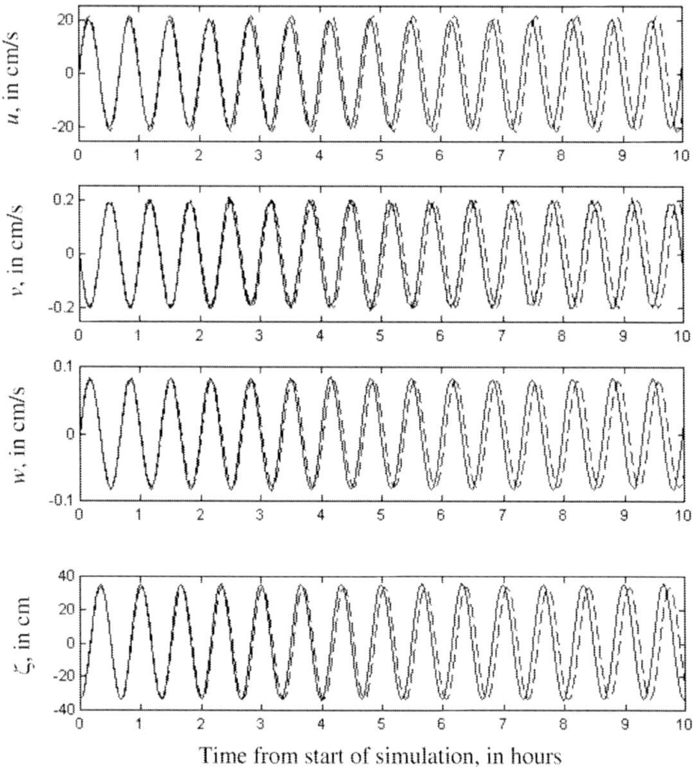

Figure 3.3.3. Test Case 2: Graphical display of analytical and numerical solutions for the high-resolution test case.

Analytical solution is dashed line; numerical solution is solid line. The results are reported for test node HR(3) at the location $x = 13{,}625$ m, $y = -125$ m, and $z = -1$ m. The numerical calculations were done with the SI3D model using 1 leapfrog and 1 trapezoidal iteration for time stepping.

Table 3.3.1 Test Case 2: Error measures for the low- and high-resolution solutions.

LR(1)	***u***	***v***	***w***	ζ
d	0.3353	0.3353	0.3352	0.3322
ℓ **1**	1.4542	1.4542	1.4550	1.4438
ℓ **2**	1.5678	1.5678	1.5686	1.5524
$\ell\,\infty$	2.0619	2.0619	2.0681	2.0285
RMSE	2.9944	2.9944	0.2634	109.64

LR(2)	*u*	*v*	*w*	ζ
d	0.3358	0.3357	0.3292	0.3015
$\ell 1$	1.4258	1.4385	1.3018	1.3026
$\ell 2$	1.5369	1.5508	1.3894	1.3925
$\ell \infty$	1.9850	2.0157	1.9738	1.9986
RMSE	54.3147	1.4813	0.0237	9.9991

LR(3)	*u*	*v*	*w*	ζ
d	0.3298	0.3296	0.3355	0.3317
$\ell 1$	1.2961	1.6406	1.4675	1.4574
$\ell 2$	1.3824	1.7591	1.5814	1.5661
$\ell \infty$	1.8013	1.4156	2.0664	2.0631
RMSE	15.3356	0.34234	0.0975	40.6272

HR(1)	*u*	*v*	*w*	ζ
d	0.5988	0.5988	0.5988	0.5882
$\ell 1$	1.0598	1.0598	1.0600	1.0716
$\ell 2$	1.1991	1.1991	1.1993	1.2089
$\ell \infty$	1.8869	1.8869	1.818	1.8777
RMSE	1.146	1.146	0.2017	85.4852

HR(2)	*u*	*v*	*w*	ζ
d	0.5991	0.5992	0.5966	0.5820
$\ell 1$	1.0500	1.0505	1.0117	1.0266
$\ell 2$	1.1886	1.1889	1.1373	1.1512
$\ell \infty$	1.8456	1.8450	1.8456	1.8481
RMSE	42.3534	0.5803	0.0227	9.6471

HR(3)	*u*	*v*	*w*	ζ
d	0.5983	0.5987	0.5989	0.5882
$\ell 1$	1.0230	1.0572	1.0726	1.0853
$\ell 2$	1.1544	1.1960	1.2129	1.2232
$\ell \infty$	1.7951	1.9061	1.8784	1.9013
RMSE	17.3833	0.16523	0.0685	29.0653

3.3.6 References

Lamb, H., 1945, Hydrodynamics, Dover Publications, New York.

Press, W. H., Teukolsky, S. A., Vetterling, W. T., and Flannery, B. P., 1992, Numerical Recipes in Fortran—The Art of Scientific Computing (2nd edition): Cambridge University Press, 963 p.

3.4 TEST CASE 3: TIDAL FORCING IN A RECTANGULAR BASIN WITH A HORIZONTAL BOTTOM

Contributors: Francisco J. Rueda and Peter E. Smith

3.4.1 Description and Objectives

The computational domain for most real problems in hydrodynamic modeling often includes one or more open connections to water bodies outside the domain. These connections are called "open boundaries," and implementing a proper numerical treatment for these boundaries is one of the most challenging problems in hydrodynamic modeling. Computational cell faces that coincide with an open boundary require special treatment to incorporate the influence of forcing from the outside water body. This test case attempts to check the numerical formulation and implementation of open boundary conditions for the case when a time series of free-surface elevation is known at an open boundary. No additional complications are introduced from friction or geometry. The geometry, governing equations, and assumptions are the same as those for test case 1, the only difference resulting from an open boundary subject to periodic oscillations of the free surface elevation. This test case is especially relevant for verifying models aimed at simulating estuaries or coastal embayments affected by tidal oscillations at the boundary that connects them to the ocean.

The variation in the water surface elevation ζ at the open boundary for this test case is defined by a simple harmonic tide (eq. 3.59), where the time t is expressed in hours from the start of the simulation. As in test cases 1 and 2, the fluid is assumed to be frictionless (inviscid). The lateral (y-direction) velocity component v and the variation in the water surface elevation in the lateral direction remain everywhere equal to zero. The tidal period for the proposed boundary tide is 12.4 hours, which corresponds with the principal lunar constituent of the astronomical tide known as the M_2 tide. The proposed dimensions for the channel length and depth were chosen so that the wavelength of the M_2 tide is close to four times the length of the channel. For these conditions, the traveling wave entering the channel is added to the wave reflected from the wall boundary, and a resonant oscillation is set up in the channel. The analytical solution defines a tidal range in water surface elevation that is greater at the closed boundary than at the open boundary.

As in test case 1, the terms $\partial v/\partial y$, $\partial v/\partial t$, and $\partial\zeta/\partial y$, although zero in magnitude, are included in the mathematical statement of the problem. This is a reminder in testing a fully 3-D code that it is preferable that these terms not be hard-wired to zero. It should be checked in the numerical solution that these terms do indeed remain equal to zero (or negligibly small).

3.4.2 The Mathematical Problem

3.4.2.1 Governing Equations

Continuity

$$u_x + v_y + w_z = 0 \quad (3.51)$$

$$\zeta_t + u_x H + v_y H = 0 \quad (3.52)$$

Assumptions:

- *u, v* are not functions of the vertical coordinate *z* (note, however, that the vertical velocity *w* is a function of *z)*
- $\zeta << H$ (allows a linearization of the code)
- *H* is constant

Momentum

$$u_t + g\zeta_x = 0 \quad (3.53)$$

$$v_t + g\zeta_y = 0 \quad (3.54)$$

Assumptions:

- Inviscid fluid (interfacial and bottom friction are neglected)
- Homogeneous fluid (ρ = constant)
- Nonrotating reference frame ($f = 0$)
- $uu_x, vu_y << u_t, g\zeta_x$ (the *x*-direction advection terms are neglected)
- $uv_x, vv_y << v_t, g\zeta_y$ (the *y*-direction advection terms are neglected)

3.4.2.2 Domain

The domain for test case 3 is shown in figure 3.4.1.

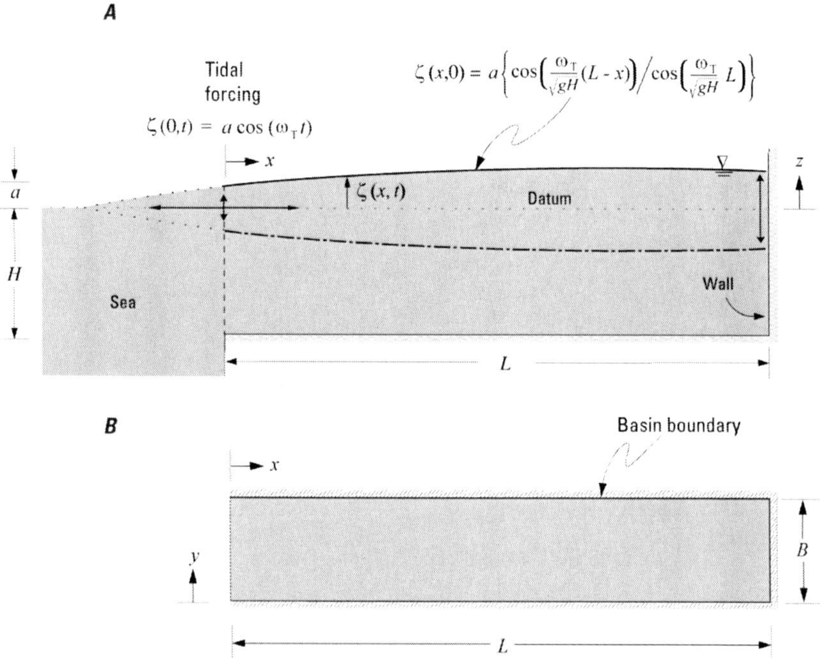

Figure 3.4.1 Test Case 3: Diagram of the domain for the test problem of tidal forcing in a rectangular basin with horizontal bottom.

(***A***) Cross section of the rectangular basin showing initial position of the free surface, and (***B***) top view. The vertical scale in (***A***) is greatly exaggerated and distorted.

3.4.2.3 Boundary Conditions

$$\left.\frac{\partial u}{\partial y}\right|_{y=0} = \left.\frac{\partial u}{\partial y}\right|_{y=B} = \left.u\right|_{x=L} = 0 \tag{3.55}$$

$$\left.\frac{\partial v}{\partial x}\right|_{x=0} = \left.\frac{\partial v}{\partial x}\right|_{x=L} = \left.v\right|_{y=B} = 0 \tag{3.56}$$

$$\left.\frac{\partial u}{\partial z}\right|_{z=-H} = \left.\frac{\partial v}{\partial z}\right|_{z=-H} = \left.\frac{\partial u}{\partial z}\right|_{z=\zeta} = \left.\frac{\partial v}{\partial z}\right|_{z=\zeta} = 0 \tag{3.57}$$

$$\left.w\right|_{z=-H} = 0 \tag{3.58}$$

$$\left.\zeta\right|_{x=0} = a\cos(\omega_T t) \tag{3.59}$$

3.4.2.4. Initial Conditions

$$\zeta\big|_{t=0} = a\left[\cos\left(\frac{\omega_T}{\sqrt{gH}}(L-x)\right)\bigg/\cos\left(\frac{\omega_T}{\sqrt{gH}}L\right)\right] \tag{3.60}$$

$$u, v, w\big|_{t=0} = 0 \tag{3.61}$$

3.4.3 Analytical Solution (Neuman and Pierson, 1966, p.291)

$$\zeta(x,t) = a\left[\cos\left(\frac{\omega_T}{\sqrt{gH}}(L-x)\right)\bigg/\cos\left(\frac{\omega_T}{\sqrt{gH}}L\right)\right]\cos(\omega_T t) \tag{3.62}$$

$$u(x,t) = \left[\left(a\frac{\sqrt{gH}}{H}\right)\bigg/\cos\left(\frac{\omega_T}{\sqrt{gH}}L\right)\right]\sin\left(\frac{\omega_T}{\sqrt{gH}}(L-x)\right)\sin(\omega_T t) \tag{3.63}$$

$$v(x,t) = 0 \tag{3.64}$$

$$w(x,z,t) = -\left[\left(a\omega_T\frac{z+H}{H}\right)\bigg/\cos\left(\frac{\omega_T}{\sqrt{gH}}L\right)\right]\cos\left(\frac{\omega_T}{\sqrt{gH}}(L-x)\right)\sin(\omega_T t) \tag{3.65}$$

3.4.4 Test Case

3.4.4.1 Physical Parameters

L = 100,000 m
B = 20,000 m
H = 12 m
a = 0.25 m
ω_T = $2\pi/(12.4$ hours) (tidal wave frequency)

3.4.4.2 Numerical Parameters

Low resolution

A regular mesh of 480 (20 x 4 x 6) computational cells, or an equivalent mesh of 1,520 degrees of freedom (3 degrees of freedom per computational cell plus 1 for each water column), can be used for a low-resolution test case. The calculated total number for the degrees of freedom includes all nodes on the basin boundary. The regular mesh cells in the horizontal plane are squares ($\Delta x = \Delta y$). It is assumed that a mesh of twelve vertical layers is used. Only 1/12 of the count for degrees of freedom should be used to discretize the rectangular basin horizontally. Of course, the user is not restricted to using regular meshes; irregular finite difference, finite element, or body-fitted meshes are encouraged, subject to the specifications on degrees of freedom.

For the regular mesh, the suggested values for the discretization parameters are

$\Delta x \quad = \Delta y = 5{,}000$ m
$\Delta z \quad = 2$ m
$T_s \quad = 24$ hours

Modelers are asked to report the numerical treatment of the time domain in full reproducible detail. For time-stepping schemes, the following time step is proposed, which guarantees a stable solution for explicit methods:

$\Delta t \quad = 360$ seconds leading to $\sqrt{gH}\Delta t/\Delta x$ (Courant Number) = 0.78

High resolution

A regular mesh of 1,920 (40 x 8 x 6) computational cells, or an equivalent mesh of 6,080 degrees of freedom (3 degrees of freedom per computational cell plus 1 for each water column), can be used for the high-resolution test case. For the regular mesh, the suggested values for the discretization parameters are

$\Delta x \quad = \Delta y = 2{,}500$ m
$\Delta z \quad = 2$ m
$T_s \quad = 24$ hours

For time stepping schemes a step size Δt of 180 seconds is proposed, leading to a Courant number ($\sqrt{gH}\Delta t/\Delta x$) of 0.78.

3.4.4.3 Reporting of Model Results

Modelers should report the $\ell 1$, $\ell 2$ and $\ell \infty$ error norms, RMSE, and the index of agreement d for the longitudinal velocity (u, in cm/s), vertical velocity (w, in cm/s), and the free-surface elevation (ζ, in cm) at the following nodal locations

Low resolution:

LR(1) $x =$ 2,500 m $y =$ 2500 m $z = -1$ m
LR(2) $x =$ 52,500 m $y =$ 2500 m $z = -1$ m
LR(3) $x =$ 92,500 m $y =$ 2500 m $z = -1$ m

High resolution:

HR(1) $x =$ 1,250 m $y =$ 1250 m $z = -1$ m
HR(2) $x =$ 51,250 m $y =$ 1250 m $z = -1$ m
HR(3) $x =$ 96,250 m $y =$ 1250 m $z = -1$ m

3.4.5 Example Verification: Low Resolution Test Case

A comparison of the analytical and numerical solutions for node LR(3) of the low-resolution test case is given in figure 3.4.2. The computed error measures for the three test nodes of the low- resolution numerical solution are given in table 3.4.1.

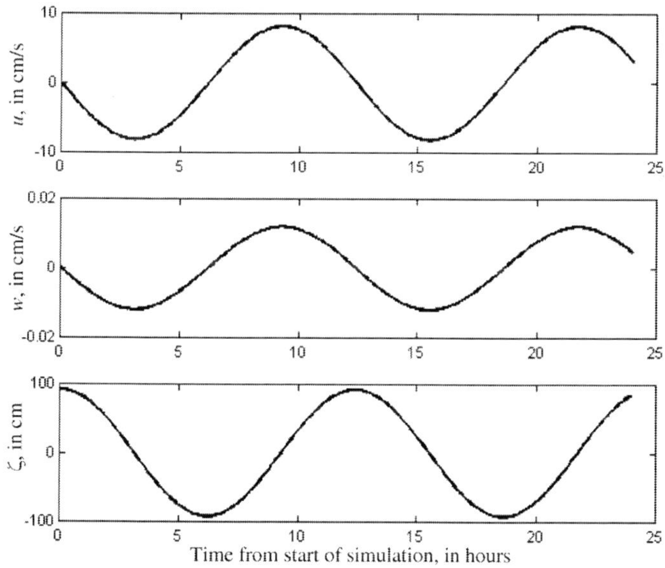

Figure 3.4.2. Test Case 3: Graphical display of analytical and numerical solutions for the low-resolution test case.

The analytical and numerical solutions plot on top of one another. The results are graphed for test node LR(3) at location x = 92,500 m, y = 2,500 m, and z = – 1.0 m. The numerical calculations were done with the SI3D model using 1 leapfrog and 1 trapezoidal iteration for time stepping.

Table 3.4.1 Test Case 3: Error measures for the low-resolution solutions.

LR(1)	u	w	ζ
d	1.0000	0.998	1.0000
ℓ **1**	1.7042×10^{-3}	2.6073×10^{-2}	1.9951×10^{-4}
ℓ **2**	1.7104×10^{-3}	2.7054×10^{-2}	2.2390×10^{-4}
$\ell\ \infty$	1.9405×10^{-3}	3.7805×10^{-2}	3.3448×10^{-4}
RMSE	0.097126	6.9126×10^{-5}	4.3954×10^{-3}

LR(2)	u	w	ζ

d	1.0000	1.0000	1.0000
ℓ_1	1.8602×10^{-3}	4.5133×10^{-3}	1.3945×10^{-3}
ℓ_2	1.8681×10^{-3}	4.6678×10^{-3}	1.3943×10^{-3}
ℓ_∞	1.9883×10^{-3}	6.7379×10^{-3}	1.5638×10^{-3}
RMSE	0.06431	3.2373×10^{-5}	7.4229×10^{-2}

LR(3)	u	w	ζ
d	1.0000	1.0000	1.0000
ℓ_1	1.9953×10^{-3}	2.9575×10^{-3}	1.5826×10^{-3}
ℓ_2	2.0248×10^{-3}	3.4832×10^{-3}	1.5756×10^{-3}
ℓ_∞	2.5550×10^{-3}	5.7619×10^{-3}	1.7340×10^{-3}
RMSE	1.1717×10^{-2}	2.9468×10^{-5}	1.0241×10^{-1}

3.4.6 References

Neuman, G., and W. J. Pierson, 1966, Principles of Physical Oceanography: Prentice-Hall, Englewood Cliffs, N.J.

3.5 TEST CASE 4: TIDAL FORCING IN A RECTANGULAR BASIN WITH A VARYING BOTTOM SLOPE AND LINEAR BOTTOM FRICTION

Contributors: Francisco J. Rueda and Peter E. Smith

3.5.1 Description and Objectives

In test case 3 the correct implementation of a tidal boundary condition was checked without additional difficulties from a complicated basin geometry or the inclusion of bottom friction or vertical momentum transport. Some of these complicating effects were incorporated in this test case, which is probably the most general of those considered in this chapter. The representation of varying bathymetry and its effects on the velocity solution can be tested. The bathymetry was assumed to vary quadratically along the longitudinal axis of the basin with maximum depth at the mouth; the plan form of the basin is rectangular. A rectangular plan form was chosen instead of a circular geometry (as presented by Lynch and Officer, 1985) to avoid the problems associated with the representation of curved lateral boundaries, already addressed in test case 2. A significant part of the error that can occur in a numerical solution when using curved open boundaries can result from the difficulty in specifying a proper open boundary condition on a curved surface.

A further important aspect of this test case is that it does not assume that the fluid is frictionless (inviscid). The vertical turbulent momentum transfer terms (terms **C** in equations 3.2 and 3.3) are included in the governing equations, and a linear frictional stress law ($\tau_{bx} = ku$; $\tau_{by} = kv$) is applied at the bottom boundary. The vertical turbulent momentum transfer coefficient in this problem is assumed to be a function of the depth squared ($K_v = \mu_0(H(x))^2$) as defined below. The linear frictional stress law is incorporated in the boundary condition given by equation 3.73.

3.5.2 The Mathematical Problem

3.5.2.1 Governing Equations

Continuity

$$u_x + v_y + w_z = 0 \tag{3.66}$$

$$\zeta_t + \left(\int_{z=-H}^{z=0} u\,dz\right)_x + \left(\int_{z=-H}^{z=0} v\,dz\right)_y = 0 \tag{3.67}$$

Assumptions:

- $\zeta << H(x)$ (allows a linearization of the code)
- $H(x)$ is not a function of time

Momentum

$$u_t = -g\zeta_x + (K_v u_z)_z \tag{3.68}$$

$$v_t = -g\zeta_y + (K_v v_z)_z \tag{3.69}$$

Assumptions:

- Homogeneous fluid (ρ = constant)
- Nonrotating reference frame ($f = 0$)
- $uu_x, vu_y \ll u_t, g\zeta_x, (K_v u_z)_z$ (the x-direction advection terms are neglected)
- $uv_x, vv_y \ll v_t, g\zeta_y, (K_v v_z)_z$ (the y-direction advection terms are neglected)
- $K_v(x, y) = \mu_0 (H(x))^2$ and μ_0 = constant (the vertical eddy viscosity is a function of the local depth squared $\{H(x)\}^2$)

3.5.2.2 Domain

The domain for test case 4 is shown in figure 3.5.1.

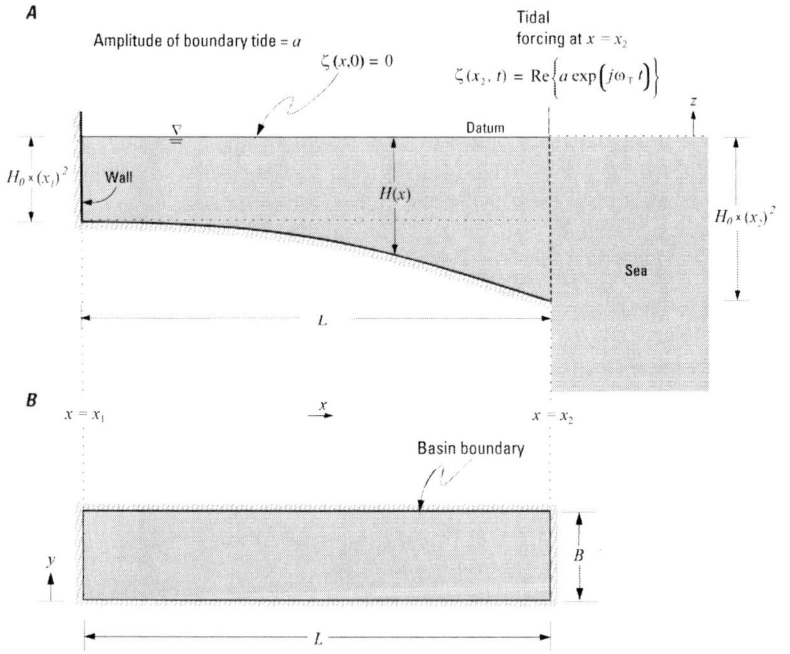

Figure 3.5.1. Test Case 4: Diagram of the domain for the test problem of tidal forcing in a rectangular basin with a variable bottom slope and linear bottom friction.
(***A***) Cross section of the rectangular basin showing the initial position of the free surface, and (***B***) top view. The vertical scale in (***A***) is greatly exaggerated and distorted.

The location of the basin bottom in figure 3.5.1 (***A***) is measured downward from the datum and is defined by the parabolic expression

$$v_t = -g\zeta_y + (K_v v_z)_z \tag{3.69}$$

$$H(x) = H_0 \times (x)^2 \tag{3.70}$$

where H_0 is a constant and x represents the distance in the horizontal direction (positive towards the right) measured from a point defined as $x = 0$. It is assumed that $x_1 < x_2$.

3.5.2.3 Boundary Conditions

$$\left.\frac{\partial u}{\partial y}\right|_{y=0} = \left.\frac{\partial u}{\partial y}\right|_{y=B} = u\big|_{x=x_1} = 0 \tag{3.71}$$

$$\left.\frac{\partial v}{\partial x}\right|_{x=x_1} = v\big|_{y=0} = v\big|_{y=B} = 0 \tag{3.72}$$

$$\left.\left(K_v \frac{\partial u}{\partial z}\right)\right|_{z=-H} = ku\,;\ \left.\left(K_v \frac{\partial v}{\partial z}\right)\right|_{z=-H} = kv \tag{3.73}$$

$$\left.\frac{\partial u}{\partial z}\right|_{z=\zeta} = \left.\frac{\partial v}{\partial z}\right|_{z=\zeta} = 0 \tag{3.74}$$

$$w\big|_{z=-H} = -u\frac{\partial H}{\partial x} - v\frac{\partial H}{\partial y} \tag{3.75}$$

$$\zeta\big|_{x=x_2} = \mathrm{Re}\left\{ a \exp(j\omega_T t) \right\} \tag{3.76}$$

Assumption:

- $kH/K_v = c_D = \text{constant}$

3.5.2.4 Initial Conditions

It is assumed that both the free surface elevation and the velocity components are zero at the beginning of the simulations, i.e.,

$$\zeta\big|_{t=0} = 0 \tag{3.77}$$

$$u, v, w\big|_{t=0} = 0 \tag{3.78}$$

To avoid the effects of this "cold start" when comparing the numerical and analytical solutions, the solutions can be compared after a sufficient period has elapsed following the start of the simulation. For the physical parameters proposed here, the observed initial transients in the numerical solution caused by the cold start were limited to two to three tidal cycles; the results were compared only after 100 hours (approximately 8 tidal cycles).

3.5.3 Analytical Solution (Lynch and Officer, 1985; Lynch and Gray, 1978)

$$\zeta = \mathrm{Re}\left\{[A' x^{s_1} + B' x^{s_2}]\exp(j\omega_T t)\right\} \tag{3.79}$$

where

$$A' = \frac{a s_2 x_1^{s2}}{(s_2 x_2^{s1} x_1^{s2} - s_1 x_1^{s1} x_2^{s2})} \tag{3.80}$$

$$B' = \frac{-a s_1 x_1^{s1}}{(s_2 x_2^{s1} x_1^{s2} - s_1 x_1^{s1} x_2^{s2})} \tag{3.81}$$

$$s_1, s_2 = -\frac{1}{2} \pm \sqrt{\frac{1}{4} - \beta^2} \text{ and } \beta^2 = \frac{(\omega_T^{\ 2} - j\omega_T \tau)}{gH_0} \tag{3.82}$$

$$\lambda = \sqrt{\frac{j\omega_T H^2}{K_v}} \tag{3.84}$$

$$u = U_0\left[1 - \frac{\cosh(\lambda Z)}{\cosh(\lambda)\left(1 + \frac{\lambda}{K}\tanh(\lambda)\right)}\right] \tag{3.85}$$

$$v = 0 \tag{3.86}$$

Where

$$U_0 = -\frac{g}{j\omega_T}\frac{\partial \zeta}{\partial x} \tag{3.87}$$

$$Z = z/H \tag{3.88}$$

$$K = (kH/K_v)\big|_{Z=-1} \tag{3.89}$$

3.5.4 Test Case

3.5.4.1 Physical Parameters

L	= 100,000 m	
B	= 20,000 m	
H_0	= 1.E-9	
x_1	= 100,000 m	
x_2	= 200,000 m	
a	= 0.25 m	
ω_T	= 2π/(12.4 hours)	(tidal wave frequency)
μ_0	= 0.01 s^{-1}	
c_D	= 1.E3	

3.5.4.2 Numerical Parameters

Low resolution

A regular mesh of 972 computational cells, or an equivalent mesh of 2,996 degrees of freedom (3 degrees of freedom per computational cell plus 1 for each water column), can be used for a low-resolution test case. The calculated total number for the degrees of freedom includes all nodes on the basin boundary. The regular mesh cells in the horizontal plane are squares ($\Delta x = \Delta y$). In the vertical dimension this problem was solved using a horizontal layering scheme (a *z*-coordinate model) so the number of vertical computational cells varied in the *x*-coordinate direction with the varying depth. Approximately 1/12 of the count for degrees of freedom are used to discretize the rectangular basin horizontally. Of course, as in previous test cases, the user is not restricted to using regular meshes; irregular finite difference, finite element, or body-fitted meshes are encouraged, subject to the specifications on degrees of freedom. Various ways of discretizing the vertical dimension are also encouraged.

For the regular mesh, the suggested values for the discretization parameters are

Δx	$= \Delta y = 5{,}000$ m
Δz	= 2 m
T_s	= 240 hours

where T_s is the duration of the simulation in hours.

Modelers are asked to report the numerical treatment of the time domain in full reproducible detail. For time-stepping schemes the following time step is proposed, which guarantees a stable solution for explicit methods:

Δt = 225 seconds leading to $\sqrt{gH}\,\Delta t / \Delta x$ (Courant Number) = 0.89

High resolution

The regular mesh of 3,808 computational cells, or an equivalent mesh of 11,744 degrees of freedom (3 degrees of freedom per computational cell plus 1 for each water column), can be used for a high-resolution test case. For the regular mesh, the suggested values for the discretization parameters are

$$\Delta x = \Delta y = 2{,}500 \text{ m}$$
$$\Delta z = 1 \text{ m}$$
$$T_s = 240 \text{ hours}$$

For time stepping schemes a step size Δt of 50 seconds is proposed, leading to a Courant number ($\sqrt{gH}\Delta t/\Delta x$) of 0.40.

3.5.4.3 Reporting of Model Results

Modelers should report the $\ell 1$, $\ell 2$ and $\ell \infty$ error norms, RMSE, and the index of agreement d using the time series for the longitudinal velocity (u, in cm/s) and the free surface elevation (ζ, in cm) at following test nodes:

Low resolution:

LR(1) $x\,(-x_1) =$	2,500 m	$y =$ 2,500 m	$z = -1$ m
LR(2) $x\,(-x_1) =$	52,500 m	$y =$ 2,500 m	$z = -1$ m
LR(3) $x\,(-x_1) =$	92,500 m	$y =$ 2,500 m	$z = -1$ m

High resolution:

HR(1) $x\,(-x_1) =$	1,250 m	$y =$ 1,250 m	$z = -0.5$ m
HR(2) $x\,(-x_1) =$	51,250 m	$y =$ 1,250 m	$z = -0.5$ m
HR(3) $x\,(-x_1) =$	96,250 m	$y =$ 1,250 m	$z = -0.5$ m

Additionally, plots comparing analytical and numerical velocity profiles should be compared for the horizontal location of nodes LR(2) and HR(2) and at four times approximately separated by ¼ of the tidal-forcing period (approximately 3 hours). The following four times are suggested for the comparison: t = 124, 127, 130, 133 hours.

A FORTRAN computer program that generates the solution for water surface elevation and velocity amplitude throughout the computational domain is given in an Appendix for this test case (section 3.5.7). In each column a fixed number of nodes was assumed (NNV), and the solution was given at NN equally spaced horizontal locations, including the boundaries of the basin. The program generates the amplitude of the velocity solution 1 m below the free surface at NN equally spaced nodes.

3.5.5 Example Verification: Low Resolution Test Case

A comparison of the analytical and numerical solutions for node LR(2) of the low-resolution test case is given in figure 3.5.2. The computed error measures for the three test nodes of the low- resolution numerical solution are given in table 3.5.1. Figure 3.5.3 provides comparisons of velocity profiles for the horizontal column of water at node LR(2).

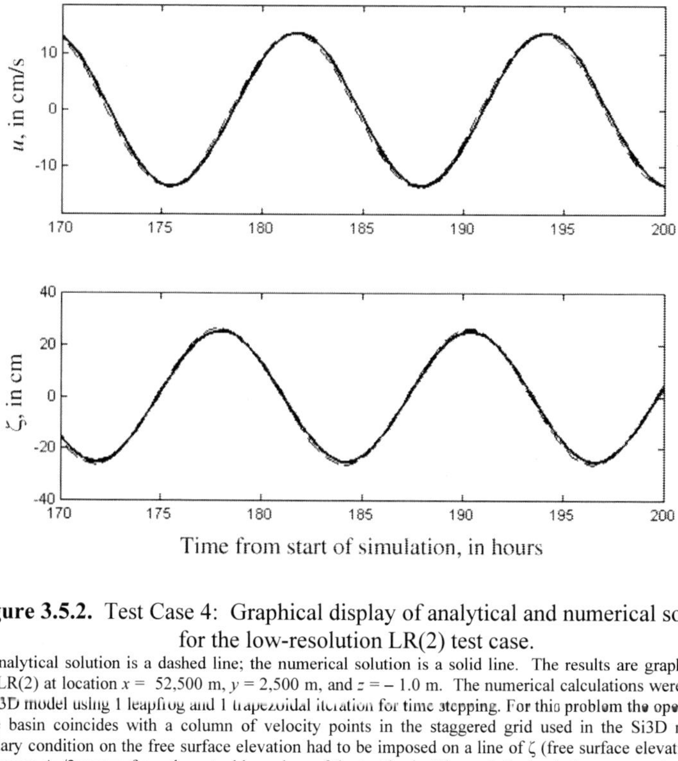

Figure 3.5.2. Test Case 4: Graphical display of analytical and numerical solutions for the low-resolution LR(2) test case.

The analytical solution is a dashed line; the numerical solution is a solid line. The results are graphed for test node LR(2) at location x = 52,500 m, y = 2,500 m, and z = – 1.0 m. The numerical calculations were done with the SI3D model using 1 leapfrog and 1 trapezoidal iteration for time stepping. For this problem the open boundary of the basin coincides with a column of velocity points in the staggered grid used in the Si3D model. The boundary condition on the free surface elevation had to be imposed on a line of ζ (free surface elevation) points, which were $\Delta x/2$ meters from the actual boundary of the test basin. The analytical solution was used to determine the time series for ζ at the location $x = x_2 - \Delta x/2$, and that time series was actually used for the model boundary condition.

Table 3.5.1. Test Case 4: Error measures for the low-resolution solutions.

LR(1)	u	ζ
d	0.9960	0.9959
ℓ **1**	0.1217	0.1255
ℓ **2**	0.1221	0.1250
$\ell\,\infty$	0.1262	0.1237
RMSE	0.132	2.6941

LR(2)	*u*	ζ
d	0.9982	0.9988
$L1$	8.3971×10^{-2}	6.6700×10^{-2}
$L2$	8.4204×10^{-2}	6.6800×10^{-2}
$L\infty$	8.5098×10^{-2}	6.6496×10^{-2}
RMSE	0.8112	1.2400

LR(3)	*u*	ζ
d	0.9991	1.0000
$L1$	6.0598×10^{-2}	5.2864×10^{-3}
$L2$	6.0709×10^{-2}	5.2896×10^{-3}
$L\infty$	6.1397×10^{-2}	5.4063×10^{-3}
RMSE	0.59201	0.09335

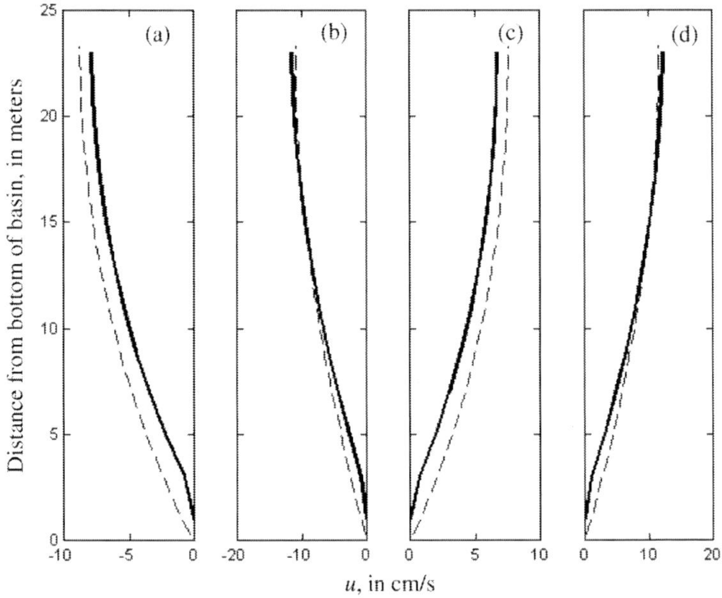

Figure 3.5.3. Test Case 4: Graphical display of analytical and numerical solutions for velocity profiles calculated for the low-resolution LR(2) test case.

The analytical solutions are dashed lines; the numerical solutions are solid lines. The results are graphed for test node LR(2) at location x = 52,500 m, y = 2,500 m, and z = – 1.0 m for times (a) *t* = 124 hours, (b) *t* = 127 hours, (c) *t* = 130 hours, and (d) *t* = 133 hours

3.5.6. References

Lynch, D. R., and G. Gray (1978). "Analytic solutions for computer flow model testing." *ASCE Journal of the Hydraulic Division*, 104(HY10), 1409-1428.

Lynch, D. R., and C. B. Officer (1985). "Analytic test cases for three-dimensional hydrodynamic models." *International Journal for Numerical Methods in Fluids*, 5, 529-543.

3.5.7 Appendix: FORTRAN Code with Implementation of Analytical Solution

```
      PROGRAM EXAMPLE
C
C EXAMPLE PROGRAM SHOWING USE OF SUBROUTINES VERT AND
HORIZ
C 3-D ANALYTIC SOLUTION FOR ASCE VERIFICATION AND VALIDATION
C TASK COMMITTEE
C CONSTANT VISCOSITY PROFILE
C FRANCISCO J. RUEDA          MAY 16, 2001
C SUBROUTINE HORIZ HAS GRAVITY=9.808 (MKS).
C
      PARAMETER (NNV=41,NN=41)
      IMPLICIT DOUBLE PRECISION (A-H, O-Z)
      COMPLEX*16 V1(NNV),TAU,HH(NN),VR(NN,NNV), VS(NN)
      COMPLEX*16 HHH,V0,EYE,T,TT,VS2
      COMPLEX*16 AMP
      REAL*8 R(NN)
      REAL*8 MU, MU0, LITK, BIGK, TAU0, R1, R2, W, DEPTH
      EYE=(0.0,1.0)
      ZERO=0.0
     AMP = (0.0, -2.50E-1)
C
C VERTICAL DATA
C
      MU0 = 0.01
      H=10.
      PERIOD = 12.4 * 3600.
      OMEGA= 2. * 3.1415926535897932 / PERIOD
      LITK=1.0
C
C HORIZONTAL DATA
C
      R1=100.E3
      R2=200.E3
      H0=1.E-9
C
C SAMPLING POINTS
C
      DO I=1,NN
        R(I)=R1+(I-1.)/(NN-1.)*(R2-R1)
      END DO
C
C DIMENSIONLESS PARAMETERS
C
      MU=MU0
```

```
      BIGK=LITK*H/MU
      W=OMEGA*H*H/MU
      TAU0=3.*MU/(H*H)
C
C GET  THE VERTICAL STRUCTURE AND THE BOTTOM STRESS PARAM.
TAU
C
      CALL VERT(NNV,W,MU,BIGK,TAU0,V1,TAU)
C
C NOW LOOP OVER HORIZ. NODES AND GET HORIZ. SOLUTION
C
      DO I=1,NN
       RI=R(I)
          DEPTH = H0 * RI ** 2.
          CALL VZ1M (DEPTH,W,BIGK,VS2)
       CALL HORIZ(R1,R2,H0,OMEGA,TAU,RI,HHH,V0, AMP)
       HH(I)=HHH
          V_MEAN(I) = V0 * V2
          VS(I) = V0 * VS2
       DO J=1,NNV
        VR(I,J) = V0 * V1(J)
       END DO
      END DO
C
C       WRITE RESULTS
C

C
C       DEFINITION OF OUTPUT VARIABLES
C
C       1.- I : HORIZONTAL NODE (I = 1 TO NN)
C       2.- R : X VALUE CORRESPONDING TO Ith HORIZONTAL NODE
C       3.  HH: AMPLITUDE OF FREE SURFACE OSCILLATIONS
C       4.- VS: AMPLITUDE OF LONGITUDINAL VELOCITY COMPONENT
1M
C       BELOW FREE SURFACE
C       5.- J : VERTICAL NODE (J= 1 TO NNV) WITH 1 BEING THE BOTTOM
C       NODE
C       6.- VR: AMPLITUDE OF LONGITUDINAL VELOCITY OSCILLATION
AT
C       NODE I,J.
C
C       TO OBTAIN THE ANALYTICAL SOLUTION AT TIME  T  MULTIPLY
THE
C       AMPLITUDES TIMES EXP(EYE * OMEGA * T) AND RETAIN THE
REAL
```

```
C       PART.
C

    OPEN(UNIT=6,FILE='example.out',STATUS='UNKNOWN')
    DO I=1,NN
          WRITE(6,101) I, R(I),HH(I), VS(I)
     DO J=1,NNV
       WRITE(6,102) I,J,VR(I,J)
     END DO
        END DO
 101  FORMAT(1X,I4,5(1X,E10.4))
 102  FORMAT(1X,2(1X,I4),2(1X,E10.4))
    CLOSE(6)
    STOP
    END
C
C-----------------------------------------------------------------
C
    SUBROUTINE VZ1M(DEPTH,W,BIGK,VS2)
C
C VERTICAL ANALYTIC SOLUTION 1 M BELOW FREE SURFACE
C FOR ASCE VERIFICATION AND VALIDATION TASK COMMITTEE
C CONSTANT VISCOSITY PROFILE
C FRANCISCO J. RUEDA
C JULY 6, 2001
C
C
    IMPLICIT DOUBLE PRECISION (A-H, O-Z)
        COMPLEX*16 VS2
    COMPLEX*16 EYE, XLAMBDA,A,B,COSHLZ,
   &          COSHL, SINHL, TANHL, XZ
    REAL*8 BIGK, W
        REAL*8 ZETA, DEPTH
    EYE=(0.0,1.0)
    ZERO=0.0
C
C DEFINITION OF CONSTANTS
C
    XLAMBDA = CDSQRT(EYE*W)
        SINHL =  (CDEXP(XLAMBDA) - CDEXP(-1.* XLAMBDA))/2.0
        COSHL =  (CDEXP(XLAMBDA) + CDEXP(-1.* XLAMBDA))/2.0
        TANHL =  SINHL / COSHL
    ZETA = -1.D0/DEPTH
        A = (0.0, 0.0)
        B = (0.0, 0.0)
        XZ = XLAMBDA * ZETA
```

```
    COSHLZ = (CDEXP(XZ) + CDEXP(-1.* XZ))/2.0
        A = COSHLZ
        B = COSHL * (1.+ (XLAMBDA / BIGK) * TANHL)
    VS2= 1.- (A / B)

        RETURN
        END

C
C-------------------------------------------------------------
C
    SUBROUTINE VERT(NNV,W,MU,BIGK,TAU0,V1,T)
C
C 1-D VERTICAL ANALYTIC SOLUTION FOR ASCE VERIFICATION AND
C VALIDATION TASK COMMITTEE
C CONSTANT VISCOSITY PROFILE
C FRANCISCO J. RUEDA
C MAY 16, 2001
C

    IMPLICIT DOUBLE PRECISION (A-H, O-Z)
    COMPLEX*16 V1(NNV)
    COMPLEX*16 T
    COMPLEX*16 EYE, XLAMBDA,A,B,COSHLZ,
   &         COSHL, SINHL, TANHL, XZ
    REAL*8 MU, TAU0, BIGK, W
        REAL*8 ZETA, DZ
    EYE=(0.0,1.0)
    ZERO=0.0
C
C DEFINITION OF CONSTANTS
C
    XLAMBDA = CDSQRT(EYE*W)
        SINHL =  (CDEXP(XLAMBDA) - CDEXP(-1.* XLAMBDA))/2.0
        COSHL =  (CDEXP(XLAMBDA) + CDEXP(-1.* XLAMBDA))/2.0
        TANHL =  SINHL / COSHL

C    COMPUTE VELOCITY PROFILE
    DZ=1.D0/(NNV - 1.D0)
    DO I=1,NNV
         A = (0.0, 0.0)
         B = (0.0, 0.0)
     ZETA= -1.D0 + (I-1.D0) * DZ
         XZ = XLAMBDA * ZETA
     COSHLZ = (CDEXP(XZ) + CDEXP(-1.* XZ))/2.0
         A = COSHLZ
```

```
        B = COSHL * (1.+ (XLAMBDA / BIGK) * TANHL)
      V1(I)= 1.- (A / B)
    END DO
C
C    COMPUTE TAU
C
    A = XLAMBDA * XLAMBDA * TANHL
        B = XLAMBDA + (XLAMBDA*XLAMBDA/BIGK - 1.) * TANHL
    T= TAU0/3. * (A/B)
    RETURN
    END
C
C——————————————————————————————————————————————————
C
      SUBROUTINE HORIZ(R1,R2,H0,OMEGA,TAU,R,HH,V0, AMP)
C
C 2-D HORIZONTAL ANALYTIC SOLUTION FOR ASCE VERIFICATION
AND
C VALIDATION TASK COMMITTEE
C CONSTANT TAU
C GRAVITY GIVEN IN MKS UNITS
C FRANCISCO J. RUEDA
C MAY 16, 2001
C
      IMPLICIT DOUBLE PRECISION (A-H,O-Z)
      COMPLEX*16 S1,S2,EYE,TAU,HH,V0,A,B,DENOM
     COMPLEX*16 AMP
     REAL*8 R1, R2
      EYE=(0.,1.)
      G=9.806
      S1=-0.5+CDSQRT(0.25E0 -(OMEGA**2-EYE*OMEGA*TAU)/G/H0)
      S2=-0.5-CDSQRT(0.25E0 -(OMEGA**2-EYE*OMEGA*TAU)/G/H0)
      DENOM=S2*(R2**S1)*(R1**S2) - S1*(R1**S1)*(R2**S2)
      A= S2*(R1**S2)/DENOM
      B=-S1*(R1**S1)/DENOM
      HH= AMP * ( A*(R**S1) + B*(R**S2) )
      V0=-AMP * G/(EYE*OMEGA*R)*( S1*A*(R**S1) + S2*B*(R**S2) )
      RETURN
      END
C
C——————————————————————————————————————————————————
C
```

3.6 TEST CASE 5: WIND-DRIVEN FLOW IN A CLOSED RECTANGULAR BASIN WITH A HORIZONTAL BOTTOM

Contributors: Yafei Jia and Sam S.Y. Wang

3.6.1 Description and Objectives

Wind is an important forcing to free surface flows when the movement of the air is significant relative to that of the water flow. The phenomenon that the flow is driven by wind at the air-water interface is very complicated: waves can be generated by the wind; the flow near the free surface moves in the wind direction, and the flow near the bed must then move in the opposite direction to offset the surface flow. Analytical solutions have been obtained for a simplified problem which excludes the waves and the associated processes. However, the solutions can be used to examine the correctness of the numerical model's boundary condition at the free surface and the accuracy of some terms in the model's governing equations. The analytical solutions by Koutitas and O'Connor (1980) and Huang (1993) for wind-driven flow in a straight shallow cavity are used in this test case. Turbulent shear stresses were assumed to be proportional to a constant eddy viscosity. One may use these solutions, however, to determine a numerical model's capability in predicting a linearized, vertical 2D flow field, its consistency with the mathematical model, convergence to the exact solution in a grid refinement process, and its quantitative accuracy.

3.6.2 The Mathematical Problem

3.6.2.1 Governing Equations

Continuity

$$\frac{\partial u}{\partial x}+\frac{\partial v}{\partial y}+\frac{\partial w}{\partial z}=0 \tag{3.90}$$

Free Surface

$$\frac{\partial \zeta}{\partial t}+u_{\zeta}\frac{\partial \zeta}{\partial x}+v_{\zeta}\frac{\partial \zeta}{\partial y}-w_{\zeta}=0 \tag{3.91}$$

Momentum

$$\frac{\partial u}{\partial t}+K_V\frac{\partial^2 u}{\partial z^2}+g\frac{\partial \zeta}{\partial x}=0 \tag{3.92}$$

$$\frac{\partial v}{\partial t}+K_V\frac{\partial^2 v}{\partial z^2}+g\frac{\partial \zeta}{\partial y}=0 \tag{3.93}$$

where u, v, and w are velocity components in Cartesian coordinates, g is the gravitational acceleration, K_V is the vertical eddy viscosity, ζ is the free surface elevation measured vertically from the reference plane (datum) where $z = 0$, and u_ζ, v_ζ, and w_ζ are the velocity components at the free surface.

Because the following are assumed in the analytic solution, the selection of numerical parameters and the setup of the numerical simulation domain should also satisfy these assumptions:

- Hydrostatic pressure
- Homogeneous fluid (ρ = constant)
- Constant vertical eddy viscosity (K_V = constant)
- $|u\frac{\partial u}{\partial x}|,|v\frac{\partial u}{\partial y}|,|w\frac{\partial u}{\partial z}|,|K_H\frac{\partial^2 u}{\partial x^2}|,|K_H\frac{\partial^2 u}{\partial y^2}| << |K_V\frac{\partial^2 u}{\partial z^2}|$
- $|u\frac{\partial v}{\partial x}|,|v\frac{\partial v}{\partial y}|,|w\frac{\partial v}{\partial z}|,|K_H\frac{\partial^2 v}{\partial x^2}|,|K_H\frac{\partial^2 v}{\partial y^2}| << |K_V\frac{\partial^2 v}{\partial z^2}|$
- Steady state flow $u_t = v_t = 0$

3.6.2.2 Domain

The wind shear driven flow definition sketch is shown in Figure 3.6.1. The problem is set in a domain so the water depth h is significantly less than the length L of the basin. The analytical solutions were compared to the numerical solution in the central portion of the basin where the simplifications to the mathematical model hold. The upward and downward flows near the ends of the basin are results of the recirculation. They will be computed as part of the numerical solution but will not be used for the verification.

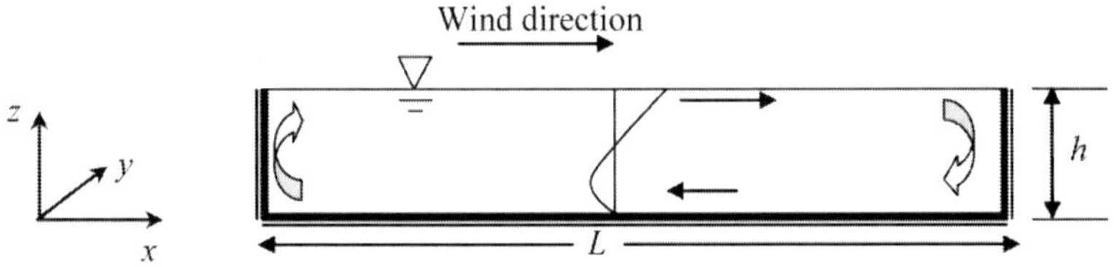

Figure 3.6.1. Test case 5: Sketch of wind driven flow in a closed rectangular basin with a horizontal bottom.

3.6.2.3 Boundary Conditions

The boundary conditions prescribed on the free surface are

$$\begin{aligned}\tau_{sx} &= \rho_a C_w U_w u_w \\ \tau_{sy} &= \rho_a C_w U_w v_w \\ U_w &= \sqrt{u_w^2 + v_w^2}\end{aligned} \tag{3.94}$$

$$\left.\frac{\partial u}{\partial z}\right|_{z=\zeta}=\frac{\tau_{sx}}{K_V} \tag{3.95}$$

$$\left.\frac{\partial v}{\partial z}\right|_{z=\zeta}=\frac{\tau_{sy}}{K_V} \tag{3.96}$$

where (u_w, v_w) are the wind velocity components in the x- and y- directions, respectively; ρ_a is the air density; C_w is the air-water drag coefficient; and U_w is the resultant wind speed. The boundary conditions prescribed on the bed surface are

$$u|_{z=0}=v|_{z=0}=w|_{z=0}=0\ . \tag{3.97}$$

At the end walls

$$u|_{x=0}=v|_{x=0}=\left.\frac{\partial w}{\partial x}\right|_{x=0}=u|_{x=L}=v|_{y=L}=\left.\frac{\partial w}{\partial x}\right|_{x=L}=0 \tag{3.98}$$

and on the side walls,

$$\left.\frac{\partial u}{\partial y}\right|_{y=side\ wall}=v|_{y=side\ wall}=\left.\frac{\partial w}{\partial y}\right|_{y=side\ wall}=0\,. \tag{3.99}$$

3.6.2.4 Initial Conditions

For steady state flow models an initial estimate of variables to begin the iteration process is needed; for unsteady flow models, initial conditions must be specified. In the case of this simplified problem, a “cold start” shown below can be applied to either steady or unsteady models.

$$u|_{t=0}=v|_{t=0}=w|_{t=0}=0 \tag{3.100}$$

3.6.3 Analytical Solution

The analytical solutions by Koutitas and O'Connor (1980), Huang (1993) and Huang and Spaulding (1995) were used here as examples of model verification. The main difference between these two solutions is the boundary condition applied to the basin floor. A non-slip boundary condition was used by Koutitas and O'Connor (1980), and a partial slip boundary condition was used by Huang (1993).

The analytical solution obtained by Koutitas and O'Connor (1980) is relatively simple:

$$u=u_{\max}\delta(3\delta-2) \tag{3.101}$$

$$u_{\max}=\frac{\tau_{sx}h}{4\rho_0 K_V} \tag{3.102}$$

$$\frac{\partial \zeta}{\partial x} = \frac{3\tau_{sx}}{2\rho_0 gh} \tag{3.103}$$

where u_{max} is the velocity at the free surface, $\delta = z/h$ is the non-dimensional or normalized vertical coordinate measured from the bed (z = 0) to the water surface (z = h), h is the total water depth (in this test case $h = \zeta$), ρ_0 is the constant reference density of water, K_V is the constant vertical eddy viscosity, and τ_{sx} is the wind shear stress acting on the surface in the x-direction. It is easy to verify that the velocity u is a maximum at the surface ($u_{max} = u_{\delta=1}$) and is zero at the bed. At the depth $\delta = 2/3$, the flow direction changes from positive to negative. The local maximum negative velocity ($u = -u_{max}/3$) is located at $\delta = 1/3$. The solid line in figure 3.6.1 shows this solution. The analytical solution obtained by Huang (1993) with a prescribed bed shear and a partial-slip boundary condition is different from equation (3.101) and is given below:

$$u = \frac{1}{6K_V} g \frac{\partial \zeta}{\partial x} h^2 [3(\delta - 1)^2 - 1] + \frac{\tau_{sx} h}{2\rho_0 K_V} (2\delta - 1) \tag{3.104}$$

$$\frac{\partial \zeta}{\partial x} = \frac{3}{2} \frac{\tau_{sx}}{\rho_0 gh} \frac{2K_V + kh}{3K_V + kh} \tag{3.105}$$

The definitions of the variables in equations 3.104 and 3.105 are the same as those for equations 3.101 through 3.103; k is an additional variable representing the linearized bottom friction coefficient. As one can see, the u velocity can not be zero on the bed surface ($\delta = 0$) because of the friction at the bed surface.

3.6.4 Test Case

Users of this test case should modify their mathematical and numerical models to eliminate or switch off the terms in the governing equations, so that they are the same as the ones used to obtain the analytical solutions. Then the user should apply the same boundary conditions and adopt the same geometric, physical and numerical parameters. In this particular case, the momentum equations are simplified so that only the second order term for the vertical variation in the longitudinal momentum equation remains. Minimal effort is needed to generate a grid. Analytical solutions for the flow field induced by wind stress acting at the free surface can be obtained by simplifying the three-dimensional momentum equations for these simplified test cases. The channel is straight, long and shallow and is closed at both ends by vertical walls. No net discharge is present; the wind stress is the only force that drives the flow. The motion of the flow in the channel is of the interest. Neglecting the advection, Coriolis forcing, horizontal diffusion terms, and the cross-sectional and vertical momentum equations, analytical solutions can then be obtained in a vertical two-dimensional domain using one momentum equation and the continuity equation.

3.6.4.1 Physical Parameters

Wind shear stress: $\tau_{sx} = 0.1$ N/m^2 ($\tau_{sy} = 0.0$)

Water depth: $h = 40$ m
Eddy viscosity: $K_V = 0.03$ m^2/s
Gravitational acceleration: $g = 9.817$ m/s^2
Density of water: $\rho_0 = 1000$ kg/m^3
Linearized bed friction coefficient: $k = 0.005$ m/s

3.6.4.2 Boundary Conditions

On the free surface the shear stress τ_{sx} can be applied directly for numerical models using a control volume or multi-layer type of method (Kodama et al. 1993). For models using natural boundary conditions the velocity gradient $\partial u / \partial z$ may be used to reflect the boundary shear force. The vertical gradient of velocity may be obtained from equations (3.101) and (3.104).

$$\frac{\partial u}{\partial z} = \frac{1}{h}\frac{\partial u}{\partial \delta} = \frac{1}{h} u_{\max} (6\delta - 2) \tag{3.106}$$

or

$$\frac{\partial u}{\partial z} = \frac{g}{K_V}\frac{\partial \zeta}{\partial x}(\delta - 1)h + \frac{\tau_{sx}}{\rho_0 K_V} \tag{3.107}$$

The values of the boundary velocity gradient may be evaluated by equations 3.106 and 3.107 using the given physical parameters:

$$\left.\frac{\partial u}{\partial z}\right|_{\delta=1} = \frac{4}{h} u_{\max} = \frac{\tau_{sx}}{\rho_0 K_V} = 0.0033333 \tag{3.108}$$

$$\left.\frac{\partial u}{\partial z}\right|_{\delta=1} = \frac{\tau_{sx}}{\rho_0 K_V} = 0.0033333 \tag{3.109}$$

If the analytical solution of Koutitas and O'Connor (1980) is used, $u = 0$ is prescribed on the basin floor. If one uses the solution of Huang and Spaulding (1995), equation 3.107 should be used to calculate the velocity gradient, $\partial u/\partial z|_{\delta=0}$. Users are advised to compute this boundary condition with a second order method.

3.6.4.3 Reporting of Model Results

Because both the numerical model and the analytical solution solve the same equation with the same boundary conditions, the numerical errors can easily be detected and the accuracy of the numerical solution can be determined quantitatively. The rate of the error reduction or the order of accuracy enhancement of the numerical model during a grid refinement process can be used to estimate the rate of convergence. A model tester should report not only the magnitudes of numerical errors from different grid resolutions but also the rate of convergence.

3.6.5 Example Verification

Figure 3.6.2 compares the computed results with the analytical solution (eq. 3.101). Three uniform grids with 6, 11 and 21 nodes in the vertical direction were used to simulate the wind driven flow in a cavity. Eleven nodes in the horizontal direction of the basin were used. All of the results using the three meshes are in good agreement with the analytical solution, and the differences between these solutions are not apparent in figure 3.6.2. This agreement occurs because the interpolation functions used in the CCHE3D model are quadratic, the same as in the analytical solution. In the computation, the surface elevation was computed by using the velocity solution and the continuity equation. The driving force for balancing the wind shear stress, represented by the surface slope, was the solution of the computation rather than being given by analytical solutions. Although the velocity solutions near the center point of the computational domain behave linearly, they are non-linear near the two ends of the channel which will affect the accuracy and convergence at the center point where the convergence is being observed. From table 3.6.1, however, one may find that the solution accuracy improves with the mesh density. The error variation with the mesh density at locations $\delta = 0.4$, 0.6, 0.8 and 1.0 shows that the errors are reduced dramatically as the mesh size is reduced.

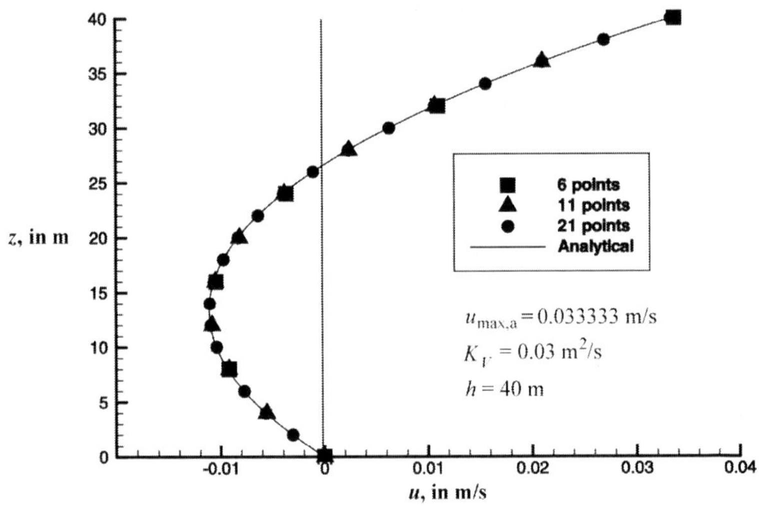

Figure 3.6.2. Test Case 5: Graphical display of analytical and numerical solutions for a velocity profile calculated using the non-slip bed boundary condition.
The analytical solution is by Koutitas and O'Connor (1980) and is plotted at the center of the basin.

Table 3.6.1 Test Case 5: Convergence test of the CCHE3D model.

No. of Nodes		6	11	21
Analytical solutions (equation 3.101)	$u_{max,a}$ = 0.0333333	0.0337177	0.033430	0.033378
	$u_{\delta=0.8}$ = 0.010666 m/s	0.0110339	0.01075797	0.01070932
	$u_{\delta=0.6}$ = 0.004 m/s	0.00367872	0.00387108	0.00396358
	$u_{\delta=0.4}$ = -0.010666 m/s	-0.0104219	-0.01052547	-0.0106407
Difference between analytical solution and numerical solution	$\Delta f_{\delta=1} = u_{max,s} - u_{max,a}$	3.844×10^{-4}	9.667×10^{-5}	4.47×10^{-5}
	$\Delta f_{\delta=0.8} = u_s - u_{\delta=0.8,a}$	3.672×10^{-4}	9.130×10^{-5}	4.27×10^{-5}
	$\Delta f_{\delta=0.6} = u_s - u_{\delta=0.6,a}$	3.212×10^{-4}	1.289×10^{-4}	3.641×10^{-5}
	$\Delta f_{\delta=0.4} = u_s - u_{\delta=0.4,a}$	-2.447×10^{-4}	-1.412×10^{-4}	-2.594×10^{-5}
	Δz	8	4	2
	Δz^2	64	16	4
Coefficient C of equation 3.111	$\Delta f_{\delta=1} / \Delta z^2$	6.00×10^{-6}	6.04×10^{-6}	1.12×10^{-5}
	$\Delta f_{\delta=0.8} / \Delta z^2$	5.74×10^{-6}	5.71×10^{-6}	1.07×10^{-5}
	$\Delta f_{\delta=0.6} / \Delta z^2$	5.02×10^{-6}	8.06×10^{-6}	9.10×10^{-6}
	$\Delta f_{\delta=0.4} / \Delta z^2$	-3.82×10^{-6}	-8.82×10^{-6}	-6.48×10^{-6}

Following Roache (1998), the error here is defined as the difference between the numerical solution and the analytical solution at a point:

$$E = f(\Delta) - f_a \tag{3.110}$$

As suggested by Roache (1998), numerical error would be proportional to Δ^p, where p is the order of the convergence, and the error can be expressed as

$$E = f(\Delta) - f_a = C\Delta^p + H.O.T. \tag{3.111}$$

The near-constancy of the ratio E/Δ^2 ($p = 2$) for all three meshes at each of the four locations (table 3.6.1) indicates the CCHE3D model is second-order accurate.

Figure 3.6.3 shows the comparison of the computed results with the analytical solution (equations 3.104). Similar to figure 3.6.2, three uniform grids with 6, 11 and 21 nodes have been used to simulate the wind driven cavity flow. Eleven nodes are used in the horizontal direction in the basin. The results are improved as the mesh is refined. Although the improvements cannot be seen clearly in the figure, they are similar to the ones shown in table 3.6.1.

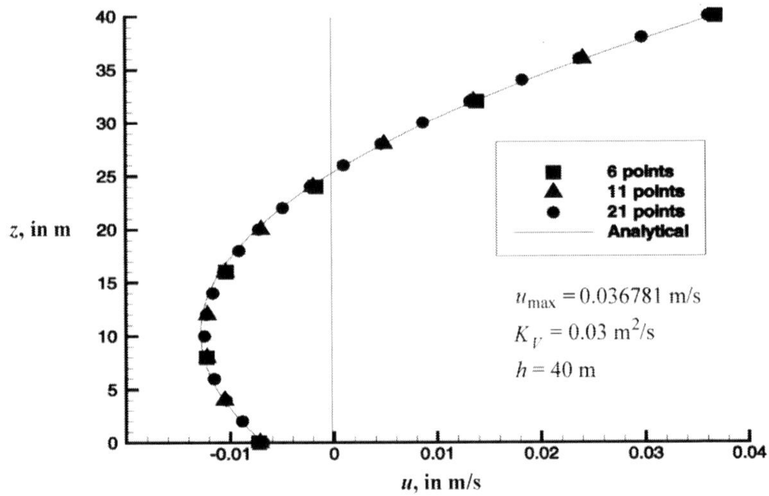

Figure 3.6.3. Test Case 5: Graphical display of analytical and numerical solutions for a velocity profile calculated using the slip bed boundary condition.
The analytical solution is by Huang and Spaulding (1995) and is plotted at the center of the basin.

3.6.6 Concluding Summary

Even though the analytical solutions in this test were obtained with several simplifications, they play a very important role in numerical model verification. Because the numerical and analytical solutions are obtained from the same set of equations having the same boundary/initial conditions and physical/geometrical parameters, the numerical solution must converge to the analytic solution. Otherwise computer program coding errors, computational algorithm errors, discretizing errors, and/or mathematical formulation errors are indicated. Furthermore, since an exact solution can be used as a measure of accuracy, the order of magnitude of numerical errors can be determined quantitatively. The order of convergence of a numerical model can be tested in a systematic grid refinement process. In the present example the accuracy of the longitudinal velocity at various water depth levels increased as the grid was refined. The rate of the error reduction indicated the terms involved were second-order accurate. Therefore, the test, if successful, can give the user a great deal of confidence in the numerical model tested. These verification results should give the model developer and the user confidence in the correctness of the model's numerical algorithm and computer coding. Additional validation tests using the laboratory experimental data and field measurements can be done to complete a comprehensive model verification and validation procedure.

3.6.7 References

Huang, W., 1993, Three-dimensional numerical modeling of circulation and water quality induced by combined sewage overflow discharge: Ph.D dissertation, Department of Ocean Engineering, Univ. of Rhode Island, Kingston, R.I.

Huang, W., and Spaulding, M., 1995, 3D model of estuarine circulation and water quality induced by surface discharge: American Society of Civil Engineers, Journal of Hydraulic Engineering, v. 121, no. 4, p. 300-311.

Kodama, T., Wang, S.S.Y., and Kawahara, M., 1993, Numerical model verification for wind-induced flow: Advances in Hydroscience and Engineering, Sam, S.Y. Wang (ed.), v. I, p. 2138-2143.

Koutitas, C., and O'Connor, B., 1980, Modeling three-dimensional wind-induced flows: American Society of Civil Engineers, Journal of Hydraulic Division, v. 106, no. 11, p. 1843-1865.

Roache, P.J., 1998, Verification and Validation in Computational Science and Engineering, Hermosa Publishers, Albuquerque, New Mexico, 446 p.

3.7 TEST CASE 6: INTERNAL-WAVE SEICHING IN A RECTANGULAR BASIN WITH A HORIZONTAL BOTTOM

Contributors: Francisco J. Rueda and Peter E. Smith

3.7.1 Description and Objectives

Stratification in natural surface-water bodies significantly modifies and controls the hydrodynamic processes of mixing and transport. In stratified water bodies, oscillatory motions with well-defined frequencies cause isopycnals to tilt and fluctuate around an equilibrium position (Imberger, 1998). These oscillatory phenomena are referred to as internal waves. They are the result of the balance between gravity and inertial forces, and to simulate numerically internal waves at the basin scale requires the baroclinic and scalar advection terms to be represented accurately. The goal of the next problem is to test the ability of a 3-D model to simulate basin-scale internal waves. The test problem simulates a baroclinic or internal seiche in an enclosed rectangular and narrow flat-bottom basin using a background linear density profile in a non-rotating coordinate system. The water motions in the test problem are caused by gradients in the water density. An analytical solution to this problem was published by Eliason and Bourgeois (1997) and referred to as "the baroclinic mode solution." It requires that (1) the motion is effectively 2-D in a vertical plane, (2) the fluid is assumed inviscid, and (3) non-linear terms are neglected. This problem is appropriate for verifying models that simulate the hydrodynamics of water bodies having gradients in the density, such as lakes, estuaries, and shallow coastal seas.

3.7.2 The Mathematical Problem

3.7.2.1 Governing equations

Continuity

$$u_x + v_y + w_z = 0 \qquad (3.112)$$

$$\zeta_t + \left(\int_{z=-H}^{z=0} u\,dz\right)_x + \left(\int_{z=-H}^{z=0} v\,dz\right)_y = 0 \qquad (3.113)$$

Assumptions:

- $\zeta << H$ (allows a linearization of the code)
- H is constant

Momentum

$$u_t + g\zeta_x + g\frac{1}{\rho_0}\int_z^0 \frac{\partial \rho}{\partial x}dz' = 0 \qquad (3.114)$$

$$v_t + g\zeta_y + g\frac{1}{\rho_0}\int_z^0 \frac{\partial\rho}{\partial y}dz' = 0 \tag{3.115}$$

Assumptions:

- Inviscid fluid (interfacial and bottom friction are neglected)
- Nonrotating reference frame ($f = 0$)
- $uu_x, vu_y << u_t, P_x/\rho_0$ (the x-direction advection terms are neglected)
- $uv_x, vv_y << v_t, P_y/\rho_0$ (the y-direction advection terms are neglected)
- where P_x and P_y represent the x and y pressure-gradient terms, including the barotropic and the baroclinic parts

Scalar transport

$$\frac{\partial S}{\partial t} + w\frac{\partial \hat{S}}{\partial z} = 0 \tag{3.116}$$

Assumption:

- Immiscible fluid
- $uS_x, vS_y << S_t, w\hat{S}_z$ (the horizontal scalar advection terms are neglected)
- $\hat{S}$ represents a time-invariant linear background vertical profile of the scalar variable (this assumption neglects the nonlinear effects that arise from the vertical advection of the scalar variable)
- $\hat{S} = \hat{S}_{z=0} - z(\hat{S}_{z=0} - \hat{S}_{z=-H})$ (the background profile is linear) (3.117)

Equation of state

$$\rho = \rho\,(S) \tag{3.118}$$

For this test problem, a small-amplitude (barotropic) seiching in the free-surface and velocity solution can be observed after the simulation is begun from a cold start. The free-surface seiche (identical to test case 1) is caused by the water body adjusting to the longitudinal gradient in density, and the seiche continues unabated once it is set in motion. The amplitude of the free-surface seiche is considered negligible and can be filtered out of the numerical solution for comparison with the analytical solution.

3.7.2.2 Domain

The domain for test case 6 is shown in figure 3.7.1.

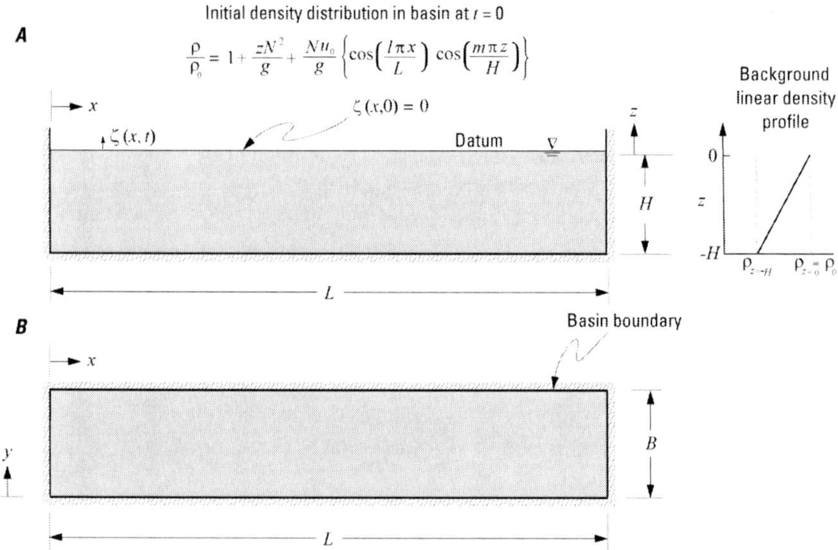

Figure 3.7.1 Test Case 6: Diagram of the domain for the test problem of internal-wave seiching in a rectangular basin with a horizontal bottom.

(***A***) Cross section of the rectangular basin showing the initial position of the free surface, and (***B***) top view. The vertical scale in (***A***) is greatly exaggerated and distorted.

3.7.2.3 Boundary conditions

$$\left.\frac{\partial u}{\partial y}\right|_{y=0} = \left.\frac{\partial u}{\partial y}\right|_{y=B} = u\big|_{x=0} = u\big|_{x=L} = 0 \tag{3.119}$$

$$\left.\frac{\partial v}{\partial x}\right|_{x=0} = \left.\frac{\partial v}{\partial x}\right|_{x=L} = v\big|_{y=0} = v\big|_{y=B} = 0 \tag{3.120}$$

$$\left.\frac{\partial u}{\partial z}\right|_{z=-H} = \left.\frac{\partial v}{\partial z}\right|_{z=-H} = \left.\frac{\partial u}{\partial z}\right|_{z=\zeta} = \left.\frac{\partial v}{\partial z}\right|_{z=\zeta} = 0 \tag{3.121}$$

$$w\big|_{z=-H} = 0 \tag{3.122}$$

In addition, the flux of the scalar variable S is zero everywhere along the basin boundary.

3.7.2.4 Initial conditions

$$\rho = \rho_0\left(1+\frac{zN^2}{g}\right)+\frac{\rho_0 N u_0}{g}\cos\left(\frac{l\pi x}{L}\right)\sin\left(\frac{m\pi z}{H}\right) \tag{3.123}$$

$$\zeta, u, v, w\big|_{t=0} = 0 \tag{3.124}$$

where

- N is the Brunt-Väisäla frequency (a measure of water column stability) for the constant background density profile, $N^2 = -(g/\rho_0)\,\partial\hat{\rho}/\partial z)$
- ρ_0 is the reference surface density
- l is a horizontal mode number (integer) that can be chosen by the user
- m is a vertical mode number (integer) that can be chosen by the user

3.7.3 Analytical solution (Eliason and Bourgeois 1997)

$$u = u_0 \sin\left(\frac{l\pi x}{L}\right)\cos\left(\frac{m\pi z}{H}\right)\sin(\omega t) \tag{3.125}$$

$$v = 0 \tag{3.126}$$

$$w = -\frac{lH}{mL}u_0\cos\left(\frac{l\pi x}{L}\right)\sin\left(\frac{m\pi z}{H}\right)\sin(\omega t) \tag{3.127}$$

$$\rho = \rho_0\left(1+\frac{zN^2}{g}\right)+\frac{\rho_0 N u_0}{g}\cos\left(\frac{l\pi x}{L}\right)\sin\left(\frac{m\pi z}{H}\right)\cos(\omega t) \tag{3.128}$$

where

ω is the frequency of the baroclinic oscillation defined by

$$\omega = \frac{lHN}{mL} \tag{3.129}$$

3.7.4 Test case

3.7.4.1 Physical parameters

L = 20,000 m
B = 4,000 m
H = 12 m

For our calculations the temperature was assumed to be the only active scalar S. The following equation of state was used to relate temperature (°C) to density (kg/m^3):

$$\rho\ (S) = 516 \times (1.939933 + S(5.88599 \times 10^{-5} - 1.108539 \times 10^{-5} \times S))$$

This approximates the equation given by Mellor (1991) and should be used in any code being tested. A code formulated in terms of salinity, or another active scalar, can be used on this test problem so long as the equation of state is suitably modified.

Additional specific parameters for this test problem are

$\hat{S}\|_{z=0}$	$= 25\ ^{o}$C	(constant background temperature at the basin free surface used in equation 3.116)
$\hat{S}\|_{z=-H}$	$= 15\ ^{o}$C	(constant background temperature at the basin bottom used in equation 3.116)
$\hat{\rho}\|_{z=0} = \rho_0$	$= 998.18$ Kg/m^3	(constant background density at the basin free surface corresponding to $\hat{S}\|_{z=0} = 25\ ^{o}$C.
$\hat{\rho}\|_{z=-H}$	$= 1000.18$ Kg/m^3	(constant background density at the basin bottom corresponding to $\hat{S}\|_{z=-H} = 15\ ^{o}$C.
u_0	$= 0.10$ m/s	(amplitude of the baroclinic horizontal speed)
N	$= 0.040448$ s^{-1}	(Brunt-Väisäla frequency)
l, m	$= 1$	(first horizontal and vertical modes)
ω	$= .000024269$ s^{-1}	(baroclinic seiche frequency)
T'	$= 71.9158$ hours	(baroclinic seiche period, $T' = 2\pi/\omega$)

3.7.4.2 Numerical parameters

High resolution

A regular mesh of 960 (20 x 4 x 12) computational cells, or an equivalent mesh of 3,920 degrees of freedom (four degrees of freedom per computational cell, u, v, w, S ; plus one, ζ, for each water column), can be used for a high-resolution test case. The total number calculated for the degrees of freedom includes all nodes on the basin boundary. The regular mesh cells in the horizontal plane are squares ($\Delta x = \Delta y$). It is assumed that a mesh of twelve vertical layers is used. Only 1/12 of the count for degrees of freedom should be used to discretize the rectangular basin horizontally. Of course, the user is not restricted to using regular meshes; irregular finite difference, finite element, or body-fitted meshes are encouraged, subject to the specifications on degrees of freedom.

For the regular mesh, the suggested values for the discretization parameters are

$\Delta x \quad = \Delta y = 1000$ m

$\Delta z \quad = 1$ m

$T_s \quad = 144$ hours (6 days)

where T_s is the duration of the simulation.

Modelers are asked to report the numerical treatment of the time domain in full reproducible detail. For time-stepping schemes the following time step is proposed, which guarantees a stable solution for methods that discretize the baroclinic term explicitly:

Δt = 720 seconds

Low resolution

The regular mesh of 120 (10 x 2 x 6) computational cells, or an equivalent mesh of 500 degrees of freedom (four per computational cell, u, v, w, S ; plus one, ζ, for each water column), can be used for a low-resolution test case. For the regular mesh, the suggested values for the discretization parameters are

Δx = Δy = 4,000 m
Δz = 2 m
T_s = 144 hours (6 days)

For time stepping schemes a step size Δt of 720 seconds is proposed.

3.7.4.3 Reporting of model results

Modelers should report the $\ell 1$, $\ell 2$ and $\ell \infty$ error norms, RMSE, and the index of agreement d for the longitudinal (u, in cm/s) and vertical (w, in cm/s) velocity solutions and the free surface elevation (ζ, in cm) at the following nodal locations

Low resolution:

Longitudinal velocity	x = 9,000 m	y = 1,000 m	z = – 1 m
Vertical velocity	x = 1,000 m	y = 1,000 m	z = – 5 m
Density	x = 1,000 m	y = 1,000 m	z = – 5 m

High resolution:

Longitudinal velocity	x = 9,500 m	y = 500 m	z = – 0.5 m
Vertical velocity	x = 500 m	y = 500 m	z = – 5.5 m
Density	x = 500 m	y = 500 m	z = – 5.5 m

As an option, modelers are also invited to submit phase and dissipation errors in the form of the celerity ratio Q and the dissipation factor ε_A.

3.7.5 Example Verification: Low Resolution Test Case

A comparison of the analytical and numerical solutions for the three locations of the low-resolution test case is given in figure 3.7.2. The computed error measures for the low- resolution numerical solution are given in table 3.7.1. Figure 3.5.3 provides comparisons of velocity profiles for the horizontal column of water at node LR(2).

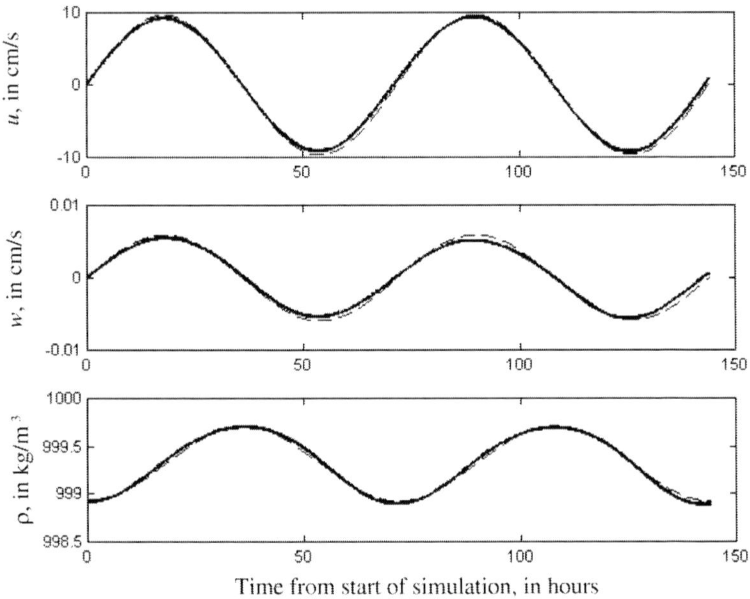

Figure 3.7.2 Test case 6: Graphical display of analytical and numerical solutions for the low-resolution test case.

The analytical solution is a dashed line; the numerical solution is a solid line. The results are graphed for the *u*-velocity component at location x = 9,000 m, y = 1,000 m, and z = – 1.0 m; for the *w*-velocity component at location x = 1,000 m, y = 1,000 m, and z = – 5.0 m; and for the density ρ at location x = 1,000 m, y = 1,000 m, and z = – 5.0 m. The numerical calculations were done with the SI3D model using 1 leapfrog and 2 trapezoidal iterations for time stepping. The velocity solution from the model in this test case exhibited small oscillations having the frequency of barotropic seiches. For the purpose of comparing the analytical and numerical solutions, the oscillations were eliminated by smoothing the numerical velocity results with a 6-point moving average filter.

Table 3.7.1 Test Case 6: Error measures for the low-resolution solution.

	u **(cm/s)**	***w*** **(cm/s)**	**ρ (kg/m^3)**
ε_A	1.0000	1.0000	1.0000
Q	1.0019	1.0095	1.0679
d	0.9990	0.9967	—
ℓ **1**	0.0612	0.1049	—
ℓ **2**	0.0616	0.1099	—
$\ell\,\infty$	0.0716	0.1456	—
RMSE	0.4199	4.6004×10^{-4}	—

3.7.6. References

Eliason, D. E., and A. J. Bourgeois, 1997, Validation of numerical shallow water models for stratified seiches: International Journal for Numerical Methods in Fluids, v. 24, p. 771-786.

Imberger, J., 1998, Flux paths in a stratified lake—a review: Physical Processes in Lakes and Oceans, J. Imberger, ed., American geophysical Union, p. 1-17.

Mellor, G. L., 1991, An equation of state for numerical models of oceans and estuaries: Journal of Atmospheric and Oceanic Technology, v. 8, p. 609-611.

3.8 TEST CASE 7: DENSITY-DRIVEN FLOW IN A RECTANGULAR BASIN WITH A HORIZONTAL BOTTOM

Contributors: Wenrui Huang, Malcolm Spaulding, and Tien-Shuenn Wu

3.8.1 Description and Objectives

Baroclincity is one of the important forcing mechanisms in estuarine flow. In this test case we considered the vertical variations in current velocity induced by a constant, horizontal gradient of density in a flat rectangular basin with no friction. This situation is similar to the circulation condition in a well mixed estuary. At steady state the horizontal density gradient is balanced by the surface slope and the vertical diffusion of momentum. The analytical solution from this test case can be used to test the ability and accuracy of a three-dimensional model in predicting the vertical variations of velocity. In addition, the numerical scheme used to determine the vertical velocity structure in a 3-D model can be examined though this simplified test case.

3.8.2 The Mathematical Problem

3.8.2.1 Governing equations

Continuity

$$u_x + w_z = 0 \tag{3.130}$$

Assumption:

- u is constant as a function of the horizontal distance x.

Momentum

$$\frac{\partial p}{\partial x} = \rho K_V \frac{d^2 u}{dx^2} \tag{3.131}$$

$$p = \rho g(\zeta - z) \text{ or, } \frac{\partial p}{\partial x} = \rho g \frac{\partial \zeta}{\partial x} - gz \frac{\partial \rho}{\partial x} \tag{3.132}$$

Assumptions:

- Density ρ linearly increases from the estuarine head to the ocean
- Density is constant as a function of the vertical coordinate z
- Constant vertical eddy viscosity K_V
- Steady state: $u_t = 0$
- Current velocity is constant as a function of the horizontal distance: $u_x = 0$

3.8.2.2 Domain

The domain for test case 7 is shown in figure 3.8.1.

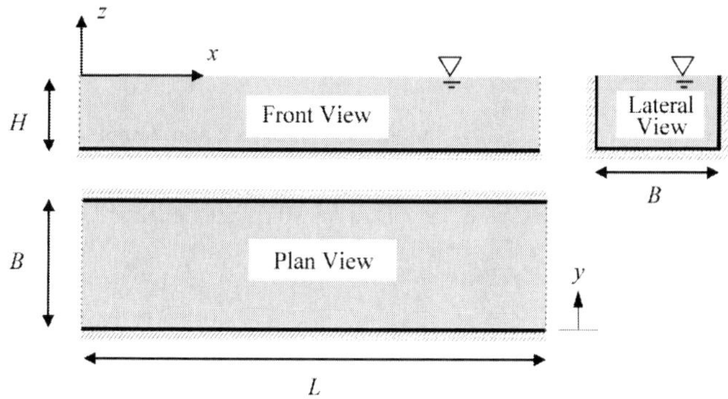

Figure 3.8.1. Test Case 7: Diagram of the domain for the test problem of density-driven flow in a rectangular basin with horizontal bottom.
Note: The drawing is not to scale.

3.8.2.3 Boundary conditions

$$\left.\frac{\partial u}{\partial z}\right|_{z=-H} = \left.\frac{\partial u}{\partial z}\right|_{z=\zeta} = 0 \tag{3.133}$$

$$\left.w\right|_{z=-H} = 0 \tag{3.134}$$

The surface wind stress and the bottom friction are equal to zero.

3.8.2.4 Initial conditions

$$\left.\zeta\right|_{t=0} = 0 \tag{3.135}$$

$$\left.u, v, w\right|_{t=0} = 0 \tag{3.136}$$

3.8.3 Analytical solution (Officer, 1976 - p.122)

Using the coordinate system given in figure 3.8.1 the analytical solution can be written as follows:

$$u(z) = \frac{1}{24}\left(\frac{gH^3}{\rho K_V}\right)\left(\frac{d\rho}{dx}\right)\left[1 - 6\left(\frac{z}{H}\right)^2 - 4\left(\frac{z}{H}\right)^3\right] \tag{3.137}$$

3.8.4 Test case

3.8.4.1 Physical parameters

L = 15,000 m
B = 5,000 m

H = 20 m
K_V = 50 cm^2/s
$d\rho/dx$ = 3 x 10^{-3} g/cm^3/(10 km).

3.8.4.2 Numerical parameters

The horizontal domain is equally divided into 1 km x 1 km computational cells. In the vertical direction, two tests using 5 and 16 layers were used to examine the model sensitivity to vertical grid resolutions. The time step for the internal mode is 2 minutes. Model simulations were conducted for 30 days to reach steady state.

3.8.4.3 Reporting of model results

Model predictions of the vertical profile of the horizontal velocity at the middle of the basin can be compared with the analytical solution given by equation 3.137. For the domain shown in figure 3.8.1, the model outputs were obtained at location x = 7.5 km. Modelers should report the maximum errors and root-mean-square errors (RMSE) between model predictions and the analytic solution.

3.8.5 Example verification

The comparison of the model predictions and the analytical solution are given in figure 3.8.2.

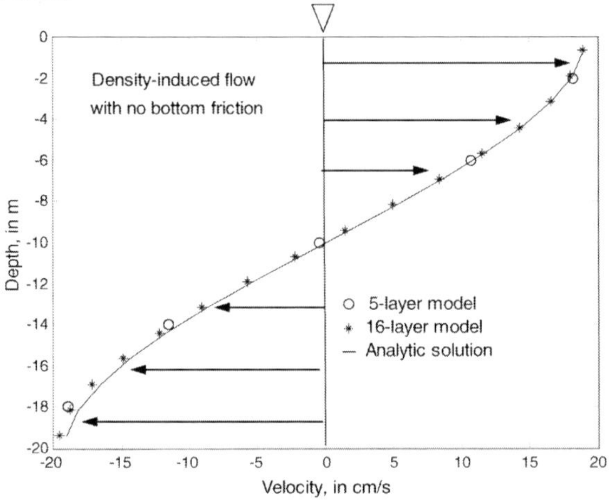

Figure 3.8.2 Test Case 7: Graphical display of analytical and numerical solutions for the low- and high-resolution solutions. The graph shows the vertical profile of the horizontal velocity component. Currents in the upper layer are in the down-estuary direction, and those in the lower layer are in the up-estuary direction. The maximum currents occur near the surface and the bottom boundaries.

The results are graphed for the u-velocity component at location x = 7,500 m. The numerical calculations were done with the Huang/Spaulding model described in section 3.1.4.

As given in table 3.8.1, the maximum error and root-mean-square error (RMSE) were used to evaluate the agreement between the model predictions and the analytical solution.

Table 3.8.1 Test Case 7: Error measures for the low- and high-resolution solutions.

Errors	*5 layers*	*16 layers*
Maximum error (%)	5.5	3.0
RMSE (cm/s)	0.24	0.14
Correlation	0.98	0.99

In general, the model predictions agree well with the analytical solution. For the low-resolution case with five vertical layers, the maximum error is five percent near the bottom. The RMSE value is 0.24 cm/s. Increasing the vertical resolution improved model accuracy. For the high-resolution case with sixteen vertical layers, the maximum error decreased to three percent near the bottom, and the RMSE value fell to 0.14 cm/s.

3.8.6 References

Officer, C. B., 1976. Physical Oceanography of Estuaries (and Associated Coastal Waters). John Wiley and Sons, New York, N.Y.

Huang, W., M. Spaulding, 1995, 3D model of estuarine circulation and water quality induced by surface discharges: American Society of Civil Engineers, Journal of Hydraulic Engineering, v.. 121, no. 4, p. 300-311.

3.9 TEST CASE 8: A THREE-STEP PROCEDURE TO CHECK NONLINEAR EFFECTS FROM TIDAL CIRCULATION IN A SQUARE BASIN WITH A HORIZONTAL BOTTOM

Contributors:* *Tien-Shuenn Wu, Gerald S. Janowitz, Wenrui Huang, and Steven C. McCutcheon

3.9.1 Description and Objectives

Verifying nonlinear numerical models by comparing their results with those of exact analytical solutions is not possible, as no such analytical solutions exist. An analytical solution can be obtained, however, as a perturbation series utilizing the ratio of tidal amplitude to water depth as the perturbation parameter. The analytical solution is typically truncated after the first few terms in the perturbation expansion, and hence it is not exact. The analytical solution obtained by perturbation expansion can closely approximate the solution of the nonlinear governing equations.

We placed no restriction on the tidal amplitude but rather considered "short" basins or harbors, i.e., those for which the seaward extent of the harbor is small compared with the wavelength of a free surface gravity wave of tidal frequency or equivalently those for which the time needed for a free surface gravity wave to propagate across the harbor is small compared to the tidal period. More precisely, our perturbation parameter is the square of the product of the seaward dimension of the harbor *B* and the free surface wave number ω_T/c.

3.9.2 The Mathematical Problem

3.9.2.1 Governing Equations

Continuity

$$\frac{\partial \zeta}{\partial t}+\frac{\partial hV}{\partial y}=0 \qquad (3.138)$$

Assumptions:

- The total water depth ($h = H + \zeta$) is not constant
- The problem is 2-D in the horizontal plane with horizontal velocities *U* and *V*

Momentum

$$\frac{\partial V}{\partial t}+V\frac{\partial V}{\partial y}+g\frac{\partial \zeta}{\partial y}=0 \qquad (3.139)$$

Assumptions:

- Inviscid fluid (bottom friction is neglected)
- Homogeneous fluid (ρ = constant)

- Non-rotating reference frame ($f = 0$)

3.9.2.2 Domain

The domain for test case 8 is shown in figure 3.9.1.

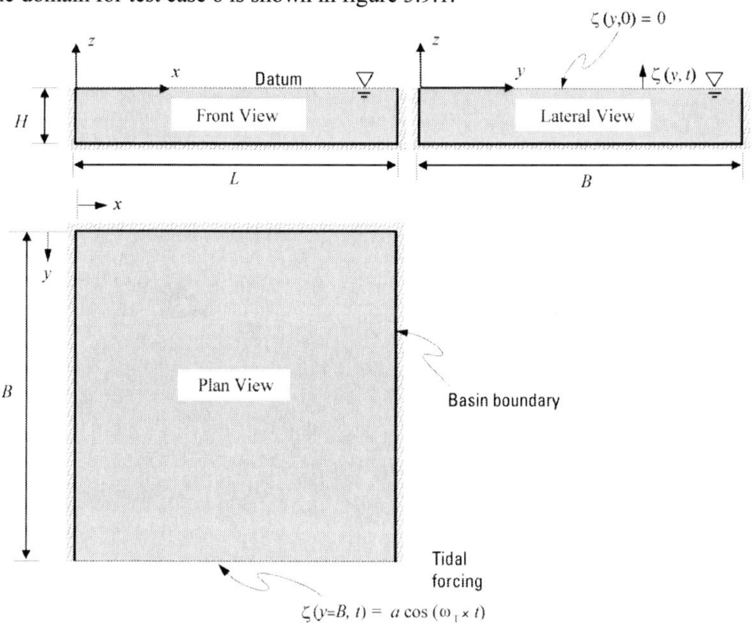

Figure 3.9.1 Test Case 8: Diagram of the domain for the test problem of tidal circulation in a square basin with a horizontal bottom.
Note: The drawing is not to scale.

3.9.2.3 Boundary conditions

$$V\big|_{y=0} = 0 \tag{3.140}$$

$$\zeta\big|_{y=B} = a\cos(\omega_T \times t) \tag{3.141}$$

where

- $\omega_T = 2\pi/T$ = tidal frequency
- T is the tidal period

3.9.2.4 Initial conditions

$$\zeta\big|_{t=0} = 0 \tag{3.142}$$

$$V\big|_{t=0} = 0 \tag{3.143}$$

3.9.3 Analytical Solution (Wu, 1987, pp. 34-40)

$$\begin{aligned}\zeta(y,t) &= a \times \left[\zeta_0 + \in \zeta_1 + O(\in^2)\right] \\ &= \mathrm{a}\cos(t') + \frac{-a[\cos(t') + (a/h) + (a/h)\sin^2(t')](y'^2 - 1)}{2[1 + (a/h)\cos(t')]^2} + aO(\in^2)\end{aligned} \tag{3.144}$$

where

- $y' = y/B$
- $t' = t \times \omega_T$
- $\in = \omega_T^2 B^2 / (gh)$ = perturbation parameter

The zero-order solution and first-order solution are shown below. Note the increase in complexity from the zero-order solution to the first-order solution:

$$\zeta_0 = \cos(t') \tag{3.145}$$

$$\zeta_1 = \frac{[\cos(t') + (a/h) + (a/h)\sin^2(t')](1 - y'^2)}{2[1 + (a/h)\cos(t')]^2} \tag{3.146}$$

3.9.4 Test case

3.9.4.1 Physical parameters

L	= 40 km for low resolution	
	= 20 km for high resolution	
B	= 40 km for low resolution	
	= 20 km for high resolution	
$\in$	= 0.345372 for low resolution	
	= 0.086343 for high resolution	
H	= 10 m	
a	= 2 m	(amplitude of the boundary tidal forcing)
T	= 12 hours	(period of the boundary tidal forcing)

3.9.4.2 Numerical parameters

Low resolution for larger (40 km x 40 km) harbor:

A regular mesh of 16 (4 x 4) computational cells can be used for a low-resolution test case. The regular mesh cells in the horizontal plane are squares ($\Delta x = \Delta y$). Because this problem is two-dimensional, there is only one layer in the vertical dimension. Of course, the user is not restricted to using regular meshes; irregular finite difference, finite element, or body-fitted meshes are encouraged.

For the regular mesh, the suggested values for the discretization parameters are

$\Delta x \quad = \Delta y = 5$ km
$T_0 \quad = 36$ hours
$T_S \quad = 72$ hours

where T_S is the duration of the simulation and T_0 is the start time at which the analytical and numerical solutions are compared.

Modelers are asked to report the numerical treatment of the time domain in full reproducible detail. For time-stepping schemes the following time step is proposed, which will require an implicit numerical method (that is not subject to a Courant stability condition on the speed of gravity waves) to achieve a stable solution:

$\Delta t = 900$ seconds

High resolution for smaller (20 km x 20 km) harbor:

A regular mesh of 16 (4 x 4) computational cells can be used for a high-resolution test case. The regular mesh cells in the horizontal plane are squares ($\Delta x = \Delta y$). Because this problem is two-dimensional, there is only one layer in the vertical dimension. Of course, the user is not restricted to using regular meshes; irregular finite difference, finite element, or body-fitted meshes are encouraged.

For the regular mesh, the suggested values for the discretization parameters are

$\Delta x \quad = \Delta y = 2.5$ km
$T_0 \quad = 36$ hours
$T_S \quad = 72$ hours

For time-stepping schemes the following time step is proposed:

$\Delta t \quad = 900$ seconds

3.9.4.3 Reporting of model results

Modelers should report the $\ell 1$, $\ell 2$ and $\ell \infty$ error norms, RMSE, and the index of agreement d for the free-surface elevation (ζ, in cm) at the following nodal locations

Low resolution:

$x = 20{,}000$ m $\quad y = 35{,}000$ m

High resolution:

$x = 10{,}000$ m $\quad y = 17{,}500$ m

3.9.5 Example verification

A comparison of the analytical and numerical solutions for the low-resolution test case is given in figure 3.9.2. A comparison of analytical and numerical solutions for the high-resolution test case is given in figure 3.9.3. The computed error measures for both the low- and high-resolution numerical solutions are given in table 3.9.1.

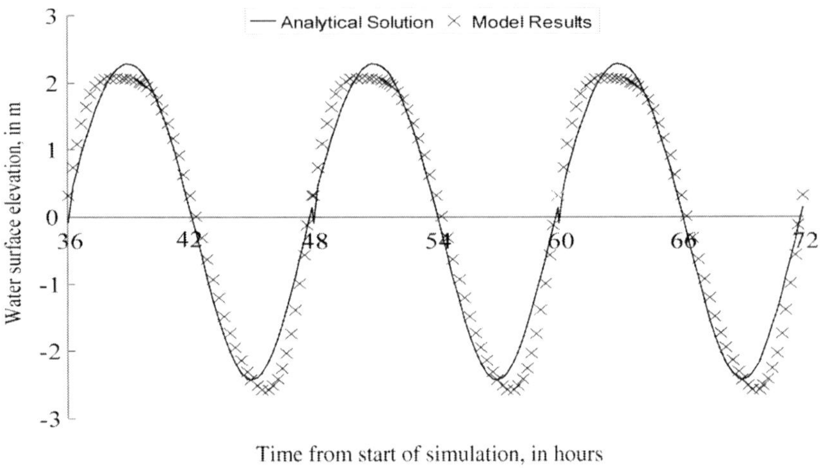

Figure 3.9.2 Test Case 8: Graphical display of analytical and numerical solutions for the low-resolution test case using a 40 km x 40 km harbor.
The analytical solution is the solid line; the numerical solution is the set of plotted points marked by an (x). The results are graphed for the node at location x = 20,000 m, y = 35,000 m. The numerical calculations were done with the SHULIN model using a fully implicit numerical method on a staggered grid.

Table 3.9.1 Test Case 8: Error measures for the low-resolution and high-resolution solutions

	Low Resolution ζ	***High Resolution*** ζ
ε_A	—	—
$\boldsymbol{Q}$	—	—
d	-0.025231	-0.072553
L1	0.178149	0.070283
L2	0.187540	0.081338
L∞	0.263178	0.126625
RMSE	0.309858	0.118159

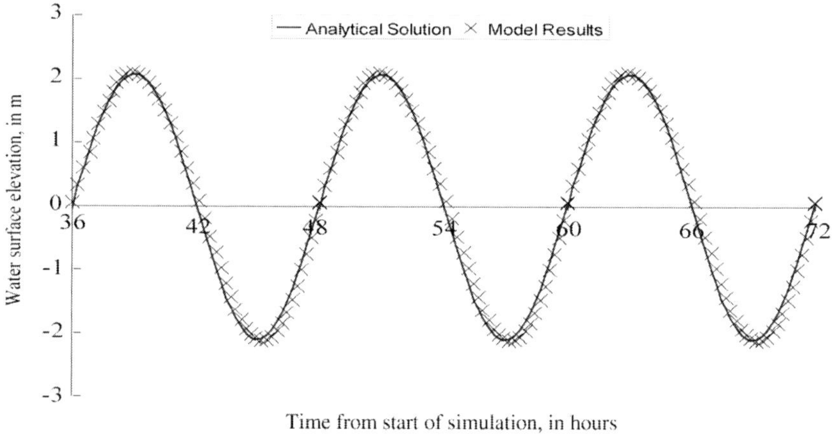

Figure 3.9.3 Test Case 8: Graphical display of analytical and numerical solutions for the high-resolution test case using a 20 km x 20 km harbor.

The analytical solution is the solid line; the numerical solution is the set of plotted points marked by an (x). The results are graphed for the node at location x = 10,000 m, y = 17,500 m. The numerical calculations were done with the SHULIN model using a fully implicit numerical method on a staggered grid.

3.9.6 References

Wu, T. S., 1987, The direct computation of tidal circulation in harbors: University Microfilm International, 300, pg. 3-63.

CHAPTER 4

MATHEMATICAL VERIFICATION USING PRESCRIBED OR MANUFACTURED SOLUTIONS

Sam S.Y. Wang and Yafei Jia

4.1 INTRODUCTION

Contributor: Sam S.Y. Wang

Numerical models have been applied to studies of more and more realistic hydrodynamic systems requiring higher and higher level of accuracy and reliability in recent years. Because the complex phenomena involved in the real world hydrodynamic systems are nonlinear, the ability of numerical model to mimic nonlinear behaviors is of a greater concern than before. The quality assurance of the nonlinear models by mathematical verification and physical validation has become more and more important.

The test cases presented in Chapter 3 were intended for the verification of the correctness in mathematics of linear free surface flow models or the linearized non-linear models and the correctness of computational codes for producing numerical simulations, because almost all analytic solutions used as basis of mathematical verification in that chapter are linear. At present, due to the limitation of the state of the art in mathematical solution of nonlinear partial differential equations, their analytic solutions are usually unavailable. In order to carry out the verification of nonlinear numerical models, alternative methodologies are needed.

The Prescribed Solution Forcing (PSF) by Dee and Da Silva (1986) and the Method of Manufactured Solution (MMS) by Roache and Steinberg (1984) and Roache (1998, 2002) can be used to verify whether a numerical model (code) is mathematically correct in solving a system of nonlinear differential equations governing more realistic nonlinear physical systems.

In general, the arbitrarily selected analytic function being used in either PSF or MMS does not satisfy the original governing nonlinear partial differential equations, and it may not even have anything to do with the physics of the system. It can serve the purpose of Mathematical Verification to determine whether the nonlinear equations governing a boundary value problem of interest have been solved correctly, because the verification of mathematical correctness of a numerical model analytically is meant to be a mathematical exercise. Once such a nonlinear analytic function is prescribed or manufactured, it can be substituted into the governing nonlinear differential equations of a boundary value problem which a numerical model has been developed to solve. Due to the fact that this arbitrarily prescribed analytic function is not a solution, the original differential equation is not satisfied, and an extra non-zero term is resulted on the right-hand side of the equation, which usually is referred to as a forcing function or source term. The resulting new differential equation including the added forcing or source term together with the original boundary and initial conditions forms a new boundary and initial value problem with a known analytic solution which is the one having been prescribed or manufactured. Detailed description and mathematical representation are to be given in the following sections.

Both approaches provide a straightforward procedure, which is to use a prescribed or manufactured analytic function to test whether a numerical model can obtain a correct solution to a nonlinear boundary and initial value problem. It can not only confirm that the model is free of model formulation mistakes, discretization mistakes, coding errors, etc., but also determine whether the model is convergent and consistent. Furthermore, it is capable of determining the order of convergence or accuracy quantitatively.

Besides, this mathematical verification process can be used to determine the relative effects of each term or each group of terms, say advective, diffusive, inertial terms, etc. on the accuracy of the solution, by inclusion or exclusion of individual or group of term(s) in or from the governing equation in a series of verification tests.

In this Chapter, the basic ideas of the application of a Prescribed Solution Forcing and Manufactured Solution methods for the verification of a numerical model simulating a nonlinear free surface flow system, and a step by step procedure to conduct such verification are presented. In addition, simple examples are given to clarify the descriptions and demonstrate the procedure. The examples presented are by no means exclusive. In fact, users are encouraged to prescribe or manufacture analytic function(s) of their own choice for conducting their mathematical verification of the model(s) or code(s), which they are developing or planning to adopt. Once again, the users are reminded that this verification test is or could be a purely mathematical exercise to confirm whether the nonlinear partial differential equations with or without the prescribed boundary conditions and/or initial conditions are solved correctly, whether the solution is consistent with the equations and whether there are mathematical and/or program coding mistakes or errors. A test case should be designed such that all the terms and related calculations are exercised non-trivially to ensure the completeness of the verification.

4.2 METHOD OF PRESCRIBED SOLUTION FORCING (PSF)

Contributors: Sam S.Y. Wang and Weiming Wu

4.2.1 Basic Concept and Approach

The Prescribed Solution Forcing (PSF) is a mathematical technique that can be used to verify numerical model's correctness in solving the governing differential equation(s) (linear or nonlinear) and boundary/initial conditions. Its basic concept and approach are briefly presented in this section.

Let's suppose that a nonlinear boundary value problem is governed by:

$$A(\overline{u}) = 0 \quad \text{in } \Omega \tag{4.2.1}$$

$$B(\overline{u}) = G(\overline{x}_\Gamma, t) \quad \text{on } \Gamma \tag{4.2.2}$$

where A and B are operators, linear or nonlinear, defined in the study domain Ω, G is a function prescribed on the boundary Γ, $\overline{x}_\Gamma$ is a coordinate of a point on the boundary, $\overline{u}$ is the unknown variable, and t is time. Because of the fact that the state of the art of mathematics does not always provide us with a technique to obtain an analytic solution of nonlinear boundary value problems, the mathematical verification of the numerical solution of the boundary value problem governed by (4.2.1) and (4.2.2) requires a special technique. The PSF has been developed to provide one with an alterative to conduct a mathematical verification of a numerical model.

Based on the concept of PSF (Dee and Da Silva, 1986, Dee, 1991, and Dee et al., 1992), an analytic function, say $\overline{\Phi}(\overline{x},t)$, can be chosen or found to automatically satisfy the boundary conditions (4.2.2) on Γ and all its partial derivatives can have finite and non-trivial values in the study domain Ω. Then, one can substitute $\overline{\Phi}(\overline{x},t)$ into (4.2.1) to obtain

$$A(\overline{\Phi}) = \overline{F}(\overline{x},t) \quad \text{in } \Omega \tag{4.2.3}$$

$$B(\overline{\Phi}) = G(\overline{x}_\Gamma,t) \quad \text{on } \Gamma \tag{4.2.4}$$

where the inhomogeneous term, $\overline{F}$, is the result obtained by substituting $\overline{\Phi}$ into (4.2.1). One should realize that the resulted $\overline{F}$ is also an analytic function of $\overline{x}$ and t. From physical interpretation, $\overline{F}$ is usually called as a forcing function, and sometimes it is also called as a source term.

One sees that by the use of the prescribed analytic function, $\overline{\Phi}$, the original boundary value problem, (4.2.1) and (4.2.2) have been changed to a new boundary value problem, represented by (4.2.3) and (4.2.4). The interesting consequence is that this new boundary value problem governed by (4.2.3) and (4.2.4) has, now, an analytic solution, $\overline{\Phi}$. Another desirable outcome is the fact that the only difference between these two boundary value problems is the term, $\overline{F}$, in (4.2.3) that itself is an analytic function. Because of this outcome, one can modify the numerical model having already been developed for simulating the original boundary value problem (4.2.1) and (4.2.2) into a new model, by adding a forcing term. Then, the modified code can be used to solve the new boundary value problem (4.2.3) and (4.2.4).

Since $\overline{F}(\overline{x},t)$ has a unique numerical value at each point (or node) and time in the discretized solution domain, it can be added into the code easily. After this is done, the modified nonlinear numerical model can be verified analytically by comparing the solutions of the modified model code to the values of the prescribed analytical solution, $\overline{\Phi}(\overline{x},t)$ at the corresponding points. If good convergence can be achieved, the numerical model is accepted as having been verified.

For conducting a complete mathematical verification, one has to perform a grid convergence analysis to determine the order of convergence of the numerical model, as well as to quantitatively assess the order of magnitude of errors of numerical solutions. These aspects are to be presented later in this chapter.

4.2.2 Procedure for Model Verification Using PSF

A more specific presentation of the basic procedure of mathematical verification of a 3D free surface flow model using PSF is given as follows.

The first step is to select an analytic function, which satisfies the boundary conditions (4.2.2). Because the boundary conditions are of lower order and simpler form than the governing differential equations, it is, thus, easier to choose an analytic function to satisfy them. In general, one can choose

$$\overline{\Phi} = \overline{\Phi}_1 \overline{\Phi}_b \tag{4.2.5}$$

where both $\overline{\Phi}_1$ and $\overline{\Phi}_b$ are analytic functions in the solution domain $\Omega$and the only difference is that $\overline{\Phi}_b$ satisfies (4.2.2), but $\overline{\Phi}_1$ does not.

For example, a simple one-dimensional differential equation in a domain between $x =$ 0 and 1, denoted as Ω[0, 1]:

$$T'' - aT = 0, \ (0 < x < 1) \tag{4.2.6}$$

and the boundary conditions at $x = 0$ and 1 are:

$$T(0) = 0$$
$$T(1) = 0. \tag{4.2.7}$$

where T is the unknown field variable of the problem. One may choose

$$\overline{\Phi}_1 = x^2 + bx + c$$
$$\overline{\Phi}_b = x(x-1) \tag{4.2.8}$$

where b and c are arbitrary coefficients determined by the boundary value problem. Thus yielding

$$\overline{\Phi} = x(x-1)(x^2 + bx + c) \tag{4.2.9}$$

One finds that $\overline{\Phi}$ is analytic in $\Omega[0,1]$ and satisfies the boundary conditions (4.2.7). Of course, when one is dealing with a 3D unsteady problem, the selection of $\overline{\Phi}$ can be tedious, but not impossible.

The second step is to substitute the selected analytic function $\overline{\Phi}$ into (4.2.6) and (4.2.7), yielding

$$T''(\Phi) - aT(\Phi) = F(x) \tag{4.2.10}$$

with

$$F(x) = -ax^4 - (ab-1)x^3 + (ab - ac + 1)x^2 + (ac + 6b - b)x - 2(b-c) \tag{4.2.11}$$

Equation (4.2.10) is the resulting new boundary value problem (BVP), and (4.2.11) is a forcing (or inhomogeneous) term, which is itself an analytic function and has a unique value at each point x in the solution domain.

The third step is to modify the numerical model developed to solve the original BVP (4.2.1) and (4.2.2), or (4.2.6) and (4.2.7), by adding the forcing function of $F(x)$ into the code, to solve the new BVP governed by (4.2.10) and (4.2.7). Because the boundary condition (4.2.7) is automatically satisfied, no change of boundary conditions in the code is needed. In the case when the user does not have the source code of the original model, the user can request the model developer to assist in carrying out this step, or include a forcing term in each equation in the code for the purpose of verification based on PSF. This forcing term should be set to zero by default for normal use of the model. Only during Mathematical Verification, a value shall be provided to this term in each equation.

The fourth step is to execute the modified model or code to produce solutions of the modified new BVP. The verification can be carried out by comparing the numerical results with the analytic solution, just like what has been done in Chapter 3.

4.2.3 An Example

Toro (1994) applied the PSF method to verify a numerical model for solving a 3D flow in an open-channel (a flume with rough bed). The physical data of Coleman (1991) was approximately fitted by a truncated Fourier Series, which is selected as the analytic function for numerical model verification. Since Fourier series expansion is a popular special function used to represent nonlinear function, users may want to use it for other cases as well. This example shall show that it is possible to achieve both mathematical verification and physical process validation by conducting one test, because the numerical model results can be compared to both the analytic solution and the experimental measurements at the same time. The interested user may refer to Dr. Toro's dissertation (Toro, 1994) for details.

Toro adopted a 3D free surface flow model, which was the early version of CCHE3D model, and used the following governing equations and boundary conditions:

X-momentum equation

$$\begin{aligned}\frac{\partial u}{\partial t}+\left(u-2\frac{\partial \nu_t}{\partial x}\right)\frac{\partial u}{\partial x}+\left(v-\frac{\partial \nu_t}{\partial y}\right)\frac{\partial u}{\partial y}+\left(w-\frac{\partial \nu_t}{\partial z}\right)\frac{\partial u}{\partial z}+g\frac{\partial \eta}{\partial x}\\ =\nu_t\frac{\partial^2 u}{\partial x^2}+\nu_t\frac{\partial^2 u}{\partial y^2}+\nu_t\frac{\partial^2 u}{\partial z^2}+\frac{\partial \nu_t}{\partial y}\frac{\partial v}{\partial x}+\frac{\partial \nu_t}{\partial z}\frac{\partial w}{\partial x}\end{aligned} \tag{4.2.12}$$

where u, v, and w are the velocity components in the x, y, and z directions, respectively, η is the free surface elevation, ν_t is the eddy viscosity, and g is the gravitational acceleration.

Y-momentum equation

$$\begin{aligned}\frac{\partial v}{\partial t}+\left(u-\frac{\partial \nu_t}{\partial x}\right)\frac{\partial v}{\partial x}+\left(v-2\frac{\partial \nu_t}{\partial y}\right)\frac{\partial v}{\partial y}+\left(w-\frac{\partial \nu_t}{\partial z}\right)\frac{\partial v}{\partial z}+g\frac{\partial \eta}{\partial y}\\ =\nu_t\frac{\partial^2 v}{\partial x^2}+\nu_t\frac{\partial^2 v}{\partial y^2}+\nu_t\frac{\partial^2 v}{\partial z^2}+\frac{\partial \nu_t}{\partial x}\frac{\partial u}{\partial y}+\frac{\partial \nu_t}{\partial z}\frac{\partial w}{\partial y}\end{aligned} \tag{4.2.13}$$

Continuity equation

$$w(z)=w_0-\int_{z_0}^{z}\left(\frac{\partial u}{\partial x}+\frac{\partial v}{\partial y}\right)dz \tag{4.2.14}$$

where w_0 is the known velocity at level z_0.

The free surface is determined by the depth-integrated continuity equation

$$\frac{\partial(\eta-\xi)}{\partial t}+\frac{\partial(\eta-\xi)U}{\partial x}+\frac{\partial(\eta-\xi)V}{\partial y}=0 \tag{4.2.15}$$

where ξ is the channel bottom elevation, and U and V are the depth-averaged velocities given by

$$U=\frac{1}{\eta-\xi}\int_{\xi}^{\eta}udz\,;\qquad V=\frac{1}{\eta-\xi}\int_{\xi}^{\eta}vdz \tag{4.2.16}$$

At the outlet, the free surface elevation is specified.

At the inlet, the following velocity profile is prescribed:

$$\frac{u}{u_*}=\frac{1}{\kappa}\ln\left(\frac{z}{z_0}\right)+2\frac{\Pi}{\kappa}\sin^2\left(\frac{\pi z}{2h}\right) \tag{4.2.17}$$

where u is the longitudinal velocity parallel to the channel bottom, z is the normal distance from the bottom, κ=0.41 is the von Karman constant, z_0 is the characteristic length of the bottom irregularities, Π=0.2 is Coles' parameter verified by Nezu and Rodi (1986) in their experiments in straight channels, and h is the flow depth. In order to use (4.2.17), the shear velocity u_* is computed from the depth-averaged velocity U obtained from the prescribed unit discharge using the following relation:

$$\frac{U}{u_*}=\frac{1}{\kappa}\left[\ln\left(1+\frac{h}{z_0}\right)+\frac{z_0}{h}\ln\left(1+\frac{h}{z_0}\right)-1\right]+2\frac{\Pi}{\kappa}\left[1+\frac{1}{\pi}\sin\left(\frac{\pi z_0}{h}\right)\right] \tag{4.2.18}$$

and the characteristic length of the bottom irregularities, z_0, is given by

$$z_0=\begin{cases} k_s/30 & u_*k_s/\nu>70 \\ k_s/30+0.1186\nu/u_* & 5\le u_*k_s/\nu\le 70 \\ 0.1186\nu/u_* & u_*k_s/\nu<5 \end{cases} \tag{4.2.19}$$

where k_s represents the equivalent roughness height on the bed, and ν is the kinematic viscosity.

At rigid boundaries (bed and side walls), the log-law is used to compute the velocity parallel to the boundary. A point inside the domain is used to iterate Equation (4.2.17) to obtain the shear velocity, and then the log-law is applied again to a point on the boundary to obtain the velocity parallel to the boundary. The details can be found in Toro (1994).

In principle, an analytical function to be used in Mathematical Verification can be chosen arbitrarily as long as it and its derivatives in the governing differential equations are not automatically trivial in the solution domain and it satisfies the given boundary conditions. Toro (1994) obtained a prescribed solution from a set of experimental data measured by Dr. Neil Coleman (1993) of the USDA-ARS National Sedimentation Laboratory, Oxford, Mississippi. The flume was 25 m long and 1 m wide. The flow was steady and uniform. The streamwise velocity at a cross-section sufficiently far away from the entrance was measured. The measurements were taken at 10 positions from the flume bed along 5 verticals located between the wall and the centerline of the flume. The prescribed solution was constructed by using the following truncated Fourier series to approximate the measured streamwise velocity:

$$W(y,\varsigma)=\left[a_0+a_1\ln(\varsigma)+a_2\sin^2(\theta)\right]\left[c_0+\sum_{k=1}^{3}b_k\sin(k\alpha)+\sum_{k=1}^{3}c_k\cos(k\alpha)\right] \tag{4.2.20}$$

where $\theta=\pi\varsigma/2h_0$, $\alpha=\pi y/L$, $h_0=0.209$ m is the measured water depth, L=0.997 m is the width of the computational domain, y denotes the vertical coordinate, and ς denotes the coordinate in the direction normal to the flume bed, as shown in Fig. 4.2.1.

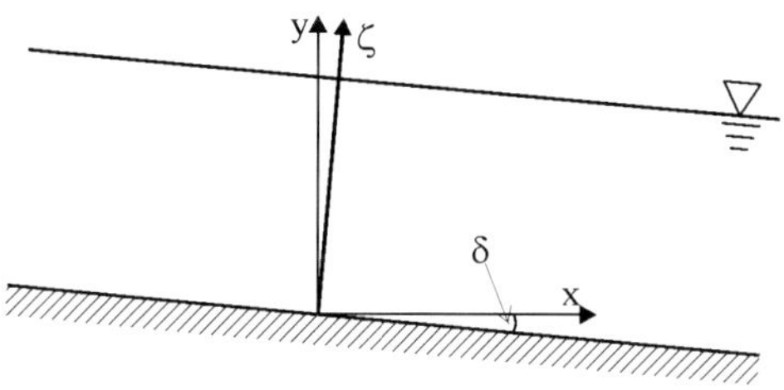

Figure 4.2.1 Coordinate Systems in the Vertical Plane

Equation (4.2.20) is rewritten as

$$\begin{aligned}W(y,\varsigma)&=a_1+\sum_{k=1}^{3}a_{3k+1}\sin(k\alpha)+\sum_{k=1}^{3}a_{3k+10}\cos(k\alpha)\\&+\left[a_2+\sum_{k=1}^{3}a_{3k+2}\sin(k\alpha)+\sum_{k=1}^{3}a_{3k+11}\cos(k\alpha)\right]\ln(\varsigma)\\&+\left[a_3+\sum_{k=1}^{3}a_{3k+3}\sin(k\alpha)+\sum_{k=1}^{3}a_{3k+12}\cos(k\alpha)\right]\sin^2(\theta)\end{aligned} \tag{4.2.21}$$

where a_1, a_2, ..., a_{21} are coefficients, whose values are determined by applying the least-square fitting and given below:

$$a_1 = -0.180404\text{E}-01;\ \ a_2 = 0.589028\text{E}-01;\ \ a_3 = -0.325773\text{E}+00$$
$$a_4 = 0.101126\text{E}+01;\ \ a_5 = -0.174058\text{E}-01;\ \ a_6 = 0.505340\text{E}+00$$
$$a_7 = 0.483502\text{E}-02;\ \ a_8 = -0.591909\text{E}-04;\ \ a_9 = 0.235567\text{E}-02$$
$$a_{10} = -0.742364\text{E}-01;\ \ a_{11} = 0.461995\text{E}-02;\ \ a_{12} = -0.646052\text{E}-01$$
$$a_{13} = -0.506114\text{E}-02;\ \ a_{14} = 0.871122\text{E}-04;\ \ a_{15} = -0.252912\text{E}-02$$
$$a_{16} = 0.483028\text{E}+00;\ \ a_{17} = -0.590340\text{E}-02;\ \ a_{18} = 0.235336\text{E}+00$$
$$a_{19} = 0.111469\text{E}-02;\ \ a_{20} = -0.693702\text{E}-04;\ \ a_{21} = 0.970070\text{E}-03$$

Figure 4.2.2 shows the agreements between the Fourier series approximates and the measurements. The root mean square error (RMS) corresponding to this fit is 0.0054876 m/s.

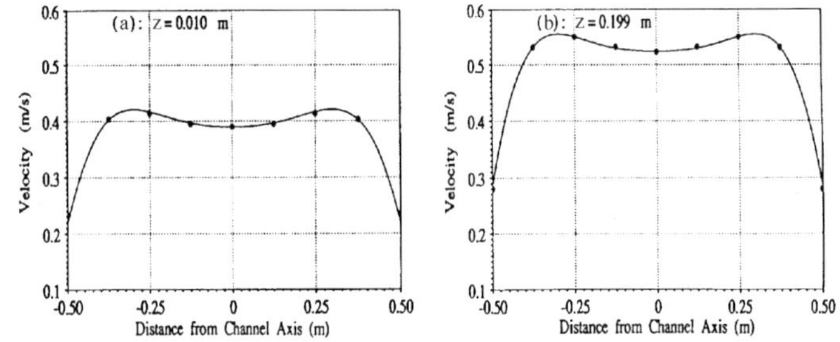

Figure 4.2.2 Comparison between Measurements (·) and the Prescribed Solution (—) at two levels of a channel cross-section

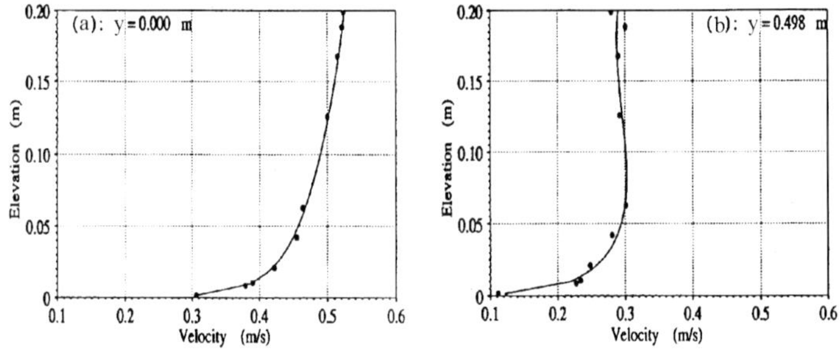

Figure 4.2.3 Comparison between Measurements (·) and the Prescribed Solution (—) along two vertical lines of a channel cross-section

Eq. (4.2.21) is converted to the velocity components in the Cartesian coordinate system as

$$\begin{pmatrix} \hat{u} \\ \hat{v} \\ \hat{w} \end{pmatrix} = \begin{pmatrix} W\cos(\delta) \\ 0 \\ -W\sin(\delta) \end{pmatrix} \tag{4.2.22}$$

where δ is the flume slope angle. A positive value of δ denotes for a downslope bed as shown in Fig. 4.2.1.

By substituting the prescribed solution (4.2.22) into (4.2.12), (4.2.13), and (4.2.15), one can obtain the set of modified inhomogeneous governing differential equations

$$\begin{aligned} &\frac{\partial u}{\partial t} + \left(u - 2\frac{\partial \nu_t}{\partial x}\right)\frac{\partial u}{\partial x} + \left(v - \frac{\partial \nu_t}{\partial y}\right)\frac{\partial u}{\partial y} + \left(w - \frac{\partial \nu_t}{\partial z}\right)\frac{\partial u}{\partial z} + g\frac{\partial \eta}{\partial x} \\ &\quad - \left[\nu_t\frac{\partial^2 u}{\partial x^2} + \nu_t\frac{\partial^2 u}{\partial y^2} + \nu_t\frac{\partial^2 u}{\partial z^2} + \frac{\partial \nu_t}{\partial y}\frac{\partial v}{\partial x} + \frac{\partial \nu_t}{\partial z}\frac{\partial w}{\partial x}\right] = P_x \end{aligned} \tag{4.2.23}$$

$$\begin{aligned} &\frac{\partial v}{\partial t} + \left(u - \frac{\partial \nu_t}{\partial x}\right)\frac{\partial v}{\partial x} + \left(v - 2\frac{\partial \nu_t}{\partial y}\right)\frac{\partial v}{\partial y} + \left(w - \frac{\partial \nu_t}{\partial z}\right)\frac{\partial v}{\partial z} + g\frac{\partial \eta}{\partial y} \\ &\quad - \left[\nu_t\frac{\partial^2 v}{\partial x^2} + \nu_t\frac{\partial^2 v}{\partial y^2} + \nu_t\frac{\partial^2 v}{\partial z^2} + \frac{\partial \nu_t}{\partial x}\frac{\partial u}{\partial y} + \frac{\partial \nu_t}{\partial z}\frac{\partial w}{\partial y}\right] = P_y \end{aligned} \tag{4.2.24}$$

$$\frac{\partial(\eta - \xi)}{\partial t} + \frac{\partial(\eta - \xi)U}{\partial x} + \frac{\partial(\eta - \xi)V}{\partial y} = P_\eta \tag{4.2.25}$$

where P_x, P_y, and P_η are the added Forcing Functions, whose expressions can be derived by substituting (4.2.22) into (4.2.12), (4.2.13), and (4.2.15). For the specific simple case, the corresponding forcing functions are (Toro, 1994)

$$\begin{aligned} P_x = &\left(\hat{u} - 2\frac{\partial \hat{\nu}_t}{\partial x}\right)\frac{\partial \hat{u}}{\partial x} - \frac{\partial \hat{\nu}_t}{\partial y}\frac{\partial \hat{u}}{\partial y} + \left(\hat{w} - \frac{\partial \hat{\nu}_t}{\partial z}\right)\frac{\partial \hat{u}}{\partial z} + gS \\ &- \left[\hat{\nu}_t\frac{\partial^2 \hat{u}}{\partial x^2} + \hat{\nu}_t\frac{\partial^2 \hat{u}}{\partial y^2} + \hat{\nu}_t\frac{\partial^2 \hat{u}}{\partial z^2} + \frac{\partial \hat{\nu}_t}{\partial z}\frac{\partial \hat{w}}{\partial x}\right] \end{aligned} \tag{4.2.26}$$

$$P_y = -\frac{\partial \hat{\nu}_t}{\partial x}\frac{\partial \hat{u}}{\partial y} - \frac{\partial \hat{\nu}_t}{\partial z}\frac{\partial \hat{w}}{\partial y} \tag{4.2.27}$$

$$P_\eta = \frac{\partial}{\partial x}\int_\xi^\eta u\,dz \tag{4.2.28}$$

where S is the water surface slope $\partial\eta/\partial x$, and $\hat{\nu}_t$ is the eddy viscosity evaluated using the prescribed solution (4.2.21).

To conduct the Mathematical Verification, one is to add the Forcing Functions into the computer code (e.g., CCHE3D), execute the modified code using the same domain geometry, physical and numerical parameters and boundary conditions, and to compare the results. The comparison of the modified CCHE3D and the prescribed solution are shown in Fig. 4.2.4.

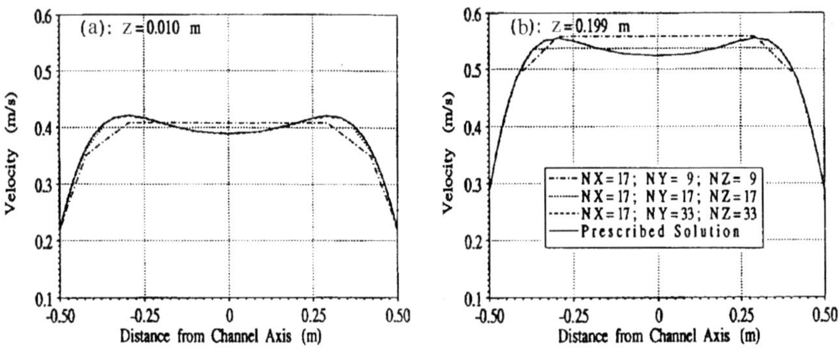

Figure 4.2.4 Comparison between computed u-velocity and the prescribed solution with grid refinement in transversal and vertical directions at two levels of a channel cross-section

To assess the Grid Resolution effects on the accuracy of the numerical model's results, three meshes with grid sizes refined in the transversal and vertical directions were used. It can be seen that the computational results obtained by using the finest grid had the best agreement with both the analytic solution and the measured data, and that the numerical results of the coarse grid failed to reproduce the streamwise velocity deficit near the center of the channel. It has been found that further refinement of the grid beyond $NY = NZ = 33$ did not significantly change the accuracy. This proves that the converged solution can be obtained. Beyond this point, the solution is insensitive to the mesh refinement, or grid resolution independence has been established.

It is of interest to point out that by applying the approach adopted in this example, one can not only accomplish the Mathematical Verification, but also a Physical Process Validation at the same time. In addition, a Calculation Verification by successive grid refinement can be carried out. Both model developers and users may want to advance this methodology further.

The convergence tests were conducted by Toro (1994) for u, v, and water surface elevation using the three mesh resolutions. Fig. 4.2.5 shows the root-mean-square (RMS) errors for u and v momentum equations (4.2.23) and (4.2.24), and η equation (4.2.25). These errors were defined as

$$RMS = \sqrt{\sum_{k=1}^{N}\left[\left(P_{model}\right)_k - \left(P_{psf}\right)_k\right]^2 \Big/ N} \tag{4.2.29}$$

where P_{psf} is the right-hand side of (4.2.23)–(4.2.25) computed using the prescribed solution; P_{model} is the left-hand side of the same equations computed by the model results; k is the computational node index; and N is the total number of nodes inside the domain.

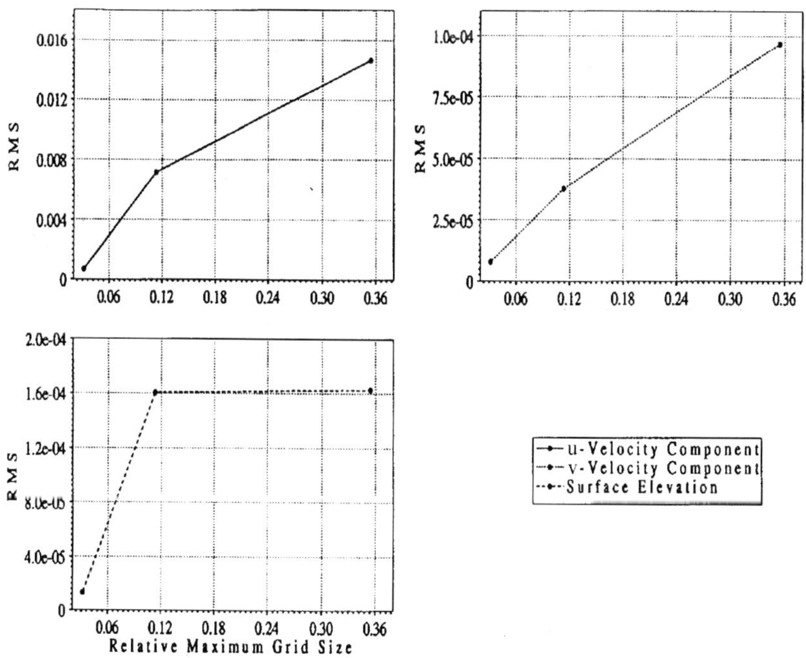

Figure 4.2.5 Convergence test for u, v, and η with grid refinement in transverse and vertical directions

From these results, it is seen that the convergence for u, v, and η is observed when the grid are refined, which means the Grid Convergence Test has been satisfied.

4.3 METHOD OF MANUFACTURED SOLUTION (MMS)

Contributor: Yafei Jia

4.3.1 Basic Concept

Application of a numerical model to a practical problem is a cost effective approach that has been adopted by engineering communities. On the other hand, it is a complicated process, and the final results inevitably involve errors which are contributed from many sources including the numerical models, measurements, and unknown or un-measurable parameters in the field. There are many ways to evaluate the errors and uncertainties resulted from a numerical model. However, when a numerical simulation is completed, one may have to answer several important questions: a) if the mathematical differential equations involved are solved correctly, b) if these equations are feasible for the particular problem, and c) what level of errors in the final solutions are due to the accuracy of a numerical model? The numerical solutions will not be reliable if these questions are left un-answered. The Method of Manufactured Solution (MMS, Steinberg and Roache 1985, Roache, 1998, 2002, Knupp and Salari, 2002,) is a general approach to provide a certain answer to the question of a model's correctness and it can verify non-linear numerical models rigorously. Although examples have been given also for checking code accuracy with linear and/or non-linear problems in many text books and research (Celia, 1992; Cao et al., 2001), the test cases provided in this section enable one to verify his/her complete Navier-Stokes equation solver for free surface flows without making assumptions like those for obtaining linear solutions. Instead, one could exercise all the terms in the differential equations and make sure they will be correct in any way the model is used.

It is ideal to test a model's correctness by comparing the model's numerical solution with its analytical solution; however, there is no non-linear solution available for this purpose. Since only the numerical procedure's (not physics) correctness is interested, it would be very effective if arbitrarily made linear or nonlinear functions can be used in model verification. The principle of MMS technique is to verify the correctness of differential equation solvers and computing codes of numerical models using arbitrarily manufactured solutions.
Let the differential equation be expressed as

$$L(u) = 0 \tag{4.3.1}$$

where L denotes differential operators and u is the variable to be solved. When a manufactured function ϕ is substituted into Eq. (4.3.1), one would have

$$L(\phi) = F \tag{4.3.2}$$

Since ϕ is not the solution of the Eq. (4.3.1), the non-zero F is obtained *analytically*. When solved *numerically*, the solution of this equation would be forced to converge to ϕ with the analytical forcing term F being added to the mathematical equations of the numerical model.

The verification of a numerical model is straightforward because the solution of Eq. (4.3.2) is known: one needs only compare the difference of the manufactured analytical function, ϕ, and the numerical solution of Eq. (4.3.2). Although the function ϕ can be manufactured arbitrarily, it has to be non-trivial for all the terms of the involved mathematic equations to make meaningful verification tests. Unlike the PSF approach, the MMS does not require the satisfaction of any particular boundary conditions other than those defined by ϕ along the boundaries of the computational domain.

To demonstrate the procedure in its entirety, a 3D free surface turbulent model is needed. The model developed at the National Center for Computational Hydroscience and Engineering, CCHE3D, was used for this purpose (Jia, et al, 2005). The governing equations are Reynolds' equations, continuity equation and the free surface kinematic equation:

$$\frac{\partial u_i}{\partial t} + u_j \frac{\partial u_i}{\partial x_j} + f_i + \frac{1}{\rho}\frac{\partial p}{\partial x_i} - \frac{\partial}{\partial x_j}[\nu_t(\frac{\partial u_i}{\partial x_j} + \frac{\partial u_j}{\partial x_i})] = 0 \tag{4.3.3}$$

$$\frac{\partial u_i}{\partial x_i} = 0 \tag{4.3.4}$$

$$\frac{\partial h}{\partial t} + u_h \frac{\partial h}{\partial x} + v_h \frac{\partial h}{\partial y} - w_h = 0 \tag{4.3.5}$$

where u_i are velocity components in the Cartesian coordinate system with coordinates x_i (i=1,2,3), p is pressure, h denotes free surface elevation, ρ is fluid density, and ν_t is viscosity. It is noted that the notation $u=u_1$, $v=u_2$ and $w=u_3$ are also adopted for convenience. The mathematical model of free surface flows includes not only the above equations, but also a complete set of boundary conditions for applications to physical problems. They may be in the form of:

$$\begin{aligned}
&u_i(x_{i,t}) = U_{i,wall} \text{ on solid surface, } \Gamma_{wall} \\
&u_i(x_{i,t}) - U_{in} \text{ at inlets, } \Gamma_{in} \\
&\frac{\partial u_i}{\partial z} = 0 \text{ at free surface } h(x_i,t) \\
&h\ (x_i,t) = H_{Ex} \text{ at outlets, } \Gamma_{Ex} \\
&\frac{\partial h}{\partial n} = 0 \text{ at walls, } \Gamma_{wall} \\
&p(x_i,t) = p_o \text{ at free surface, } \Gamma_o \\
&\frac{\partial p}{\partial n} = 0 \text{ at } \Gamma_{wall}
\end{aligned} \tag{4.3.6}$$

In more general cases, the eddy viscosity, ν_t, is a function of dependent variables of the flow field, additional equations are needed to close the mathematical model. The zero-equation approach, an algebraic form for the eddy viscosity, is used in the examples in this section. This simplification enables us to verify all the terms in the equation (4.3.3) except for those related to $\frac{\partial \nu_t}{\partial x_i}$. If one would like to verify two-equation turbulence closure schemes (k-ε model, for instance), one may need to manufacture solutions for these additional equations.

4.3.2 Manufactured Solutions for a Free Surface Flows

To apply the Method of Manufactured Solution to conduct the mathematical verification of the nonlinear system governed by (4.3.3)–(4.3.5), two sets of analytic functions have been manufactured. They are given in (4.3.7) and (4.3.10) for unsteady and steady cases, respectively.

The first set, *Function I,* reads

$$u = \sin y \cos^2 \frac{x}{2} \cos \frac{\pi}{2}(1 - \frac{z}{h}) \sin t \tag{4.3.7a}$$

$$v = -\sin x \cos^2 \frac{y}{2} \cos \frac{\pi}{2}(1 - \frac{z}{h}) \sin t \tag{4.3.7b}$$

$$w = -A \cos \frac{x}{2} \cos \frac{y}{2} \cos \frac{\pi}{2}(1 - \frac{z}{h}) \sin t \tag{4.3.7c}$$

$$h = A \cos \frac{x}{2} \cos \frac{y}{2} \cos t + h_0 \tag{4.3.7d}$$

$$p = C_p \cos(x) \cos(y) \cos(\frac{\pi}{2} \frac{z}{h}) \tag{4.3.7e}$$

where A is an amplification parameter. The boundary conditions for the computational domain are:

$$\begin{aligned} & u_i(x_i, t) = 0 \text{ on } \Gamma_{wall} : x = \pm\pi,\ y = \pm\pi, \text{ and bed : } z=0 \\ & \frac{\partial u_i}{\partial z} = 0 \text{ at free surface } h(x_i, t) \\ & \frac{\partial h}{\partial t} + u_h \frac{\partial h}{\partial x} + v_h \frac{\partial h}{\partial y} - w_h = 0 \\ & h = h_0 \text{ on } \Gamma_{wall} \end{aligned} \tag{4.3.8}$$

and those for the pressure:

$$\frac{\partial p}{\partial x}\Big|_{wall}=0,\ \frac{\partial p}{\partial y}\Big|_{wall}=0,\ \frac{\partial p}{\partial z}\Big|_{bed}=0,\text{ and } p_h=0 \tag{4.3.9}$$

For this particular case, the computational domain is confined in Γ_{wall}, the virtual "bed" and the "free surface". The words "wall" and "bed" are used for convenience to represent surrounding and bottom boundaries of the computational domain rather than physical ones. The boundaries are located such that the Dirichlet boundary conditions from Eq. (4.3.7) for velocity ("non-slip") and Neumann boundary conditions for pressure at walls and the bed are satisfied (Eqs. 4.3.8 and 4.3.9). It is obvious that these Dirichlet boundary conditions can be given wherever the boundaries of the computation domain, Γ_{wall}, are located. At the free surface, the free surface boundary condition is always satisfied, and Neumann boundary condition for velocities is applied. For the pressure, Neumann boundary conditions are applied on walls and bed, and Dirichlet boundary condition is applied at surface. Although the functions in the domain and along the boundaries have no physical meanings, satisfying these conditions in (4.3.8) and (4.3.9) makes it possible that some hard coded general physical boundary conditions implemented in a model can be tested and one takes little effort to modify his/her code for verification. Since Eq. 4.3.7 is time dependent, one may use it for verifying unsteady models. The distributions of the vector field and free surface represented by *Function I* are depicted in Figures 4.3.1 and 4.3.2.

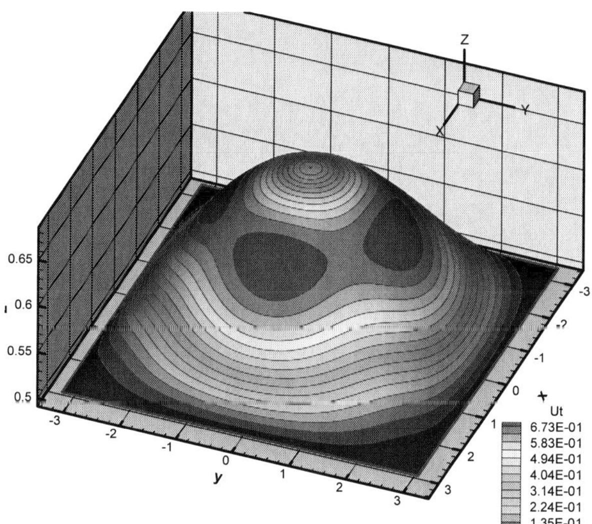

Figure 4.3.1 The 3D view of the surface shape and velocity magnitude distribution on the surface (*Function I*).

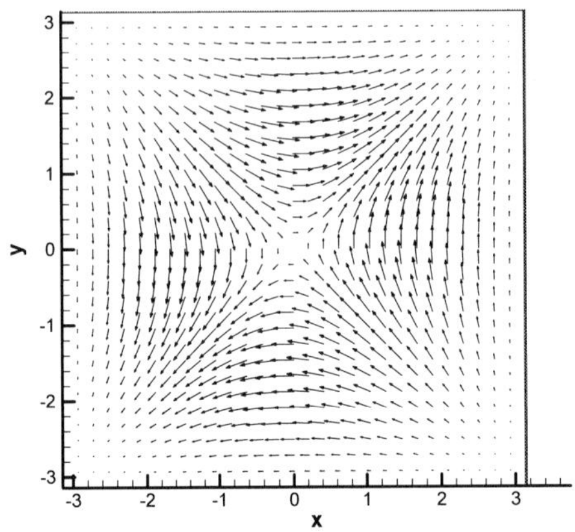

Figure 4.3.2 Velocity vector field at $z=0.5h$ (*Function I*).

The second set, *Function II*, reads:

$$u = \sin(x)\cos(y)\sin(\frac{\pi}{2}\frac{z}{h}) - \cos(x)\sin(y)[\cos(2\pi\frac{z}{h}) - 1] \tag{4.3.10a}$$

$$v = -\cos(x)\sin(y)\sin(\frac{\pi}{2}\frac{z}{h}) + \sin(x)\cos(y)[\cos(2\pi\frac{z}{h}) - 1] \tag{4.3.10b}$$

$$w = -\frac{A}{2\pi}[\sin^2(x)\cos^2(y) - \cos^2(x)\sin^2(y)][\sin(2\pi\frac{z}{h}) - 2\pi\frac{z}{h}\cos(2\pi\frac{z}{h}) + 2\pi] \tag{4.3.10c}$$

$$h = A\sin(x)\sin(y) + h_0 \tag{4.3.10d}$$

Boundary conditions:

One the surround vertical (wall) boundary:

$$\Gamma_{wall} : x = 0, x = 2\pi, y = 0, y = 2\pi$$

Vectors on the walls were set: $u_i(x_i, t)\big|_{wall}$

$$\left.\frac{\partial u_i}{\partial z}\right|_h = 0$$

On the "bed": $z=0$

$$u_i(x_i, t)\big|_{bed} \tag{4.3.11}$$

On the "free surface":

$$\frac{\partial h}{\partial t} + u_h \frac{\partial h}{\partial x} + v_h \frac{\partial h}{\partial y} - w_h = 0$$

with $h\big|_{wall} = h_0$

This function is manufactured to satisfy both the continuity equation and free surface condition. The advantage over *Function I* of being divergence free is that the velocity correction procedure will result in the form of a Laplacian equation rather than Poisson equation for the pressure solver's verification. Because of this, the dynamic pressure solution will not be affected by the manufactured solution if the velocity correction type of methods is used in the numerical model. It will only be affected by the pressure boundary conditions. Since the pressure solution is not formulated in *Function II*, pressure solvers could be used to check if the numerical solution is truly divergence free. One may note that the "velocities" on the surround boundaries are non-zero, they are not even parallel or normal to the "walls". The analytical derivatives for formulating forcing terms of these two functions are included in Appendix B.

Figure 4.3.3 depicts the shape of the free surface and the velocity field in the domain of $0 \le x \le 2\pi$, $0 \le y \le 2\pi$, z=0, and h (h_0 = 2m, A =0.5). The free surface looks like standing waves, and the flow field indicated by the vector is very complex. In Figure 4.3.3, the color on the top level indicates the surface elevation and those on the lower levels represent the magnitude of flow velocity. The w velocity is zero at the free surface level, u and v velocity are cycling around the wave centers. At the bed level, the u and v velocities are zero, and w velocity varies at different locations. The surface elevation and the velocity magnitude are represented with two color bar legends. Figure 4.3.4 shows two cross-sections of the flow field cut normal to x direction; the vector field is highly non-linear.

Because these functions are continuous, they can be used for testing models using any shape of computational domain with any mesh system. Coordinate transformation is a fundamental procedure for any numerical model, when non-Cartesian meshes are used, the transformation for irregular mesh must be activated and exercised. Additional error in the transformation could then be detected and the mistake in derivation or coding could be identified.

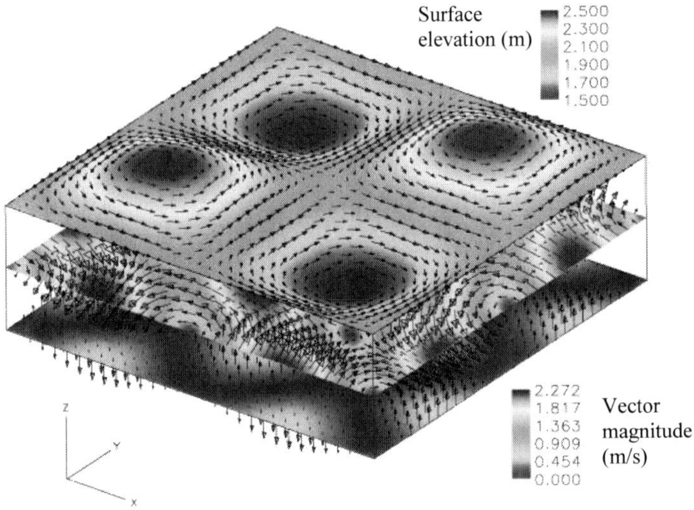

Figure 4.3.3 A 3D view of the manufactured flow field (*Function II*). The shading of the top surface is free surface elevation. The second and the third level are velocity magnitude and vectors at $z = 0.5h$ and $z = 0$. Vectors represent velocity field.

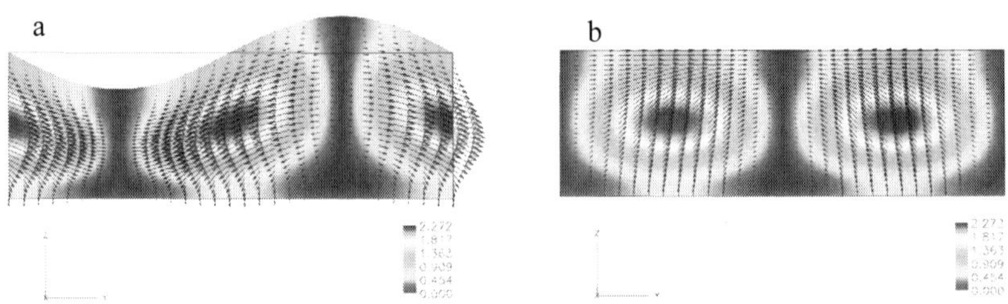

Figure 4.3.4 Velocity magnitude and vectors in vertical sections in x direction (*Function II*). a: $x = \pi / 2$, b: $x = \pi$

By substituting either of these manufactured solutions into the differential equations (4.3.3) – (4.3.5), one obtains a set of new differential equations (4.3.12), assuming a constant eddy viscosity. The difference between the differential equations before and after substituting the manufactured solution is a forcing function, F, appeared in each of the new equations.

$$\frac{\partial u}{\partial t}+u\frac{\partial u}{\partial x}+v\frac{\partial u}{\partial y}+w\frac{\partial u}{\partial z}-\nu_t(\frac{\partial^2 u}{\partial x^2}+\frac{\partial^2 u}{\partial y^2}+\frac{\partial^2 u}{\partial z^2})-\nu_t(\frac{\partial^2 u}{\partial x^2}+\frac{\partial^2 v}{\partial y\partial x}+\frac{\partial^2 w}{\partial z\partial x})+\frac{1}{\rho}\frac{\partial p}{\partial x}=F_u$$

$$\frac{\partial v}{\partial t}+u\frac{\partial v}{\partial x}+v\frac{\partial v}{\partial y}+w\frac{\partial v}{\partial z}-\nu_t(\frac{\partial^2 v}{\partial x^2}+\frac{\partial^2 v}{\partial y^2}+\frac{\partial^2 v}{\partial z^2})-\nu_t(\frac{\partial^2 u}{\partial x\partial y}+\frac{\partial^2 v}{\partial y^2}+\frac{\partial^2 w}{\partial z\partial y})+\frac{1}{\rho}\frac{\partial p}{\partial y}=F_v \quad (4.3.12a)$$

$$\frac{\partial w}{\partial t}+u\frac{\partial w}{\partial x}+v\frac{\partial w}{\partial y}+w\frac{\partial w}{\partial z}-\nu_t(\frac{\partial^2 w}{\partial x^2}+\frac{\partial^2 w}{\partial y^2}+\frac{\partial^2 w}{\partial z^2})-\nu_t(\frac{\partial^2 u}{\partial x\partial z}+\frac{\partial^2 v}{\partial y\partial z}+\frac{\partial^2 w}{\partial z^2})+\frac{1}{\rho}\frac{\partial p}{\partial z}=F_w$$

$$\frac{\partial u}{\partial x}+\frac{\partial v}{\partial y}+\frac{\partial w}{\partial z}=F_c \quad (4.3.12b)$$

$$\frac{\partial h}{\partial t}+u_h\frac{\partial h}{\partial x}+v_h\frac{\partial h}{\partial y}-w_h=F_h \quad (4.3.12c)$$

To simplify the presentation, the analytic expressions of the derivatives of these forcing functions, F_i, are given in Appendix B.

In both functions, one does not need to manufacture the static pressure field, since the static pressure terms for free surface flows is related to surface elevation: $g\frac{\partial h}{\partial x}$ and $g\frac{\partial h}{\partial y}$. They could be verified because the distribution of pressure is known: $\rho g(h-z)$; however, this is a minor task in comparison to other terms involved. In fluid dynamics models, the dynamic pressure solution is a function of the flow field. The pressure distribution of the *Function I* (4.3.7e) is manufactured and it is not related to the flow. Like many CFD codes, in CCHE3D model, the equation system is solved with the velocity correction method, and pressure is solved using the Poisson equation obtained from the discretized momentum equations with the mass conservation condition demanded. Since the manufactured *Function I* does not satisfy the mass conservation condition, non-zero divergence of the numerical solution and associated source terms from the manufactured solutions will appear in the Poisson equation. It would be, therefore, much easier to verify the pressure solver separately than couple it to the solution of the vector field.

4.3.3 Grid Convergence Analysis

It is well known that the truncation error in formulating a numerical model can be determined by Taylor's series analysis. A central difference scheme is second-order accurate. Its leading truncation error has an order of magnitude of Δ^2, where is the mesh spacing (Celia and Gray, 1992). According to Roache (1998), the error of a numerical model can be expressed as a function of Δ^{p_c}:

$$E=f(\Delta)-f_a=C\Delta^{p_c}+H.O.T. \quad (4.3.13)$$

where $f(\Delta)$ and f_a are numerical and exact solutions for the mathematical model respectively, p_c is the order of the convergence (or accuracy), and Δ is mesh size or time step size. The "near"-constancy of the ratio E/Δ^{p_c} or C is the evidence of a model's convergence with a rate p_c. In the following analysis, the error of numerical solutions and mesh spacing Δ will be used to establish such a relation and the coefficient C and p_c will be identified. Since the numerical model CCHE3D, was developed to be of second-order accuracy, a Grid Convergence Analysis is performed as a part of the Mathematical Verification to confirm that the numerical solution does exhibit a second-order convergence.

The following discussion is based on tests using *Function I*. To conduct a grid convergence test of a numerical model, it is more convenient to test it under steady state conditions. Even if the numerical model was developed for unsteady simulation, the grid convergence test can be performed when the solution has reached the steady state. A very small time step $\Delta t = 0.001\pi$ was used to avoid the convergence being affected by the error in time integration. The initial condition or guessed value for starting the test run can be arbitrary. In this test, the vector field u, v, and w at all internal nodes were set to be zero (cold start). The solutions at $t_0=\frac{1}{4}\pi$ were used as the steady solution. The amplitude parameter A=0.5m was set. To perform a steady state test, one should exclude the time derivatives of the manufactured solution from the source functions. Since the exact solution at any point in the computational domain is known, the computational error can be calculated at any mesh point and any time. The error at boundaries of vertical wall and bed are zero since exact boundary values are specified. To avoid round off errors affecting the analysis, double precision was set for the computer code. For simplicity, the errors of the three computed velocities, pressure and free surface are evaluated with error norm on the internal nodes.

$$Err = \sqrt{\frac{\sum (V_a - V_n)^2}{(I_{max} - 2)(J_{max} - 2)(K_{max} - 2)}} \qquad (4.3.14)$$

where Err indicates the mean error of a dependent variable, V. I_{max}, J_{max} and K_{max} are numbers of nodes in the x, y and z directions, respectively. Subscripts a and n indicate analytical value and numerical solution, respectively. One notes that only errors at internal nodes are evaluated. In all tests, four sets of meshes were used:

$$I_{max} = J_{max} = K_{max} = 11, \quad I_{max} \times J_{max} \times K_{max} = 1{,}331$$
$$I_{max} = J_{max} = K_{max} = 21, \quad I_{max} \times J_{max} \times K_{max} = 9{,}261$$
$$I_{max} = J_{max} = K_{max} = 31, \quad I_{max} \times J_{max} \times K_{max} = 29{,}791$$
$$I_{max} = J_{max} = K_{max} = 61, \quad I_{max} \times J_{max} \times K_{max} = 226{,}981$$

In this study, the model is unsteady. The time stepping is used to achieve the steady state solution from the arbitrarily assumed initial condition. Figures 4.3.7a and 4.3.7b show the rate of convergence of the tested CCHE3D model. A second-order accurate upwinding scheme was applied. The results for time step $\Delta t = 0.01\pi$ are shown in Figure 4.3.7a and those for $\Delta t = 0.001\pi$ are shown in Figure 4.3.7b. From these two figures, one sees that the error norms of all variables exhibit second order convergence as expected. The regression coefficients for these convergence curves are listed in Table 4.3.1. The correlations for all curves are very strong. The coefficients for the test with smaller time step: $\Delta t = 0.001\pi$ are significantly lower than those for $\Delta t = 0.01\pi$, which indicate that the error is also affected by the time step size.

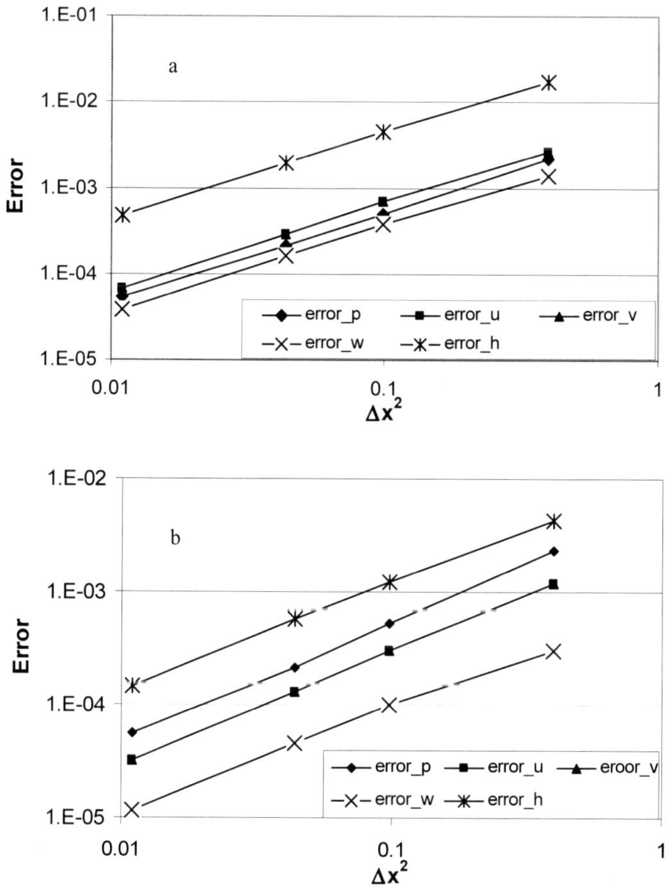

Figure 4.3.7 Convergence test for the CCHE3D free surface model with steady state source terms: (a) $\Delta t = 0.01\pi$ and (b) $\Delta t = 0.001\pi$

Table 4.3.1 Coefficients of the convergence curves for steady state test (the convergence error equations are $Error = C\Delta x^2$)

Parameters	A=0.5m, $t=t_0$ $\Delta t=0.01\pi$		A=0.5m, $t=t_0$ $\Delta t=0.001\pi$	
	C	R^2	C	R^2
Error-u	0.0068	0.9999	0.003	1.
Error-v	0.0068	0.9999	0.003	1.
Error-w	0.0035	0.9996	0.0007	0.9937
Error-p	0.0055	0.9995	0.0059	0.9995
Error-h	0.0437	0.9999	0.0107	0.9989

In CFD codes, non-linear advection terms normally have to be handled with upwinding techniques, such as the first-order upwind, second-order upwind, and QUICK schemes, etc. These schemes have different accuracy which should be reflected by the convergence tests. Tests were conducted for several schemes in CCHE3D model and Table 4.3.2 summarized these verification cases and the coefficients of the related error equations. Equation 4.3.12 was used for testing with all linear and non-linear terms except for the pressure terms. The results of testing convergence with only linear terms are also listed as a reference.

Table 4.3.2 Test cases regard to non-linear terms using *Function I*

F	Linear terms only	1st order upwinding		Convective interpolation 1.6 order upwinding		2nd order upwinding		QUICK scheme			
	A=0.5	A=0.5		A=0.5		A=0.5		A=0.5		A=0.0	
	C	C	R^2	C	R^2	C	R^2	C	R^2	C	R^2
Error-u	0.0017	0.0041	0.9999	0.0021	0.9998	0.0021	1.0	0.0017	1.0	0.0011	1.0
Error-v	0.0017	0.0041	0.9999	0.0021	0.9998	0.0021	1.0	0.0017	1.0	0.0011	1.0
Error-w	0.0002	0.0008	1.0	0.0003	0.9999	0.0002	0.9996	0.0002	0.9999	0.00005	0.9994

The coefficients of the first-order upwind scheme is 0.0041 for u and v components and 0.0008 for w; much higher than those for linear terms, indicating the absolute error induced by the first-order scheme is dominant and the overall convergence is reduced from second order for only linear terms to first order. When the upwinding scheme is computed by the convective interpolation function (Jia and Wang, 1999), the convergence curve exhibits a p_c value between 1 and 2. When p_c =1.6 was used by trial and error, the convergence exhibits straight lines. The convergence of this upwinding scheme is therefore identified to be of order 1.6.

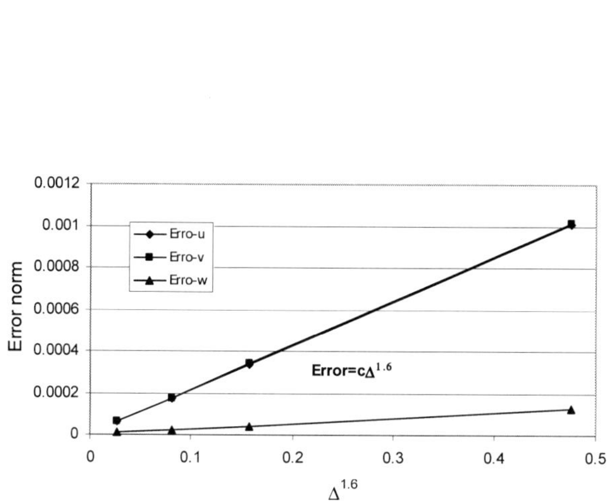

Figure 4.3.8 Convective interpolation function is of the order of 1.6

Test of the second-order upwind scheme is straightforward; the second-order convergence is obtained for all three vectors. The coefficients of the regression error equations are somewhat higher than those for linear terms, but lower than those of the first-order upwind scheme. The test results also indicate the QUICK scheme is of second-order accuracy. Its error norms are linearly correlated to Δ^2 as shown in Figure 4.3.9. However, QUICK scheme has the smallest coefficients for its convergence curves which indicate the lowest error norm and the highest accuracy. In the two cases tested using QUICK scheme, the error for the case with A=0 is less than that for A=0.5m. When A>0, the free surface is curved, mesh deformation would introduce a higher level of errors to the calculation, but the convergences were still of second order. It is also noted, the coefficients of the QUICK scheme with A=0.5 is the same as those of linear terms, indicating the errors from non-linear terms are even lower than those of linear terms. It should be pointed out that derivation or coding mistakes of a numerical scheme can be easily detected by MMS. A coding error was found in the terms of QUICK scheme handling mesh vertical deformation. The error would show only when $A \neq 0$, because that small term could not be exercised (no mesh vertical deformation with $A = 0$). Good convergence trends were obtained after the derivation and coding for this scheme were reviewed and the error was removed.

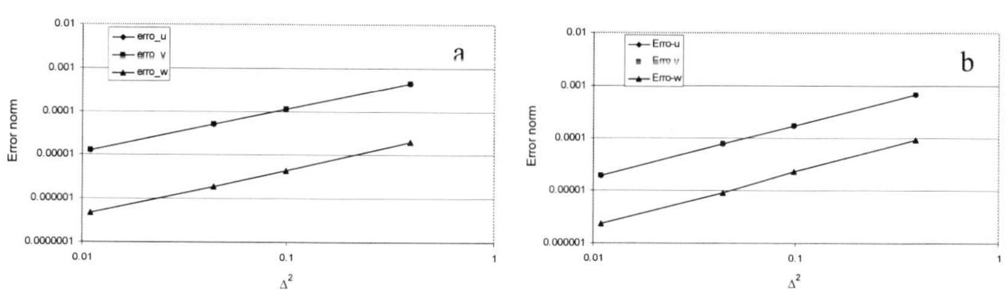

Figure 4.3.9 Convergence curves for the QUICK scheme with A=0 (a) and A=0.5 (b)

Since *Function I* is time dependent, it is possible to use it for verification of time integration schemes directly. When an unsteady case is concerned, it is possible to compute error norms with equation (4.3.14) because the errors also vary in time. Figure 4.3.10 shows the periodic variations of the error norms calculated by the first-order Euler scheme and QUICK scheme.

However, to make it easy for evaluating the error norms at the end of calculations with different time steps, the time averaged error norms (TAEN), an overall estimator of the convergence performance of a model, was used:

$$Err_{u_i,t} = \frac{1}{N_t}\sum_n Err_{u_i} \tag{4.3.15}$$

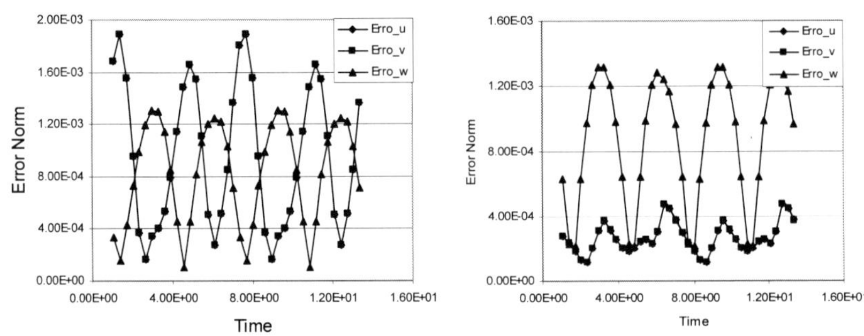

Figure 4.3.10 Error norm for verifying Eq. 4.3.3 with all the spatial derivatives and the first order Euler and QUICK scheme

When spatial derivatives are involved in the differential equation, the errors induced by the spatial discretization will be mixed with those by the time steps. The time averaged error norm will show a trend of decreasing with time step size if the error due to time step is significantly larger than that due to mesh size. For this reason, A=0 was set for the test for preventing the influence of mesh deformation on time integration.

Table 4.3.3 shows the time steps used for this unsteady test example. All the tests were run for a time period of 2π to make the averaged error norm accurate. The mesh used for these cases was 31x31x31. The convergence test results are shown in Figure 4.3.11. The curve of the time averaged error norm exhibits a second order convergence when $\Delta t \geq 0.0025\pi$, and the convergence rate drops rapidly when the time step is smaller than 0.0025π. The TAEN level where the curve turns is close to that of the steady test cases (the steady cases were computed with a very small time step: $\Delta t = 0.001\pi$, the error due to time step is minimal). This indicates the errors due to the spatial descritization (Δ) are equal or higher than that due to time stepping.

Table 4.3.3 Unsteady momentum equation (A=0.0, t=t_0)

Δt	0.00125π	0.001875π	0.0025π	0.00375π	0.005π	0.0075π	0.01π

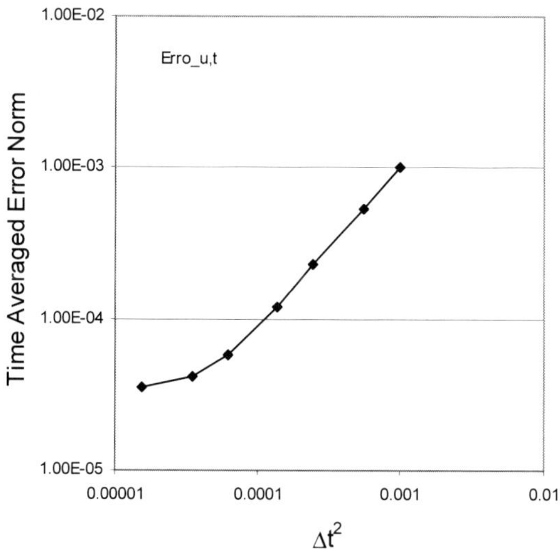

Figure 4.3.11 Time averaged error norm using a second order corrected Euler time integration scheme

Function II was also used to test the CCHE3D model. Convergence curves similar to those for *Function I* were produced. Figure 4.3.12 shows the comparison of the results of the numerical model and the manufactured solution, *Function II*, indicated in Fig. 4.3.3. The upper part of the figure compares the solutions at the top level, and the lower part compares the middle level. The mesh for the computation was 11x11x11. The general pattern of the manufactured solution was reproduced by the numerical model. One can see, however, some slight differences in the domain.

Figure 4.3.13 shows that the comparison of the manufactured solution and the results of the numerical model are similar to those of Figure 4.3.12, with the upper part of the figure comparing the solutions at the top level and the lower part comparing the middle level. The mesh for the computation was 21x21x21. The agreement between the numerical and analytic solutions has improved, as one can hardly tell a difference visually.

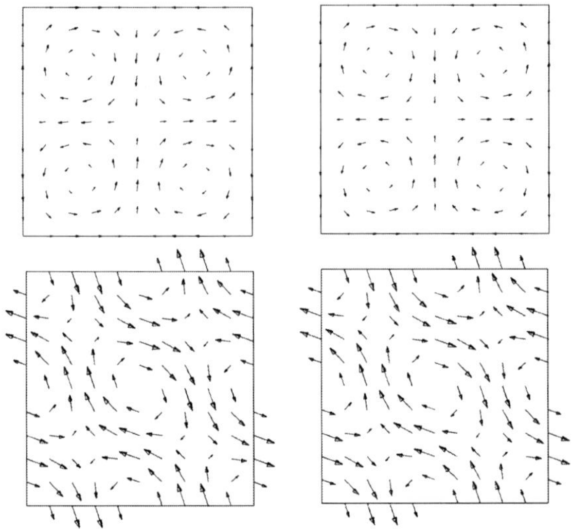

Figure 4.3.12 Comparisons of MMS solution (Left) and analytic (manufactured) solution (Right) at top level (Top) and middle level (bottom).
The mesh for this result was 11x11x11.

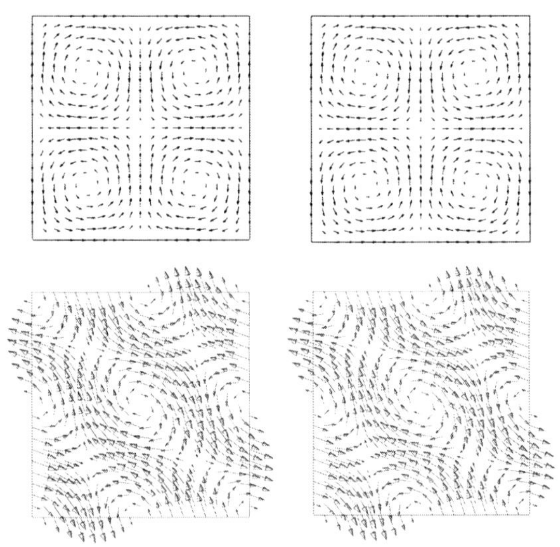

Figure 4.3.13 Comparisons of MMS solution (Left) and analytic (manufactured) solution (Right) at top level (Top) and middle level (bottom).
The mesh for this result was 21x21x21.

4.4 Model Verification Procedure of MMS

Contributor: Yafei Jia

This section is to outline the essential steps needed to carry out a mathematical verification of a developed numerical model in order to determine whether it can solve a set of nonlinear differential equations correctly as well as whether it can show to have a pre-designed order of accuracy or rate of convergence.

These goals are achieved by comparing the results of the numerical model to the values of the prescribed or manufactured analytic solutions of the boundary value problem. The verification procedures for applying the MMS are outlined below.

The mathematical verification of nonlinear models using the method of manufactured solution is suggested to follow the procedure below:

(1) Manufacture a set of non-linear analytic functions, one for each unknown variable, which may or may not satisfy the boundary conditions.

(2) By substituting the manufactured solutions into the governing differential equations, a set of forcing or source functions are obtained analytically.

(3) Modify the computer code of the numerical model developed for solving original differential equations for initial/boundary value problems (without the new forcing terms) by adding the analytic forcing terms. The forcing term for one equation is a combination of terms generated from each of the terms in the differential equation. If one would like to verify some of the terms, others in the equation and their associated source terms should be canceled.

(4) Select a computational domain and make sure that the manufactured solution is valid in the domain.

(5) Generate several numerical grid systems with appropriate gird intervals in all three dimensions. This is prepared to conduct the Grid Convergence Test and/or Calculation Convergence Verification.

(6) Specify boundary conditions on all boundaries consistent with the manufactured solutions. All boundary conditions could be time-dependent. At the boundaries of this domain, boundary conditions of these functions should be satisfied (Eq. 4.3.8, 4.3.9, and 4.3.10). In general, Dirichlet, Neumann, or mixed or radiation boundary conditions can be specified using the manufactured solution. Dirichlet type boundary conditions of the manufactured solution in particular can be specified easily on arbitrarily shaped boundaries.

(7) Execute the modified numerical model once for each of the generated grid systems.

(8) Compare the numerical model results to the corresponding manufactured solutions, and perform a convergence test to determine the rate of convergence or the

order of accuracy. Since the exact solution is known, one could check errors at points, in certain areas or the entire domain.

One should note that the key difference in verification procedures between the PSF and MMS (described above) is the fact that the prescribed analytic solutions are required to satisfy the prescribed boundary conditions of the boundary value problem before it is substituted into the governing differential equations. As a result it is not as easy to obtain as the analytical function required by the MMS.

4.5 Concluding Remarks

Contributors: Sam S.Y. Wang and Yafei Jia

Two alternatives, the PSF and the MMS, have been presented in this chapter for model developers and/or users to conduct mathematical verification of numerical models for solving nonlinear free surface flow problems. Both methodologies require model testers to select (prescribe or manufacture) a nonlinear analytic function, which is the analytic solution of a new boundary/initial value problem. The new boundary value problem represented by (4.2.3) and (4.2.4) is obtained by substituting the selected analytic function, $\overline{\Phi}(\overline{x},t)$ into the original boundary value problem defined by (4.2.1) and (4.2.2). The difference between the new and the original boundary value problems is the addition of a new (or an extra) forcing function in the new boundary value problem. Since this forcing function, $\overline{F}(\overline{x},t)$, is an analytic function of independent variables, it has a unique value at each numerical node and a particular time. Therefore, each forcing value can be easily added to each numerical equation of the original numerical model to result in a new numerical model for solving the new boundary value problem. Consequently, one can compare the solutions of this new numerical model or code to the prescribed or manufactured solution for the Mathematical Verification of the numerical model. The detailed step by step procedure has been outlined in the previous sections of this chapter.

To complete the mathematical verification, one needs to perform a grid convergence test to obtain the rate of convergence or to assess the order of accuracy quantitatively. This step is conducted by refining the grid size and obtaining additional results for each grid size. Once the order of convergence obtained from calculation verification or convergence test is the same as the one designed for the model, the mathematical verification is completed. This also assures that the numerical model tested does not have mathematical mistakes in derivation, discretization, solution, and coding; it is consistent with the boundary value problem (PSF) and has achieved the designed order of the rate of convergence.

The tester is reminded that the MMS approach in general does not require the inclusion of the satisfaction of the prescribed boundary conditions of the original boundary value problem. Instead, it uses the values or gradients of the manufactured solutions as Dirichlet or Neumann type condition(s) at the geometric boundaries of the computational domain. However, the MMS does confirm that the numerical model has solved the differential equations correctly. By adopting this approach, the model (code) tester needs to specify these boundary conditions to the numerical model before the execution of the model. On the other hand, in the PSF approach the selected (or prescribed) analytic function satisfies automatically all the prescribed boundary conditions of the original boundary value problem. A user choosing to use the PSF approach is recommended to select or derive his/her own prescribed analytic solution. However, the selection of a Fourier series or one of the special functions may not be a bad idea. It is expected that additional analytic functions for this purpose are to be published in the open literatures in the near future.

From the above summary, one sees that a straightforward approach has been developed to mathematically verify a nonlinear numerical model's capability and accuracy for solving three-dimensional free surface flows. A prohibitive task, not too long ago, has become a routine mathematical exercise. In addition, if the tester chooses to use one of the manufactured solutions as suggested, he/she does not need to perform the tedious mathematical manipulations to obtain the forcing functions. To facilitate model testers to use the manufactured solutions to conduct the verification test of a numerical model, not only all the derivative terms needed to assemble the forcing functions, but also computer subroutines written in Fortran are provided in Appendix B as well as on CD ROM attached to this report. A model tester may simply use the subroutine to generate the values of all forcing functions and add them to each numerical equation at each interior node and the boundary conditions to the boundary nodes before executing the numerical model. The numerical solutions obtained can then be compared to those evaluated from the manufactured solution at corresponding points to draw the conclusion of the mathematical (code) verification.

Again, the tester are reminded that Mathematical Verification is necessary but not sufficient to complete a comprehensive verification and validation test, one should continue to perform the required physical process validation to insure that the model has the capability of predicting all basic physical processes important to the system to be modeled, and the application site validation to make sure that the model can indeed predict the real world problem realistically. Only after these three major steps are completed, the numerical model can be applied to the study of a real world problem. These last two steps are described in the next two chapters.

4.6. REFERENCES

Cao, W., Huang, W., and Russell, R.D., 2001, "An error indicator monitor function for an r-adaptive finite-element method", *J. of Computational Physics*, 170, 871-892.

Coleman, N.L., 1991, "Experiments in alluvial bed channel flow resistance", Experiments in Fluids, Springer Berlin/Heidelberg, Vol. 1, No.4, pp205-208.

Coleman, Neil, 1993. Former research geologist of National Sedimentation Laboratory of USDA Agricultural Research Service, Private communication and notes.

Celia, M.A. and W.G. Gray, 1992, *Numerical Methods for Differential Equations – Fundamental Concepts for Scientific and Engineering Applications*, Prentice-Hall International (UK) Limited, London.

Dee, D.P., and A.M. Da. Silva, 1986, "Using hough harmonics to validate and assess non-linear shallow water models". Monthly Weather Review, 114, p2191-2196.

Dee, D.P., 1991, "Prescribed solution forcing method for model verification in hydraulic engineering", Proceedings of 1991 National Conference of the Hydraulic Division, ASCE, Nashville.

Dee, D.P., Toro, F.M., and Wang, S.S.Y., 1992, "Numerical model verification by prescribed solution forcing — a test case", Proceedings of the Hydraulic Engineering Sessions at the Water Forum 92, ASCE, Baltimore, p416-421.

Jia, Y., Scott, S., Xu, Y.C., Huang, S.L., and Wang, S.S.Y., 2005, "Three-Dimensional Numerical Simulation and Analysis of Flows around a Submerged Weir in a Channel Bendway", *Journal of Hydraulic Engineering*, Vol. 131, No. 8:682-693, August 1, 2005.

Jia, Y and Wang, S.S.Y., 1999, "Numerical model for channel flow and morphological change studies", *Journal of Hydraulic Engineering*, ASCE, Vol. 125, No. 9, pp. 924-933

Knupp, P. and Salari, K., 2002, *Verification of Computer Codes in Computational Science and Engineering*, CRC Press, Boca Raton, FL.

Nezu, I. and Rodi, W., 1986, "Open-channel flow measurements with a laser Doppler anemometer", J. Hydraulic Engineering, ASCE, Vol. 112, No. 5, pp. 335-355.

Roache P.J., and Steinberg, S. 1984, "Symbolic manipulation and computational fluid dynamics", AIAA Journal, Vol. 22, No. 14. Paper No. 83-1952. October, 1984, pp.

1390-1394.

Roache, P,J., 1998, *Verification and Validation in Computational Science and Engineering,* Hermosa Publishers, PO Box 9110, Albuquerque, New Mexico 87119-9110.

Roache, P. J., 2002, "Code verification by the method of manufactured solutions", ASME, Journal of Fluids Engineering, Vol. 114, No. 1, March 2002, pp. 4-10.

Steinberg, S. and Roache P.J., 1985, "Symbolic manipulation and computational fluid dynamics", Journal of Computational Physics, Vol, 57, pp. 251-284.

Toro, P.M., 1994, "Verification of the CCHE3D hydrodynamic model for open channel flow", dissertation, the University of Mississippi.

CHAPTER 5

PHYSICAL PROCESS VALIDATION

Yafei Jia and Sam S.Y. Wang

5.1 INTRODUCTION

Contributor: Sam S.Y. Wang

After a numerical model has been verified to be mathematically correct, it implies that the model is free of mathematical errors in the process of model formulation, numerical solution, computer program coding, and computational algorithms. It is also confirmed that the model solution is convergent and consistent to the analytic or prescribed (or manufactured) solution of the governing differential equations. Furthermore, it can determine whether the order of magnitude of errors of model's numerical approximation is that intended in the numerical solution scheme. With this confirmation, the next question to be asked is: whether this numerical model is capable of simulating the basic physical processes, especially those most important to the problem to be investigated.

This chapter is to present an approach to the answer to this question. In addition, several examples are given to demonstrate how this validation step is to be carried out. One shall quickly find out that all examples given are based on experiments performed in laboratories and most of them are indoor experiments, because the laboratory experiments are performed in controllable environments. Therefore, numerous non-essential and insignificant complexities of forcings can be eliminated and the system can be modeled better by prescribed (controlled) forcings and boundary conditions. In addition, the measurements obtained in the laboratories are usually more accurate and reliable. Besides, the spatial distribution and resolution of data collection points can be better and easier to design than that of field measurements. More importantly, the entire experiment can be repeated, if needed. As a result, the functional relationship between system's responses to each forcing changes can be tested and established individually.

To carry out Physical Process Validation for confirming a numerical model's capabilities in reproducing or predicting an observable physical process (or mechanism) in a free surface flow system, one should select a laboratory experiment with high quality measured data of sufficient amount. These data sets may be from the test cases provided in this chapter, published literature, unpublished reports of some research institution, or self-designed test. If time and fund are available, the self-designed tests are recommended, because they can be specially designed for the problem to be investigated. After having obtained such a set of data or test cases, the model developer or user can input all physical dimensions, conditions and values of parameters into the computer and execute a run to simulate the laboratory experiment numerically for making comparisons with the measured data.

Sometimes, it may be necessary to use a part of the laboratory measurements to calibrate the values of the physical parameters in the numerical model, because each experimental facility set up has some unique characteristics, which may affect the parameter's values, such as the bed and wall resistance, etc. Those data used for parametric calibration can not be used again for validation. If there are two or more physical processes in the flow system, which are important to the intended investigation, one needs to select or design one test for each of these processes. The simulated results are to be compared with the experimental measurement.

If a reasonable agreement between the results predicted by the numerical model and those measured form the laboratory experiment(s) could not be obtained either in values or in trends (spatial and temporal); it could be caused by the deficiency of the numerical model. Either the model user or most likely the model developer needs to examine the model's assumptions and simplifications closely to identify the improvement(s) needed. In one example, during a test of a 3D free surface flow in a sharp curved channel, the small vortices in the upper corner of the outer bank could not be simulated. By reasoning and testing, it was found that the deficiency was caused by the turbulence closure adopted rather than by insufficient grid resolution as commonly reported. After switching off the simple k-ε turbulence closure and switching on the nonlinear k-ε turbulence closure included in the model, the agreement was achieved. Should the model users have no access to the source code nor the possibility of selecting an appropriate option among the ones included in the model, it may be advisable for the user to select a different model, which has the required capability to conduct his/her studies.

One is reminded that there are uncertainty and errors in both measurements and numerical calculations. The Task Committee has recognized their importance and recommended a study of these and other related problem by a new task committee. The result shall be reported shortly. All users and developers of 3D free surface flow models are highly advised to refer to Section 2.18, Basic Methodology of Calculation Verification and conduct a Calculation Verification to estimate the overall errors due to all calculations.

5.2 OVERVIEW OF TEST CASES

Contributor: Sam S.Y. Wang

In this chapter, six test cases are provided to the prospective model testers for their use to construct a series of tests to determine whether or not a numerical model is capable of predicting the basic physical processes of a three-dimensional free-surface flow problem. How many test cases are needed and which ones are to be selected depend upon the problem to be investigated. In case of a simple problem which the essential physical process happens to be the same as or similar to that of a particular test case given in this chapter, that test case alone may be sufficient to serve the purpose. On the other hand, if the problem to be studied is complex, which involves two or more basic physical processes, the tester may need to select two or more test cases to do the physical process validation. For example, if the study is to simulate the hydrodynamics of the flow field in a river bendway with submerged bendway weirs, one may need to select at least two test cases, say the Delft U-Shaped Channel Test Case in Section 5.3 and the Flow in a Channel with a Spur Dike Test Case in Section 5.5. These two cases can test the model's capabilities of simulating the 3D free surface flow along a curved channel, more specifically the generation of helical secondary motions and superelevation along the outer bank, as well as the complicated 3D flow with complex vortices around an in-stream obstruction, respectively.

Each test case is presented with its clearly defined objectives, experimental setup, geometric dimensions, physical parameters and boundary conditions. Most of the required boundary and initial conditions are given in Appendix A of this report. The prospective testers are advised to read the "README" file of each data file for additional information and instructions. After the test run(s), the numerical results obtained from the model are compared to the experimental measurements, which are also provided in Appendix A. Disagreement(s) may be due to the errors in the governing equations of the model and/or in the input data. If it is confirmed that the errors or incapability are in the governing equations of the model due to over simplifications or unrealistic assumptions, then the model's governing equations have to be corrected. Agreement validates the model's capability in reproducing the basic physical processes of key importance to the flow field to be investigated.

After a satisfactory agreement is obtained the model tester should conduct a Calculation Verification to estimate the errors of all calculations, which is important to model user, especially when the model is to be applied to conduct a field study of large scope with significant long term implications. The examples of Calculation Verification process are included in Sections 5.4.10 and 5.5.7.

The objectives of each of the six test cases are briefly summarized below:

- *Free Overfall Flow Test Case*

This case can be used to determine whether a three-dimensional flow model is capable of (1) handling two free-surface boundaries with an adaptive numerical grid, (2) predict a pressure field with both hydrostatic and non-hydrostatic pressure regions, and (3) other physical characteristics properties such as free-surface locations, velocity field, and pressure distribution measured in the experiment.

- *Delft U-Shaped Smooth Channel Test Case*
  If one is interested in determining whether a three-dimensional numerical model can predict secondary helical currents and their gradual increase and decrease in strength as they move through a smooth channel bend, this test may serve the purpose. In addition, it has identified the deficiency of the linear $\kappa\varepsilon$ turbulence closure in predicting small vortices observed in the upper corners of the outer wall in the curved reach of the channel bend. As a result, the tester should decide to adopt the nonlinear $\kappa\varepsilon$ turbulence closure, if those small vortices are important to the proposed study.

- *WES Riprap Curve-Channel Test Case*
  This case can test the numerical model's capability in predicting the velocity field affected by four successive curved bends with different radius of curvature, sloped (rather than vertical) banks, and rough surfaces of channel bed and banks. (Note that this is an outdoor test facility).

- *Recirculation Flow around a Spur Dike Test Case*
  Before one plans to apply a three-dimensional free-surface flow model to investigate the flow and local scour phenomena around obstructions and/or in-stream hydraulic structures such as spur dikes, bridge piers and abutments, break waters, jetties, and others, he/she may use this case to insure that the selected model has the capabilities in simulating several complex flow mechanisms including flow deceleration, stagnation, downwash, formation of a recirculation zone, reattachment, and induced vortex-like secondary currents, etc.

- *Flow Around a Submerged Dike Test Case*
  A series of experiments has been conducted at the USDA National Sedimentation Laboratory in Oxford, Mississippi in cooperation with the National Center for Computational Hydroscience and Engineering of The University of Mississippi (Kuhnle et al., 2002). The three-dimensional flow field around a trapezoidal spur dike submerged in the flow was measured at well-located data collection points. The detailed and complex mean turbulent flow field was recorded for the purpose of model validation. This is an example of studying a physical system by a joint effort of physical and numerical modeling. This test case provides a large set of detailed velocity field data for testing numerical model's capacity and accuracy in simulating a highly complex three-dimensional flow field. It proves that to perform a

successful model validation, one must have sufficient amount of high quality data collected at well-selected data locations.

- *Flow in the Vicinity of a Harbor Model Test Case*
  Both steady and unsteady data collected at the Franzius Institute in Germany is provided for the validation of three-dimensional models' capability in simulating flows in the vicinity of a simplified harbor (a rectangular cavity) with or without a groyne attached to the harbor.

In summary, one can choose one or more of the test cases from this chapter, or the published literature, or the experiment(s) designed by the testers to validate a numerical model by comparing the numerical simulations to laboratory experimental measurements. If significant discrepancies in both value and the trend of the flow field variables are found, the model developer needs to correct the mistakes, and/or improve the capability of the numerical model; or the user may select to test and find an alternative model to use. The model tester is reminded again to try his/her best in carrying out a Calculation Verification to estimate the accumulated calculation errors and determine the grid resolution convergence.

Once the model tested passes a set of well-selected test(s), one should have higher confidence in the models' capability to predict the basic physical processes of the three-dimensional free-surface flow system intended to investigate.

Models pass the Mathematical (Code) Verification and Physical Validation described in Chapters 3, 4 and 5 should not be applied to study the real-life problems without going through the Application Site Validation step. Details of the Application Site Validation are presented in the next chapter.

5.2.1 References

Kuhnle, R., Jia, Y., Alonso, C., 2002, "3-Dimensional Measured and Simulated Flow for Scour Near Spur Dikes", First International Conference on Scour of Foundations, ICSF-1, Texas A&M University, College Station, Texas, USA, November 17-20, 2002, Vol.1 pp349-363

5.3 FREE OVERFALL FLOW TEST CASE

Contributors: Yafei Jia and Sam S. Y. Wang

5.3.1 Introduction

Free overfall flows have been observed at nick point or head-cut of natural river channels and at manmade in-stream structures such as weir or drop structures. Numerous laboratory experiments have been conducted to study this phenomenon. Free over-falls have at least three special features, i.e. the flows turn from subcritical to supercritical near the brink of the channel; they are bounded by strongly curved free surfaces and the vertical pressure distribution near the free fall is non-hydrostatic. The results obtained from laboratory experiments by Rajaratnam and Muralidur (1968) are chosen as a validation test case for free surface flow models. It provides both users and developers of 3D numerical models with a simple way to determine the model's capabilities in predicting a non-hydrostatic pressure field in the supercritical free surface flow and in handling two free surface boundaries.

5.3.2 Objectives

To carry out this test with a 3D free-surface flow model, one can evaluate whether the model being considered for application has the following capabilities of

- handling two free surface boundaries without prior information of their locations;
- predicting non-hydrostatic pressure field in and near the free fall region of the flow;
- solving both subcritical and supercritical flows and
- obtaining a reasonable velocity field

5.3.3 Approach

For simplicity, this test case is designed as a vertically two-dimensional numerical simulation, because the flow is symmetrical and the data were taken along the centerline of the flume. Users of this test case need to simplify their mathematical models to eliminate the lateral variations of the flow or apply a symmetric boundary condition in the lateral direction, solving only a strip of computational domain along the central line of the flow with appropriate boundary conditions imposed on the free surface, bed surface, inlet and downstream side.

5.3.4 Description of Experiment

The free overfall flow physical model experiments performed by Rajaratnam and Muralidhar (1968) are sketched in Figure 5.3.1, and the flow condition selected for this test case is listed in Table 5.3.1.

Among many data sets collected and published in their paper, the most comprehensive set, Test 1A, was selected for numerical simulation. The experiment was conducted in a rectangular flume, 0.457m wide and 6.1m long, with a flat and horizontal bed. The approaching flow from upstream is subcritical, which reaches the critical condition before the brink. L is the so-called "Length of Overfall", the length between the critical flow point and the brink. The flow depth at the brink (h_e) is actually less than the critical depth (h_c). Velocity and shear stress were measured by a calibrated pitch probe and a Prandtl tube with an external diameter of 3mm. Since the flow is symmetric to the center line, its variation in transverse direction can be neglected by simulating just the flow in the central vertical plan. As a result, one only needs to specify at the inlet the unit discharge q for the simulation.

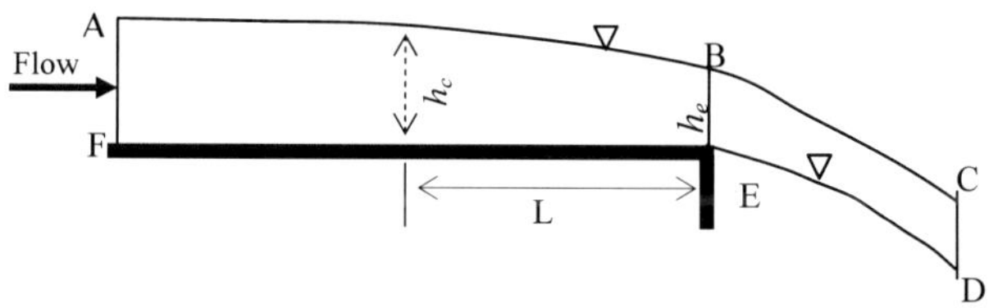

Figure 5.3.1 Sketch of the free overfall flow experiment, ($FE>L$)

Table 5.3.1 Flow condition of test case 1A, Rajaratnam and Muralidhar (1968)

Run No.	Bed slope, So	Unit discharge q (m^2/s)	Critical depth, h_c (m)	End depth, h_e (m)	Length of overfall L(m)
1A	0.0	0.143	0.128	0.0945	0.286

5.3.5 Boundary Conditions

Boundary conditions for each portion of the flow boundary were specified as shown in Figure 5.3.2. Namely, (1) the atmospheric pressure (p_a) should be specified along the surface AC, DE, and the end section CD; (2) hydrostatic pressure should be prescribed along the upstream inlet section AF; and (3) over the solid bed surface, EF, $\partial^2 p/\partial z^2=0$ is prescribed, which approximates the pressure variation very close to the bed as linear. If one sets the simulation channel longer than the length of the free fall, L (Figure 5.3.1), one can simply prescribe subcritical flow boundary conditions at the inlet AF; a supercritical flow boundary condition would have to be used otherwise. The dynamic pressure at the brink is non-zero, but it vanishes at the section CD (Rouse 1936, Strelkoff and Moayeri, 1970). The length of the free fall part is long enough to ensure that the atmospheric pressure can be applied at CD. The vertical location of this section and its water depth depend on the final profile of the overfall flow which are therefore unknown prior to the simulation. In the simulation example, this section is located along the trajectory of the flow, and depth of this section is assumed to be equal to the adjacent upstream section ($\partial(\eta_t - \eta_l)/\partial s = 0$). Figure 5.3.2 shows the details of boundary conditions prescribed for the test case from which one sees that the pressure boundary conditions as well as the log law assumed at the smooth bed boundary and at the inlet to approximate fully developed approach flow. x and z are horizontal and vertical coordinates for u and w velocity component, respectively, s denotes the direction along the flow, η is the free surface with subscript t and l indicating the top and lower free surface, respectively. u_*, k are shear velocity, and Karman Constant (=0.41), $z_0 = 0.11\nu / u_*$ and is the kinematic viscosity of water. ρ is the density of water, g is gravitational acceleration, p is pressure and p_a is the atmospheric pressure.

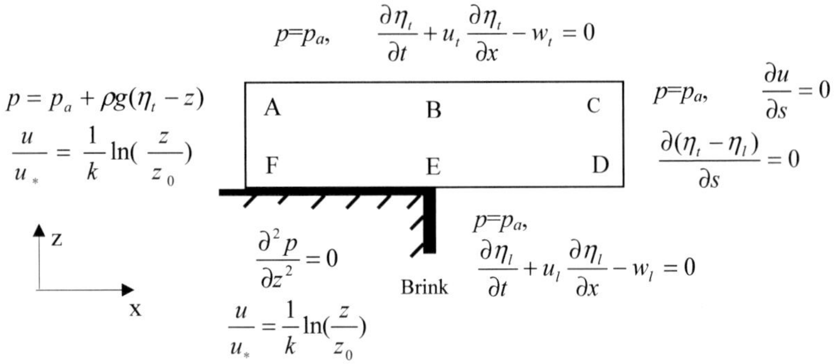

Figure 5.3.2 Boundary condition for free overfall flow simulation

5.3.6 Test Example

To demonstrate the test procedure, an unsteady free surface flow model, CCHE3D (Jia and Wang, 1997, Jia, Kitamura, and Wang, 2001) was tested according to the suggested approach. Boundary conditions shown in Figure 5.3.2 were prescribed, and all geometric and physical parameters were provided as input data. Due to the fact that the initial locations of the two free-surface boundaries are unknown, analytic solutions (Marchi, 1993) were used as guessed initial conditions. It was found later that an arbitrarily shaped initial profile can be used also to achieve the same results. A numerical grid was generated to discretize the computational domain, which has finer resolution near the brink, near the lower boundary of bed and lower free surface. The two free surfaces are moving boundaries and they have to be located dynamically; the actual computational domain or the locations of the free-surface boundaries were modified in each time step until the steady state flow is reached. The steady state solutions of velocity and pressure fields were obtained. The numerically simulated results were compared to the experiment measurements of Rajaratnam and Muralidhar (1968) as given in Figures 5.3.3-5.3.6.

Figure 5.3.3 shows comparisons of the computed and measured pressure. It can be seen that the pressure distribution is hydrostatic in the region upstream of the singularity (the brink) and changes gradually to non-hydrostatic pressure in the region close to the brink, and finally to a constant or atmospheric, downstream of the brink. In the upstream of the brink, pressure is linearly distributed in the vertical direction. However, near the brink, the pressure rapidly changes to parabolic and it gradually vanishes further downstream.

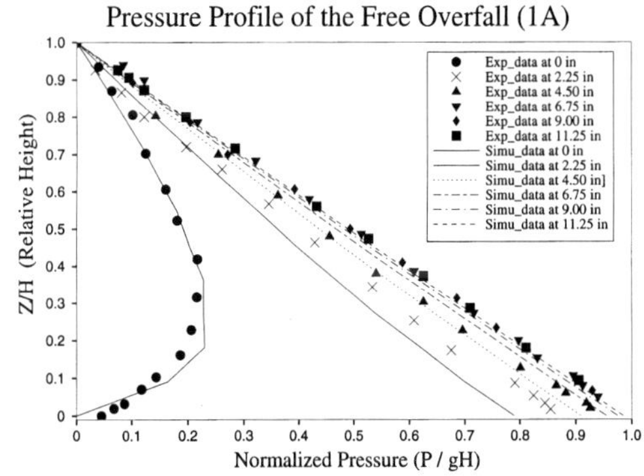

Figure 5.3.3 Comparison of measured and simulated pressure
(The numbers in inches in the legend indicate the distances from the brink upstream)

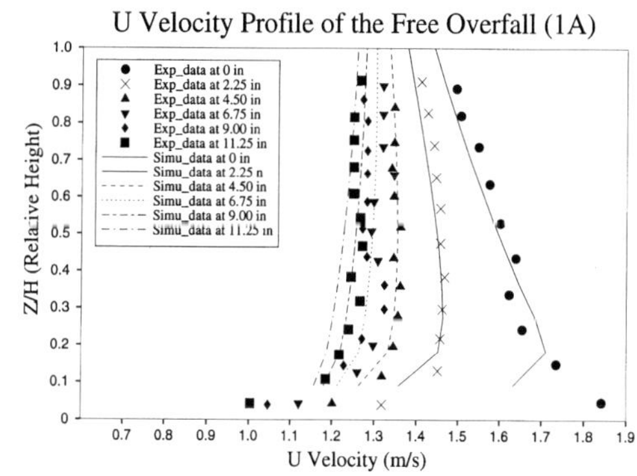

Figure 5.3.4 Comparison of the measured and simulated horizontal velocity
(The numbers in inches in the legend indicate the distances from the brink upstream)

The agreement appears to be very good. Figure 5.3.4 shows the computed and the measured horizontal velocities. The overall agreement is good except near the brink where the differences may be attributed to the singularity. One should note that near the brink the actual flow may oscillate in time and aeration may take place. As a result, the assumptions made in prescribed boundary conditions over the solid surface near the brink may be unrealistic. For example, the validity of the Law of Wall may be questionable and the pressure boundary condition $\partial^2 p/\partial z^2=0$ may not be accurate enough. Of course, if the discrepancy is significant, the model user should refine the grid locally to enhance the accuracy of the numerical solution near the singularity. Figures 5.3.5 and 5.3.6 show comparisons of shear stress and water surface profile, respectively. The comparisons indicate the simulation results are quite good.

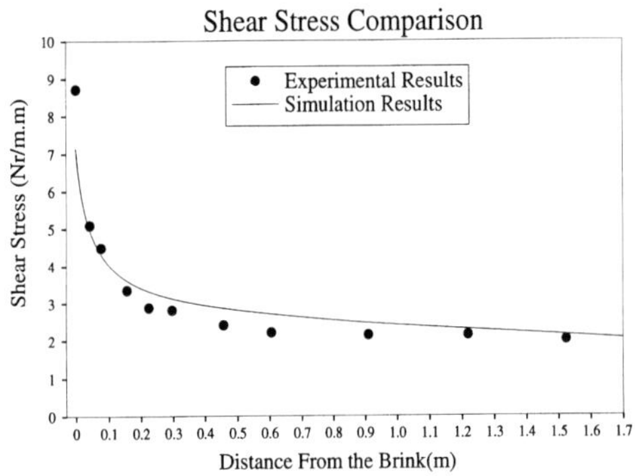

Figure 5.3.5 Comparison of measured and simulated shear stress

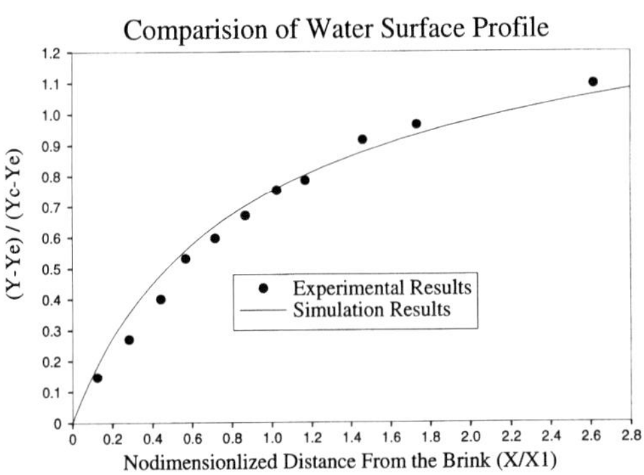

Figure 5.3.6 Comparison of measured and simulated (upper) free surface. $Y(=h)$ is water depth, $Y_e\ (=h_e)$ is depth at the brink, $Y_c\ (=h_c)$ is the critical depth, and $x_1=x$ at which $z-h_e=0.75(h_c-h_e)$

5.3.7 Conclusions

The information presented above shows that the Overfall Flow Test Case can be used to validate a model's capabilities of predicting realistic velocity and pressure fields, free-surface locations and bed shear stress distribution of a three-dimensional free-surface flow, especially the non-hydrostatic pressure distribution in the flow field near and downstream of the brink, which includes both subcritical and supercritical flow regimes. It can also identify deficiencies of a numerical model such as the discrepancy in velocities near the bed surface, especially very close to the brink. If the accuracy of the velocity field near this singular point (the brink) is very important, one may enhance the capability of the numerical model, refine the grids of numerical solution locally and/or apply more realistic boundary conditions.

5.3.8 References

Jia, Y., Kitamura, T., and Wang, S.S.Y., 2001, "Simulation scour process in a plunge pool with loose material", ASCE, *Journal of Hydraulic Engineering*, Vol. 127, No. 3, pp219-229.

Jia, Y. and Wang, S.S.Y., 1997, "Numerical Simulation of Free Overfall", Proceedings of the Conference on Management of Landscapes Disturbed by Channel Incision, held in Oxford, Mississippi, pp 171-176.

Marchi, E., 1993, "On the free overfall", *Journal of Hydraulic Research*, Vol. 31, No. 6, pp777-790, Vol. 32, No.5, pp 792,796.

Rajaratnam, N., and Muralidhar, D., 1968, "Characteristics of the rectangular free over fall", *Journal of Hydraulic Research*, Vol. 6, No. 3, pp233-258.

Rouse, H., 1936, "Discharge characters of the free overfall", *Civil Engineering*, Vol. 6, No. 4, pp 257-260.

Strelkoff, T., and Moayeri. M., 1970, "Pattern of potential flow in a free overfall", ASCE, *Journal of Hydraulic Division*, Vol. 96, No. HY4, pp 879-901.

5.4 DELFT U-SHAPED CHANNEL FLOW (INDOOR) TEST CASE

Contributor: Yafei Jia

5.4.1 Introduction

Some distinctive characteristics have been observed in the free surface flows in curved channels in natural river and streams, irrigation canals, aqueducts, laboratory flumes, etc. In these flows, surface superelevations are observed along channel bends; the flow near the free surface is toward the concave bank and that near the channel bed is moving in the opposite direction. The resulting flow is a helical secondary current combined with the primary flow in the longitudinal direction along the channel. Often times, more than one helical secondary current has been observed, with the smaller secondary helical motion occurring in the upper corner of the outer bank. Secondary currents cause a unique morphological evolution of the channel cross-section shape and bank migration of meandering rivers. Investigations of these complex three-dimensional flow and morphodynamic processes are often needed in soil conservation, river stabilization and restoration/rehabilitation, waterway infrastructure, ecological and environmental research and engineering. The applications of numerical simulation to the studies of the curved channel flows have been emphasized more and more in recent years.

In order to insure the quality of a computational simulation model selected for the study of the flows in curved channels, its capability of predicting the aforementioned physical characteristics has to be confirmed. As discussed in the introduction section of this chapter, it is advised to test the model using test cases based on both field data and laboratory experiments. The test case presented here is the one based on the laboratory experiment of De Vriend (1979).

5.4.2 Objectives

Since the data set is a realistic representations of a three-dimensional velocity field measured at a large number of cross-sections, this test case can be applied to validate three-dimensional free surface flow simulation models for validating a model's capability in predicting the following three-dimensional flow characteristics.

- The distribution of the secondary helical flow in a channel bend, the transition from a straight channel approaching to a channel bend, and the weakening of these currents as the flow is moving away from bended portion of the curved channel and returning to the straight exit section.
- A smaller helical current in the upper corner near the outer bank.
- The vertical distribution of longitudinal velocity with the maximum velocity below water surface.
- The free surface elevation variation longitudinally and transversally in the entire channel, especially in the channel bend.

5.4.3 Description of the Experiment

The configuration sketch of the physical model is shown in Figure 5.4.1, and the flow condition of the experiment is listed in Table 5.4.1. The bed and vertical walls were of concrete. Some thin iron plates that strengthen the outer bank glass panels were roughened with fine grains with a mean diameter of 0.006m to achieve the equivalent roughness.

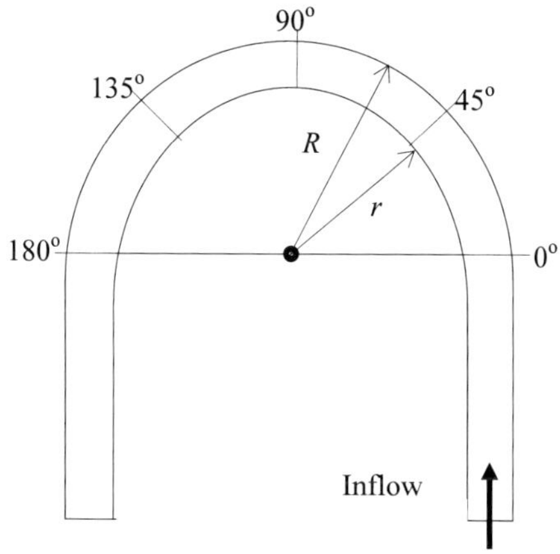

Figure 5.4.1 Configuration sketch of the physical experiment.

Table 5.4.1 Flow condition of the 180° U-shaped channel, De Vriend (1978)

Bed slope *So*	Discharge Q (m^3/s)	Depth (m)	Width (m)	Channel Length (m)	Inner radius r(m)	Outer radius *R*(m)
0.0	0.18	0.189	1.7	25.35	3.4	5.1

Velocity data were measured in 21 cross-sections including 13 sections in the curved part of the channel, which were equally spaced at 15° interval from 0° to 180°. Six more sections in the downstream straight channel at one meter intervals and two other sections were measured in the upstream straight segment; one was 1 meter and the other was 4 meters upstream of the first curved channel section (at 0°).

At each of the cross-sections, measurements were made along 11 vertical lines and on each vertical line, there were 9 measuring points. The total number of points is therefore 2079 (21x11x9). Figure 5.4.2 shows the grid of the measuring points. The total horizontal velocity component V and the direction angle α were measured, but the vertical velocity components were not measured. The velocity component in tangential ($u_t = V\cos\alpha$) and radial direction ($u_r = V\sin\alpha$) can be computed by using these measurements.

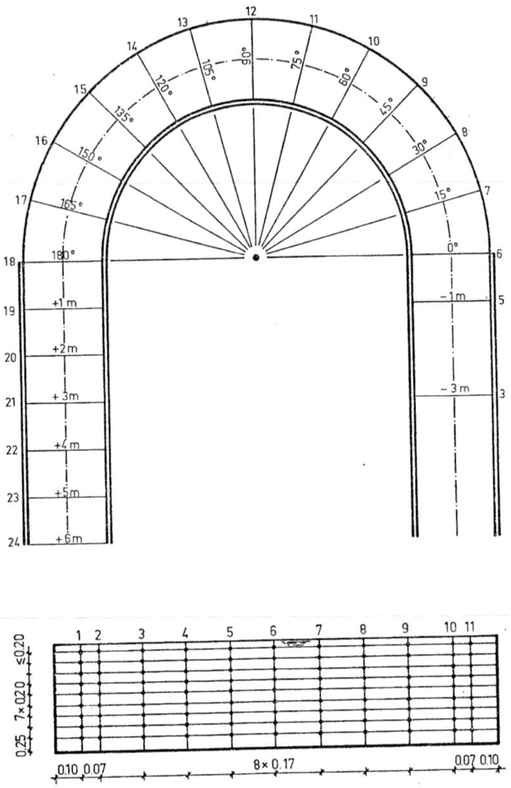

Figure 5.4.2 Meshes of the velocity measuring points

5.4.4 Approach

To test a numerical model using this case, one is advised to select a computation domain of the same configuration and dimensions in full-scale. It would be convenient, for comparing the numerical results to the measurements, to locate some of the nodes of the numerical grid at the measuring points thus avoiding additional errors due to interpolation. The boundary conditions and physical parameters to be used for conducting the test runs should be the same as those used in the laboratory experiments.

5.4.5 Data Organization

The collected data have been tabulated and drawn in figures using Microsoft Excel. These data files are provided in Appendix A of this report. Some examples of the data plots are shown in Figures 5.4.3 and 5.4.4. In the experiment, the total horizontal velocities (V_{total}) and the angle between this total velocity and the longitudinal direction (α) were measured, and the longitudinal ($V_{\Phi}=V_{total}\cos\alpha$) and transverse ($V_R=V_{total}\sin a$) velocities were then computed, which are separately presented in different figures.

The vertical lines in these plots represent the locations of measurement; the distance between the vertical lines and the corresponding velocity points is proportional to the magnitude of velocity. One can see that the longitudinal velocity magnitude does not vary too much along the vertical direction, and the maximum velocities often appear below the free surface, especially near the banks. This is because the transverse and vertical momentum transfer due to the secondary flow greatly changed the momentum distributions in the bend. The measured secondary flow pattern (Figure 5.4.4) is classical: the upper part of the flow moves towards the concave bank and the lower part of the flow moves toward the convex bank. The upper corner of the outer bank has a small secondary flow cell with a counter-rotational direction to the primary one. This small secondary flow cell appears where the bend starts and continues in the straight reach downstream of the bend.

5.4.6 Boundary Conditions

The boundary conditions for simulating this case are simple. One only needs to prescribe the surface elevation at the exit of the channel and flow discharge at the inlet. One could even use the measured velocity at the first section as the upstream boundary condition. However, the location of the upstream boundary should be set reasonably far away from the channel bend. The wall functions may be used on vertical walls as well as on the bed surface. A free surface solver can be used for predicting water surface elevation. The experiment flume was said to be "smooth", and the roughness height k_s was found to be in the range of (0.001~0.0005m) with the numerical simulation.

5.4.7 Test Example

Simulation results of this test case (Jia and Wang, 1992) using CCHE3D model, developed at the National Center for Computational Hydroscience and Engineering of The University of Mississippi, are presented as the first test example. A finite element mesh, designed for conducting this test, has a total of 49x15x7 nodes distributed at 49, 15 and 7 stations along longitudinal, transversal and vertical directions respectively. To improve accuracy and resolution, modification or refinement of this mesh according to one's model validation needs is advised.

The flow boundary conditions and parameters are chosen to be the same as those used in conducting the laboratory experiments.

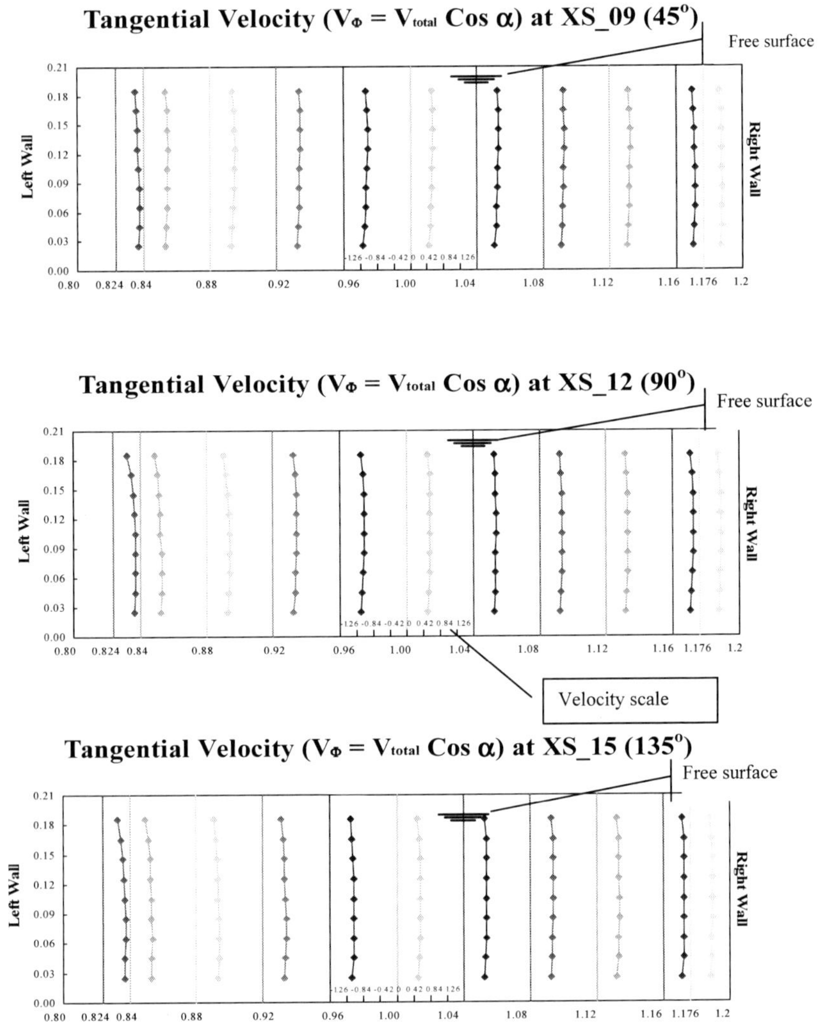

Figure 5.4.3 Measured longitudinal velocity in section 45^0, 90^0, and 135^0 or in section XS_09, 12 and 15 (See Figure 5.4.2.). α is the angle the total velocity diverges from the longitudinal direction.

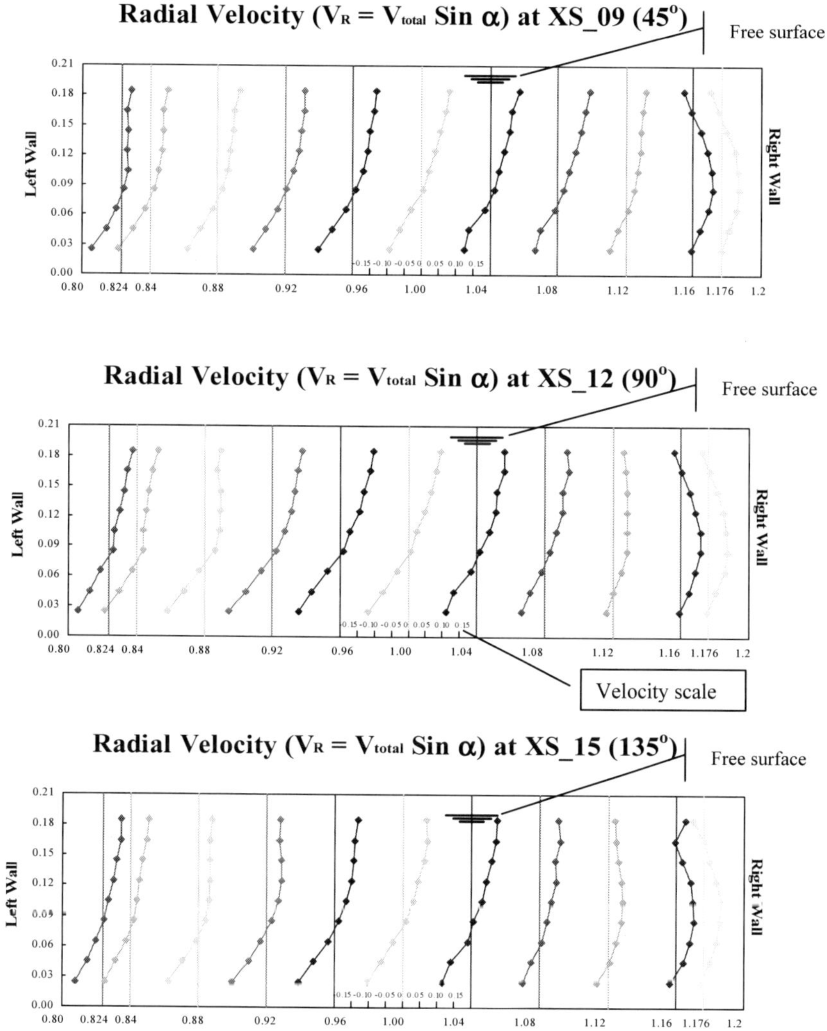

Figure 5.4.4 Measured transverse velocity in section 45^0, 90^0, 135^0 or in section XS_09, 12 and 15 (See Figure 5.4.2.). α is the angle the total velocity diverges from the longitudinal direction.

5.4.8 Numerical Solution

The comparisons of the simulated and measured longitudinal and transversal velocity components at selected cross-sections No. 6 (0°), No. 12 (90°), No. 18 (180°) and No. 21 (+3m), are presented in Figures 5.4.5 and 5.4.6. These show that for engineering purposes the agreements are quite acceptable. It appears that the discrepancies between the measured and simulated transversal velocity components, especially near the side walls and the bed in the channel bend, are more pronounced than those of the longitudinal components. One should note that the scale of the measured and simulated transversal velocity components is five times larger than that of the main velocity.

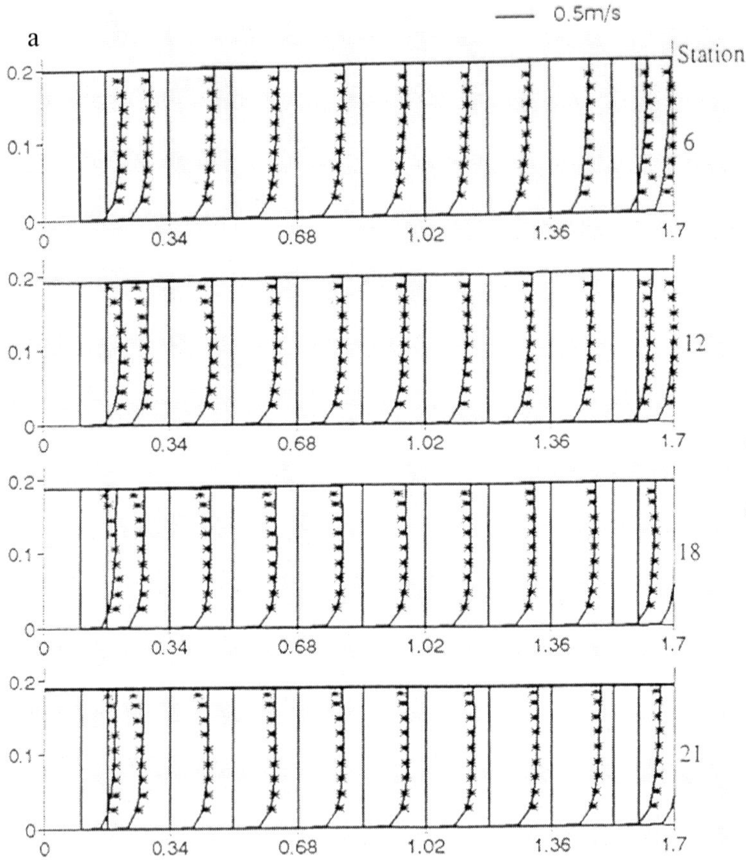

Figure 5.4.5 Simulated and measured longitudinal velocity in selected sections.

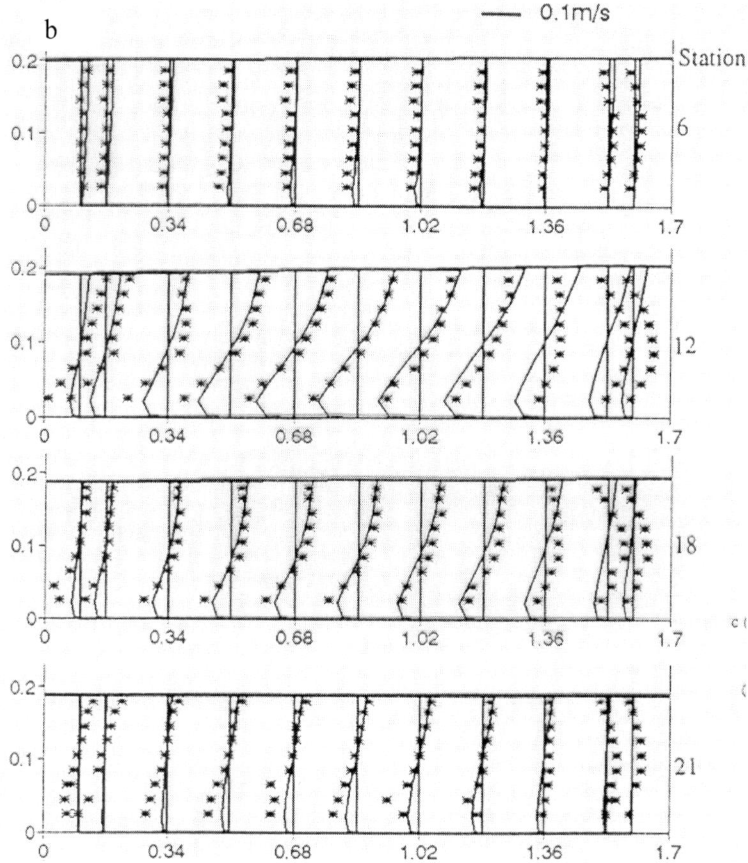

Figure 5.4.6 Simulated and measured transverse velocity in selected sections.

A discrepancy that should be pointed out is the fact that the vertical profiles of simulated longitudinal velocity components show that the maximum value along each vertical is always at the free surface. This is different from the measurements, in which the maximum is mostly beneath the free surface, especially near the side walls where the maximum can even be in the lower half of the water depth. From Figure 5.4.6, one sees that the transversal velocity components are essentially zero in the straight reach up to the cross-section No. 6, where the flow is entering the bend; they increase in magnitude quickly near the bend entrance, and gradually decrease in the downstream straight reach but not yet becoming zero at Section No. 21 (3 meters from the exit of the bend). The small vortex at the upper corner near the outer bank could not be reproduced by the numerical model with a zero equation closure scheme. This numerical error is considered insignificant in engineering applications, as in most of the published papers reporting computational simulations, because the most important characteristics of a curved channel flow have been predicted correctly. But one could look into the cause of some discrepancies such as those observed at Sections No. 12, 18 and 21 in Figure 5.4.6, and try to improve the model for better accuracy. Free surface elevation data and those predicted along the outer bank, central line and the inner bank are shown in the Figure 5.4.7. It is evident that the superelevation occurs only in the bend part of the channel. The elevations in both inner and outer banks as well as along the center of the channel are straight lines except in the transition zones.

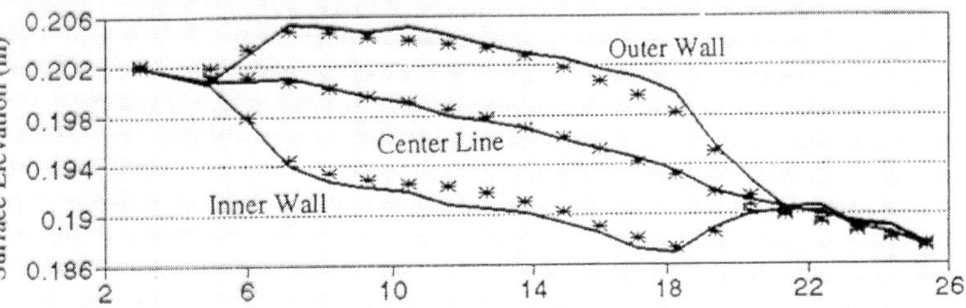

Figure 5.4.7 Simulated and measured free surface along the U-shaped channel (— Simulation, ** Measurement)

In fact, the deficiency mentioned above has been improved by implementing a nonlinear k-ε turbulence closure for the flow in the channel bend (Jia, et al, 2001). By an enhancement of the turbulence closure scheme and pressure solution, the results of the simulation of flows in the same channel and another curved channel with an eroded bed were reported and these discrepancies have been significantly reduced. Figure 5.4.8 shows the measured and simulated main flow and secondary current; one can see that the small vortex at the upper corner of the outer bank (right hand side) has been captured by the improved turbulence closure model. These results indicated that the validation test can be applied to reveal the deficiency of a numerical model, so that the model can be improved before it is applied to real-life problems.

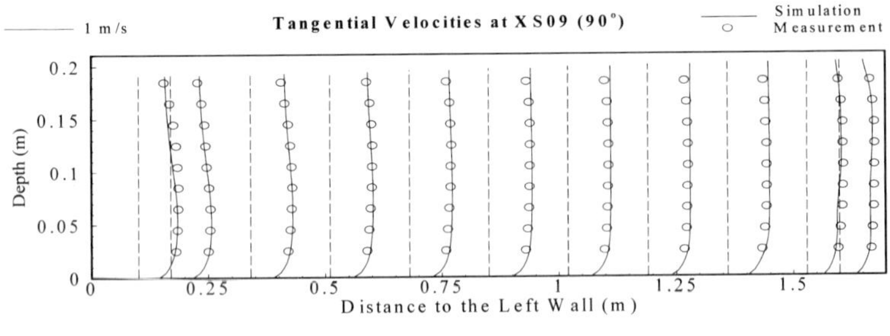

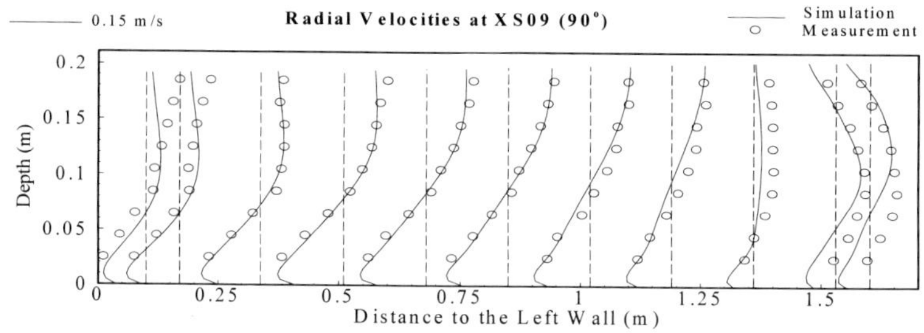

Figure 5.4.8a Computed and measured longitudinal and transverse velocity profiles in the cross-section of $\alpha = 90^0$.

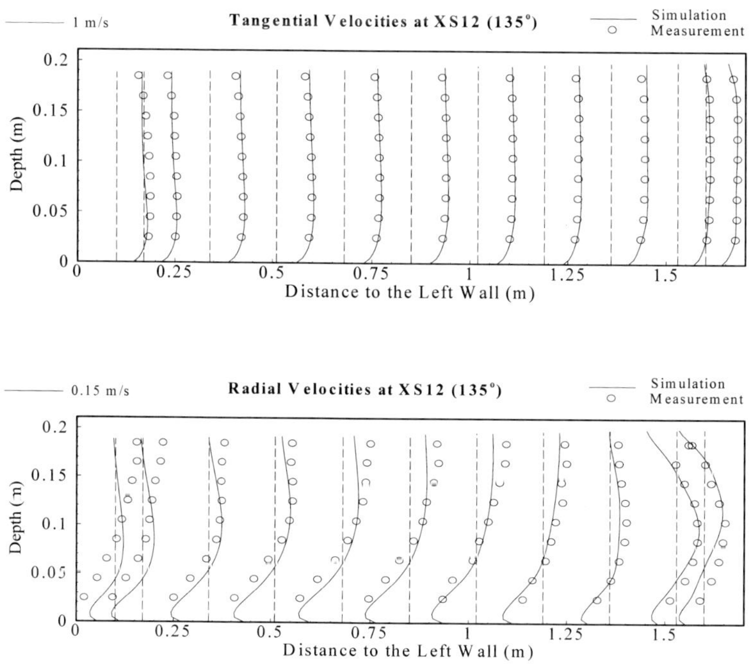

Figure 5.4.8b Computed and measured longitudinal and transverse velocity profiles in the cross-section of $\alpha = 135^0$.

5.4.9 Calculation Verification

To estimate numerical errors in a particular case of physical phenomenon, calculation verification is performed. Once one has validated a model's ability to approximate the observed flow physics, it is often necessary to evaluate the model's numerical error for this particular physical problem and the applied discretization (grids). The numerical error for the final mesh should be reasonably small and the solutions should be converged and not vary with the mesh significantly. The general approach of calculation verification has been described in Chapter 2 with simple one-dimensional examples. Extending the approach to 3D free surface flows is more involved with more factors to be considered.

Three grids were generated for conducting the calculation verification of the 180° U-shaped bend channel case. The three grids are defined with G_I (I=1,2,3) with I=1 being the finest, and I=3 the coarsest. G_1, the finest grid, was the one generated for simulating the physical experiment and validating the non-linear k-ε model. It was not obtained in the first attempt. In the processes of the validation study, several different distributions of grid density and number of nodes in three directions have been tested to confirm that the non-linear k-ε is capable of reproducing the flow recirculation patterns observed in physical model. Calibration was then conducted by fine tuning the effective roughness for the bed and for the walls. To match the computed free surface profiles with those measured reasonably, more than ten runs of computations were conducted. Coarser grids G_2 and G_3 were generated later for CV purpose. Grid G_3 was obtained by deleting one grid line from every two lines from G_1 with the first near wall grid lines preserved (almost halved). Since the three dimensional modeling case needs a minimum number of grid lines to resolve the flow and the channel geometry, one cannot afford to coarsen the grid further for the purpose of CV; on the other hand, the time and computer capacity would not allow one to increase the total number of lines either. It was determined to create another grid, G_2, with an intermediate grid density, which was also based on G_1 by deleting some lines and add some new lines.

Let $N_{Gi} = N_1 \cdot N_2 \cdot N_3 \big|_{Gi}$ be the total number of the grid G_i (i=1,2,3), where N_I, N_J and N_K are nodal numbers in transverse, longitudinal and vertical direction.

$$N_{G1} = 59 \times 97 \times 42 = 240{,}366$$
$$N_{G2} = 40 \times 65 \times 32 = 83{,}200$$
$$N_{G3} = 31 \times 49 \times 22 = 33{,}418$$

The effective ratio of mesh refinement is defined

$$r_{32} = \sqrt[3]{2.49} = 1.355$$
$$r_{21} = \sqrt[3]{2.889} = 1.424$$
$$r_{31} = \sqrt[3]{7.193} = 1.930$$

Although the grid refinement ratio, r_{21} and r_{32} are not exactly the same, they were made very close with a mean value of $r = \sqrt{r_{12} r_{23}} = \sqrt{r_{31}} = 1.389$. Due to the nature of general channel flows and that of the secondary currents, one has to use non-uniform grid to resolve the variations of the flow near the walls and in the domain. In the longitudinal direction, the grid is uniformly spaced. In the bend part, the grid spacing near the inner bank is less than those along the outer bank. In transverse direction, most of the grid points are concentrated near the banks and where the outer bank eddy would appear. In the vertical direction, the grid distribution was set uniform at each point; the actual mesh size Δz depends on local water depth. To keep the wall boundary condition valid, the first near wall (bed) grid point is the same for all grids.

Comparison of results

The same boundary conditions were applied to the three meshes and the results are compared. Since the three computational grids are different, we only compare the solution and calculate errors on the common nodes in the *x-y* plan shared by all three meshes, to avoid error of interpolation. In the vertical direction, the nodal spacing is uniform for the finest grid, while for the two coarser grids, the location of the first near bed node is kept unchanged, and other nodes are equally spaced between this near bed node and the free surface because we would like to keep the physical wall boundary condition for all three cases as close as possible. Since the locations for the nodes along vertical mesh lines and the resulted surface elevation (local flow depth) of the three cases differ, interpolation is needed for comparison of velocity results between different meshes. All the vertical velocity profiles presented are on the coarsest grid (G_3); the results of finer grids are interpolated to the coarsest one via a third order (second order near bed or surface) Lagrangian interpolation.

Using GCI as an error estimator, the (approximate) constancy of the equation (4.6.1), and assuming constant (not necessarily an integer) *r,* the rate of convergence *p* is evaluated (Roache, 1998),

$$p_g = \ln\left(\frac{\varepsilon_{23}}{\varepsilon_{12}}\right) \tag{5.4.1}$$

Where $\varepsilon_{12} = (f_2 - f_1)/f_1$ and $\varepsilon_{23} = (f_3 - f_2)/f_2$

On the other hand, by using the formulation of Richardson Extrapolation for solutions f_1: f_2 and f_2 : f_3, the convergence rate can be expressed differently involving a constant *r*:

$$p_r = \ln\left(\frac{f_3 - f_2}{f_2 - f_1}\right) / \ln(r) \tag{5.4.2}$$

These two formulations are both used to evaluate the convergence of the bend channel solutions.

Four variables are selected for analyzing convergence of the solutions using these formulas:

- Size (width) of the outer bank vortex, ℓ_v
- Total water surface slope
- Main (longitudinal) flow velocities: V_ϕ
- Transverse velocity: V_R.

Two turbulence schemes, standard k-ε model and the non-linear k-ε model were tested with these three grids and the resultant numerical errors are compared and analyzed. Steady solutions were obtained for all the cases when the flow discharge along the channel ranges $Q_j/Q_0 \sim 0.0003$.

Free surface profiles

Figure 5.4.9 shows the comparisons of free surface profiles of the computed flow with k-ε model and non-linear k-ε model. For each set of result, the surface elevations along left bank, right bank and the center line are presented. The convergence of the water surface profiles of both models is clearly seen. Since there is no grid lines located along the true center line, interpolations are used. The pattern of the free surfaces looks very similar for all cases. For the k-ε model, the two profiles from the finer grids are closer than that from the coarsest gird, G_3. They are differentiated simply by the general surface slope. For the non-linear k-ε model, the convergence trend of the surface is similar with the two profiles of the finer grids overlaping closely.

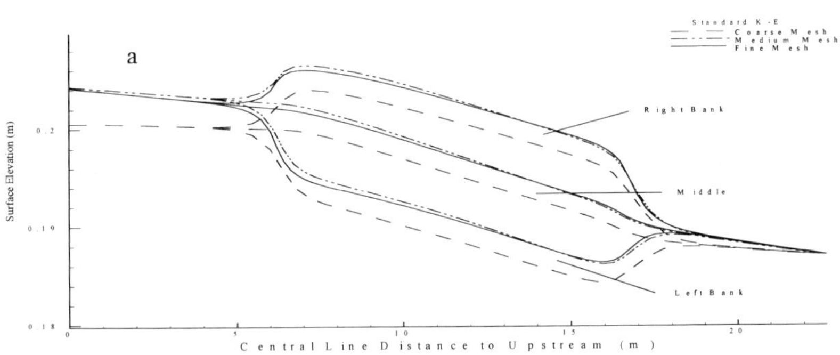

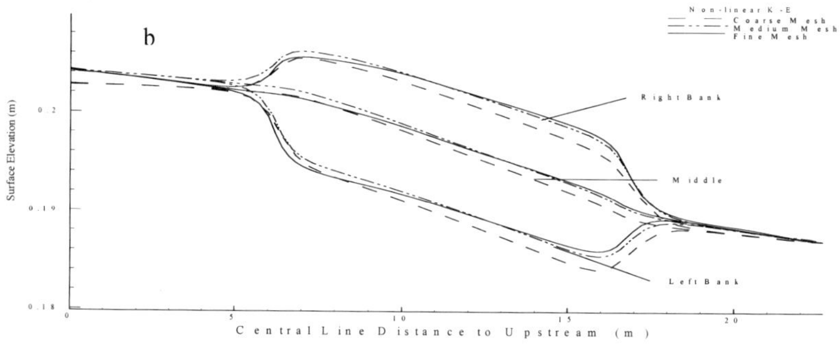

Figure 5.4.9. Free surface elevation computed using three meshes: a) k-ε model, b) non-linear k-ε model

The difference of the mean surface elevations at inlet and outlet and the actual length of the center lines from the three grids were used to compute the total slope. The slope of the three meshes converges in an oscillatory fashion as shown in Table 5.4.2. Absolute value of the ε and Δf were used to compute convergence as a result. The convergence rate computed from equations (5.4.1) and (5.4.2) are quite large, much larger than the expected order of 1~2.

Table 5.4.2 Total longitudinal slope and convergence.

Model	Grid	Slope (f)	$\|\varepsilon\|$	p_g	$\|\Delta f\|$	p_r
k-ε	G3	5.9790E-04	0.2187	3.341	1.6738E-04	10.2
	G2	7.6528E-04				
	G1	7.5940E-04	0.007743		5.8800E-06	
Non-linear k-ε	G3	6.9836E-04	0.08568	4.253	6.4450E-05	12.9
	G2	7.6380E-04				
	G1	7.6287E-04	0.001219		9.7000E-07	

Figure 5.4.10 shows the computed secondary flow vector fields near the outer bank on three grids. The size of the outer bank eddy varies with the grids. The distance of the separation point of the side vortex to the outer bank is defined as ℓ_v, and the non-dimensional variable ℓ_v / h was used to evaluate the convergence. Table 5.4.3 shows the vortex size from the three grids and the convergence rate computed by equations (5.4.1) and (5.4.2). The convergence rate computed by these two equations has quite large a difference with that from equation (5.4.1) being closer to the expected rate of convergence: order 1.

Table 5.4.3 Convergence of the near bank vortex

Model	Grid	ℓ_v / h*	ε	p_g	p_r
Non-linear k-ε	G3	1.915	0.323	1.05	0.446
	G2	1.447			
	G1	1.300	0.113		

* ℓ_v / h: Relative size of side vortex. ℓ_v is the width of the outer bank eddy at the free surface, h is water depth, the ratio is determined from the enlarged graphic visually.

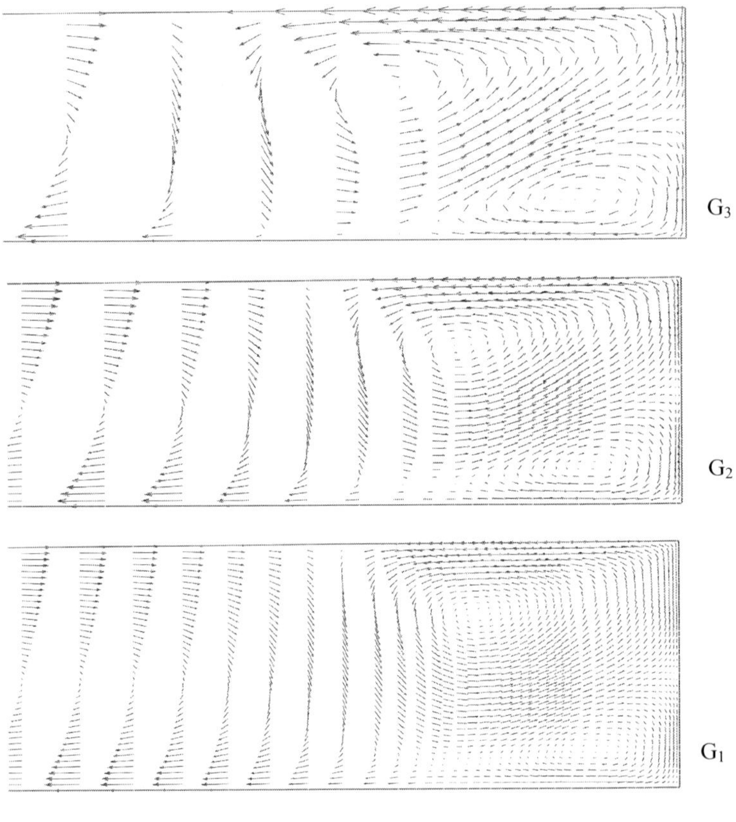

Figure 5.4.10 Computed vector field near the outer bank with three grids.

Since GCI is defined as

$$E = \text{GCI}_{\text{fine grid}} = F_s \frac{\varepsilon}{r^p - 1}, \text{ with } F_s = 1\sim3 \tag{5.4.3}$$

The GCI$_{fine21}$ (Fs=1) for the size of the side vortex is 0.27 or 0.72, computed with p_g and p_r, respectively.

Convergence of velocity field

Figure 5.4.11 shows the longitudinal velocity distribution of the k-ε model results in the 90° cross-section computed with the finest grid. The arrow heads locate the positions in this cross-section where vertical distributions of velocity solution from the three meshes are compared. Figure 5.4.12 shows several velocity profiles (point 2, 4, 6 and 8). The general trend is that the velocity profile converges as the grid is refined. However the degree of convergence at different vertical locations differs. Velocity at a vertical point of the coarser grid may be greater than that of the finer grid, but it may be less at other positions. Figure 5.4.13 shows the computed secondary current in the same locations, the trend of convergence is also obvious: the differences between the finer grids are much smaller than those between coarser ones.

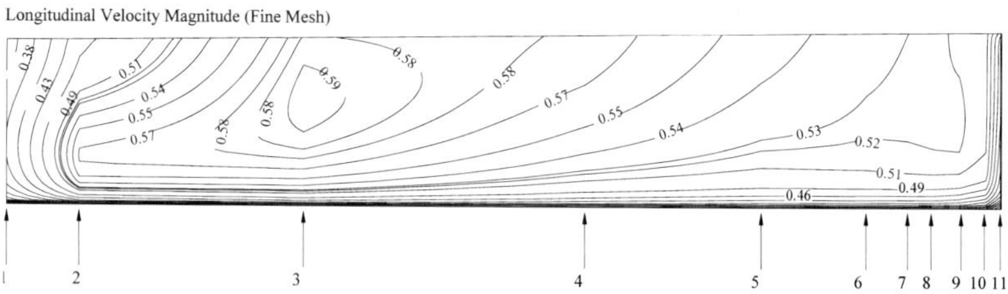

Figure 5.4.11 Main flow velocity distribution in 90° section, k-ε model.

Tables 5.4.4 and 5.4.5 show the error calculation for the longitudinal and transverse velocity profiles at point 4 (Figure 5.4.11). In these tables, vertical location, calculated velocities with the three meshes, errors and convergence rate are indicated. The convergence rate p computed from Equation 5.4.1 for each point with r=1.389, p_g, is less than those using Equation 5.4.2, p_r. It seems the range of p_g is more reasonable: [0.19-2.8] for longitudinal velocity and [0.09, 2.0] for transverse velocity In the middle of the profiles, one negative p_g and p_r are resulted because the local condition, $|\varepsilon_{32}| < |\varepsilon_{21}|$. In some other profiles, there are several negative points. Average p_g is 0.63 and 1.0 for the longitudinal and transverse velocity, respectively; the corresponding average p_r is 1.9 and 3.0. Because the code is of the second order, and the meshes for the calculations are non-uniform, it is expected that the convergence should be around the first order. Large variation of p similar to this was also reported by Celik and Karatekin (1997) studying a simpler two-dimensional turbulent fluid flow case with a backward facing step.

Errors for all the compared points are computed using Equation 5.4.3 with F_s =1, which makes GCI$_{fine\ grid}$ equal to the error of Richardson Extrapolation. The errors with the negative p_g were not computed. The errors indicated in Tables 5.4.4 and 5.4.5 vary in wide ranges of 10~0.01%, and 260~0.6%, for longitudinal and transverse velocities. The average error, however, is reasonably reflected, around 3% and 5% for these two velocity components.

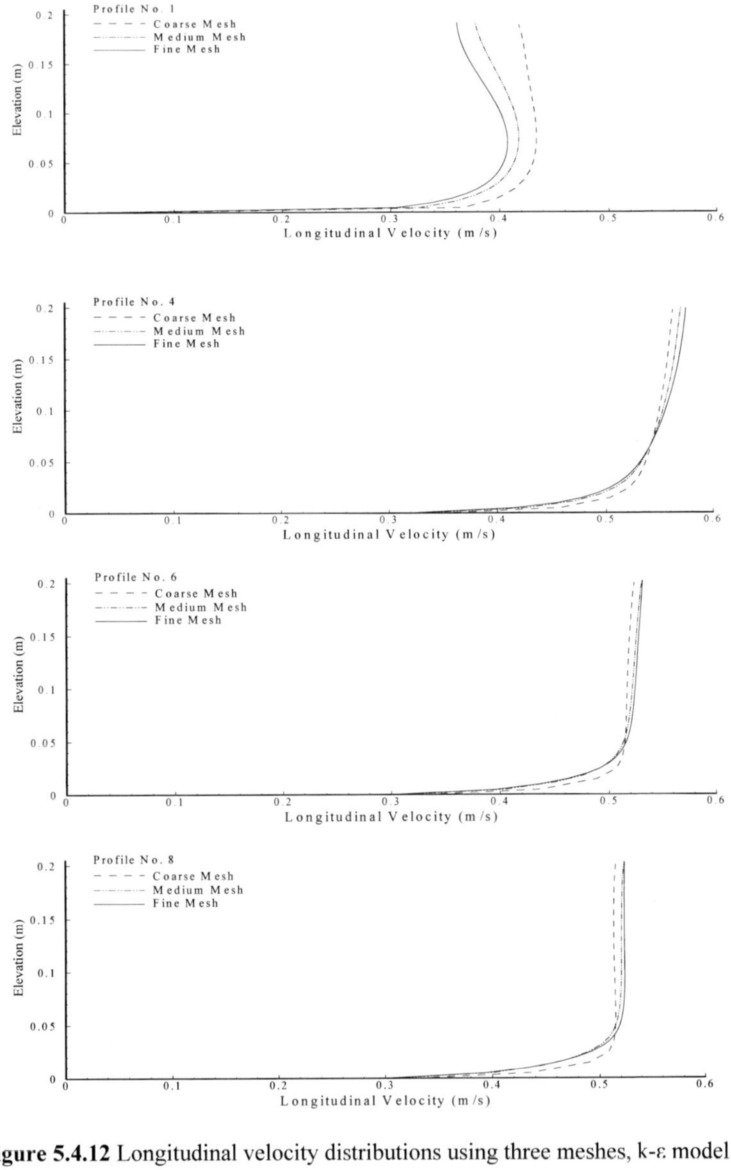

Figure 5.4.12 Longitudinal velocity distributions using three meshes, k-ε model.

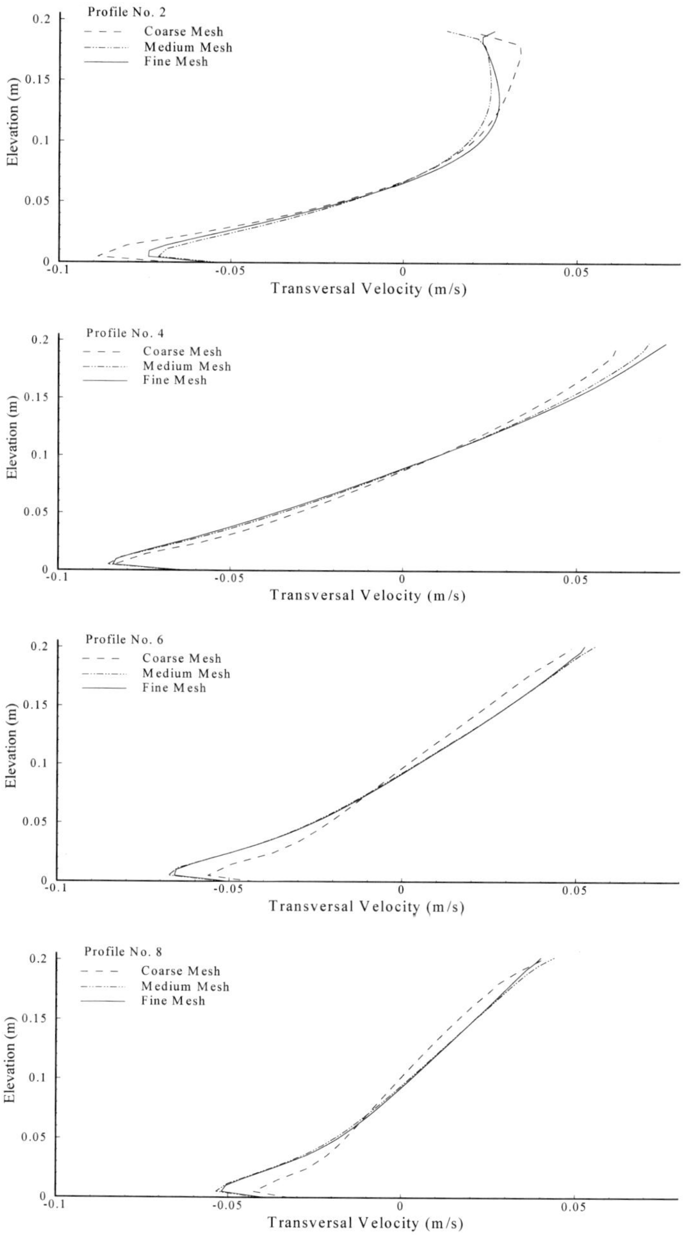

Figure 5.4.13 Transverse velocity distributions using three meshes, k-ε

Table 5.4.4 Error calculation for longitudinal velocity, k-ε model, 90° section

z	f_3	f_2	f_1	ε_{32}	ε_{21}	p_g	p_r	E_{21}
4.79E-03	4.57E-01	4.26E-01	4.15E-01	7.21E-02	2.68E-02	0.989081	3.09E+00	0.0698438
1.44E-02	5.03E-01	4.82E-01	4.77E-01	4.41E-02	1.05E-02	1.432104	4.39E+00	0.017508
2.40E-02	5.21E-01	5.07E-01	5.03E-01	2.71E-02	9.17E-03	1.082918	3.32E+00	0.0214586
3.36E-02	5.30E-01	5.20E-01	5.17E-01	1.78E-02	5.67E-03	1.145504	3.50E+00	0.0123957
4.31E-02	5.35E-01	5.29E-01	5.27E-01	1.08E-02	3.60E-03	1.103099	3.37E+00	0.0082324
5.27E-02	5.39E-01	5.35E-01	5.34E-01	6.09E-03	1.62E-03	1.324252	4.04E+00	0.0029712
6.23E-02	5.42E-01	5.40E-01	5.40E-01	2.34E-03	1.40E-04	2.812234	8.56E+00	9.237E-05
7.19E-02	5.44E-01	5.44E-01	5.45E-01	-5.14E-04	-1.22E-03	-0.86114	-2.62E+00	-
8.15E-02	5.46E-01	5.48E-01	5.49E-01	-2.85E-03	-2.35E-03	0.192624	5.79E-01	-0.035939
9.11E-02	5.48E-01	5.51E-01	5.53E-01	-4.70E-03	-3.39E-03	0.325651	9.81E-01	-0.030043
1.01E-01	5.50E-01	5.54E-01	5.56E-01	-6.24E-03	-4.32E-03	0.368192	1.11E+00	-0.033554
1.10E-01	5.52E-01	5.56E-01	5.59E-01	-7.49E-03	-5.18E-03	0.36841	1.11E+00	-0.040254
1.20E-01	5.54E-01	5.58E-01	5.62E-01	-8.53E-03	-5.97E-03	0.357442	1.07E+00	-0.047902
1.29E-01	5.55E-01	5.60E-01	5.64E-01	-9.39E-03	-6.71E-03	0.336089	1.00E+00	-0.057482
1.39E-01	5.57E-01	5.62E-01	5.66E-01	-1.01E-02	-7.39E-03	0.312368	9.28E-01	-0.068395
1.49E-01	5.58E-01	5.64E-01	5.68E-01	-1.07E-02	-8.01E-03	0.288294	8.53E-01	-0.080619
1.58E-01	5.59E-01	5.65E-01	5.70E-01	-1.12E-02	-8.54E-03	0.268089	7.90E-01	-0.092783
1.68E-01	5.60E-01	5.66E-01	5.72E-01	-1.16E-02	-8.95E-03	0.258469	7.59E-01	-0.101003
1.77E-01	5.61E-01	5.68E-01	5.73E-01	-1.20E-02	-9.18E-03	0.266713	7.84E-01	-0.100253
1.87E-01	5.62E-01	5.69E-01	5.74E-01	-1.24E-02	-9.12E-03	0.307383	9.08E-01	-0.085768
1.97E-01	5.63E-01	5.70E-01	5.75E-01	-1.31E-02	-8.27E-03	0.462409	1.38E+00	-0.05037

Table 5.4.5 Error calculation for the transverse velocity, k-ε model, 90° section

z	f_3	f_2	f_1	ε_{32}	ε_{21}	p_g	p_r	E_{21}
4.79E-03	-8.38E-02	-8.55E-02	-8.40E-02	-1.94E-02	1.76E-02	0.093699	3.38E-01	0.5638503
1.44E-02	-7.42E-02	-7.78E-02	-7.82E-02	-4.54E-02	-5.94E-03	2.03504	6.18E+00	-0.006239
2.40E-02	-5.89E-02	-6.49E-02	-6.62E-02	-9.26E-02	-2.04E-02	1.513297	4.54E+00	-0.031646
3.36E-02	-4.71E-02	-5.32E-02	-5.48E-02	-1.14E-01	-2.93E-02	1.359035	4.05E+00	-0.052065
4.31E-02	-3.72E-02	-4.25E-02	-4.42E-02	-1.25E-01	-3.75E-02	1.203338	3.55E+00	-0.077332
5.27E-02	-2.83E-02	-3.27E-02	-3.42E-02	-1.35E-01	-4.38E-02	1.129094	3.30E+00	-0.097481
6.23E-02	-2.00E-02	-2.35E-02	-2.48E-02	-1.48E-01	-5.27E-02	1.035149	2.99E+00	-0.130105
7.19E-02	-1.21E-02	-1.46E-02	-1.57E-02	-1.76E-01	-6.65E-02	0.971186	2.75E+00	-0.176993
8.15E-02	-4.42E-03	-6.08E-03	-6.84E-03	-2.73E-01	-1.12E-01	0.889284	2.34E+00	-0.330173
9.11E-02	2.96E-03	2.19E-03	1.73E-03	3.52E-01	2.64E-01	0.288394	1.59E+00	2.6527761
1.01E-01	1.01E-02	1.02E-02	1.01E-02	-1.20E-02	1.40E-02	-0.15246	-4.22E-01	-
1.10E-01	1.70E-02	1.80E-02	1.82E-02	-5.59E-02	-1.08E-02	1.643441	4.97E+00	-0.015098
1.20E-01	2.36E-02	2.55E-02	2.60E-02	-7.43E-02	-2.06E-02	1.283333	3.84E+00	-0.039253
1.29E-01	2.99E-02	3.26E-02	3.35E-02	-8.46E-02	-2.65E-02	1.160929	3.45E+00	-0.057027
1.39E-01	3.59E-02	3.95E-02	4.07E-02	-9.15E-02	-3.04E-02	1.102477	3.26E+00	-0.06957
1.49E-01	4.15E-02	4.59E-02	4.75E-02	-9.62E-02	-3.33E-02	1.06169	3.13E+00	-0.079735
1.58E-01	4.68E-02	5.20E-02	5.39E-02	-9.95E-02	-3.54E-02	1.032777	3.03E+00	-0.08763
1.68E-01	5.17E-02	5.76E-02	5.97E-02	-1.01E-01	-3.66E-02	1.014995	2.98E+00	-0.092511
1.77E-01	5.64E-02	6.27E-02	6.51E-02	-1.01E-01	-3.70E-02	1.003794	2.94E+00	-0.094694
1.87E-01	6.04E-02	6.73E-02	7.01E-02	-1.02E-01	-3.93E-02	0.952701	2.78E+00	-0.106922
1.97E-01	6.19E-02	7.10E-02	7.60E-02	-1.29E-01	-6.53E-02	0.680727	1.87E+00	-0.26041

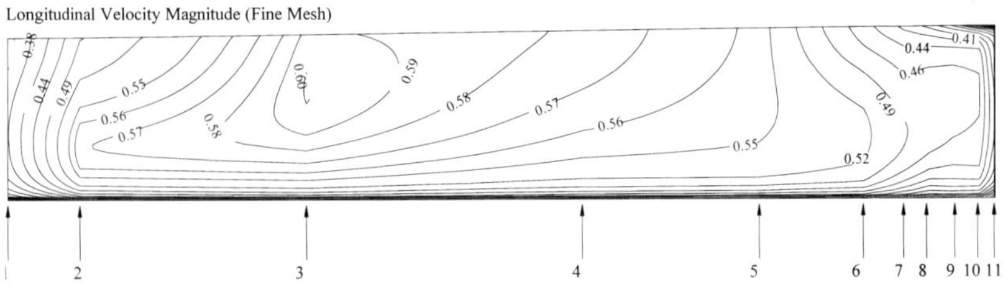

Figure 5.4.14 Main flow velocity distribution in 90° section, non-linear k-ε model.

Figure 5.4.14 shows the longitudinal velocity distribution in the 90° section resulted from the non-linear k-ε model. On the left side of the cross-section (inner bank), the general pattern of the flow is similar to that shown in Figure 5.4.11. The flow on the right side is affected by the near outer bank vortex.

Figure 5.4.15 shows vertical distributions of longitudinal velocity along point 4, 6, 8 and 10. It is seen that the distribution is not affected by the near bank vortex near the center of the channel (point 4). In fact the convergence of the longitudinal velocities is not affected as well, similar to that in the linear k-ε case. In the near bank zone, velocity profiles are changed significantly and the convergence appears to be slower, particularly around where the vortex separates from the main helical current (point 6). One of the reasons for this would be that the meshes in transverse direction in this separation area are relatively sparse and the flow pattern varies fast.

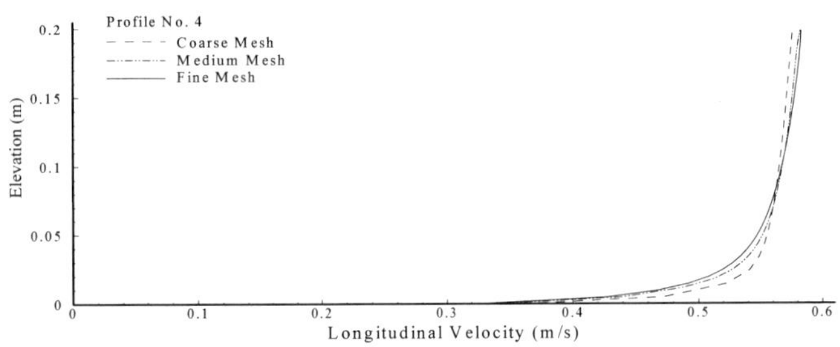

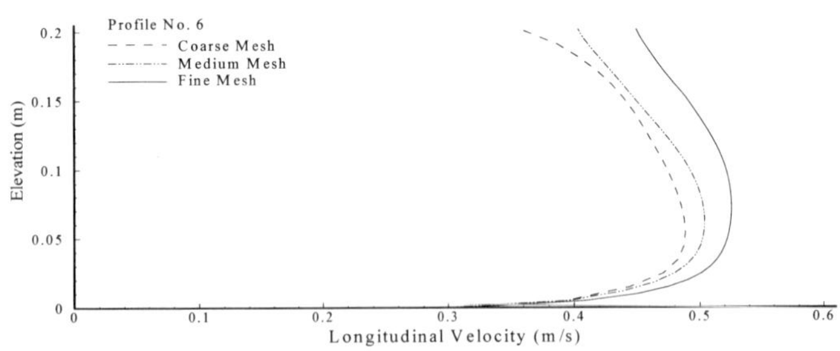

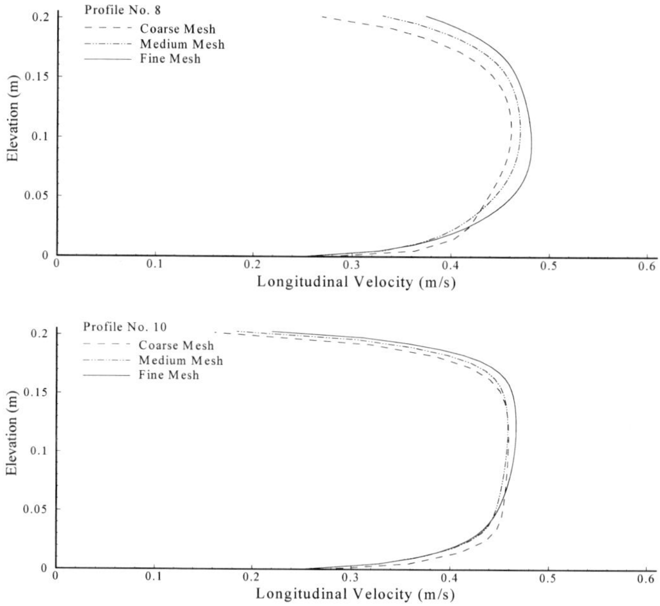

Figure 5.4.15 Longitudinal velocity distributions using three meshes, non-linear k-ε model

The comparisons of the secondary velocities at these points are shown in Figure 5.4.16. Again, the trend of convergence is obvious to the left side of the separation area and near the right bank. The largest numerical error appears also at point 6, the separation zone. The general trend of convergence near the right bank can be recognized visually, but more difficult to be expressed with a point convergence rate since it varies very much in this zone.

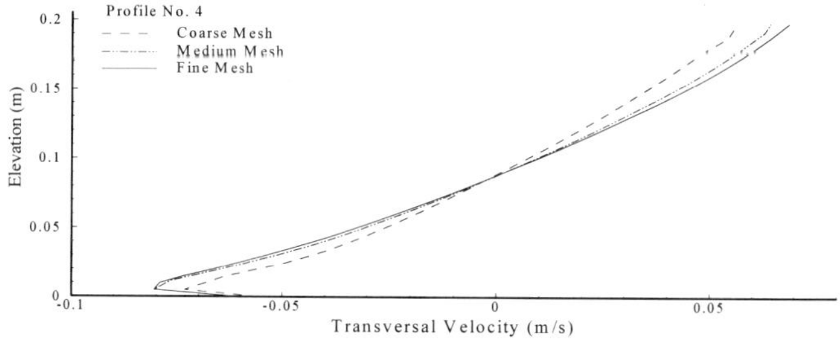

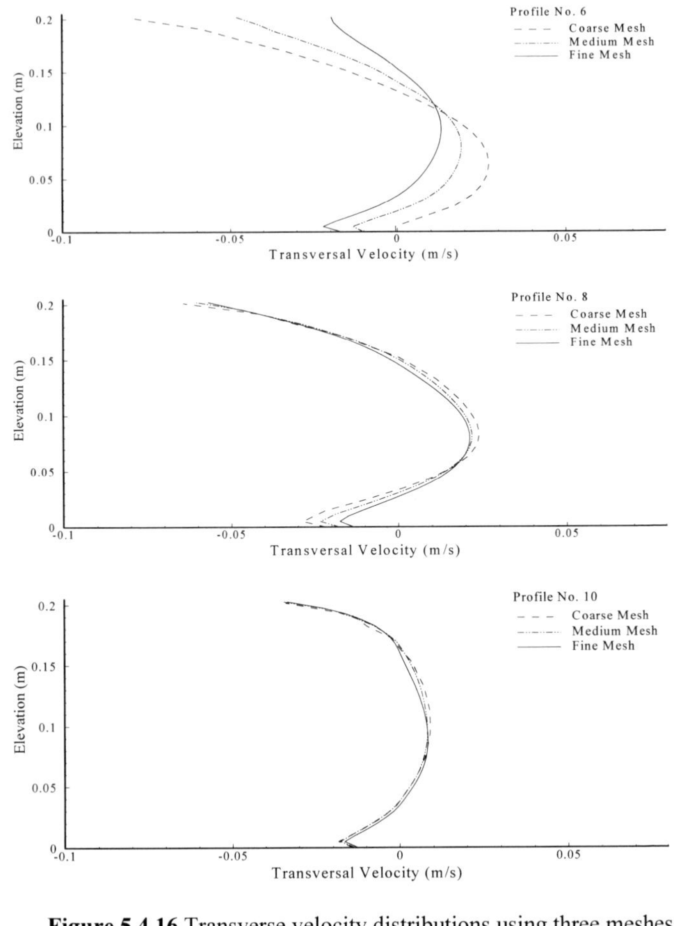

Figure 5.4.16 Transverse velocity distributions using three meshes, non-linear k-ε model

5.4.10 Several Concerns Regarding to Calculation Verification

Free surface

Free surface confines the computational domain of free surface models. The exact location and distribution of the free surface can not normally be prescribed; it is rather the solution of the model. It is expected that the computed free surface will be different if any parameter of the computation has been changed. A direct consequence therefore of using grids or meshes with different distribution or resolutions is a different free surface distribution or, in other words, a different computation domain. Since a change in free surface elevation also induces a pressure field variation and a respond to the flow field, the convergence of 3D free surface simulations is slower than those of non-free surface.

Walls

Wall boundaries appear in almost all simulations of flow physics. It is commonly handled with the so called "Law of Wall": logarithmic distribution of near wall velocity. It is required to have the first near wall grid line be placed in the range, $30<y^+<300$ ($y^+=u_*\Delta/\nu$). This requirement could not be easily achieved for near wall flow velocities of practical problems often vary spatially. This posts a difficulty in the process of refining or coarsing the near wall grid which may result in having many of the near wall nodes out of the physically correct distance range. It is possible to overcome the difficulty by using the "low Reynolds Number" modeling approaches with additional efforts.

Mesh stretching

It is well known that where flow varies rapidly in space and time, finer resolutions of mesh and small time steps are necessary respectively. Non-uniform grids have to be used for gaining the solution accuracy and more computing efficiency. This is more significant in three dimensional free surface flow modeling. Because numerical error will increase with non-uniform grid, the convergence of a problem with a non-uniform grid is likely to be of lower order. A code of the second order accuracy with uniform grids may become the first order with non-uniform grids. The observed order of convergence may vary spatially within the computation domain depending on the local solution and grid distribution. One has to check at more locations in the domain in order to insure a global convergence.

Change of boundary geometry

In general, grid refinement and coarsening should not result in changes of boundary geometry, otherwise, the problem being studied is changed, which directly jeopardizes the numerical error analysis that follows. In practice, this constrain is hard to enforce. We have seen the free surface would be changed with the grid refinement or coarsing. Boundary change is likely to occur when the grid resolution is adjusted particularly for cases where sharp turns, high curvatures and rough bed forms are present. Figure 5.4.17 indicates a real spur dike structure and three meshes representing it in the computation. There is not doubt that the solution will not only be affected by numerical error but also the boundary shape. Since everything is defined by grid lines and surfaces, change of geometric boundary is almost inevitable.

Calculation verification: before or after validation process?

Having been indicated that validation is to check weather a model can reproduce correct flow physics; calculation verification is for finding the numerical error and convergence of a model for a particular experiment or a field problem, it is suggested that CV is to be conducted as a final step after obtaining the final satisfactory results based on the following facts:

1. Reproduce the concerned flow physics is the final goal of a research project, if one could not achieve this, it is not necessary to conduct a CV;
2. Several grids with different number of nodes/lines and density distributions have to be tested to achieve 1;
3. CV has to be done if there is any significant change in parameters, boundary conditions, and grid distributions, etc; and
4. CV is not a trivial effort in comparison to the physical validation study.

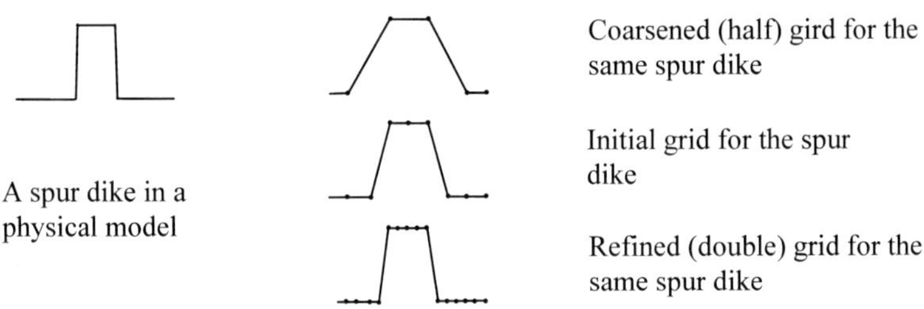

Figure 5.4.17 Boundary shape change due to grid refinement and coarsening

Unsteadiness

Since CV is to find numerical errors and convergence by comparison of the numerical solutions with different grids, it is relatively easy to do with steady problems. One has to run the computation from initial conditions given for these grids and compare the final results. The equivalence of these initial conditions is not very critical to the final comparison. If the problem is unsteady (with unsteady boundary condition: tidal wave; or naturally unsteady even with steady boundary conditions: vortex shedding, for instance), the equivalence of initial condition on these grids will be critical to the CV results.

Turbulence modeling

Turbulence modeling is needed for almost all flow simulations. The complexity of a particular model scheme represents the level of understanding of the flow physics. Normally, the turbulence scheme used does not affect the process of CV since it is considered only a mathematical exercise. However, the complexity will affect the time it takes to conduct the CV; more complex turbulence models (two equation schemes, Reynolds Stress models, for instance) involve more equations to solve and the solutions are more complicated which take more iterations or larger number of time steps with smaller step size to converge.

5.4.11 Conclusions

Most natural and manmade alluvial rivers have curved channels. The test case presented here is expected to have wide applications by both the developers and users of free surface flow models in simulating three-dimensional flows in curved channels to study not only the hydrodynamics, but also the associated sediment transport and morphodynamic processes.

As demonstrated, the test case is capable of determining a numerical model's capability in predicting the super-elevation distribution (Figure 5.4.7) the formation and reduction of secondary helical currents (Figures 5.4.5 and 5.4.6), of a three-dimensional free surface flow in a curved channel. The small vortex near the outer bank and the vertical profile of the longitudinal velocity could be simulated only with an appropriate turbulence closure scheme. The results from two versions of the CCHE3D model for this test case have demonstrated the need to further improve the numerical models.

Once the model is validated that it is capable of reproducing (approximately) the observed physics in physical experiments with a satisfactory mesh (grid), it is often questioned what is the numerical error related to the mesh used for the final results, and if the final results has converged to an asymptotical value or distribution. Confidence on the final solutions can be enhanced if one has positive answers. The step Calculation Verification is a procedure to analyze the computational data and provide answers to these questions. Instead of using an infinitesimally fine 3D mesh, it needs three sets of meshes with different resolutions. The general procedure and examples are given in Chapter 2. This test case provides a practical example. In addition to presenting the physical validation of a zero equation model and two two-equation (linear and non-linear) turbulence models in simulating experiment data, error and convergence are studied.

This example demonstrates that it is cost-effective and objective if the CV is conducted after the Physical Process Validation; general trend of convergence can be confirmed both visually from the plotted results and analytically by computing the errors and convergence rate due to mesh density. The analysis of water surface elevation, near bank vortex, longitudinal and transverse velocity distributions indicates the results from the finest mesh are converged with tolerable error. The convergence of the velocity field is around first order with mean errors around a few percent. Considering the complicity of the numerical model, irregular mesh distribution, and highly non-linear solution, this range of numerical errors is expected. In the highly non-linear part of the domain (vortex near outer bank) visual exam of convergence is more effective. The convergence rates computed by Equation 5.4.1 are more realistic for this test case. The evaluated local convergence rate and error demonstrate large variations particularly at locations where the solution is highly non-linear. It seems necessary to have a method to account for global convergence of the entire solution to enhance the current method.

5.4.12 References

Celik, I., and Karatekin, O., 1997, "Numerical experiments on Application of Richarson Extrapolation with Nonuniform Grids", ASME, Journal of Fluid Engineering, September, 1997, Vol. 119. No.9, pp584-590.

De Vriend, D.J., 1979, Flow measurements in a Curved Rectangular Channel, Laboratory of Fluid Mechanics, Department of Civil Engineering, Delft University of Technology, Internal Report No. 9-79.

Jia, Y., Blanckaert, K., and Wang, S.S.Y., 2001, "Simulation of secondary flow in curved channels". Published on the FMTM2001 International Conference, Tokyo, Japan.

Jia, Y., and Wang, S.S.Y., 1992, "Computational model verification test case using flume data". Hydraulic Engineering, pp436-441, ASCE.

Roache, P,J., 1998, *Verification and Validation in Computational Science and Engineering,* Hermosa Publishers, PO Box 9110, Albuquerque, New Mexico 87119-9110.

5.5 RIPRAP TEST FACILITY (OUTDOOR) TEST CASE

Contributors: Nigel G. Wright and H.P. Morvan

5.5.1 Introduction

Fully 3D CFD codes have been in use for nearly forty years, but it is only more recently that there has been wider interest in their use for analyzing the dynamics of river flows (Neary and Odgaard 1993, Shiono and Muto 1993, Meselhe et al. 1995, Thomas and Williams 1995, Hodkinson 1996, Sinha et al. 1998, Wu et al. 2003, Morvan et al. 2002). This interest arises because flows in river channels are often highly three-dimensional. Three-dimensional phenomena determine a channel's shape, its transformation and its conveyance.

Many complex features are present in open channel flow and each requires careful treatment in a computational model. In view of this, adequate testing and validation of a computer model is necessary before it can be generally applied to a naturally occurring channel. Benchmarks are valuable because they isolate particular features of the simulations for detailed testing and validation. In the past many flume experiments have been used to investigate the nature of open channel flow. These are restricted by their size, but measurements in actual rivers are difficult and are more prone to inaccuracies. Large-scale experimental facilities such as the Riprap Test Facility (RTF) at the Waterways Experiment Station (WES) in Vicksburg, Mississippi or the Flood Channel Facility (FCF) at HR Wallingford, UK can provide excellent data for testing numerical codes. In this section the case of the RTF is presented as a benchmark for three-dimensional simulations and some results are presented. These are used to indicate the areas of good agreement and areas where more work or particular care is required.

5.5.2. Objectives

The objectives of this test case are:

- To correctly predict flow through a series of bends of different orientation.
- To examine the prediction of secondary currents that are neglected in 2D layered models.
- To examine the implementation of roughness models in 3D CFD codes.
- To demonstrate the importance of sensitivity tests in a computational study.

5.5.3 Test Case

The Riprap Test Facility is described by Abraham (1991) who used data collected by Maynord (1998). The data were measured on the John L. Grace Riprap Test Facility (RTF) at the Waterways Experiment Station in Vicksburg, Mississippi whose main purpose was to carry out studies of riprap stability. The velocity and depth measurements provide an excellent set of benchmark data. The RTF consists of a trapezoidal channel with a bottom width of 3.66m (12ft) and a side slope of 1:2, vertical: horizontal (Figure 5.5.1). It is 237.7m (780ft) long and has a slope of 0.00216. The facility is shown in plan view in Figure 5.5.1. For these measurements the bed consisted of riprap with a range of diameters 13-57mm (0.5-2.5 inches).

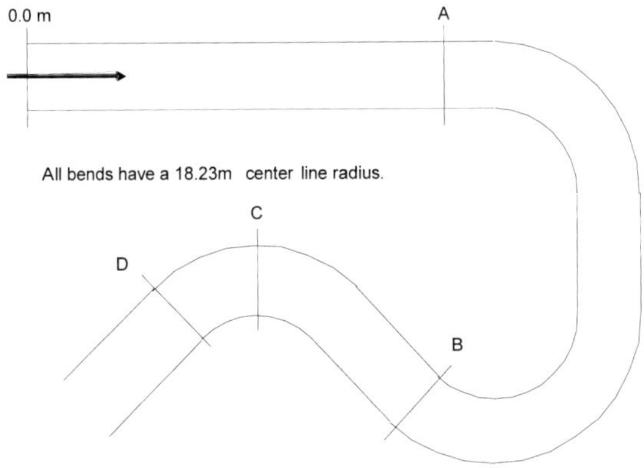

Cross-section label	Distance
A	54.24m
B	137.11m
C	163.92m
D	190.43m

Figure 5.5.1 RTF Channel Layout (based on Maynord, 1998)

Abraham (1991) used the data to evaluate a 2D layered model. He found reasonable, general agreement, but the predictions were unreliable near the walls which he attributed to poor turbulence modeling in the lateral direction. The use of a fully 3D approach in this work ought to give improvements from enhanced turbulence modeling.

5.5.4 Measured Velocity Data

At cross-section A (Figure 5.5.6) the velocity profile is symmetrical, and the isovels follow the walls. Obviously there are no secondary currents there, and the free surface profile is fairly flat.

At cross-section B (Figures 5.5.7 & 5.5.8) the maximum velocity is near the outer bank of the channel. The isovels can be seen close together near the outer wall at this section which is located after the second bend. The maximum velocity is 2.0 m/s. This imbalance is translated in a lateral velocity to the left, which generates a secondary current in the counterclockwise direction, of intensity equal to about 15% of the mean velocity.

At cross-section C (Figures 5.5.9 & 5.5.10), the apex of the third bend, the flow cuts across the bend and is mostly located on the inner side of the bend. The velocity field is slightly reduced. What is noticeable is that a reversal of the secondary helical current has occurred and that it is now turning rightward through inertia. This generates a secondary current in the clockwise direction equal to about 25% of the mean velocity.

At cross-section D (Figures 5.5.11 & 5.5.12) the flow has been transferred to the right as a result of the clockwise current transfer past the third bend. The velocity maximum is in the region of 1.75 m/s and the secondary current equal to about 15-20% of the velocity mean.

5.5.5 Description of Example Simulations

In presenting computer simulations it is important to detail the assumptions and specifications used in the model set-up. This should allow readers to evaluate the results and repeat the calculations. Care should also be taken to ensure that the quality of the numerical solution is satisfactory prior to examining the outputs in the light of comparative field measurements. This two-step process is too often reduced to just the second phase. In this context it is useful to refer to the ASME guidelines on evaluating the accuracy of CFD (Freitas 1993). In this document comparison with experimental measurements is step 10 of 10 and comes after an examination of various aspects of a numerical solution and comparison with analytic solutions. A more detailed discussion of verification and validation can be found in the book by Roache (1998)

Boundary Conditions

Correct specification of boundary conditions is an important part of modeling open channel flow. CFD codes are usually designed for applications in mechanical or process engineering and the default forms of boundary specification need to be adapted to the case of open channel flow. For the case under consideration here the following conditions were adopted.

1. Inlet

It is assumed that the distance between the inlet and the first bend is sufficient to allow the profile to settle down to equilibrium conditions. Examination of the results confirms that this is the case. To reduce stability problems in the iterative solution process a varying profile was used. The case considered here had a measured discharge of 4.248 m^3/s (150 cfs). The following equation was implemented and the constant C was adjusted to give the correct discharge:

$$U_{in} = C \ln (z/z_0) \quad \text{with } z_0 = {}^{k_s}\!/_{30}$$

This is not an exact equation for the velocity, but it is acceptable in view of the length of the reach leading up to the first bend, as stated above. The following equations were used to specify the turbulence quantities:

$$\overline{k_{in}} = 0.5\frac{u_*^2}{\sqrt{C_\mu}} = 0.0204\text{m}^2/\text{s}^2$$

$$\nu_{t\,in} = 0.067 \cdot u_* h$$

$$L = \frac{\nu_{t\,in}}{C_\mu \sqrt{2\overline{k_{in}}}} = 0.323\text{m}$$

with $u_* = 0.116\text{m/s}$.

2. Outlet

At the outlet it is assumed that a fully developed flow exists. In order to ensure this a length of straight channel (15.24m) is appended after cross-section D. The numerical procedure implements developed flow by applying zero derivatives in the streamwise direction and adjusting the velocities to ensure that the net flow in and out of the domain is zero.

The free surface is implemented by means of a rigid lid parallel to the bed at a slope of 0.00216. This common approximation has been used by others (Demuren 1993; Manson 1994, Cokljat and Younis 1995, Sinha et al. 1998, Morvan et al. 2002) to remove the computational complexity of solving for the location of the free surface and adapting the mesh at each numerical iteration. When this approach is taken the pressure field must be monitored to check that the position of free-surface is consistent, and in particular, that the lateral free-surface slope at bends is satisfactory.

3. Bed walls

At all solid walls a no-slip condition is applied so that the velocities are zero. To adequately resolve the boundary layer in the vicinity of the bed would require extremely fine grids that would make the solutions unfeasible. This is overcome in three-dimensional modeling through the use of wall functions that specify a logarithmic profile in this region. This profile is used to calculate the velocity and turbulence quantities in this region.

$$\frac{u_\tau}{u_*} = \frac{1}{\kappa}\ln(Ey^+)$$

where E is a function of the hydraulic roughness. This function is calculated from:

$$E(k_s^+) = \frac{1}{(1 + 0.3k_s^+)}$$

This has been found to compare well with the experimental data of Nikuradse (1933), and is similar to the expression used by Naot (1984). Further details on this are given in Morvan (2001).

In this work a commercially available code, CFX4, has been used. Where necessary this code allows the addition of user-defined FORTRAN subroutines to accomplish specific tasks and for more substantial alterations the University of Nottingham has access to the source code.

5.5.6 Discretization

The test case has been run with hybrid (Spalding 1972) and CCCT (Gaskell and Lau 1988) differencing for the advection terms. Figures at key cross-sections demonstrate differences between hybrid and CCCT differencing, especially in regions of sharp gradients. The solution is enhanced with CCCT, especially for the streamwise velocity isovels at cross-sections C and D (Figure 5.5.2). This difference emphasizes the importance of thorough checking of sensitivity to discretization. Although higher order solutions are more time-consuming and require greater attention to stability they often give significant improvements in solutions.

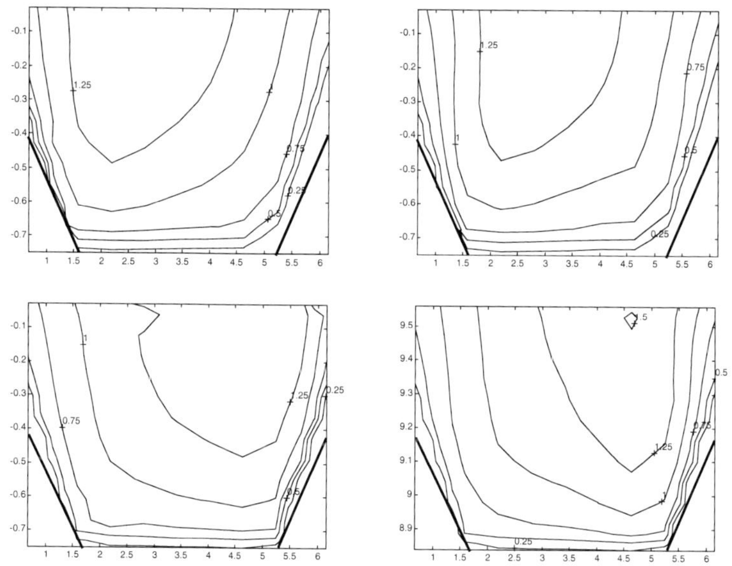

Figure 5.5.2 Comparison between Hybrid (left) and CCCT (right) solutions at Cross-Sections C and D with RSM

5.5.7 Grid Refinement

Tests have been carried out by doubling the grid resolution of the solution grid to reach 135,828 elements. The non-dimensional pressure term Cp along the channel centreline at the bed has been satisfactorily compared at the bed for both coarse and fine grid (Figure 5.5.3). A comparison of the non-dimensionalized pressure, Cp, provides a measure of the grid independence of the numerical solution because of the influence of pressure on the solution. The aim of this comparison is to demonstrate that the numerical solution presented hereafter is independent of the grid size.

The solutions obtained with both grids at cross-section C for the main velocity field have also been successfully compared.

5.5.8 Sensitivity Analysis

The wall function approach was detailed above. Within this a value of the local roughness height k_s is required. As in experimental work the model should be tested to calibrate this value within the range appropriate for the physical situation. The effect of roughness height should be monitored not only in terms of the velocity change in the domain, but also in terms of the pressure evolution on the lid.

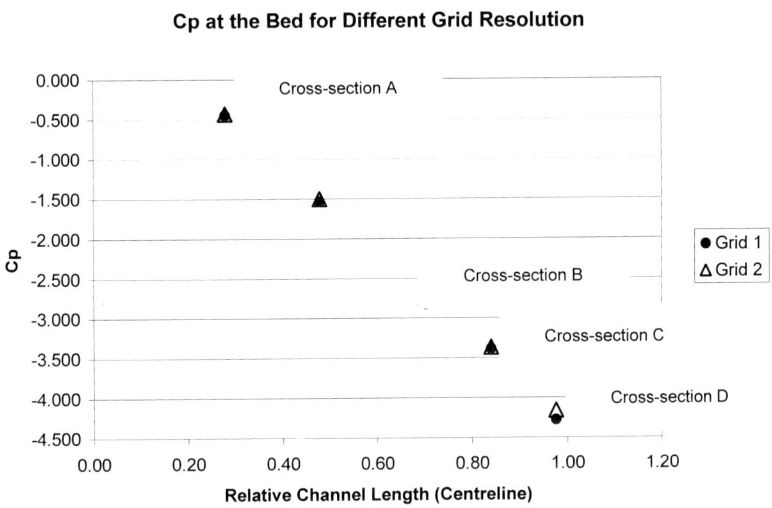

Figure 5.5.3 Grid independence Test Results

Monitoring the correct position of the fixed lid is essential to achieving a correct solution. If the fixed lid does not represent the physical situation accurately there will be a higher pressure where the lid should be higher, for example. This higher pressure will correspond to the pressure equivalent of the head of water that would be there if the lid were raised. In this way the effect on the momentum equation is only slight unless the difference is significant relative to the depth. However, the effect on the mass conservation equation can be much greater as no account is taken of the extra discharge for the extra depth. Thus the effect of a badly defined free surface will be seen throughout the domain.

Tests of sensitivity to k_s value have been carried out with a RSM turbulence closure on the basis that this model yielded a better comparison with the measurements. It was found that a value of $k_s = 0.10$ m was suited to reproduce the longitudinal free-surface profile using the pressure gradient calculated by CFX. Other values show that for lower roughness the pressure slope became less steep, which is physically consistent with open-channel flow behavior (Figure 5.5.4).

If the pressure term is very sensitive to the value of roughness, the velocity field is less affected by its choice. This is especially true when roughness is low, as in the UK Flood Channel Facility cases investigated by the authors (Morvan et al. 2002). In the present work the impact of varying roughness is again little noticeable, with the exception of an increased recirculation in the region of cross-section B (although it is impossible to say whether this constitutes an improvement as the real recirculation is not known - Maynord 1998). At cross-sections B and C the region of maximum velocity is slightly squeezed towards its centre, and the velocity at the walls is slightly reduced.

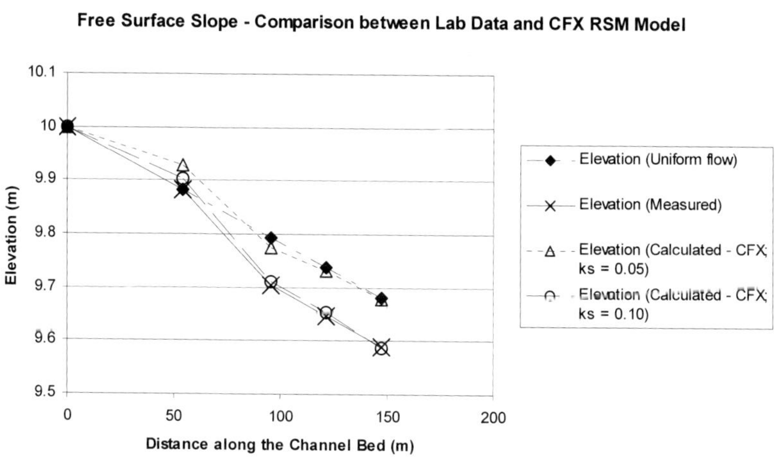

Figure 5.5.4 Longitudinal Free Surface Profile for Different Roughness Values

Earlier work conducted at the University of Nottingham (Swindale 1999) and the University of Glasgow (Morvan 2001) has indicated that rigid lid models are not very sensitive to inlet boundary conditions. A comparison of models run either with an inlet velocity profile and using periodic boundary conditions yielded almost identical results. However, periodic boundary conditions, although more in accordance with the fully developed flow condition assumption, are more difficult to utilize and prone to generate numerical instability. For this reason, an inlet flow condition was adopted here, especially as the flume was particularly suited to this in view of its long straight channel at the entry. Additionally this approach remains the more practical for application of CFD to river engineering problems.

Two turbulence models were used during the course of this work: the standard k-ε model (Launder and Spalding 1974) because of its widespread use and robustness, and a RSM (Launder et al. 1975) because of the level of detail it encompasses. In this work the simplified Launder et al. (1975) version of the Reynolds Stress model has been used with no wall-reflection terms. A converged solution was rapidly attained with the k-ε model. The Reynolds Stress simulation was started from a converged k-ε solution to aid convergence which was considerably more difficult than with the k-ε model.

The first outputs from RSM showed an improvement of the solution compared with that using k-ε, especially in the region of cross-section B. At this cross-section, the isovels close to the surface are not squeezed against the outer bank as they are with k-ε, but deflected towards the channel centre. Overall however, the solution obtained with the k-ε model is quite similar to that with RSM, and compares with the measured data.

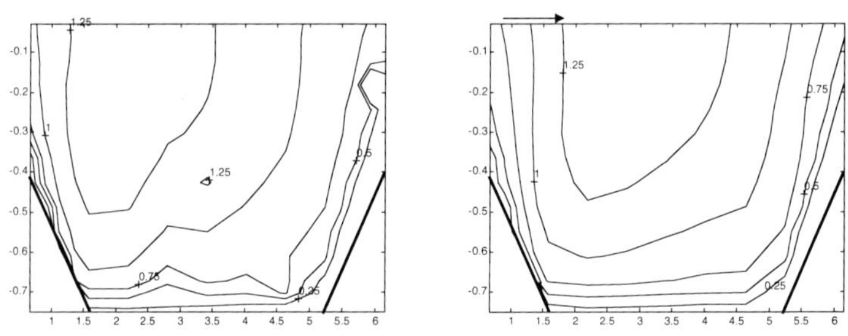

Figure 5.5.5 Outputs for Cross-Section C using k-ε (left) and RSM (right)

It can be concluded that the RSM yields a better result, as also demonstrated by Basara and Cokljat (1995). RSM is better able to cope with channel curvature as seen at cross-section C for example (Figure 5.5.5). Yet it remains costly and has problems in achieving convergence, which restricts its use. For the level of detail and the scale of the RTF -and consequently for large river application - it should be said that the k-ε performed reasonably well. It proved more robust, faster to run, and it provided a satisfactory picture of the flow. Further work should evaluate the use of the various modified forms of the k-ε model, especially its anisotropic versions. This could give a valuable compromise between robustness and accuracy.

5.5.9 Results

Four cross-sections have been analyzed and compared in detail (see Figure 5.5.1). They are located at 54.24m, 137.11m, 163.92m and 190.43m along the channel centerline. At each cross-section main velocities, secondary currents, and free surface profile are examined.

CFX Velocity Results

The modeled flow at section A (Figure 5.5.6) compares very well with the data, which demonstrates the validity of the inlet boundary conditions.

At cross-section B (Figures 5.5.7 & 5.5.8), the velocity can be clearly seen at a maximum against the outer side of the channel. However, the region of high velocity is more widely spread, which gives a reduced velocity maximum compared to the observed maximum. The velocity transfer to the left is well reproduced, although with slightly larger intensity. This generates a satisfactory secondary current picture. The intensity of the recirculation is about 10-15% of the velocity mean.

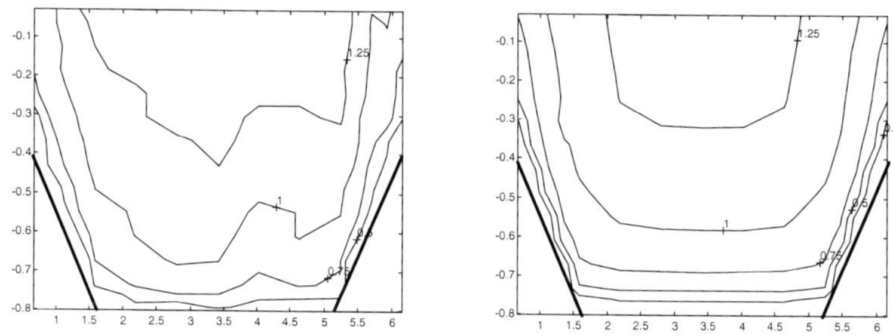

(a) Observation (b) Simulation

Figure 5.5.6 Observed and Calculated Normal Velocity at Cross-*section A (SI Units)*

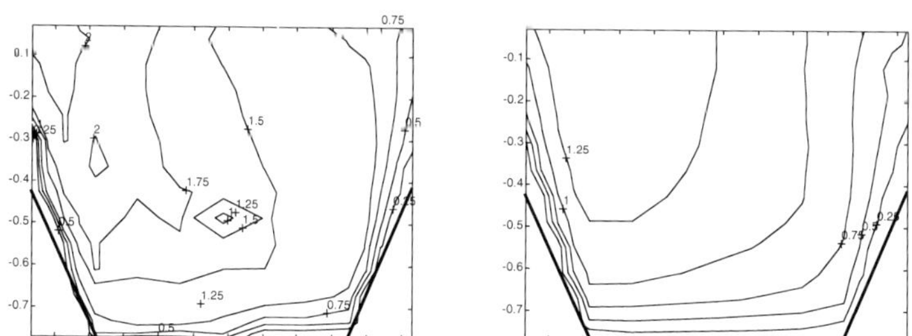

Observed and Calculated Normal Velocity at Cross-section B (SI Units)

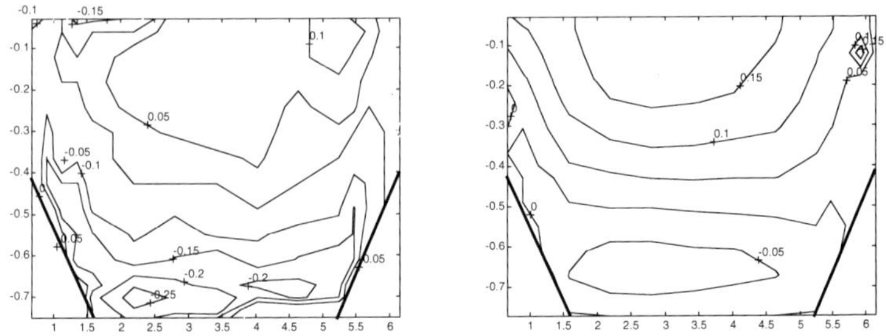

Observed and Calculated Tangential Velocity at Cross-section B
(SI Units, positive to the left)

Figure 5.5.7 Observed and Calculated Velocities at Cross-section B
(SI Units, positive to the left)

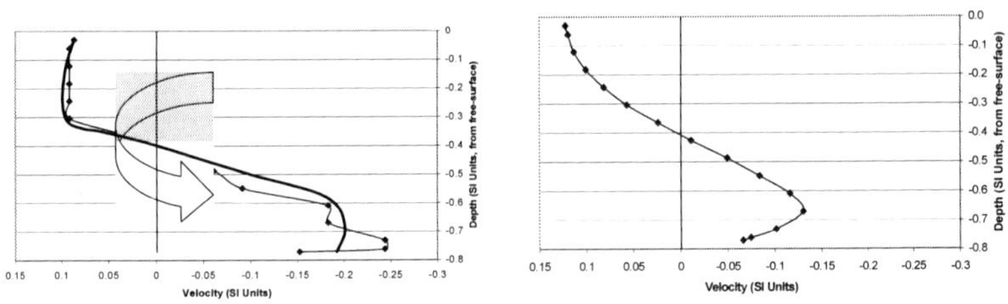

Figure 5.5.8 Observed and Calculated Secondary Currents at Section B

At cross-section C (Figures 5.5.9 & 5.5.10) the model outputs show a picture similar to that of section B. The velocity pattern is correctly reproduced, however, the main velocity core is larger and of lower intensity (-30%). The velocity transfer to the right is correctly modeled, although with a lower intensity (20% of the mean velocity instead of about 25%).

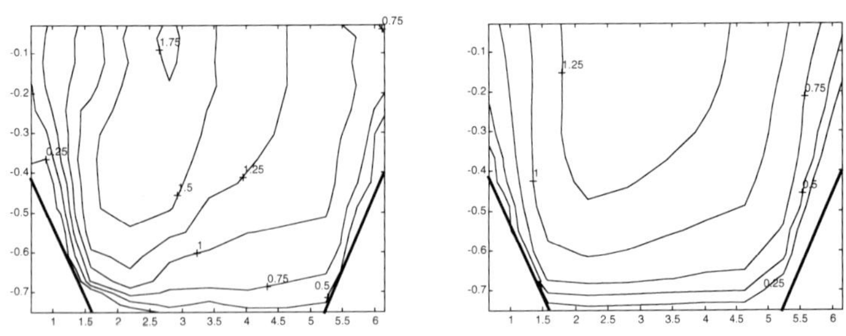

Observed and Calculated Normal Velocity at Cross-section C (SI Units)

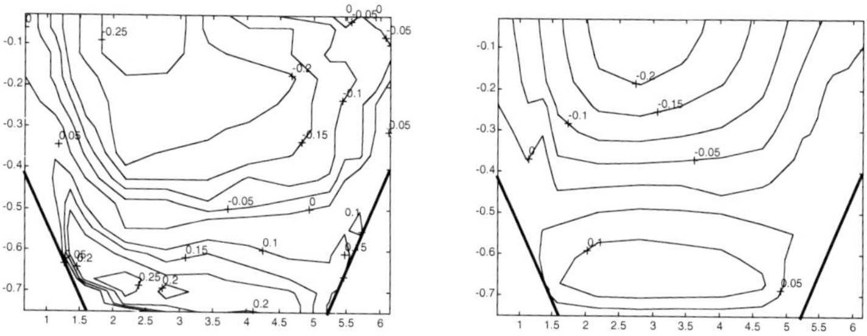

O
Observed and Calculated Tangential Velocity at Cross-section C (SI Units, positive to the left)

Figure 5.5.9 Observed and Calculated Velocities at Cross-section C (SI Units, positive to the left).

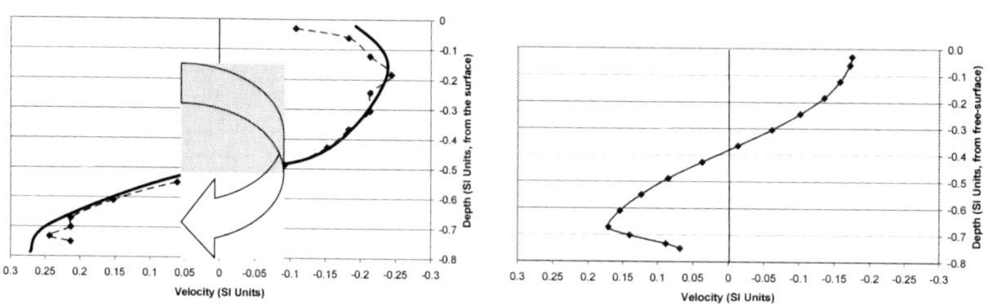

Figure 5.5.10 Observed and Calculated Secondary Currents at Cross-Section C

At cross-section D (Figures 5.5.11 & 5.5.12), the position of the velocity core is correct, but once again larger and with a lower velocity maximum (1.5 m/s). Of more concern is the fact that no secondary current is modeled at all. This may well be an effect of the boundary conditions located 15.24 m downstream.

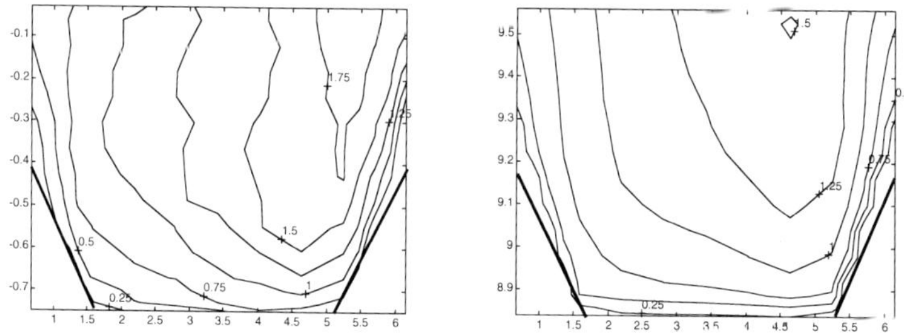

Observed and Calculated Normal Velocity at Cross-section D (SI Units)

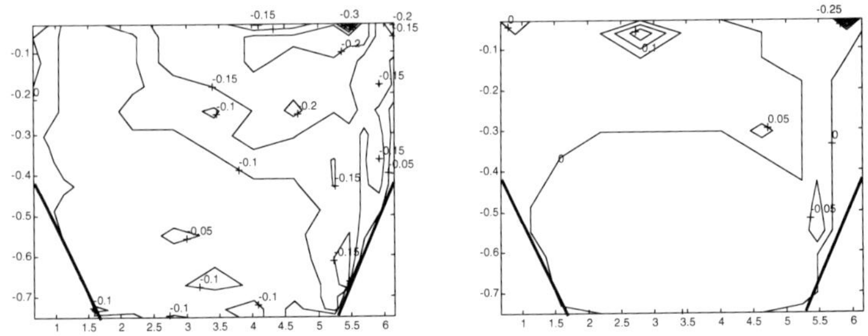

Observed and Calculated Tangential Velocity at Cross-section D (SI Units, positive to the left)

Figure 5.5.11 Observed and Calculated Velocities at Cross-section D (SI Units, positive to the left).

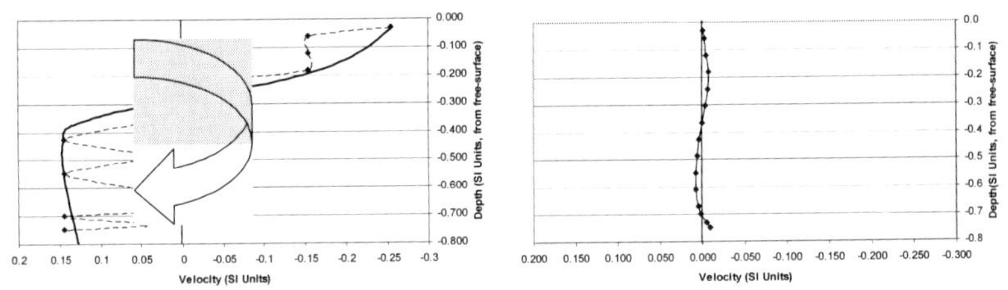

Figure 5.5.12 Observed and Calculated Secondary Currents at Cross-Section D

In general the main flow features are reasonably well modeled, although there seems to be some diffusion in the solution which reduces the velocity maximum. This entails error of about 30% in certain locations. Secondary currents and their reversal are well modeled. Their intensity is usually correct (within 20% error) although according to Maynord (1998) the experimental measurements lack some accuracy in the transverse direction.

5.5.10 Comparison of Free Surface Position and Pressure Field on the Lid

In Figure 5.5.13, three sections are examined: B, C and D. The pressure field from the numerical model is extrapolated using a hydrostatic assumption to obtain a water head to add to the lid position.

A look at the three profiles shows that the comparison between the measurements and the model data is satisfactory and consistent with the velocity data. In general the free-surface extrapolation from the model data shows a lower free-surface level at the edges. This could be related to the lower velocity maximum calculated by the model. Overall error in water level appears to be within 10 mm (as opposed to 3mm in the measurements), which means an error of 1.25% on the total flow depth, which is negligible.

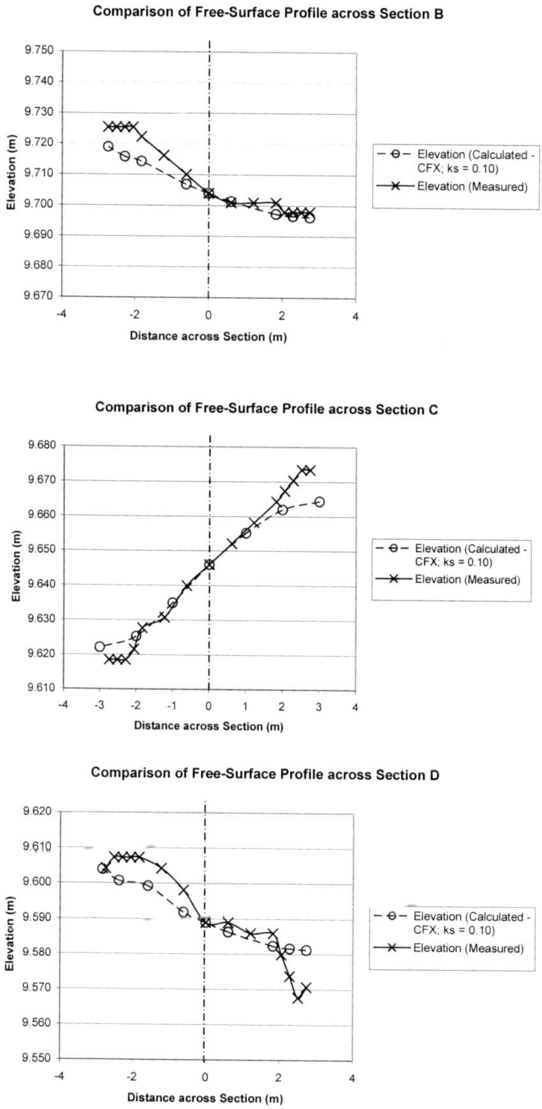

Figure 5.5.13 Comparisons of measured transverse water surface profile and computed pressure head in three cross-sections.

5.5.11 Conclusions

The Riprap Test Facility test case has been described and the application of a fully three-dimensional CFD code has been demonstrated. This application required little calibration and gave an accurate simulation of the free surface and in most cases predicted secondary circulations well (within 15%). In terms of the CFD approach it can be said that discretization and choice of turbulence model are important. RSM gives enhanced solutions and its use would appear necessary for accurate solutions in these cases. Grid independence was established with a feasible grid size. The data forms a good benchmark and test for CFD codes particularly in terms of turbulence models and representation of roughness. This work has demonstrated the need to carry out sensitivity test on various modeling parameters.

5.5.12 Acknowledgements

The second author gratefully acknowledges the receipt of a Ph.D studentship funded by EPSRC grant No. GR/L95038. Numerical simulations were conducted at the University of Glasgow on a Sun Station Ultra 10/433 with 384 MB. The authors would also like to gratefully acknowledge the contribution of data from Stephen Maynord of USACE.

5.5.13 References

Abraham, D, (1991), Analysis and Verification of a Three-dimensional Hydrodynamic Numerical Model, Masters Thesis, Texas A&M University.

Basara, B, and Cokljat, D, (1995), "Reynolds-Stress Modeling of Turbulent Flows in Meandering Open Channels", FED-Vol. 221, Industrial and Environmental Applications of Fluid Mechanics, ASME, pp. 27-32.

Cokljat, D, Younis, BA,(1995), Compound-Channel Flows - A Parametric Study Using A Reynolds-Stress Transport Closure, J Hydraul Res, 33: (3) 307-320.

Demuren A.O, 'A numerical model for flow in meandering channels with natural bed topography', Water Resources Research, Vol. 29, No. 4, 1993.

Freitas, CJ, "Editorial Policy Statement on the Control of Numerical Accuracy", Journal of Fluids Engineering, 155, pp. 339-340, 1993.

Gaskell, PH, and Lau, AKC, (1988) "Curvature Compensated Convective Transport: SMART, a New Boundedness Preserving Transport Algorithm", Int. J. Num. Meth. In Fluids, Vol. 8, pp. 617-641.

Hodkinson A., 'Computational Fluid Dynamics as a Tool for Investigating Separated Flow in River Bends', Earth Surface Processes and Landforms, Vol. 21, 1996.

Launder, B. E. and Spalding, D. B. (1974). The numerical computation of turbulent flows', Computational Methods in Applied Mechanics and Engineering, Vol. 3, pp. 269- 289.

Launder, B.E., Reece, G.J., Rodi, W., (1975), " Progress in the Development of a Reynolds Stress Turbulence Closure", J. Fluid Mech., Vol. 68, p. 537.

Manson, J. R., (1994), "The Development of a Predictive Procedure for Localised Three Dimensional River Flows" Ph.D. Thesis, The University of Glasgow.

Maynord, S., (1998), Private Communication.

Meselhe E. A., Sotiropoulos F., Patel V. C., 'Three Dimensional numerical model for open channel flows'. Proceedings of the International Conference on Hydropower, 1995.

Morvan, H., (2001), "Modelling River Flood using Computational Fluid Dynamics", PhD Thesis, The University of Glasgow, UK.

Morvan, H., Fender, G, Wright, N.G., Ervine, D.A., (2002), "Three Dimensional Hydrodynamics of Meandering Compound Channels", ASCE J. Hydr. Eng., 128(7), pp. 674-682, 2002.

Naot, D., (1984), "Response of Channel Flow to Roughness Heterogeneity", J. Hydr. Eng., Vol. 110, No. 11, ASCE.

Neary, V.S., and A.J. Odgaard. (1993) "Three-dimensional flow structure at open-channel diversions." Journal of Hydraulic Engineering. 119(11), 1223-1230.

Nikuradse, J., (1933), Laws of Flow in Rough Pipes, Verein deutscher Ingenieure, Forschungsheft No. 361, Berlin, 1933.

Roache, P.J., (1998), Verification and Validation in Computational Science and Engineering, Albuquerque, NM, Hermosa Publishers.

Shiono, K., and Muto, Y., (1993), "Secondary Flow Structure for In Bank and Over-Bank Flows in Trapezoidal Meandering Compound Channel", Refined Flow Modelling and Turbulence Measurements, Paris, 7-10 Sept., pp. 645-652.

Sinha S. K., Sotiropoulos F., Odgaard A. J., 'Three Dimensional Numerical Model for Flow through Natural Rivers', Journal of Hydraulic Engineering, Vol. 124, No. 1, 1998.

Spalding, D.B., (1972), "A Novel Finite-Difference Formulation for Differential expressions Involving both First and Second Derivatives", Int. J. Num. Meth. Eng., Vol. 4,p.551.

Swindale, N.R., Numerical Modelling of River Rehabilitation Schemes, Ph.D. Thesis, University of Nottingham, 1999.

Thomas, T.G., Williams, J.J.R., (1995), Large Eddy Simulation of a Symmetric Trapezoidal Channel at a Reynolds number of 430,000, J. of Hydr. Res., 33(6).

Wu W.M., Rodi W., Wenka T., 3D Numerical Modeling of Flow and Sediment Transport in Open Channels, J. Hydr. Eng. ASCE 126: (1) 4-15, 2000.

5.6 FLOW IN A CHANNEL WITH A SPUR DIKE TEST CASE

Contributors: Yafei Jia and Sam S. Y. Wang

5.6.1 Introduction

Manmade structures are installed in natural rivers, streams and waterways for river stabilization, restoration, rehabilitation, navigation safety, flood control, bank protection, and transportation infrastructures, etc. Flow fields in the vicinity of these structures are inevitably complicated by the presence of these obstacles. The approaching flow is decelerated and a stagnation zone is formed in front of the structure. A down wash along the front surface induces secondary vortex-like currents, the so-called horse-shoe vortex wrapping around the toe of the structure and extending toward downstream. Behind the structure, the turbulence intensity increases, and a wake or recirculation zone and vortex shedding are often observed. In short, the flow field around a manmade structure in a stream is complex, unsteady and three-dimensional. Therefore, one must make sure that a three-dimensional hydrodynamic model capable of simulating most, if not all, of the key flow mechanisms is used to carry out the investigations for research, engineering analysis and/or design of in-stream structures.

The validation test can be conducted by comparing the simulated results with laboratory experimental measurements and/or field data collected around an in-stream structure. If the field measured data is complete and reliable, it is obviously the first choice for the validation test. The fact is however that the field data available at present is mostly inadequate for model validation purposes, and laboratory data still provide the most reliable information for validating models. The laboratory experiments of Rajaratnam and Nwachukwu (1983) are adopted to the design of this validation test. Even though it is a relatively simple case of a flow around a single spur dike attached perpendicularly to one side wall of a straight open channel with uniform, rectangular cross-section, the flow field does include most of the key flow features important to realistic validations of a free-surface flow model. Furthermore, the case is simple and requires minimal effort to construct the numerical grids and conduct the tests.

5.6.2 Objectives

This test case is designed to determine whether a numerical model is capable of predicting the following flow characteristics:

- The velocity profiles at several cross-sections around a spur dike in an open channel.
- The shear stress distribution on the channel bed.
- The size and shape of the recirculation zone and/or the length of reattachment (the distance from the dike to where the flow is reattached to the wall).

- The vortices generated by the downwash in the close vicinity of the tip of the dike.

5.6.3 The Experimental Set Up and Measured Data

The set up of physical model conducted by Rajaratnam and Nwachukwu (1983) is shown in Figure 5.6.1. One spur dike of a thin plate perpendicular to the longitudinal direction was mounted vertically in a straight and uniform channel with a width of 0.914 m. The physical model dimensions and flow conditions of the experiments are listed in Table 5.6.1. It is seen that the recirculation zone length is about 12 times as long as the length of the dike, consistent with observations of several other experiments. Smooth bed and sidewalls were used in the physical model.

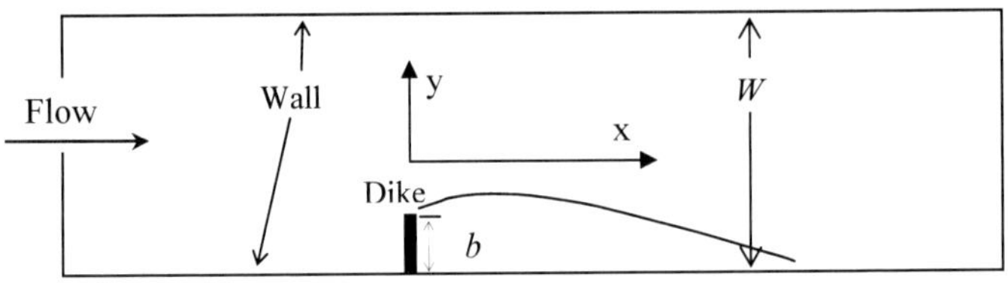

Figure 5.6.1 Sketch of the experimental flume and simulation

Table 5.6.1 Flow Conditions of the Experiment

Expt.	Q(m^3/s)	H(m)	V(m/s)	Fr	b(m)	Remark	τ_{00}(N/m^2)
A1	0.0453	0.189	0.25	0.19	0.152	Smooth bed	0.129
A2	0.0453	0.223	0.216	0.15	0.152	Smooth bed	0.105
A3	0.0453	0.256	0.192	0.12	0.152	Smooth bed	0.081
A4	0.0453	0.152	0.32	0.26	0.152	Smooth bed	0.239
A5	0.0453	0.152	0.244	0.20	0.152	Smooth bed	0.134
B1	0.0453	0.152	0.351	0.29	0.0762	Smooth bed	0.239
B2	0.0453	0.219	0.229	0.16	0.0762	Smooth bed	0.115
B3	0.0453	0.189	0.238	0.17	0.0762	Smooth bed	0.115
C1	0.0447	0.152	0.326	0.27	0.152	Sand paper (ks=0.56mm)	0.287
C2	0.0453	0.223	0.223	0.15	0.152	Sand paper (ks=0.56mm)	0.11
D1	0.0453	0.219	0.229	0.16	0.152	Hemispherical roughness (ks=6.3mm)	0.287

Q=discharge; H=depth of flow; V=mean velocity; Fr=Froude Number, b=length of dike; τ_{00}=approach bed shear stress, k_s=bed roughness.

5.6.4 Validation Test Procedure

The computational mesh size distribution should be designed such that the three-dimensional flow details, especially near the dike can be captured. It is expected that the nodal spacing near the bed, the wall, and especially the dike, should be fine enough for implementing the law of wall boundary condition. Since the experiments were conducted under steady flow conditions, constant discharge at inlet and water surface elevation at the outlet boundary are specified. One should use the standard physical parameters for water such as density, kinematic viscosity, and von Karman constant, etc. for his/her simulation.

The computational domain should be the same as the configuration of the experiment as shown in Figure 5.6.1. The velocity field and shear stress on the bed surface were measured under a steady inflow condition. The flow distribution at the inlet section is unknown; therefore the simulation channel length upstream of the spur dike should be sufficiently long to allow the simulated flow to be fully developed in this approach channel, if one applies a uniform flow distribution at the inlet section. The channel length downstream of the dike should also be sufficiently longer than the expected reattachment length so that the accuracy of the simulated recirculation zone would not be affected by the constant surface elevation boundary condition prescribed at the outlet. The spur dike in the physical model was made of a thin plate; the mesh should be refined in its vicinity. A two-dimensional mesh for discretizing the channel plane configuration has been suggested (Figure 5.6.4). The nodal number and distribution in the vertical can be determined, noting that near the channel bed the mesh should be refined as well.

5.6.5 Physical Model Data - Available Velocity Field

Velocities in the longitudinal direction at four equally spaced cross-sections behind the dike were measured. The shear layer and shape of the recirculation zone are clearly seen from these velocity profiles (Figure 5.6.2). The measured velocity profiles in the recirculation zone are similar to each other. The near bed velocity showed a strong peak in the shear layer due to the presence of the dike, which decreased downstream and upward. The magnitude of velocity near the surface (0.85H, H is the water depth) is higher than those near the bed (0.03H) in both the main flow and backward flow directions. The general velocity distribution in both levels look similar, especially the gradients of velocity in the shear layer. It is seen that a peak velocity occurs between the shear layer and the main flow. The length of the recirculation due to this thin plate dike was close to 12*b*, (12 times the dike length, *b*).

Not all the data of the experiments described in this test case (Table 5.6.1) have been published by Rajaratnam and Nwachukwu (1983). The test case contributors have obtained a rather complete data set from Prof. Rajaratnam, and the data have been tabulated and made available to the interested users.

5.6.6 Bed Shear Stress

Shear stress distribution around the spur dike is of engineering importance, because it directly causes local scouring. This measured shear stress distribution in the vicinity of the dike is shown in Figure 5.6.3. The measured shear stress was obtained using a Preston tube (Rajaratnam and Muralidhar, 1968; Rajaratnam and Nwachukwu, 1983). The data indicates that the shear stress dramatically increased to 4.9 times as much as that of the approach flow, resulted from flow acceleration due to the presence of the spur dike. Predicting the shear stress with reasonable accuracy is a pre-requisite of a numerical model's capability to simulate the scouring process.

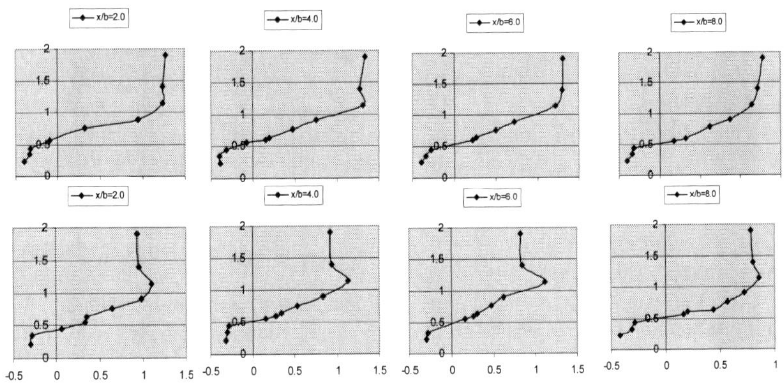

Figure 5.6.2 Measured velocity profiles across the shear layer (after Rajaratnam and Nwachukwu, 1983).

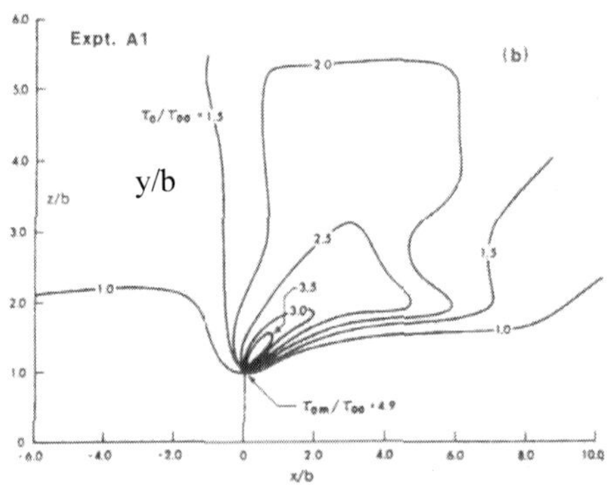

Figure 5.6.3. Measured shear stress distribution (after Rajaratnam and Nwachukwu, 1983).

5.6.7 Test Example

Jia and Wang (1996) have conducted a numerical simulation of this experiment using a three-dimensional model, CCHE3D, developed at National Center for Computational Hydroscience and Engineering. Figure 5.6.4 shows a mesh used for the computation. The nodal density near the dike was much higher than those used elsewhere in order to capture the fast variation of the flow properties. Some mesh lines have been omitted in this plot for clarity of the mesh system. This model solves three dimensional Reynolds equations with k-ε turbulence closure. The flow velocities, shear stress distribution on channel bed and the reattachment length of the recirculation zone were simulated.

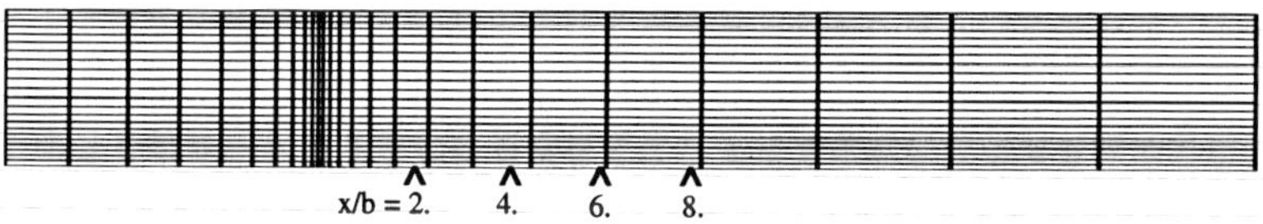

Figure 5.6.4 A computational mesh for the spur dike case (Some lines are omitted for clarity).

5.6.8 Boundary Conditions

Boundary conditions for simulating this test case are relatively simple. Steady inflow discharge, constant outlet surface elevation and wall boundary approximation (the Law of the Wall) are sufficient for the hydrodynamic model. The discharge was converted to velocity as the actual boundary condition at the inlet for momentum equations. For this simulation, a parabolic lateral distribution of specific discharge or flow discharge per unit width was specified, and the vertical distribution of longitudinal velocity was assumed to be logarithmic at the inlet section. Along the outlet cross-section, longitudinal gradients of all variables were set to zero: $\partial/\partial x=0$. On the water surface, the gradients of all variables in vertical direction were also assumed zero: $\partial/\partial z=0$.

Boundary conditions for turbulence closure schemes should be given consistent with the nature of the flow. Rastogi and Rodi (1978) used a uniform distribution at the inlet section by applying depth averaged values k and ε Leschziner and Rodi (1979) specified inlet boundary conditions of k using experimental data and $\varepsilon=C_m^{3/4}\, k^{3/2}/l_m$, where C_m=0.09, and l_m is a mixing length. The semi-empirical relationships of turbulence intensity for open channel flow (Nezu and Rodi, 1986) were adopted for estimating kinetic energy. One could obtain boundary values of kinetic energy dissipation using the kinetic energy and eddy viscosity. Turbulence eddy viscosity in the inlet cross-section was assumed to be of a parabolic distribution as suggested by experimental data (Nezu and Rodi, 1986). In regions very close to the channel bed as well as the vertical walls and dike surfaces, the law of the wall was applied for momentum equations, and local equilibrium of turbulence energy k and dissipation ε was assumed. The water surface elevation is adjusted using the pressure gradient near the surface (Leschziner and Rodi, 1979; Rastogi and Rodi, 1978). Vertical discretization does not change with water surface; the computation domain, therefore, is covered by a "rigid lid".

5.6.9 Simulation Results

Figure 5.6.5 shows comparisons of the simulated velocity field and the measured data. These comparisons are along the lines behind the dike in transversal direction (Figure 5.6.1 and 5.6.4). The agreement is reasonably good and the velocity predictions in the main channel are more accurate than those in the shear layer. For the lower level (0.03H), higher error appears along the transition zone. The difference in the recirculation zone is higher for the level 0.85H. The measured negative velocities in the recirculation region are almost constant in a very wide area, $2 \leq X_r / b \leq 8$.

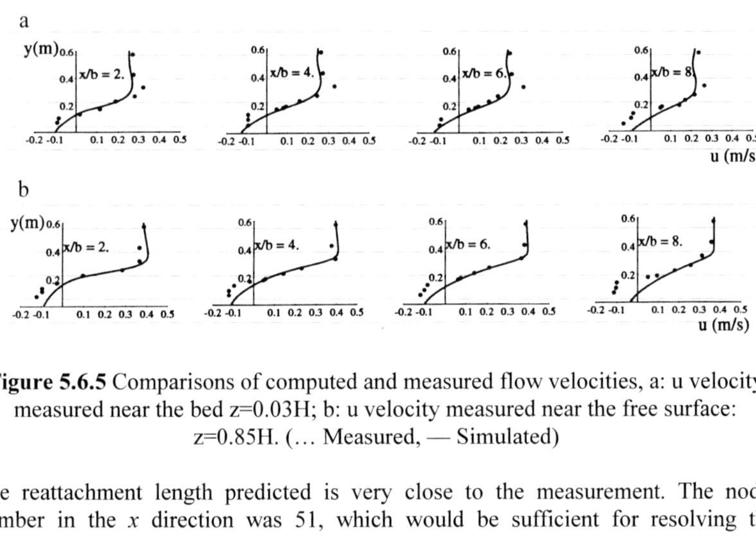

Figure 5.6.5 Comparisons of computed and measured flow velocities, a: u velocity measured near the bed z=0.03H; b: u velocity measured near the free surface: z=0.85H. (… Measured, — Simulated)

The reattachment length predicted is very close to the measurement. The nodal number in the x direction was 51, which would be sufficient for resolving the recirculation flow (Tingsanchali and Maheswaran, 1990). Three meshes with different nodal numbers in the y direction (25, 45, 61) were used to examine the sensitivity of the simulated recirculation length. It was found that the calculated reattachment could be improved when mesh density was increased in the y direction and 45 nodes seemed to be sufficient because further increase of node number had negligible effect on the resultant flow field. This indicates the numerical solution has shown a trend of convergence during a grid refinement process.

Figure 5.6.6 shows that the predicted shear stresses are generally in good agreement with the measurements, except in the small region near the tip of the dike. It is also observed that the location of the maximum shear stress obtained by the simulation are slightly upstream of the dike tip (x=0) while measured maximum values are located a little downstream of the tip.

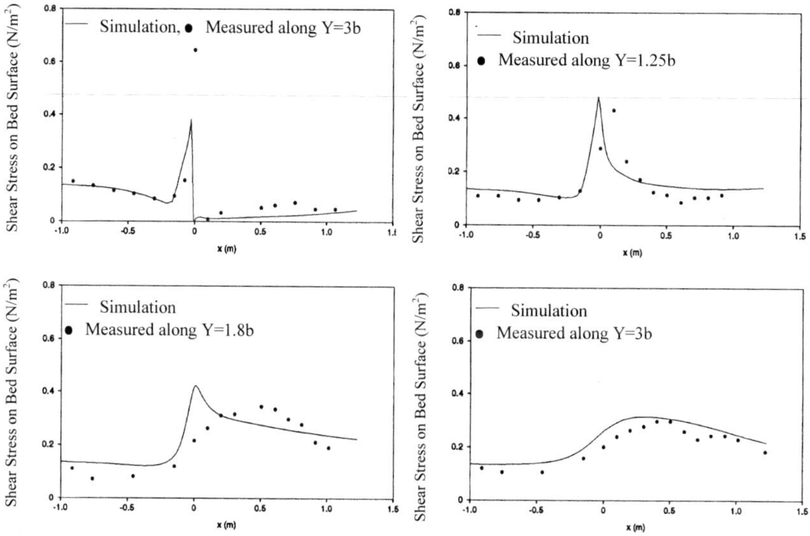

Figure 5.6.6 Comparisons of simulated and measured bed shear stresses in several longitudinal sections.

5.6.10 Conclusions

By presenting an actual example of a validation test following the suggested test procedure, the prospective user may be able to follow this approach and estimate the amount of effort required, as well as, see the usefulness of this validation test case.

The results of this test show that the velocity field predicted or reproduced by the selected model, CCHE3D, is in general agreement with the measured ones; however, one can also see the discrepancies between the two in the recirculation zone. This finding indicates that a model of higher level of sophistication to reproduce the shear flow more accurately is needed, if it is essential to the investigated problem. This simple validation test also indicates that a comparison of computed and measured reattachment length only is not sufficient to tell a model's ability of producing highly fidelity solution.

Similarly, the differences between the measured and the simulated shear stress distributions on the channel bed (Figure 5.6.6) show the limitation of a model using a certain turbulence closure scheme and the need for further mesh refinement in vertical direction, although general agreement is satisfactory.

5.6.11 References

Jia, Y. and Wang, S.S.Y., 1996, "Verification of 3D Flow Model Using Laboratory Data", Proceedings of and Presented at North American Water and Environment Congress in Anaheim, CA, CD-ROM.

Leschziner, M.A., and Rodi, W., 1979 "Calculation of Strongly Curved Open Channel Flow", *Journal of the Hydraulics Division,* ASCE, Vol. 105, No. HY10, ppl297-1313.

Nezu, I., and Rodi, W., 1986, "Open-Channel Flow Measurements with a Laser Doppler Anemometer", *Journal of the Hydraulic Engineering,* ASCE, Vol. 112, No. 5, pp335-355.

Rajaratnam, N., and Nwachukwu, B.A., 1983, "Flow Near Groin-Like Structure", *Journal of Hydraulic Engineering,* ASCE, Vol. 109, No. 3, pp463-480.

Rastogi, A.K., and Rodi, W., 1978, "Predictions of Heat and Mass Transfer in Open Channels, *Journal of the Hydraulics Division*, ASCE, Vol. 104, No. HY3, pp. 397-420.

Rajaratnam, N., and Muralidhar, D., 1968, "Characteristics of the rectangular free over fall", *Journal of Hydraulic Research*, Vol. 6, No. 3, pp233-258.

Rajaratanm, N., and Nwachukwu, B., 1983 "Flow near grion-like structures", *Journal of Hydraulic Engineering*, ASCE, Vol. 109, No. 3, pp 463-480.

Tingsanchali, T. and Maheswaran, S., 1990, "2-D Depth-Averaged Flow Computation Near Groyne", J. Hyd. Engng., ASCE, 116(1), 71-86.

5.7. FLOW AROUND A SUBMERGED TRAPEZOIDAL SPUR DIKE TEST CASE

Contributors: ***Yafei Jia, Sam S. Y. Wang, Roger A. Kuhnle and Carlos V. Alonso***

5.7.1 Introduction

Spur dikes have been widely used for bank protection, river training, and aquatic habitat rehabilitation. In the field, a practical and economical way to build such a structure is to use riprap rocks. As a result, spur dikes are often trapezoidal shaped and are often submerged during high flows. Under these conditions, the flow structure around a submerged dike is more complicated than those resulting from emergent dikes with a simpler shape (a plate for instance) reported previously. A comprehensive data set has been measured at the National Sedimentation Laboratory, of the United States Department of Agriculture, Agricultural Research Service (ARS), with three-dimensional velocities measured at 2592 points around the dike. It provides an opportunity for numerical model developers and end users to validate models being developed or applied, and enhance their confidence on the models' capabilities of simulating highly turbulent flows around hydraulic structures.

5.7.2 Objectives

Because the flow is around a trapezoidal shaped dike with high side slopes, the vertical motion and acceleration of the flow are not negligible. The turbulent flow structure and body-fitted mesh wrapping around the dike demand the model to be robust, adequate, and comprehensive in order to produce reasonable results. A three-dimensional numerical model would be able to reproduce the measured vector field reasonably if it has the capabilities such as dynamic pressure, turbulence closure model and transformation for irregular mesh. Although the case is about free surface flow, the surface elevation change in the physical model is not large; a "rigid lid" approach (if one has to use it) would be acceptable, although the capability of computing free surface is preferred.

The main objective of this case is to validate a numerical model with the measured three-dimensional velocity data. Turbulence energy and some turbulence stresses were also measured, so the performance of turbulence models can be tested. In short, the following variables and capabilities should be validated with this data set.

Variables:

- 3D velocity field
- Turbulence kinetic energy
- Turbulence stresses

Capabilities:

- Non-hydrostatic pressure
- Turbulence closure schemes

- Irregular mesh

Since the three variables were measured directly they can be compared with simulation results. The above three numerical model capabilities are required to insure correct prediction; they could only be validated indirectly when the variable comparisons are successful. One should not attempt this case if a model is incapable of these minimum requirements.

5.7.3 Approach

Because the dike is submerged and is of trapezoidal shape, the computational grid has to be either built around the dike (body fitted) or built along Cartesian coordinates. In the second case, the mesh lines intercept with the dike surface, which requires setting appropriate boundary conditions on the dike surface and blanking (deactivating) the computational nodes under the dike surface. It is a common practice that near the walls including bed and the dike surfaces, the mesh should be refined to handle the high rate of velocity variations. After the mesh is generated, boundary conditions at the inlet should be prescribed and simulations can then be started. Boundary conditions specified by a model developer for the governing equations being solved are described later in the test case.

5.7.4 Description of Experiment

All of the flow measurements were collected in an experiment flume of 30 m long, 1.2 m wide, and 0.6 m deep located at the National Sedimentation Laboratory. The physical experiments of flow around a trapezoidal shaped, submerged spur dike (Kuhnle, et al 1997, 1999) were used for the model simulation as summarized in Table. 5.7.1. The flow in the main channel was subcritical. The bed was covered with sediment (D_{50} = 0.8 mm, $[D_{84}/D_{16}]^{1/2}$ =1.35) which was immobilized with a thin layer of cement to prevent the bed from changing with time. The length and the width of the flume are 30 m and 1.2 m, respectively.

Table 5.7.1 Flow condition of the physical experiments (Kuhnle, et al, 1999)

Exp. Run	Dike length (m)	Flow rate (m^3/s)	Flow depth (m)	Froude Number
S90-3	0.152	0.129	0.302	0.206

The height of the dike is 15.2 cm and other dimensions of the dike are shown in Figures 5.7.1 and 5.7.2.

The velocity data were collected using a commercially available Acoustic Doppler Velocimeter (ADV). The measurement head of the ADV is mounted on a stainless steel mast 60 cm long and 1 cm in diameter. The sampling volume of the ADV is a cylinder 6 mm in height and 6 mm in diameter (170 mm^3) located 5 cm away from the head of the ADV. Flow velocity data at each point were collected at 50 Hz for 5 minutes. The 5 minute sample duration was determined empirically as the optimum length of time to capture the mean velocity at the sampling location within a reasonable time frame. Flow velocities were measured at 288 locations as shown in Figure 5.7.2. At each location the flow was measured at 9 vertical positions: 0.0100, 0.0225, 0.0350, 0.0475, 0.0600, 0.1000, 0.1400, 0.1800, and 0.2200 m from the bed. The vertical measurement positions were adjusted accordingly at the locations above the spur dike to arrive at nine measurement positions. A total of 2592 velocity vectors were measured. All velocity records were processed using the public domain program, WinADV. Measurements were filtered using WinADV to reject points with a correlation coefficient less than 0.7. In most files 90% or more of the data was above 0.7.

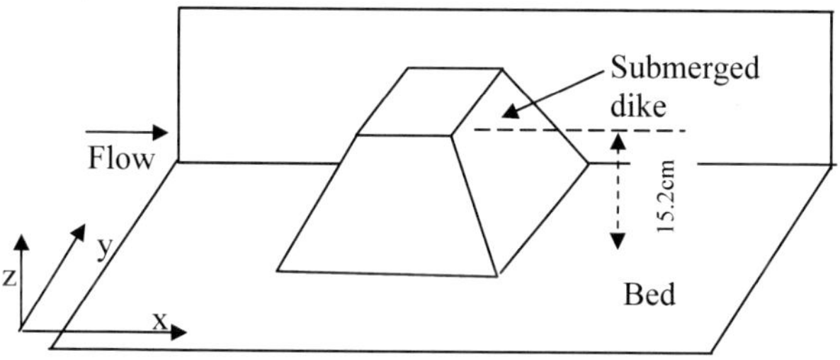

Figure 5.7.1 Sketch of model trapezoidal shaped dike (not in scale)

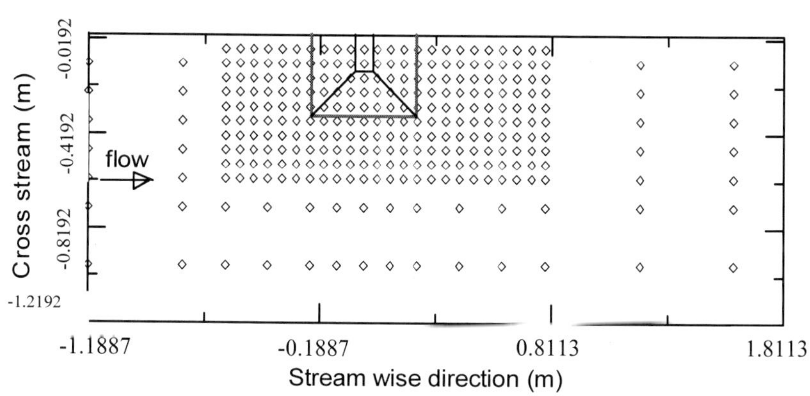

Figure 5.7.2 Plan view of experimental flume with measurement locations indicated by diamond symbols. Outline of spur dike is shown.

5.7.5 Boundary Conditions

Boundary conditions for simulating this test case are relatively easy to set, because the approach flow is steady and almost uniform. One only has to distribute the discharge in the inlet with uniform flow velocity distribution and set the inlet section reasonably far from the dike. A velocity distribution in the vertical can be specified with the logarithmic profile and it should decrease similarly toward the sidewalls. One has to make sure that the total flow discharge at the inlet section is preserved with such a velocity distribution. Similarly, the outlet section should be placed far downstream so that the flow calculation in the recirculation zone would not be affected. Normally, this section is placed at least 2-3 times the recirculation length downstream of the dike. It is commonly accepted that the zero gradient boundary condition, $\partial\partial n=0$, for all the computed variables can be applied at outlet section, where n is with the direction normal to the outlet.

On the bed surface, dike surface and vertical side walls, the wall function (log profile) can be used to specify the boundary condition under the assumption that the side walls are smooth, the bed material is of d_{50}=0.8 mm, and the dike module was made of concrete. The boundary conditions for turbulence closure models depend on the nature of the specific model being applied. Zero equation models normally do not need a boundary condition, while two equation models such as the family of k-ε models require boundary conditions at all boundaries of the computation domain because turbulence properties are computed with transport equations. For Reynolds stress models, more involved boundary conditions are needed particularly at the free surface boundary.

5.7.6 Test Example

The aforementioned data have been simulated as a test example using a three-dimensional hydrodynamic model, CCHE3D, developed at the National Center for Computational Hydroscience and Engineering, The University of Mississippi. This model has been developed for simulating open channel hydraulics and sediment transport around structures (Jia, et al. 2001, Jia and Wang, 1996, 1999, 2000). This 3D flow data set provided is an excellent case to test the numerical model's capability in simulating 3D turbulent flow around a submerged structure.

CCHE3D is a finite element based numerical simulation model that simulates time dependent free surface turbulent flows. Full Reynolds equations are solved with the options of computing either hydrostatic or dynamic pressure. Finite element techniques are used for transforming variables from local element space to physical space. The code has several turbulence closure schemes including zero equation models and two equation models. Time marching schemes are used to discetize the momentum equations and the algebraic equation system is solved implicitly. The following sketch (Figure 5.7.3) shows boundary conditions used for the flow solver which in fact are quite commonly accepted in CFD community. One should note that the log profile specified at the inlet section is from the bed to surface which is different from those specified near the wall (bed or dike) surface.

A standard k-εmodel was selected for this test case. Since the dike generates a three-dimensional recirculation and curved shear layer around it, a two-equation model is considered necessary. Near the walls, the local equilibrium condition was specified for the k and ε equation, respectively. In CCHE3D model, the "universal" function (Nezu and Nakagawa, 1993) for calculating free stream turbulent energy k is adopted.

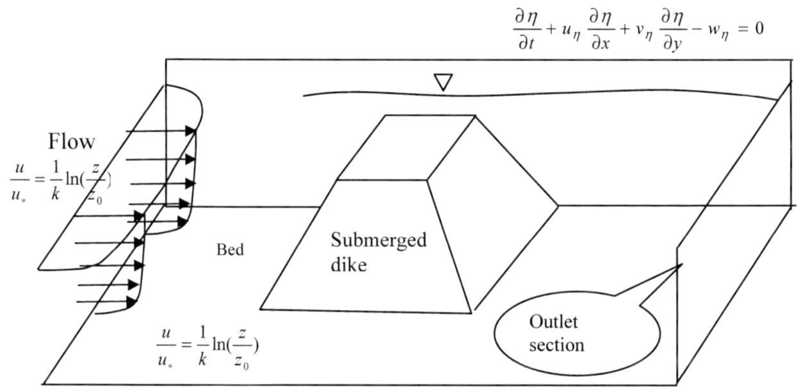

Figure 5.7.3 A sketch for boundary conditions for Reynolds stress equations

5.7.7 Velocity Distribution in Longitudinal-Sections

Figure 5.7.4 shows comparisons of measured flow (u) velocities and simulation results in the longitudinal direction (x). Velocities are presented along vertical lines at several locations in longitudinal sections. The measurements for all vertical lines started about 8 cm from the water surface to near the bed, indicated by circles, while computational solutions are from the surface to bed, represented by dark solid curves. Vertical straight lines indicate the location of the measuring profile as well as local reference for velocity profiles; positive velocity is on the right and negative ones are on the left side of the lines. The trapezoidal shape depicts the submerged dike. If a section is quite close to the attaching wall (Fig 5.7.4a), it cuts the dike on the flat top. Further away from the wall, the cross-section of the dike shown in the sections is smaller (Fig 5.7.4b) cutting on the side slope of the dike. The dike disappears from the section when it is located in the main channel (Fig 5.7.4c).

Fig. 5.7.4 shows that the general agreement is quite convincing at all locations, except near the dike surface where slight discrepancies are observed. The measurements indicate that there is no horseshoe vortex in the front side of the dike as has been observed in other physical model experiments in which dikes (Rajaratnam and Nwachukwu, 1983), cylindrical piers (Melville and Raudkivi, 1977, Istiyato and Graf, 2001), or abutment (Kwan and Melville, 1994) have a vertical front wall. Velocity profiles in the downstream side show a recirculation; the velocity above the dike top level is forward and that below the level is backward, and the shear layer that separates the flow is clearly represented by the vertical velocity profiles. An abrupt decrease of velocity in the longitudinal direction results in a small recirculation length. The flow is fully three-dimensional and the reattachment to the side wall at different levels has different lengths; the reattachment on the bed shown in this figure is about the height of the dike. Fig. 5.7.4b shows the size of recirculation in the vertical reduces rapidly at sections away from the attaching wall. At the section near the tip of the dike (Fig. 5.7.4c) the circulation disappears completely.

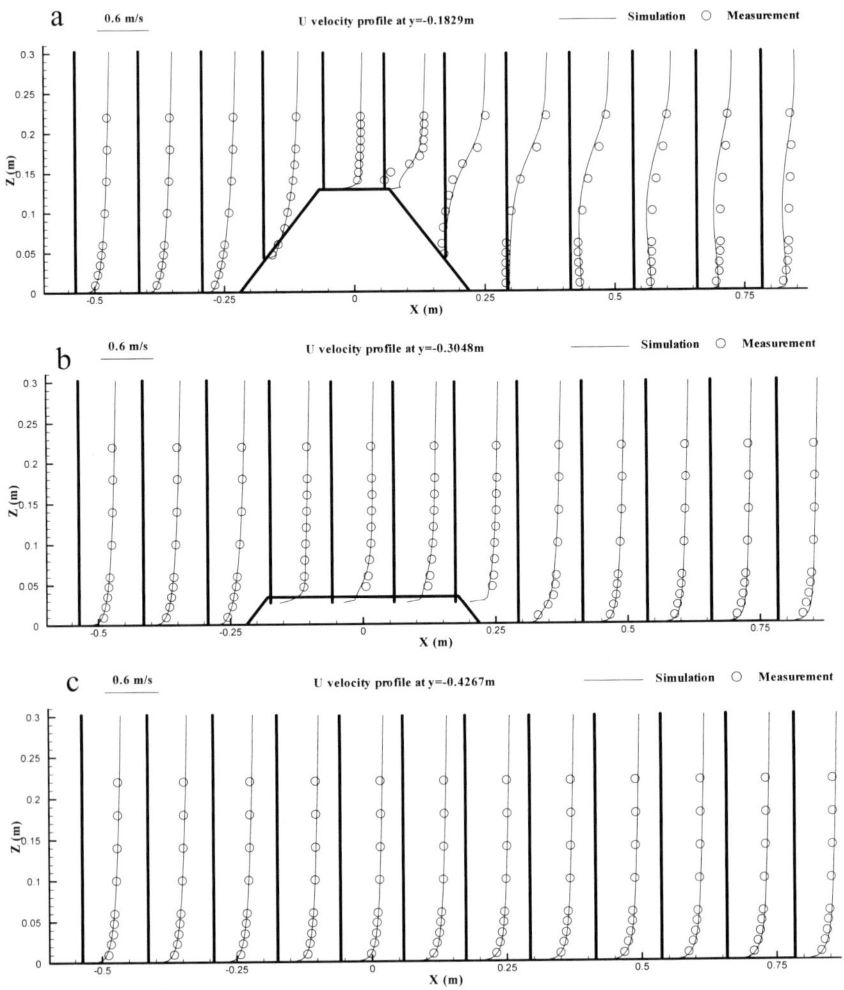

Figure 5.7.4 Comparison of simulation and measured longitudinal velocity in several longitudinal sections

Figure 5.7.5 shows a comparison of simulated and measured velocity in the *y*-direction (Fig. 5.7.1). The velocity toward the spur dike is on the right side of the solid vertical lines. The large velocity near the top corners of the dike (Fig. 5.7.5b) indicates the motion from the dike side of the channel. The distributions behind the dike (Fig.5.7.5a, and 5.7.5b) indicate a recirculation region; the upper part of the flow is toward the main channel with the lower part toward the wall to which the dike is attached. Near the immediate downstream side, the *v* velocity changes direction near the upper part of the profiles, indicating the existence of a smaller cell with an opposite rotation direction. The magnitude of the transverse velocity, *v,* decreases significantly toward the main channel (Fig. 5.7.5c). It is interesting to point out that the length of the flow rotation shown by *v* velocity profiles downstream of the dike is greater than that indicated by the *u* velocity profiles. The recirculation in Fig 5.7.4 along the *y* axis is confined tightly in the back corner of the dike. Its size is dominated by the height of the dike. The size of the circulation shown in the Figure 5.7.5 along the *x* axis is characterized by the length of the dike. This is because the shear layer created by the main flow is stronger and it extends further downstream than the circulation driven by the overtop flow.

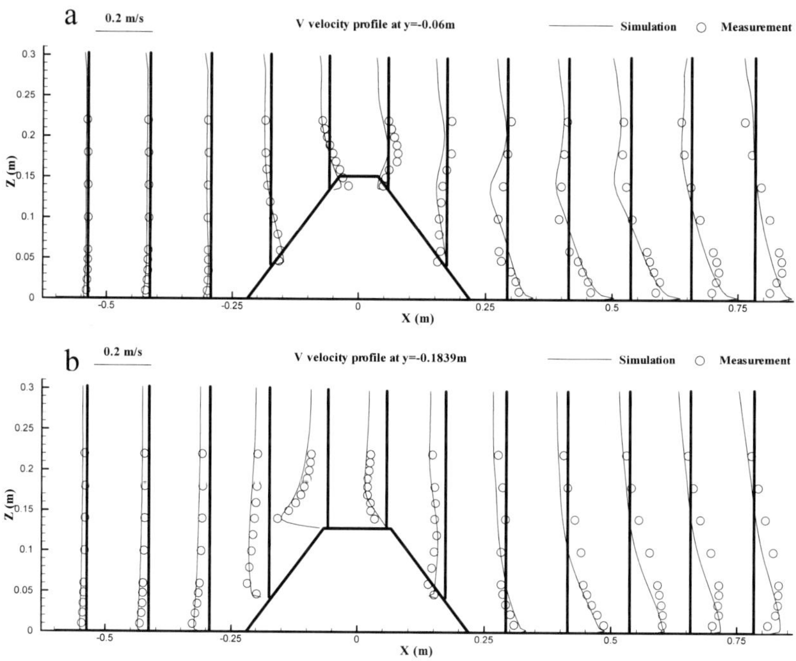

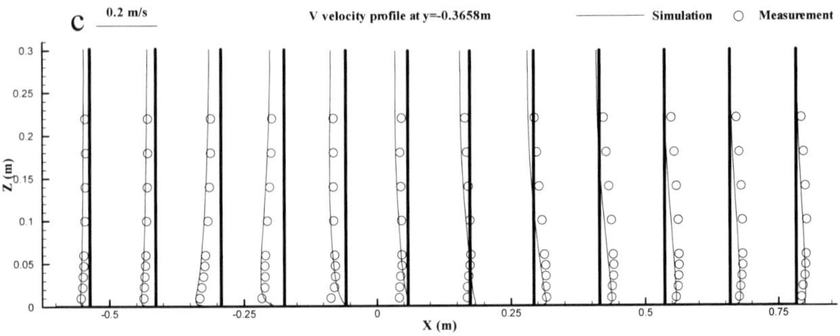

Figure 5.7.5 Comparison of simulation and measured transverse velocity in several longitudinal sections

The vertical velocity component *w* is depicted in Figure 5.7.6. Upward velocity is indicated to the right of the straight solid lines. Similar to the transverse velocity, the vertical velocity is significant only in the area close to the dike. The velocity on the dike front slope is upward, climbing the slope (Fig. 5.7.6a, 5.7.6b). The flow just downstream of the dike also has an upward motion, which is under-predicted to some extent. Further downstream, both measured and simulated velocity show a trend of decreasing strength with the simulation under-predicted somewhat.

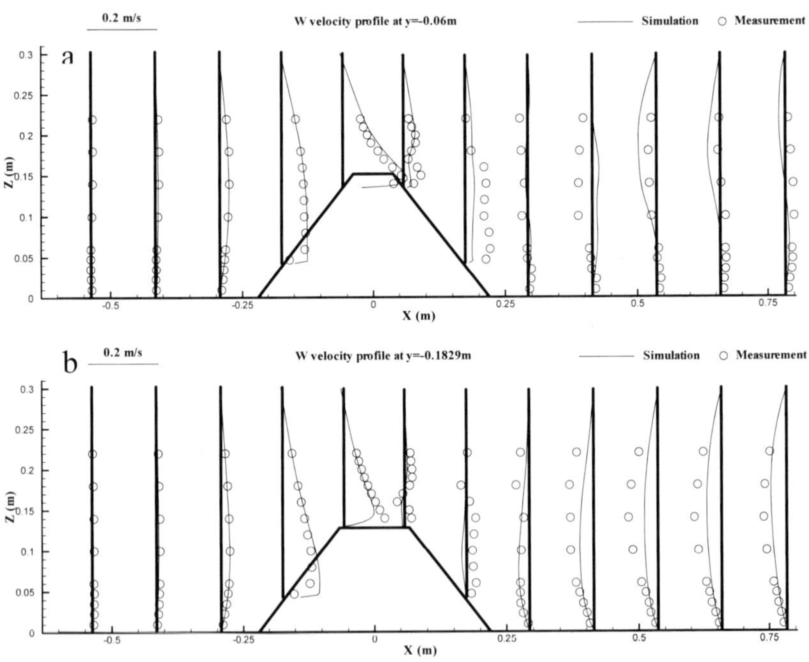

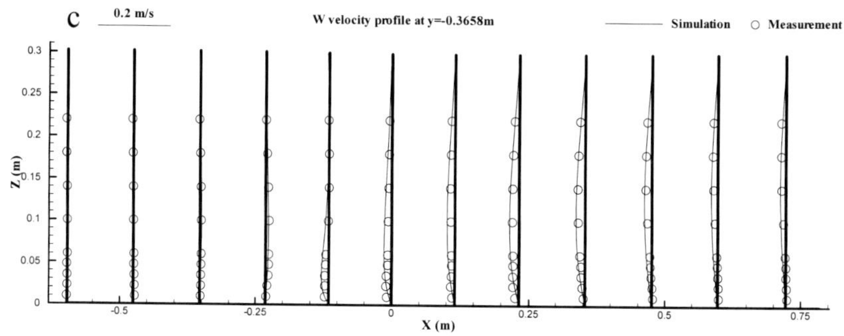

Figure 5.7.6 Comparison of simulation and measured vertical velocity (w) in several longitudinal sections

Summarizing Figures 5.7.4, 5.7.5, and 5.7.6 and the above discussions, the general flow pattern is sketched in Figure 5.7.7. One can see the three-dimensionality of the flow, particularly in the recirculation behind the dike. On the front side, the typical horse-shoe vertex often observed in physical experiments of a dike or pier with vertical wall was not seen in this data set. The vectors in Figure 5.7.7 only indicate the flow pattern rather than exact direction and magnitude.

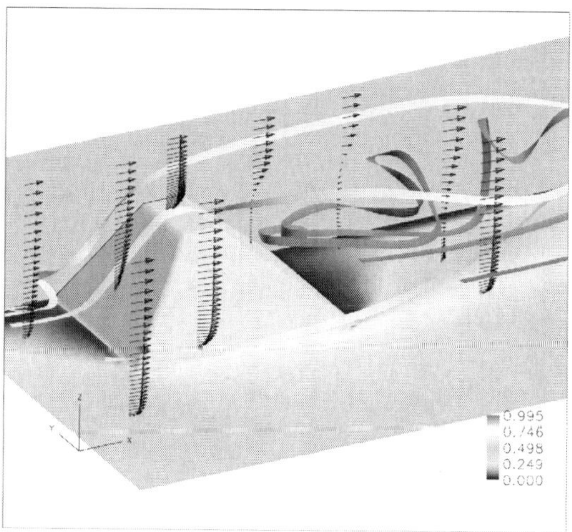

Figure 5.7.7 Flow pattern around the trapezoidal dike

5.7.8 Transverse Velocity (v) Distribution in Cross-Sections

To further examine the transverse velocity distribution and simulation results, measured and computed v velocity are presented in several cross-sections perpendicular to the mean flow direction (Figure 5.7.8). In Figure 5.7.8, the sections are aligned from upstream to downstream with the right side being the dike side, and the left side being the main channel. One section (Fig. 5.7.8a) is placed just upstream of the dike. The velocities turn to the left indicating the blockage of the dike. Figure 5.7.8b shows a section placed across the middle of the dike. Because of the trapezoidal shape, the flow coming from the front of the dike is pushed into the side space where acceleration occurs. Both data and simulation show this trend on the side slope of the dike. The simulation captured the trend of the flow but the peak velocity. Another difference is that the simulation indicates a separation on the side slope with a small depth but the data do not have this trend, due to the fact that the recirculation is very close to the surface of the dike where data is difficult to obtain.

Figures 5.7.8c and 5.7.8d show comparisons in two sections downstream of the dike. Discrepancies of simulation and measurement appear to be higher on the dike side but they are quite consistent with the main trend. Flow is moving away from the dike side in the upper part of the water depth and it moves into the backside of the dike in

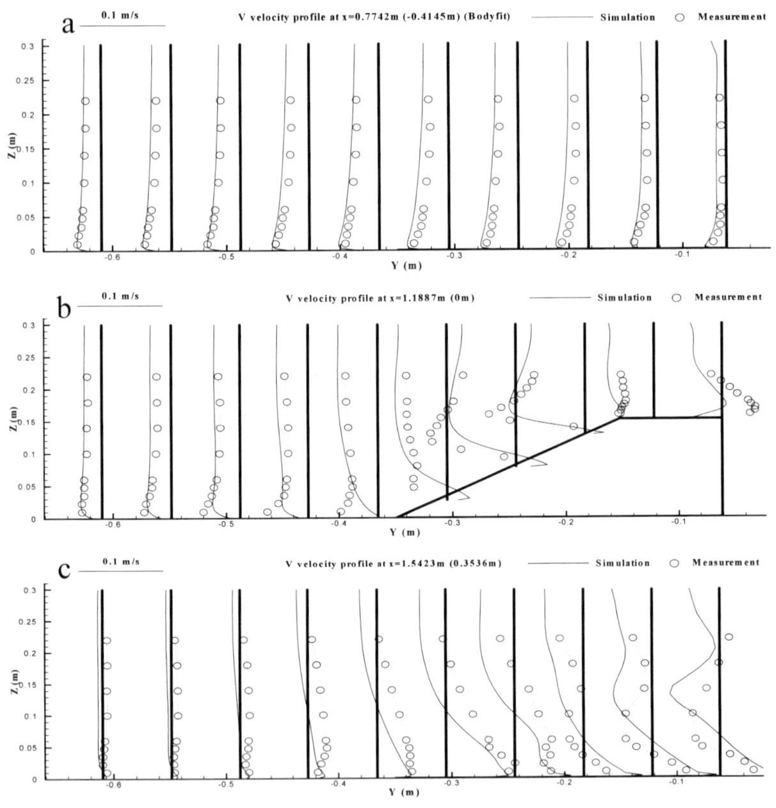

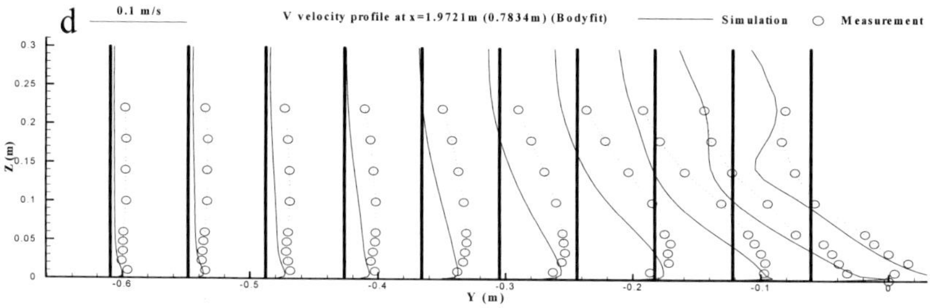

Figure 5.7.8 Comparison of simulation and measured transverse velocity (v) in several cross- sections

the lower part of the water depth. The near bed velocities agree quite well. Sediment could be transported to the backside if it were involved in the physical experiment or numerical simulation.

5.7.9 Turbulence Energy *k*

Figure 5.7.9 shows comparison of measured and simulated turbulence kinetic energy ($k = (\overline{u'}^2 + \overline{v'}^2 + \overline{w'}^2)/2$) in four cross sections. Figures 5.7.9a and 5.7.9b are on the upstream side of the dike, Figure 5.7.9c cuts across the middle of the dike and Figure 5.7.9d is downstream of the dike. One may note that the scales for Figures 5.7.9a and 5.7.9b are larger than those for 5.7.9c and 5.7.9d for display purposes. The simulation agrees with the measurement very well in the main channel and discrepancies are higher near the dike but the predicted trend of the variations is consistent with the measured data. Because k is solved using a transport equation (k-ε model), the boundary condition at the upstream is important. This comparison indicates that the boundary condition specified by Nezu's "universal function" (Nezu and Nakagawa, 1993) is very good. The highest discrepancy occurs behind the dike in the recirculation zone where the flow is highly turbulent and the turbulence energy is much higher than elsewhere.

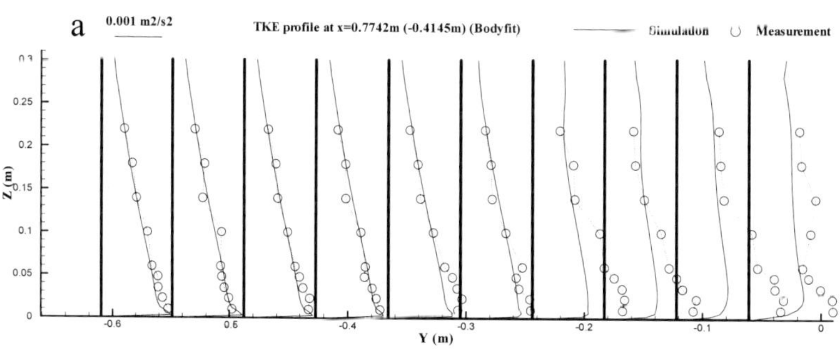

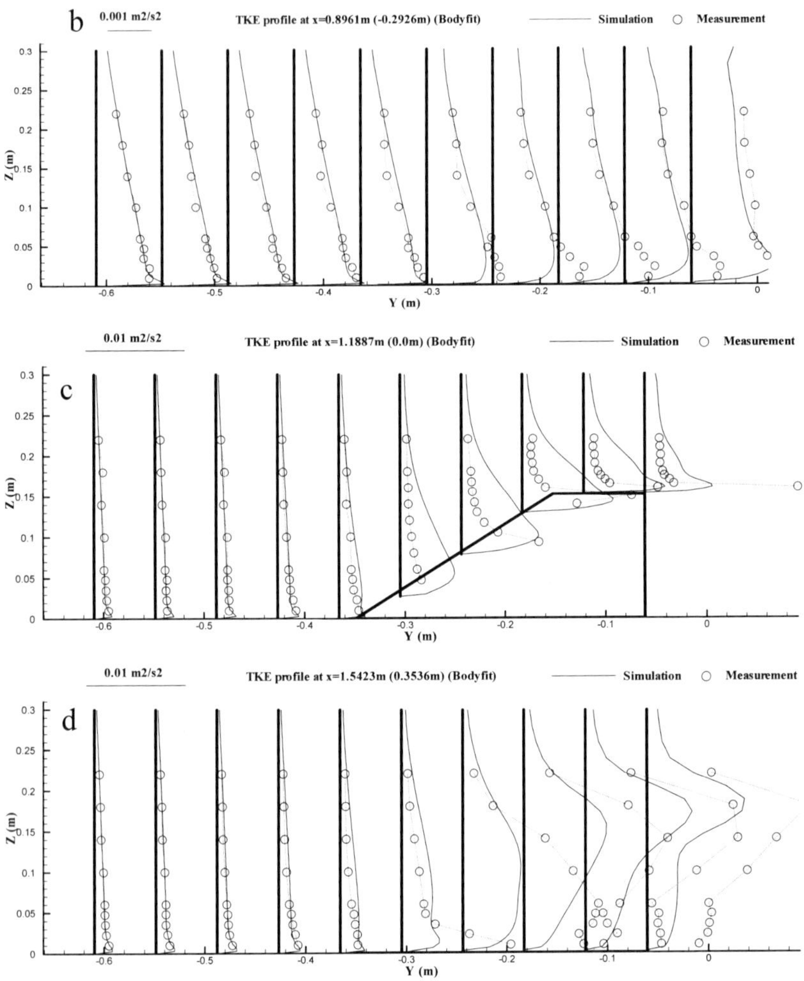

Figure 5.7.9 Comparison of simulation and measured turbulent kinetic energy (k) in several cross- sections

5.7.10 Shear Stress $-\overline{u'w'}$ Comparison

Turbulent shear stresses are important in understanding the flow, erosion and sediment transport in open channels, particularly around the dike structure. As discussed above, this submerged dike is surrounded by shear flow and creates turbulent shear layer, separation and recirculation. Shear stress distribution measured along vertical profiles and the numerical model's simulation are presented in a longitudinal section placed on top of the dike (Figure 5.7.10). The high shear stress zone is consistent with the corresponding velocity profiles (Figure 5.7.4a). The predicted shear stress however, is less than that of measurements in the shear layer. The agreement improves further downstream where the shear decreases.

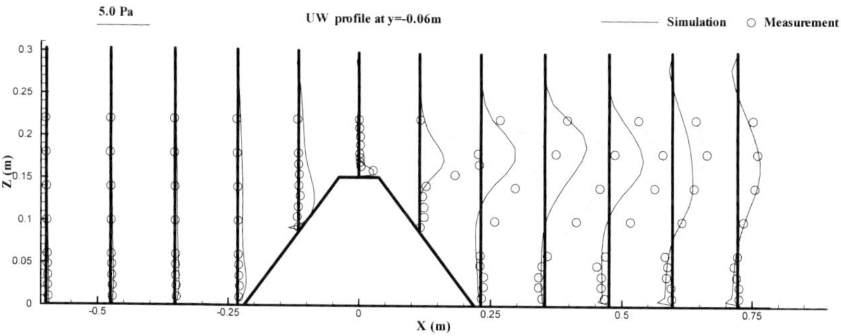

Figure 5.7.10 Comparison of simulated and measured shear stress in a longitudinal section

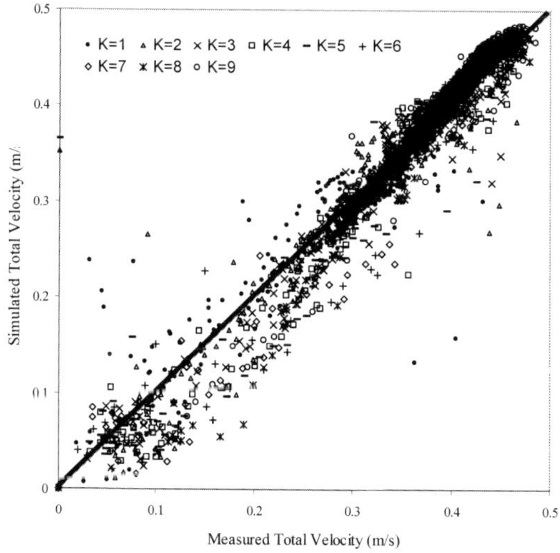

Figure 5.7.11 Overall agreement of simulated and measured total velocity magnitude

To illustrate overall agreement between the simulation and measurement, total velocity magnitude is presented in Figure 5.7.11. The diagonal line represents the perfect agreement. It can be seen that the numerical prediction reproduced the physical model data with very little systematic error (r^2=0.97). The root mean square error for the total velocity

$$\sigma = \sqrt{\frac{\Sigma(U_s - U_m)^2}{N}} \approx 0.0306 \text{ m/s} \tag{5.7.1}$$

is reasonably small with U_s and U_m representing simulated and measured total velocity and N=2592. The point scattering is slightly higher for smaller velocities (<0.3m/s). In Figure 5.7.11, the labels for the symbols K=1, K=2, ... K=9 indicate that the vertical level of measurement with K=1 represents the level near the bed and K=9 represents the level near the surface. Most of the points scattering farther away from the diagonal line are those close to the bed, dike surface or in the recirculation zone. Because most of the data points near the bottom (K=1, 2) agree very well to the diagonal line, this scattering reflects the difficulty of measuring acoustic data close to the dike surface.

5.7.11 Conclusions

The test case of a three-dimensional flow around a submerged spur dike is presented. The velocity field and turbulence stresses were measured by an Acoustic Doppler Velocimeter (ADV) at 2592 individual measurement points. The data show that the recirculation behind the dike was fully three dimensional and that it is different from that induced by a non-submerged dike. The sharp edges of the dike create separation and shear layers in the turbulent flow behind the dike. This data set is excellent for validating hydrodynamic numerical models with capabilities of dynamic pressure, irregular grid and turbulence closure models.

Simulation results indicate that the velocity field can be predicted quite well. The differences of the data and simulation in the channel across the dike are very little for all three velocity components. Upstream of the submerged dike, the simulation of the flow is also close to the measured velocity field. Although larger discrepancies between simulated and measured v and w velocities in the recirculation zone are observed, the correspondence with the u velocity is better. The overall trend of the data has been predicted quite reasonably. The mean error for the overall velocity comparison is about 0.03 m/s.

5.7.12 References

Jia, Y., Kitamura, T., and Wang, S.S.Y., 2001, "Simulation scour process in a plunge pool with loose material", ASCE, *Journal of Hydraulic Engineering*, Vol. 127, No. 3, pp219-229.

Jia, Y., and Wang, S.S.Y., 1996, "A modeling approach to predict local scour around spur dike-like structures", Proceedings of the Sixth Federal Interagency Sedimentation Conference, p. II-90-97.

Jia, Y., and Wang, S.S.Y., 1999, "Simulation of horse-shoe vortex around a bridge pier", Proceedings of the International Water Resources Engineering Conference". CD-ROM, 10 pp.

Jia, Y., and Wang, S.S.Y., 2000, "Numerical Study of Turbulent Flow around Submerged Spur Dikes", 4th International Conference for Hydroscience and Engineering, 2000, Seoul, Korea.

Istiyato, I., and Graf, W.H., 2001, "Experiments on flow around a cylinder in a scoured channel bed", International Journal of Sediment Research, Vol. 16, No. 4, pp431-444.

Kuhnle, R., Alonso, C.V., and Shields, F.D., 1997, "Geometry of scour holes around spur dikes, and experimental study", Wang, S.S.Y., Langendoen, E.J., and Shields, F.D., (ed.) Proceedings of the Conference on Management of Landscapes Disturbed by Channel Incision, Center for Computational Hydroscience and Engineering, School of Engineering, The University of Mississippi, p. 283-287.

Kuhnle, R., Alonso, C.V., and Shields, F.D., 1999,"Geometry of scour holes associated with 90° spur dikes", Journal of Hydraulic Engineering, ASCE, 125(9), p.972-978.

Kwan, T.F., and Melville, B.W., 1994, "Local scour and flow measurements at bridge abutments", Journal of Hydraulic Research, IAHR, 32(5), 661-673.

Melville, B.M., and Raudkivi, A., 1977, "Flow characteristics in local scour at bridge piers", Journal of Hydraulic Research, IAHR, 15(4), 373-380.

Nezu, I.& Nakagawa, H. (1993). *Turbulence in Open Channel Flows*. IAHR/AIRH Monograph, A.A.Balkema, Rotterdam, Netherlands.

Rajaratnam, N., and Nwachukwu, B. A. 1983. "Flow near groin-like structures." Journal of Hydraulic Engineering, ASCE, 109(3), 463-480.

5.8 FLOWS AROUND GROYNE AND IN HARBOR

Contributors: K.-Peter Holz and Peter Mewis

5.8.1. Introduction

Numerical models allow for the simulation of various complex flow conditions with hydraulic structures at a low cost. They are being applied for analysis and prediction of flow phenomena at groynes, harbor entrances, jetty geometries, as well as for the prediction of sediment transport and water exchange between harbor basins and rivers or the sea. When numerical models are applied for real world problems it is not sufficient to verify them using analytic solutions for simplified situations. In the engineering practice, their validity and limitations with respect to more complex situations have to be known. This demands comparison with physical model tests.

In 1983 the "Stiftung Volkswagenwerk" provided a research grant for a project entitled "Numerical simulation of 3D-turbulent tidal flow in harbor entrances and navigation channels" to be executed by the "Institut fur Stromungsmechanik und Elektronisches Rechnen im Bauwesen", the "Franzius-Institut fur Wasserbau und Kusteningenieurwesen" (both of the University of Hannover) and Delft Hydraulics Laboratory (Delft Hydraulics).

The objective was to obtain detailed information about various current conditions at different geometrically simple groyne and harbor configurations in a physical model and to detect test cases where three-dimensional effects become important to simulations. The experimentally obtained data on velocities and water levels are used as a basis to calibrate and validate two- and three-dimensional numerical models. The experiments comprise single groyne and several harbor entrance geometries for steady and unsteady flow conditions.

An important difference between numerical and hydraulic model investigations is the repeatability of the model experiments. A numerical model will always return the same result for the same conditions prescribed. This does not hold for hydraulic models. Thus, data obtained by measurements need to be interpreted before being compared with numerical model results. The information about the fluctuations of the measured values may reveal important hints on real physics and may not be neglected in comparative studies. For this reason, all experiments have been run twice to determine confidence measures on the obtained data. Moreover, the local variability of the flow field allowed some conclusions regarding the turbulence structure.

5.8.2 Evaluation of 3D-Effects

The objective of the performed hydraulic model investigations was to evaluate the 3D-effects in the flow field. Three criteria may be relevant for the description of the spatial structure of the flow field:

1. The changes of the horizontal flow direction over the depth
2. The deformation of the vertical velocity profiles
3. The value of the vertical velocity component relative to the horizontal flow indicating secondary flow.

These criteria may be used to answer the question whether a 2D vertical integrated numerical model sufficiently represents the physics and may be applied for flow field calculations, or whether it becomes necessary to simulate the flow field with a 3D model.

5.8.3 Physical Processes

In distinction to test cases based on analytical solutions, which can only be used to test numerically obtained results from a model in idealized situations, the physical model results give the opportunity to test the model in more complex situations involving complex physics. This way it is possible to validate the assumptions made in the formulation of the numerical model. For the test cases presented in this study, we see an interaction of the following mainly nonlinear physical processes:

- Acceleration of the flow by water level difference and continuity of mass
- Advection of momentum
- Turbulent momentum transfer in the vertical and horizontal directions
- Vertical acceleration in the vicinity of the structure (non-hydrostatic)
- Wall friction at the bottom and the side walls
- Turbulence anisotropy

In a straight undisturbed channel, the flow is driven by the water level difference. The well-known logarithmic velocity profile over the vertical, with the effect of the typical secondary currents in a rectangular channel, is observed in the undisturbed part of the flume. However, the most important process in all the test cases described in this chapter is the nonlinear advection of momentum, which is the only process causing the development of the rotational flow field. Therefore the most interesting regions are those where a balance between advection of momentum, water level gradient and momentum transfer is established, e.g. around the groyne head, behind the groyne and at the harbor entrance in the harbor basin test case.

To set up a valid model for this class of problems, one has to account for all the processes mentioned above in the numerical model. It is desirable to test different turbulence models and the wall friction term, at least at the bottom. The bottom friction and momentum exchange in the vertical influence the water level gradient in the flume. For the reattachment length in the groyne flow situation the horizontal momentum exchange is important. By an increased horizontal viscosity the eddy motion is damped, e.g. lower velocity in the harbor basin or a shorter recirculation zone is developed. For extremely high eddy viscosity this finally may lead to the disappearance of the recirculation zone.

Another important process is the vertical acceleration in the vicinity of the groyne plus the wall friction and horizontal exchange for the vertical velocity component. In most numerical models these processes are neglected for simplicity and faster computation. But with a hydrostatic pressure assumption and no wall friction for the vertical velocity component, a very strong downward velocity is computed at the last row of nodes in front of the groyne. This may be tolerated in many applications, but in general it is a serious violation of the physics.

5.8.4 Numerical Aspects

For the numerical algorithms, the test case offers, beyond the primitive check of individual terms of the mathematical formulation, a test of the interaction between the terms describing the above-mentioned processes. Common problems are:

- Overshooting by higher order accurate advection schemes resulting in unrealistically high velocities near gradients in the velocity field
- High numerical diffusion in the direction transverse to the flow
- Spurious oscillations in the results.

The well-known problem of overshooting in the advection scheme and instability of the algorithm may occur. Due to the advection schemes, spurious oscillations are often caused in the solution, which in several models are smoothed out by numerical filters. In the case of flow around a groyne, this kind of error may be observed in particular in the recirculation zone. Here the velocity vectors may alternate, in direction which is physically unreasonable. In combination with sediment transport calculations these errors may lead to a complete failure of the model.

Generally these effects are avoided in numerical models by appropriate up-winding strategies or corresponding mixed higher interpolation procedures in Finite Element Models. However, if low order advection schemes are applied, too much numerical diffusion is introduced, which of course adds to the physical viscosity formulation (as filtering and smoothing does). This results in an underprediction of the recirculation zone length and, in terms of parameters used, an underestimation of the eddy viscosity coefficients.

Other numerical difficulties arise from the implementation of the boundary conditions. Different formulations may be applied at the pronounced sharp tip of the groyne and the corners of harbor basins pointing into the stream, where the main flow velocities detach and at the downstream corner where the very important stagnation point is established. Here the influence of the formulation of the boundary conditions is most sensitive, and inappropriate formulations will deteriorate the solution and lead to physically unreasonable results.

5.8.5 Description of the Hydraulic Model and the Measuring Systems

All models were implemented in a flume with rectangular cross-section. The flume has an overall length of 32.60 m, a width of 2.50 m and permits a maximum water depth of 0.50 m. It is not inclined. In one of the sidewalls of the flume an opening to a quadratic basin with a length of 3.0 m is installed, which allows the construction of different harbor geometries (see Figure 5.8.1). The bottom consists of a horizontal layer of concrete. The vertical walls are built up in plastered brickwork. All constructions for the different test cases are made of plastics or form-boards with smooth surfaces. To guarantee turbulent current conditions, metal grids are placed at the bottom. The Manning-Strickler coefficient for these elements was measured to be 50 $m^{1/3}/s$.

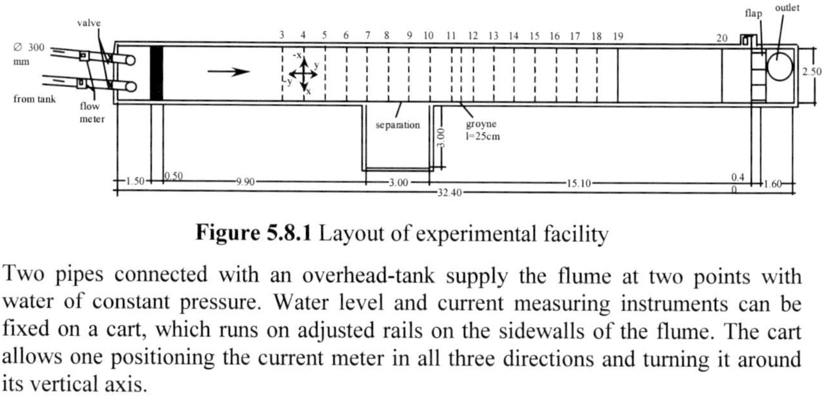

Figure 5.8.1 Layout of experimental facility

Two pipes connected with an overhead-tank supply the flume at two points with water of constant pressure. Water level and current measuring instruments can be fixed on a cart, which runs on adjusted rails on the sidewalls of the flume. The cart allows one positioning the current meter in all three directions and turning it around its vertical axis.

Discharge rates through the supply-pipes were measured by inductive flow meters (IFM) with a measuring accuracy of 1 %. The water level measurements were carried out with water level followers. The repeatability of the measurements is 0.1 mm. A 2D velocity probe NSW based on the principle of magnetic induction has been used. The probe NSW 1 ("NSW Stromungssensor S") measures two horizontal velocity components v_x and v_y and applies a 0.1 Hz RC-low pass filter. The probe NSW 2 allows a maximum data-sampling rate of 40 Hz. The accuracy of the probe is better than 1% of the scale.

5.8.6 Investigated Variants

Several geometrical variants of different groyne and harbor configurations have been investigated, including harbor basins with sloped groynes at the corners. Only the first two variants with the most common geometry are reproduced here and have been used for model testing in the past. Variant 1 is a single groyne in the flume and variant 2 consists of three different simple harbor geometries without groynes.

5.8.6.1 Variant 1: The Vertical Sharp-edged Groyne

1. Geometrical Layout and Hydraulic Boundary Conditions

In variant 1 the influence of a vertical, sharp edged groyne, positioned in a rectangular flume was investigated. The geometrical layout of this groyne test is shown in Figure 5.8.2. The length of the groyne corresponds to 10 % of the flume width.

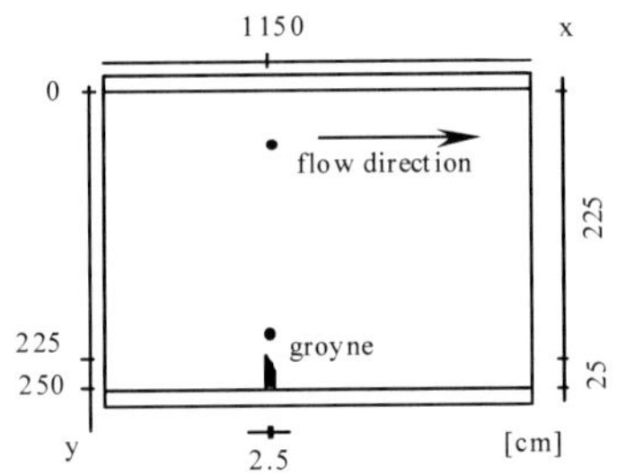

Figure 5.8.2 Layout of groyne test case. At the position of the dots, the vertical velocity profiles are given in Figure 5.8.8 and 9.

The measurements were carried out with a constant discharge of Q = 200 l/s. With a mean water depth of 23 cm, the mean velocity in the undisturbed flume can be calculated as v_f= 34.5 cm/s. This coincides well with the measured mean velocity in the undisturbed cross-section of the flume.

Mean values and standard deviations were calculated from 200 data values, sampled with a frequency of 3 Hz. Because of this low sampling rate and the low pass filter used, the calculated standard deviation describes only the long periodical fluctuations but not the real turbulent motion.

2. Water Level Measurements

The water was impounded by the groyne on the right side of the flume (Figure 5.8.2). The water level difference between upstream and downstream side of the groyne is 0.8 cm. The isolines run radial to the head of the groyne (Figure 5.8.3).

3. Results of the Velocity Measurements

Horizontal flow field

Figures 5.8.4 to 5.8.9 give more detailed information on the flow field. Behind the groyne the well-known recirculation zone establishes. The maximum extension of this zone is 12 times the groyne length (see Table 5.8.1). The back stream area is at most 20 cm wide (Figure 5.8.4). The flume current accelerates in front of the groyne. At the lee-side of the groyne the isolines run nearly parallel to the wall in short spacing to each other. This documents the high velocity gradient between the main flow and the back stream region.

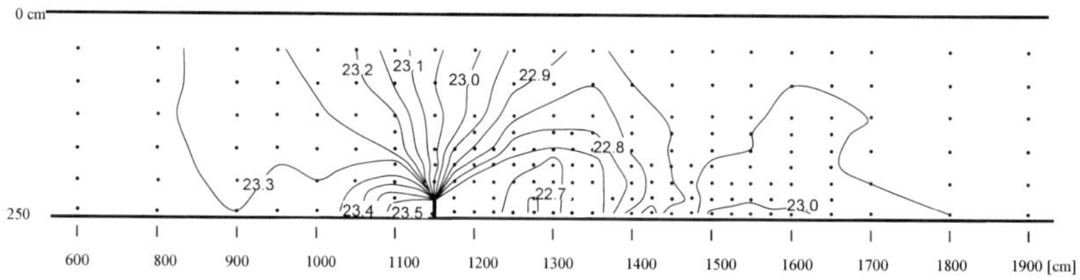

Figure 5.8.3 Water surface elevation above horizontal bottom for variant 1 in cm. The dots indicate the measuring points.

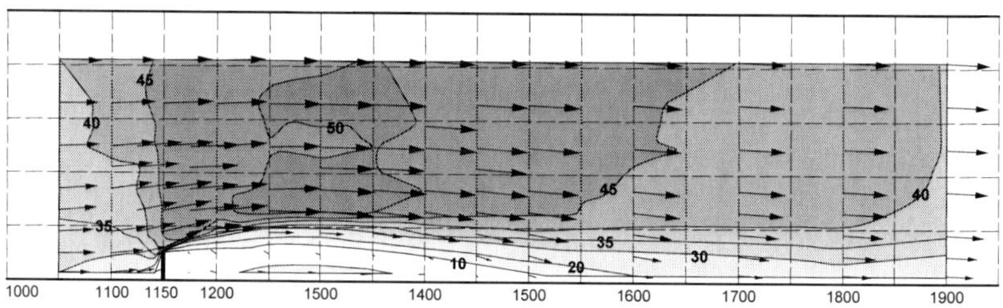

Figure 5.8.4 Flow field 13cm above the bottom

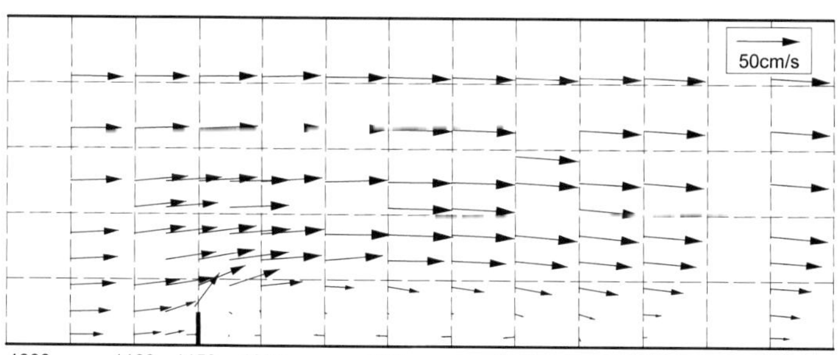

Figure 5.8.5 Flow field 21cm above the bottom.

Due to the high velocity gradient and the detaching of the current at the groyne tip, vortices are generated. The interaction between the recirculation zone and the main flume current creates highly turbulent motion in the flow field downstream of the groyne, which is indicated by higher standard deviations of the measured velocity components.

Figures 5.8.5 to 5.8.7 show the horizontal distribution close to the water surface and near the bottom. Remarkable is the strong deflection of the flow at the surface near the groyne head (Figure 5.8.5).

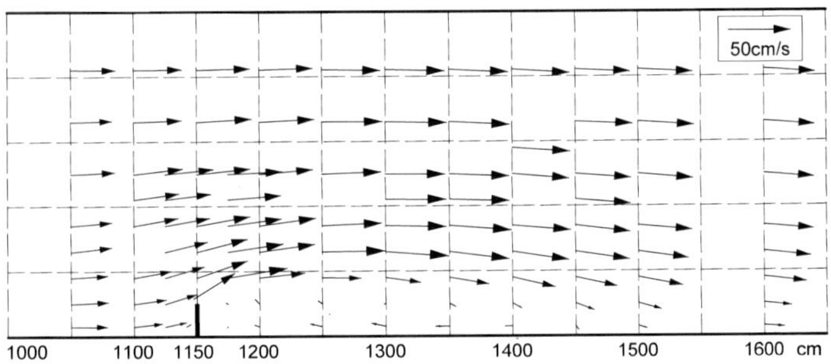

Figure 5.8.6 Flow field 8.5cm above the bottom.

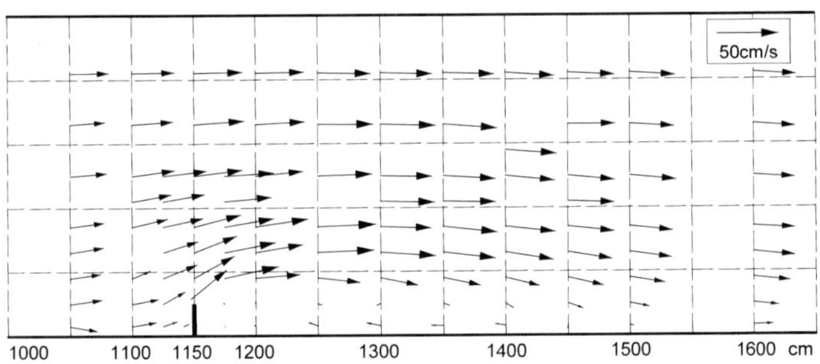

Figure 5.8.7 Flow field 5cm above the bottom.

In the undisturbed flume the vertical distribution of the velocities corresponds well to the logarithmic velocity law for hydraulic rough boundary conditions found by von Karman and Prandtl (Schlichting, 1965).

The profiles in Figure 5.8.8 show the velocity distribution at a distance of 45 cm from the wall opposite to the groyne (see dots in Figure 5.8.2). The longitudinal component $v_y(z)$ is logarithmically distributed with a small decrease close to the surface. This decrease can be explained by secondary flow of second Prandtl kind, i.e. turbulence anisotropy. The maximum velocities outside the area of influence of the groyne are located at 0.8 of the water depth. The cross channel component $v_y(z)$ is nearly zero, which indicates a constant flow direction over the depth parallel to the sidewalls of the flume.

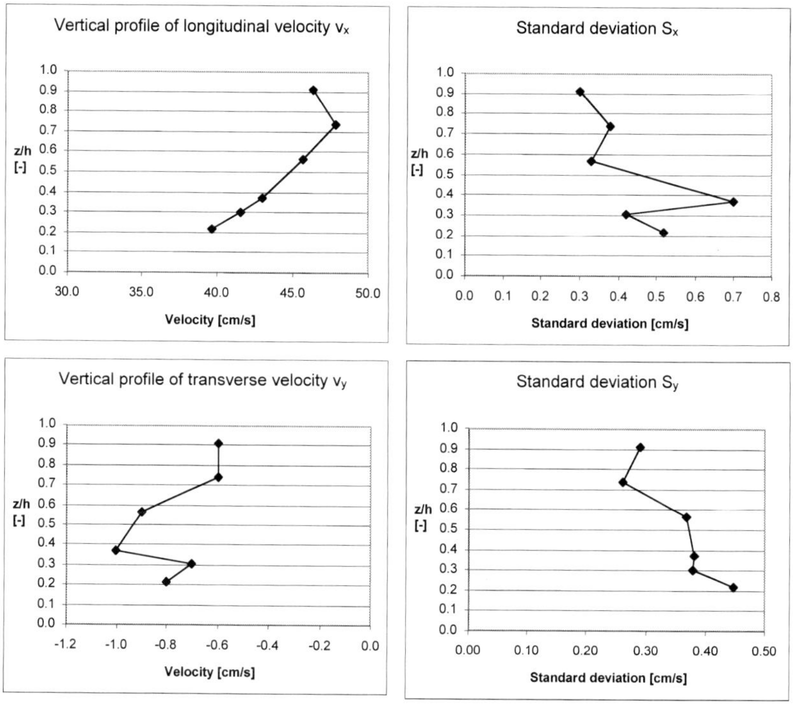

Figure 5.8.8 Vertical velocity profiles 45cm from the wall opposite to the groyne

Approaching the groyne tip the cross channel $v_y(z)$ component has significantly increased, especially in the layers close to the bottom and the surface. This indicates changes in the flow direction over the depth (Figure 5.8.9) from zero at the opposite wall to 28° at 0.2 of the flow depth and 23° at the surface 20cm away from the groyne head. It is expected that closer to the bottom the angle further increases. Unfortunately the measuring device cannot be applied at that short distance to the bottom. Right in front of the groyne head, the highest angle of 50° is observed near the surface (see Figure 5.8.5).

In the area of the eddy current, the velocity slightly increases towards the bottom. The flow direction is nearly constant over the depth. Only near the eddy center are there deviations in the flow direction, but this is mainly due to small velocities; i.e., small changes of the velocities result in strong changes in the flow direction.

Near the reattachment point (y=1450-1475 cm) the flow field is very unstable. This is due to the irregular impinging of the vortices on the sidewall causing changes in the flow direction and higher standard deviations. An interesting parameter is the so-called reattachment length, the distance behind the groyne where the flow reattaches to the sidewall. The reattachment length, divided by the length of the groyne, is a very useful dimensionless parameter for model tests.

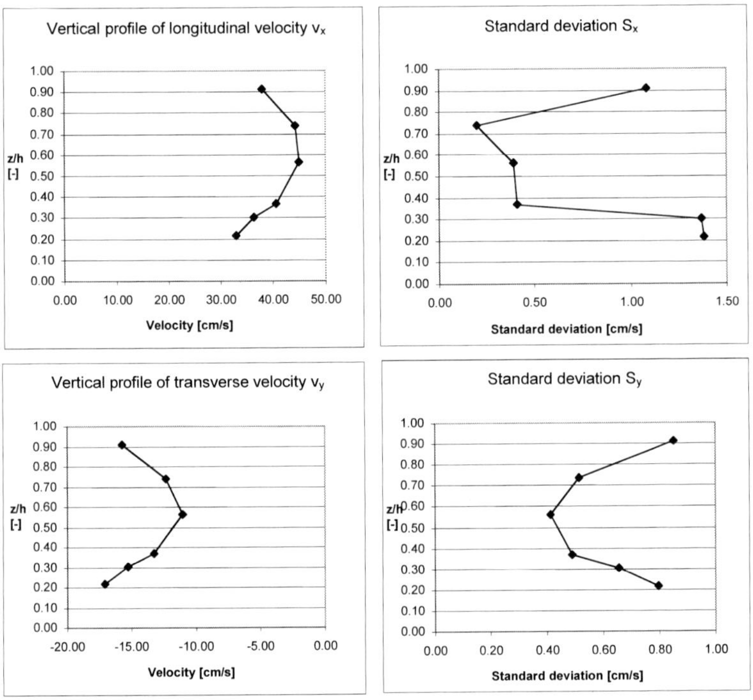

Figure 5.8.9 Vertical velocity profiles 20cm from groyne head in prolongation of the groyne.

Table 5.8.1 Reattachment length for different flow depths and shape of the groyne tip

Experiment	Groyne-length in m and shape	Water-depth in m	Froude	Reattachment versus groyne length
1	0.25 sharp-edged	0.15	0.28	12.5
2	0.25 sharp-edged	0.23	0.23	11.5
3	0.25 sharp-edged	0.3	0.2	11
4	0.25 regular	0.23	0.23	11.5

The typical measured ratio of the reattachment length to the groyne length is given in Table 5.8.1. A comparison between experiment 2 and 4 clearly shows that there is little influence of the shape of the groyne, whether sharply edged or rectangular, on the reattachment length. In the numerical model studies, the groyne may therefore be represented by a rectangular geometry.

5.8.6.2. Variant 2: Eddy Current in Three Different Rectangular Harbor Basins - General Remarks

A general situation of the current conditions in a harbor basin is shown in Figure 5.8.10. Within the shear layer at the harbor entrance an intensive vortex motion develops, which causes an exchange of the water between the flume and the basin. Part of the momentum is transferred via the turbulent mixing from the outer current to the water in the basin, where it generates the well-known eddy motion. Energy is dissipated within the turbulent mixing zone. The maximum of the turbulent fluctuations is reached at the stagnation point.

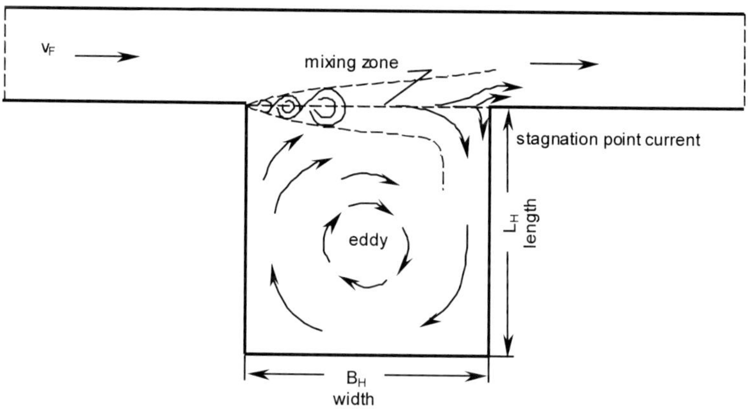

Figure 5.8.10 Flow conditions in a harbor basin (after Dursthoff, 1970)

A one-eddy system is generated in rectangular harbors with a length-to-width ratio between 0.4 and 1.7. For smaller length-to-width ratios the current in the upstream part of the basin is disordered. In this case, the eddy current is very unstable and does not occupy the whole harbor basin reach. For higher ratios from 1.7 to 2.0, a secondary eddy will be induced.

1. Geometrical Layout and Hydraulic Boundary Conditions

The three different harbor basin configurations are given in Table 5.8.2.

The discharges given in Table 5.8.2 were calculated from the measured velocities and water levels in the flume; they differ up to 10 % from the discharge of Q = 200 l/s prescribed in the channel. The mean water depth in the harbor basin is approximately h = 23.2 cm.

Table 5.8.2 Geometrical layout of Variant 2

Variant	Length, m	width, m	discharge Q, l/s	mean velocity V_f, cm/s
2.1	3.00	3.00	200	34.5
2.2	1.50	3.00	185	32.0
2.3	1.50	1.50	180	31.0

2. Water Level Measurements

The water levels are plotted in Figures 5.8.11 to 5.8.13 show that water level differences between the flume and the harbor and especially in the basin itself are very small, almost within the range of the measurement accuracy. The lowest water levels are in the middle of the harbor entrances. Close to the up- and downstream edges of the harbor mouth, the highest water levels were measured.

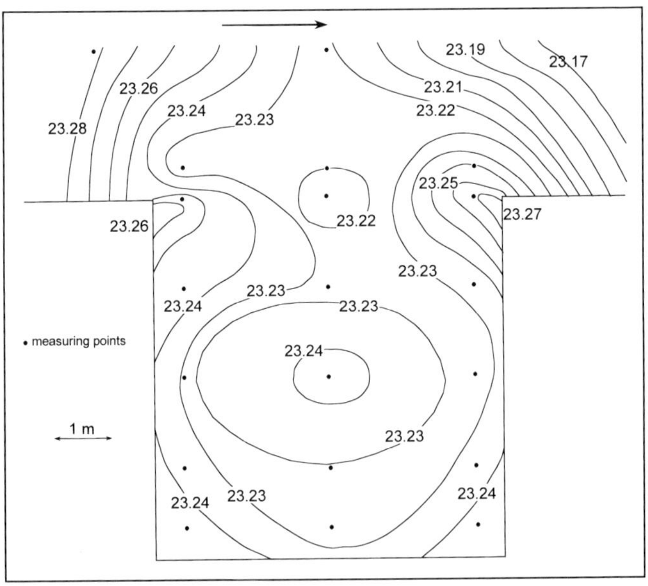

Figure 5.8.11 Water surface elevation for variant 2.1

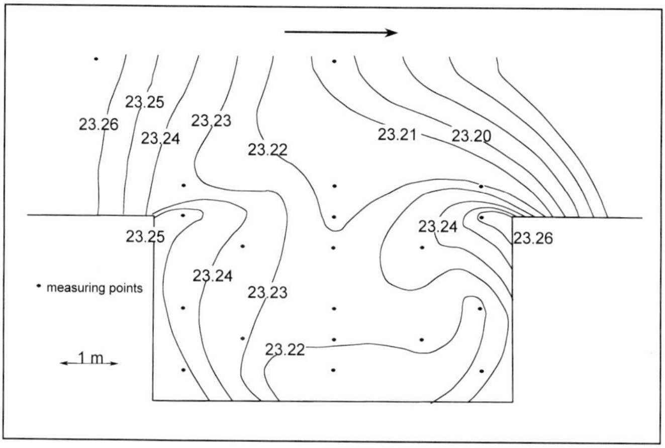

Figure 5.8.12 Water surface elevation for variant 2.2

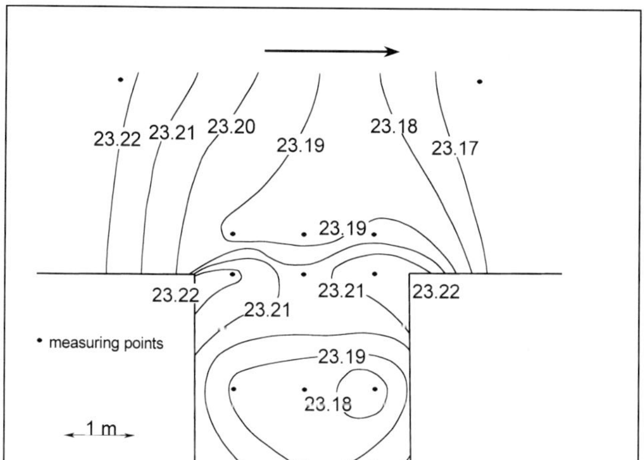

Figure 5.8.13 Water surface elevation for variant 2.3

For comparison to the flume investigations without any harbor (straight channel), the energy loss in each of the three variants, expressed in water level differences dH in the centerline of the flume, is given in Table 5.8.3.

Table 5.8.3 Height of energy loss caused by the harbor basin.

Variant	dH in mm
2.1	0.2
2.2	0.15
2.3	0.05

3. Results of the Velocity Measurements

Variant 2.1

In the quadratic basin the center of the eddy coincides well with the geometrical center. The highest velocities within the harbor basin are induced by the water level at the stagnation point (Figure 5.8.14). In the plane 5 cm or 0.2 of the water depth (Figure 5.8.15), the velocity vectors show in all three variants a pronounced convergence of the flow to the center of the eddy. This compares well with the picture given by Schlichting, 1965 (Figure 5.8.20). The separation area is not sharply shown by the isolines because of the distance of the measuring points. The velocities in the main flume area are influenced downstream and to a small extent upstream of the harbor basin.

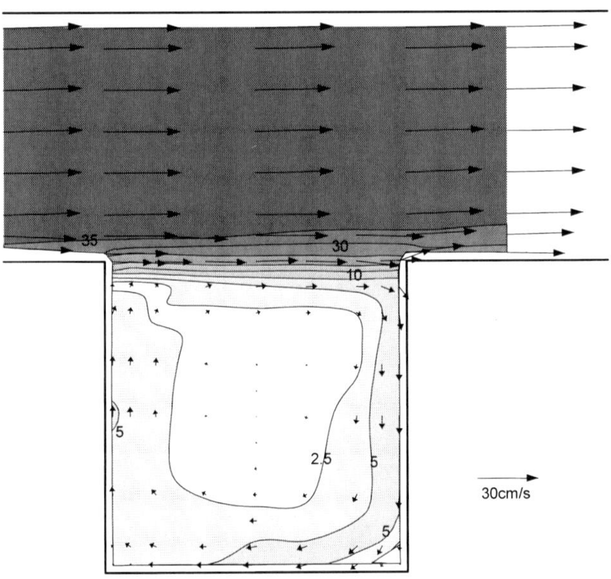

Figure 5.8.14 Measured flow field 17cm above the bottom, numbers give the velocities in cm/s.

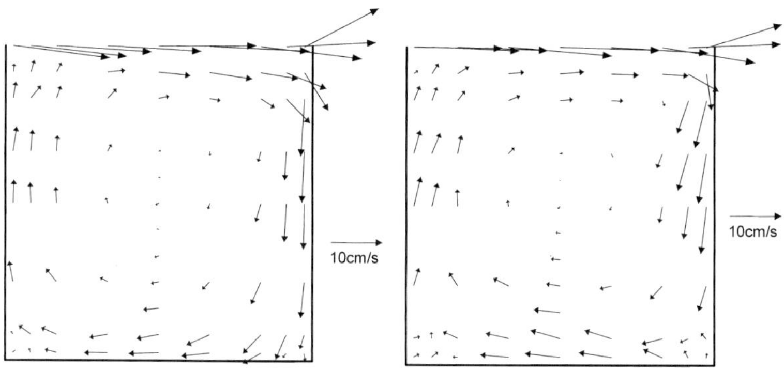

Figure 5.8.15 Flow field 5cm (left) and 21cm (right) above the bottom

Variant 2.2

The velocity distribution of variant 2.2 is shown in Figure 5.8.16. The current velocities in the harbor basin are slightly higher than those in the geometrically quadratic harbors. The reason is the smaller area of the harbor basin relative to the length of the mixing zone. In comparison to variant 2.1, the area of the rectangular harbor basin of this variant is only half size.

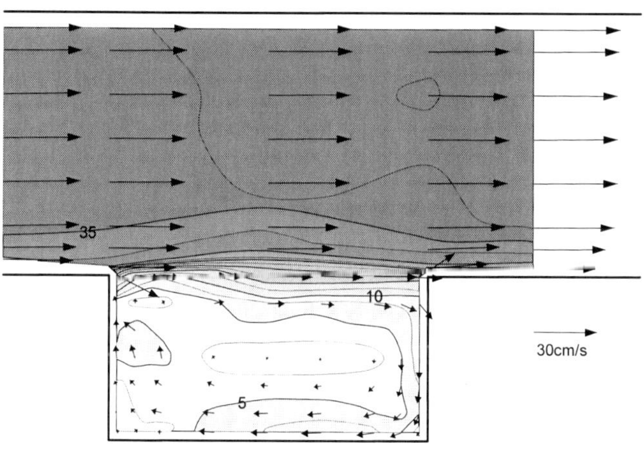

Figure 5.8.16 Flow field 17cm above the bottom

As mentioned above, the eddy is deformed in rectangular basins. The center of the eddy moves closer to the downstream wall. The eddy motion becomes more unstable. The convergence of the flow near the bottom (Figure 5.8.17) is pronounced.

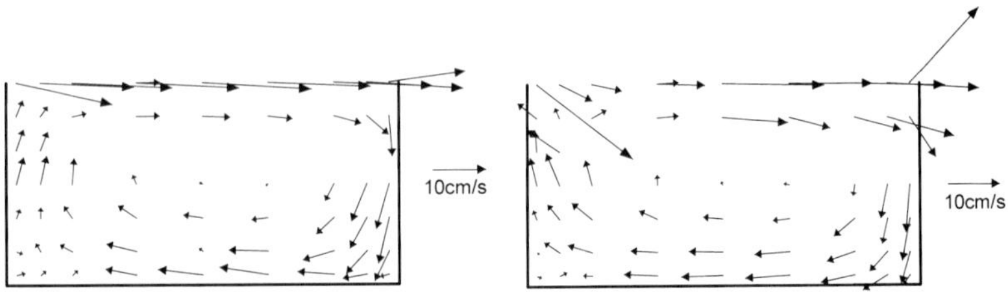

Figure 5.8.17 Flow field 5cm (left) and 21cm (right) above the bottom

Variant 2.3

In variant 2.3 the velocities in the harbor are smaller than those in variant 2.1 (Figure 5.8.18). This is due to the 10 % smaller mean flume velocity than that of variant 2.1 and the smaller size of the basin. Again the convergence of the current near the bottom is pronounced (Figure 5.8.19).

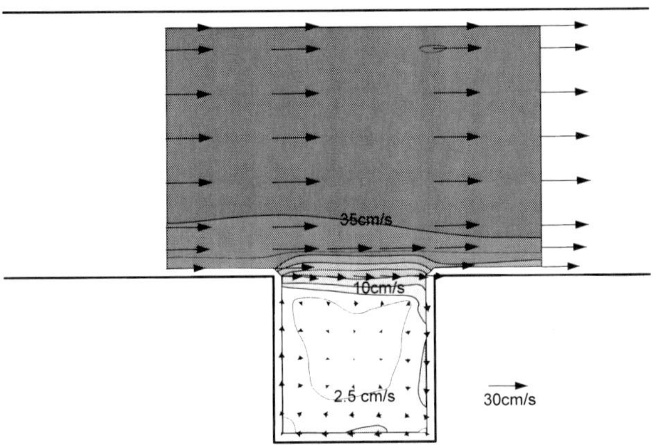

Figure 5.8.18 Measured flow field 17cm above the bottom.

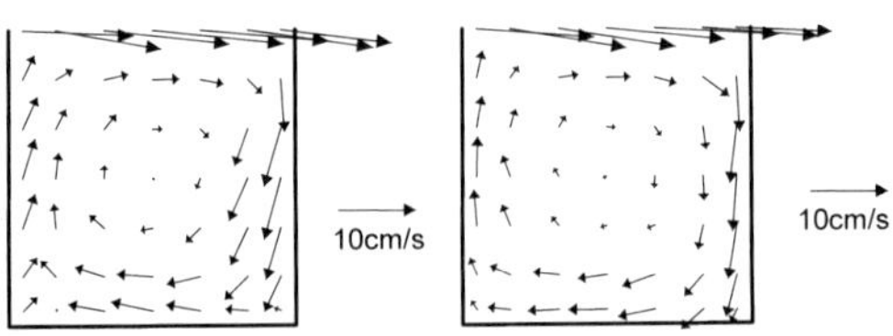

Figure 5.8.19 Measured flow field 5cm (left) and 21cm (right) above the bottom.

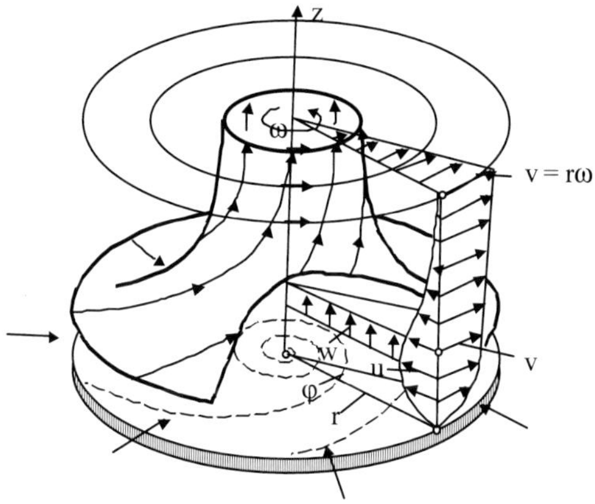

Figure 5.8.20 Rotating flow above a fixed bottom (after Schlichting, 1965)

5.8.7 Performed Tests

The physical experiments were used for the validation of several 2D and 3D models. In Flokstra's report (Flokstra, 1986), experience gained by a depth-averaged model is reported. Later a quasi-3D k-ε model developed at Delft Hydraulics was applied. This model uses the rigid-lid condition at the water surface. The results are summarized as follows. The measured and computed velocity profiles compare reasonably well. Differences however have been observed downstream of the groyne in terms of smaller computed mean flow velocity and larger computed velocities in the recirculation zone. One interesting result is a comparison of the k-ε turbulence model with a constant turbulent viscosity model. "By tuning the magnitude of the constant eddy viscosity the experimental reattachment length could be reproduced. The computed velocities for that case agree more closely to the experimental data than those by use of the k-ε model." The tuned eddy viscosity varies between 0.00025 m^2/s and 0.001 m^2/s, depending on the approach and the test case.

Another example was given by Mewis and Holz (1994), by the use of a multilayer 3D-model. The horizontal discretization is based on the FEM method using triangles, which are extended to prisms in the vertical. This model uses constant eddy viscosity in the horizontal, classical upwinding for the advection of momentum and a mixing length approach for the vertical eddy viscosity. The free surface flow was simulated by an explicit leapfrog time integration scheme. The grid or element size of the computational mesh should of course be fine enough not only to allow the eddy motion to develop, but to keep the numerical viscosity small in comparison to the physically reasonable eddy viscosity, which is not large for these test cases.

For the calibration of the bottom friction (the vertical eddy viscosity has no free parameter) the water level gradient along the centerline of the channel of 0.47cm over a distance of 20m resulting in 0.00023 has been used. In the groyne case, the water level gradient is increased due to the effect of the groyne; i.e., the energy loss in the eddy motion behind the groyne. For the harbor variants, the energy loss is much smaller (see Table 5.8.4). The mean bottom shear stress is 0.52 N/m^2. This corresponds to a friction factor $(u^*/U_m)^2$ of 0.004.

Table 5.8.4 Water level gradient along the centerline of the flume for the three variants

Variant	1	2.1	2.2	2.3
Gradient in %	0.358	0.233	0.229	0.228

Next the eddy viscosity was calibrated using the reattachment length. By variation of the horizontal eddy viscosity the model was tuned to reproduce the ratio of reattachment length to the groyne length, which was 11.5 (see Table 5.8.1). The tuned value of the horizontal eddy viscosity was 0.0003 m^2/s, which is in good agreement with the model results described above. For the harbor cases no such clear criterion for the calibration procedure exists. Here the velocities of the eddy motion in the harbor basin and the right position of the center of the eddy should be used as criteria.

Mayerle et al. (1995) used this test case for the validation of a three-dimensional numerical model. A structured grid with non-equidistant orthogonal grid lines was employed. The paper focused on or "aimed at" the testing of different eddy viscosity formulations. Comparisons were made for the surface elevation along the centerline, the reattachment length and vertical velocity distributions at different locations. A definite dependence of the reattachment length on the eddy viscosity was confirmed. A small influence of the sidewall boundary condition is stated. In this application the mixing length approach and the assumption of local equilibrium for the turbulent kinetic energy performed best with respect to the reattachment length. Near the tip of the groyne the measured flow field could not be captured because of the hydrostatic pressure assumption.

Ouillon and Dartus (1997) tested both a free surface and a rigid-lid version of their three-dimensional model for the groyne case. They used a structured grid with non-equidistant, orthogonal grid lines for a refinement at the tip of the groyne. The model was based on a hybrid finite-volume scheme based on the SIMPLE and HH-SIMPLE algorithms. The model accounts for non-hydrostatic pressure. The eddy viscosity was computed from the k- ε model. Typical computed values for the eddy viscosity were 0.00043 m^2/s in the undisturbed zone and 0.00144 m^2/s in the shear zone. Comparisons were made for the reattachment length, the distributions of the water level and the flow structure in the vicinity of the groyne. The authors stated that dramatic three-dimensional effects were observed. The model was not designed to reproduce the secondary flow effects in the flume. The maximum velocities therefore occurred at the water surface. Problems arose also with the k- ε model in the recirculation zone. The rigid-lid assumption leads to a serious underestimation of the reattachment length of 30% whereas the free surface computation with the same model gives an underestimation of about 7%. This underestimation thereby was related to the rigid lid assumption and not to the k-ε model approach in this study. Only very close to the groyne is the pressure distribution non-hydrostatic. Within a distance of a few centimeters around the groyne head, the maximum values of 3-4mm water column were computed. The authors stated, "following the type of desired spatial description, a 3D free-surface flow model may be employed with or without the hydrostatic pressure assumption."

5.8.8 References

Flokstra, C., 1986, "Computation of depth-averaged velocity distribution for channel with groyne and harbor" Delft Hydraulics Laboratory, Report H402, Oct 1986.

Mayerle R., Toro F.M. and Wang S.S.Y., 1995, "Verification of 3-dimensional numerical model simulation of the flow in the vicinity of spur dikes". Journal of Hydraulic Research, 33(2), pp. 243-256

Mewis, P. and Holz, K.-P. 1994, "3D model verification for a groyne and harbor experiment" In ASCE Proceedings of the 1994 National Conference held in Buffalo, 1994.

Ouillon S. and Dartus D., 1997, "Three-dimensional computation of flow around Groyne", Journal of Hydraulic Engineering, 123(11), pp. 962-970.

Schlichting, H., 1965, "Grenzschicht-Theorie", Karlsruhe, Verlag G. Braun

Dursthoff, W., "Uber den quantitativen Austausch zwischen FluB und Hafen", Mitteilungen des Franzius-Institutes der Universitat Hannover, Heft 34, 1970

CHAPTER 6
APPLICATION SITE VALIDATION

Richard A. Schmalz, Jr.

6.1 INTRODUCTION

Contributor: Sam S.Y. Wang

In recent years, more and more free surface flow models have been applied to the investigations of real-life problems in a variety of water resources and environmental engineering projects for performing engineering analysis and designs, assessing their ecological and environmental impact, selecting the most cost-effective designs and even supporting the optional decision making in planning, management and policies. Sometimes, the requirement of detailed flow field information demands the use of three-dimensional modeling capability at least in the near field. As the project scope becoming larger and the total cost becoming higher than ever before, concerns about the free surface flow model's validity in simulating the real-life problems have been increased. No matter how good a model has been systematically tested with successes in he Mathematical (Code) Verification and Physical Process Validation, the project directors would want to determine how realistic are the model's predictions comparing to the field measurements of the designated application site. It is important for them to know not only the accuracy of the short-term predictions, but also the reliability of longer-term effectiveness and impact. The objective of this chapter is to provide model testers a methodology to conduct the application site validation.

To conduct a successful Application Site Validation, one must first have a sufficient quantity of high quality field data collected at the study site. Preferably, the data collection locations and time are well distributed. Each field data set needs to cover all locations and time needed by the model validation. Multiple sets are needed, because one or more sets are needed for calibrating the model parameters to insure that the unique site-specific characteristics can be taken into account by the model with the calibrated physical parameters, and the data sets used for calibrations should not be re-used for validation. Due to the fact that the boundary geometry and surface roughness of the free surface flows in the natural environment are subject of changes from time to time, especially in case of flows on mobile boundaries and/or during extreme hydrological events. Therefore, the field data collections may need to be conducted also from time to time, especially for longer time flow simulations, say 1, 5, 10 or longer years.

It has happened often in the past, that the a numerical modeler was award a contract to perform a simulation project of free surface flow with a set of inadequate number of data not at all key locations, and most likely quite a few years ago. The modeler was told that those were all the data he/she could expect to have, because there was no more funds budgeted for additional field data collection. This practice caused many application site studies to have used numerical models without validation. Some of them reported excellent agreements between the model simulations and a few field measurements, but didn't mention how those agreements were achieved. It has been a common knowledge that the fine-tuning of the model parameters can do the trick. In the opinion of this Task Committee, the practice of fine-tuning is unacceptable. Therefore, the T. C. strongly suggests that all modelers and users to demand a sufficient quantity of high quality data or the funds to collect these data, before conducting real-life site study project.

Once a sufficient quantity of high quality field data is available, one can proceed to the step of Application Site Validation in two sub-steps. First, a part of the available data are used to calibrate the model parameters such as turbulence eddy viscosity, bed and bank resistance, Manning's n values, wind shear coefficients, etc. After the values of these parameters have been calibrated, they should not be modified without acceptable justifications. The model calibrated by the site-specific geo-physical, topographic and hydrologic characteristics should be able to predict realistic site-specific physical phenomena. The comparison between simulated results and the remaining field measurements (un-used for calibrations) can determine whether the model is validated or not.

The model testers are advised to perform the Calculation Verification to estimate the accumulated calculation errors of the numerical solutions. The basic methodology given in Section 2.18 and the examples presented in Sections 5.3.10 and 5.4.7 can be used as references.

The model users conducting the Application Case Validation are expected to meet a reasonable level of agreement between the model predictions and field measurements before claiming the model is validated. As stated in Chapter 2, the general (and necessarily vague) level of acceptable agreement must be determined by common practices in hydraulic engineering for the particular type of modeling project; in some cases, the level of acceptable agreement may also be affected by the accuracy of the field measurements available. At the very least, acceptable validation normally requires that the spatial and temporal variation trends in the field variables predicted by the numerical model should be in agreement with the measurements. In this report the Task Committee has avoided making any very specific recommendations regarding error tolerances for achieving a pass/fail grade in the validation process because the acceptable tolerance is dependent on the purpose for the modeling project. In the end, it is often those that must rely on model results, such as governmental agency managers, that have the final say on what the acceptable pass/fail tolerance must be for a particular project and purpose. A model pass/fail decision made by engineering management would fall under the category of Certification as discussed in Chapter 2.

6.2 OVERVIEW OF TEST CASES

Contributor: Richard A. Schmalz, Jr.

Unsteady flow test cases are presented for the following five estuaries: 1) Chesapeake Bay (East Coast, USA), 2) San Francisco Bay (West Coast, USA), 3) Apalachicola Bay (Gulf Coast, USA), 4) Meldorf Bay (North Sea Coast, Germany), and 5) Tokyo Bay (Japan). In addition, a steady flow test case for Victoria Bendway, Mississippi River is also provided. For each case a computational grid, bathymetry, initial and boundary conditions are provided. A set of field data is given for model validation. Salient features of each test case are outlined below.

The Chesapeake Bay test case extends over the one month period of September 1993. Chesapeake Bay is the largest estuary on the US East Coast with a moderate tidal range and neap-spring tidal variability. The initial conditions provided are considered in near dynamic balance and only a short 3-5 day spin-up period is used thereby allowing a near one month simulation of the neap-spring modulated residual velocity and temperature inversion during episodic wind events. Water surface elevation, temperature, salinity, and residual velocity profiles are provided to focus on the vertical mixing mechanics under wind and surface cooling and to characterize the tidal dynamics, thermohaline and residual circulation.

The San Francisco Bay test case extends over the seven month period 1 December 1997 - 30 June 1998. San Francisco Bay represents a large estuary with high and low water asymmetries and strong neap-spring tidal variability. A one-month spin-up period is used thereby allowing a six month simulation in 1998 of the neap-spring modulated residual velocity and salinity stratification development during episodic freshwater inflows. Wind, atmospheric pressure anomaly effects and temperature gradients are not included as their effects are considered small. Salinity and residual velocity profiles are provided to focus on the vertical mixing mechanics under tidal forcing and freshwater inflows. Water surface elevation and salinity data provide for a validation of the model's ability to simulate the tidal dynamics and the gravitational circulation.

Apalachicola Bay is a shallow water system similar to many Gulf Coast estuaries. The tidal range is small and meteorological forcings are often dominant. The test case extends over the period May through November 1993. A one-month spin-up period is used thereby allowing a six month simulation in 1993 of the salinity associated with the Spring-Summer freshwater inflow patterns. A uniform temperature distribution and negligible atmospheric pressure anomaly are assumed. Water surface elevations and salinity data are provided to assess the ability to simulate the horizontal and vertical salinity distributions under the influence of wind and freshwater inflows.

Meldorf Bay represents an extremely complex shallow estuary with braided tidal channels. The tidal range is large with significant neap-spring variability and an extensive area of tidal flats occupies the system even under neap tide conditions. Two separate cases are presented. The first case focuses on the simulation of the astronomical tide over May 1990, during which stratification effects are small. Water surface elevations are provided to test the ability of the model to simulate tidal flats under both spring and neap tide conditions. The spin-up period is limited to only the first day. The second case considers complete meteorological forcing and provides 28 vertical velocity sections along one repeated north-south transect over a complete tidal cycle. The test case encompasses the period 15-21 May 1999 and assumes a one day spin-up due to the weak stratification. The second case tests the ability of the model to simulate the three-dimensional velocity structure over a section with extreme bathymetric variability (two channels cut through the section).

Tokyo Bay represents an estuarine system with an irregularly shaped deep natural channel system (25-35m). The tidal range is large order 2 m with a large diurnal inequality resulting in appreciable residual currents of over 20 cm/s. The computational boundary is directly across the entrance and necessitates a reflective boundary condition. It is this feature that makes the test case so intriguing. The user has the option of comparing results for a directly forced elevation condition with a reflective boundary condition. Both tidal elevations at eight gauges and currents at ten gauges are provided for model validation. Freshwater inflows for four rivers are provided and a uniform density structure is imposed over a six day simulation period based on data collected during the period 25 August through 25 October 1983. A representative wind field is also provided to allow for the computation of residual currents.

Victoria Bendway on the Mississippi River represents an extremely complex curved channel section with a very comprehensive three-dimensional velocity measurement set at 34 cross-sections. At each cross-section, order thirty vertical profiles were taken. Several secondary helical flow patterns were well represented and serve as an excellent test of the turbulence scheme and numerical model formulation. Results from a two-dimensional flow model are provided to set the upstream and downstream boundary conditions.

Input data for each test case dataset is provided along with appropriate read me files. Users may also directly contact the first contributor of each test case for further information on alternate forcings, data accuracy issues, and updated datasets.

6.3 CHESAPEAKE BAY TEST CASE

Contributors: Billy H. Johnson, Harry V. Wang and Mac Sisson

6.3.1 Background

Hydrodynamic modeling of the Chesapeake Bay has been conducted at the U.S. Army Engineer Research and Development Center (ERDIC) at Vicksburg, MS and at the Virginia Institute of Marine Sciences (VIMS) at Gloucester Point, VA (Johnson, et al., 1993 and Wang and Johnson, 2000) to provide flow fields to a three-dimensional (3D) water quality model (Cerco and Cole, 1993 and Cerco and Meyer, 2000). Thus, the modeling study was driven by the need to predict the impact on Chesapeake Bay water quality due to various scenarios, e.g., reducing loads into the bay. Such modeling efforts require long-term (years) simulations. Therefore, in addition to requiring that the numerical hydrodynamic model reproduce basic processes such as temperature inversions, wind mixing, and residual circulation, the computational time required to produce yearlong solutions must be reasonable. For example, with the numerical grid to be presented, yearlong simulations of the hydrodynamics should not take longer than about 1 day of computational time on a state-of-the-art computer in order for such water quality studies to be feasible.

6.3.2 Objectives

In a large water body, the objectives are to determine a 3D free-surface numerical hydrodynamic model's

- ability to simulate temperature inversions,
- ability to simulate mixing over the water column due to wind events,
- accuracy in the computation of water surface elevations, and 3D fields of velocity, salinity, and temperature,
- ability to accurately compute residual currents throughout the water body since they are crucial in water quality studies, and
- computational efficiency.

6.3.3 Approach

A structured boundary-fitted grid is provided along with initial fields of salinity and temperature that are cell-centered. In addition, the bathymetry of the bay is represented as cell-centered water depths on the numerical grid. If the user modifies the numerical grid provided, these data files must also be modified. For example, the initial salinity and temperature fields, along with the water depths, must be interpolated to provide appropriate values on the users' numerical grid.

Boundary condition data are provided for a month long simulation of flow conditions during September 1983 in Chesapeake Bay. These consist of water surface elevations, salinity, and temperature data at the bay mouth; freshwater flow on the various rivers, wind data from which the surface stress is computed, and surface heat exchange data in the form of an exchange coefficient and an equilibrium temperature. Unless the user can convert the surface heat exchange coefficients and equilibrium temperatures into the type data required by his/her model, the model should be modified to compute surface heat exchange using the concept of an equilibrium temperature.

Observed interior data in both graphical (postscript files) and digital form are provided at several stations for comparison with model results. These consist of water surface elevations, water velocity, salinity, and temperature. These data were provided by Blumberg, et. al. (1991).

6.3.4 The Physical Domain

As shown in Figure 6.3.1 (file STATM.PS), Chesapeake Bay is located on the East Coast of the United States, and is the largest estuary in the United States. The bay is about 300 km long from its mouth to its northern boundary, with its width varying from about 6 km to about 48 km. The length of the bay allows for a complete M2 tide to be contained within the bay. The average water depth in the bay is about 8 m; however, a deep natural channel with an average depth of about 25 m traverses much of the bay. Several rivers feed freshwater (see Figures 6.3.2 and 6.3.3) (files PRIV_Q1.PS and PRIV_Q2.PS) into the bay, with about half of the annual inflow of 2000 m^3/s coming from the Susquehanna River at the northern end of the bay. Wind forcing accounts for much of the energy in the bay.

6.3.5 The Numerical Grid and Bathymetry

The planform numerical grid is shown in Figure 6.3.4 (file CHESBAY.PS). There are 1965 surface cells. In the numerical modeling work conducted by ERDIC and VIMS, the water column is represented by a maximum of 20 cells, each 1.52 m thick below the top layer. However, the top layer was set to be 2.13 m relative to mean sea level (MSL) in the ERDIC / VIMS model to prevent cells from drying out. With this vertical resolution, there are a total of 7485 computational cells. The numerical grid is listed in Virginia State Plane Coordinates in feet in the file labeled CHESBAY.GRD in Appendix A. Instructions for reading the data are given in the READ_ME.GRD file in Appendix A. The cell-centered water depths are presented in file CHESBAY.DEP, with read instructions located in READ-ME.DEP.

6.3.6 Initial Conditions

The initial salinity and temperature fields in the ERDIC/VIMS model were constructed to be as close to reality as possible. These were established by using the available field data on 1 September, 1983. The value for each individual cell of the 3D grid was first set to be that of the nearest field data point. The resulting 3D fields of salinity and temperature were then smoothed several times. The initial velocity field was taken to be zero and the water surface was taken to have a zero elevation relative to mean sea level. Initial conditions in the user's model should reflect these conditions. Files SEP83.SA15V and SEP83.TE15V list the initial salinity and temperature data applied on the ERDIC /VIMS 3D grid. Read information is given in files READ_ME.SA15V and READ_ME.TE15V.

6.3.7 Boundary Conditions

On the open ocean, the water surface elevation must be prescribed along with the salinity and the temperature. The water surface elevations presented in file SEP83.TIDE are observed data at the mouth of the bay. Read instructions can be found in READ_ME.TIDE. Figure 6.3.5 (file PTIDE1.PS) presents a graphical presentation of these data. The values of the salinity and temperature are observed data from the moored station at the bay mouth labeled 40 on Figure 6.3.6 file (STATWT.PS). These data are listed in file SEP83.TSATE, with read instructions given in file READ_ME.TSATE. Graphical presentations of these data are given in Figures 6.3.7 and 6.3.8 (files PSAL040.PS and PTEM040.PS).

At river boundaries, daily values for freshwater inflows are prescribed along with the temperature of the inflow. Salinity is assumed to be zero. File SEP83.RIVR contains the various river inflows previously shown in Figures 6.3.2 and 6.3.3. The water temperatures of those inflows are listed in file SEP83.RIVRT. Read instructions are provided in files READ_ME.RIVR and READ_ME.RIVRT, respectively.

At the surface, wind stress in the lower to middle bay is computed from linearly interpolated wind data from the Norfolk and Patuxent stations. A similar interpolation is performed between the Patuxent and Baltimore-Washington International Airport (BWI) stations. North of the BWI station, only wind data from the BWI station are used. The locations of these stations are shown on Figure 6.3.1. Wind vectors from these stations are presented in Figure 6.3.9 (file PWIND.PS). File SEP83.WIND contains hourly values for the (x,y) components of the wind velocity at the three stations. All wind data have been adjusted to reflect over water winds. Read instructions are given in READ_ME.WIND.

The computation of the surface heat exchange is based on the concept of an equilibrium temperature. Therefore, the boundary condition on temperature at the water surface becomes:

$$\frac{\partial T}{\partial Z} = K(T - T_e), \tag{6.3.1}$$

where K is the surface heat exchange coefficient and T_e is the equilibrium temperature.

The daily-averaged equilibrium temperatures and surface heat exchange coefficients were computed from meteorological data collected at the Patuxent weather station (Figure 6.3.1) and are listed in file SEP83.TEK. Read instructions are given in

READ_ME.TEK. As previously noted, if the user's model computes surface heat exchange differently, the model must be changed to use these data.

6.3.8 Validation Data

Interior data are available at the stations shown on Figures 6.3.1 and 6.3.6. Water surface elevations at the stations shown on Figure 6.3.1 are listed in file TID_OBS.DAT and presented graphically in files PTIDE1.PS, PTIDE2.PS, PTIDE3.PS, and PTIDE4.PS. Instructions for reading the data file are given in file READ_OBS.TID. Water velocity data in the form of (x, y) components at the four stations shown on Figure 6.3.6 are presented in files VEL036.DAT, VEL040.DAT, VEL065.DAT, and VEL121.DAT. Plots of these data are presented in files PUV036.PS, PUV040.PS, PUV065.PS, and PUV121.PS. Read instructions for the data files are given in files READ_036.VEL, READ_040.VEL, READ_065.VEL, and READ_121.VEL. Similarly, salinity and temperature data at those stations are listed in files SAL_OBS.DAT and TEM_OBS.DAT, and are presented graphically in files PSAL036.PS, PSAL040.PS, PSAL065.PS, and PSAL121.PS. Read instructions for the data files are given in files READ_OBS.SAL and READ_OBS.TEM.

As noted, one objective of this field test case is to demonstrate the numerical model's ability to compute temperature inversions due to surface cooling and mixing of the water column due to wind events. Figures 6.3.10 and 6.3.11 (files PSAL036.PS and PTEM036.PS) show near surface and near bottom salinity and temperature, respectively, at the station labeled 36 on Figure 6.3.6. An inspection of the wind data at the Patuxent station presented in Figure 6.3.9 clearly shows a correlation between the water column mixing (Figure 6.3.10) and the wind event that occurred around the 20th of the month. The mixing was likely aided by the temperature inversion (Figure 6.3.11) that occurred around the 15th of the month. Reproduction of these processes will provide a good indicator of the adequacy of the vertical turbulence closure algorithm in the user's numerical model.

Quantitative data aren't available on residual currents throughout the bay during the month of September 1983. However, near-surface and near-bottom residual currents generated from data collected during 1977-83 are shown in Figure 6.3.12. These provide qualitative information on whether the user's numerical model is doing a good job of reproducing residual currents throughout the bay.

6.3.9 Remarks

Various concerns can result in the need to compute 3D flow fields in large bodies of water such as Chesapeake Bay. The need to provide flow fields to conduct 3D modeling of the bay's water quality was the driving force in construction of the databases presented here. In addition to accurately computing tides, tidal currents, salinity, and temperature under normal tidal conditions, such models should be able to accurately reproduce episodic events, e.g., water column mixing and set ups/set downs of the bay's water surface due to meteorological forcing associated with wind events. In addition, the numerical hydrodynamic model must be able to accurately compute residual currents resulting from the process of gravitational circulation in partially stratified large water bodies if model results are to be used in water quality studies. Finally, the computational efficiency of the model on a state-of-the-art computer must be such that yearlong computations on grids containing 7,500 to 10,000 computational cells can be accomplished within 1day. Finally, the computational efficiency of the model on a state-of-the-art computer must be such that yearlong computations on grids containing 7,500 to 10,000 computational cells can be accomplished within one day. (Comment: This may be an operational requirement, but it is not necessarily a requirement for the reader/modeler. E.g., a researcher may be exploring what resolution is required to reach a x level of accuracy, or if x is possible at any resolution.) It is believed that the Chesapeake Bay Test Case presented here provides a basis for aiding a user in determining if his/her model can adequately satisfy these requirements.

6.3.10 References

Blumberg, A.F., Johnson, B.H., Heath, R.E., Hsieh, B.B., Pankow, V.R., and Kim, K.W. 19991. "Data Employed in the Development of a Three-Dimensional, Time-Varying Numerical Hydrodynamic Model of Chesapeake Bay," TR HL-91-1, US Army Engineer Waterways Experiment Station, Vicksburg, MS.

Cerco, C.F. and Cole, T. 1993. "Three-Dimensional Eutrophication Model of Chesapeake Bay," Journal of Environmental Engineering, ASCE, Vol. 119, No. 6, pp1006-1025.

Cerco, C.F. and Meyer, M. 2000. "Tributary Refinements to Chesapeake Bay Model," Journal of Environmental Engineering, ASCE, Vol. 126, No 2, pp164-174.

Johnson, B.H., Kim, K.W., Heath, R.E., Hsieh, B.B., and Butler, H.L. 1993. "Validation of A Three-Dimensional Hydrodynamic Model of Chesapeake Bay," Journal of Hydraulic Engineering, ASCE, Vol 119, pp2-20.

Wang, H.V. and Johnson, B.H. 2000. "Validation and application of the second generation three dimensional hydrodynamic model of Chesapeake Bay," Water Quality and Ecosystem Modeling, Vol. 1, pp 51-90, Kluwer Academic Publishers, The Netherlands.

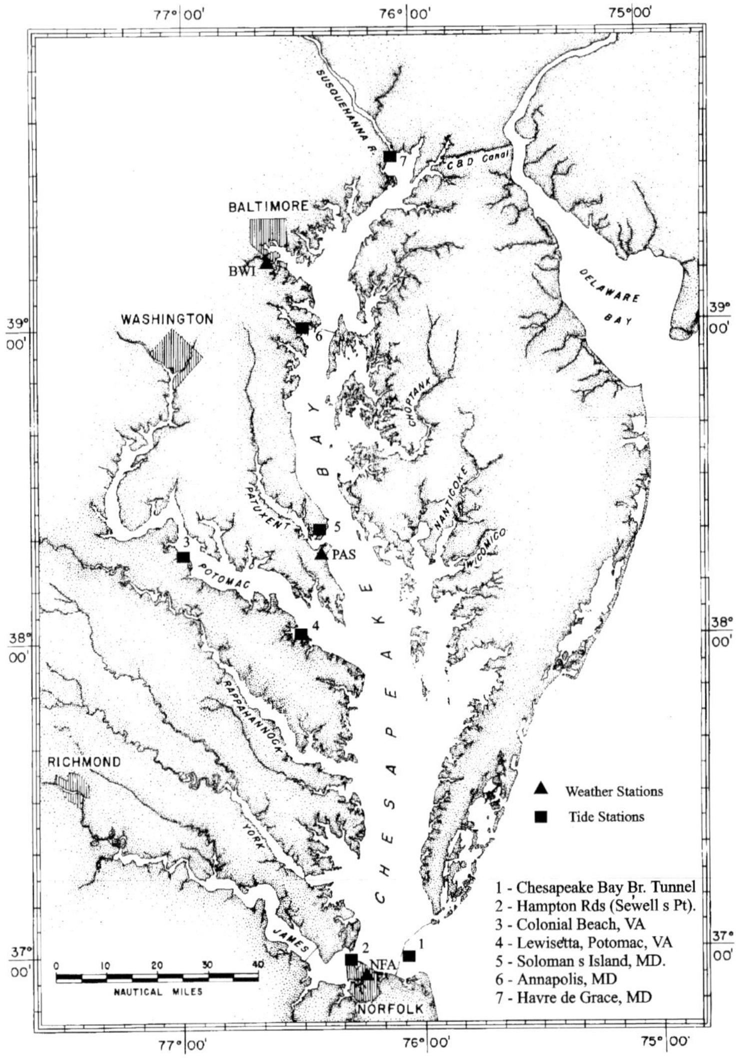

Figure 6.3.1 Chesapeake Bay Test Case: Base Map

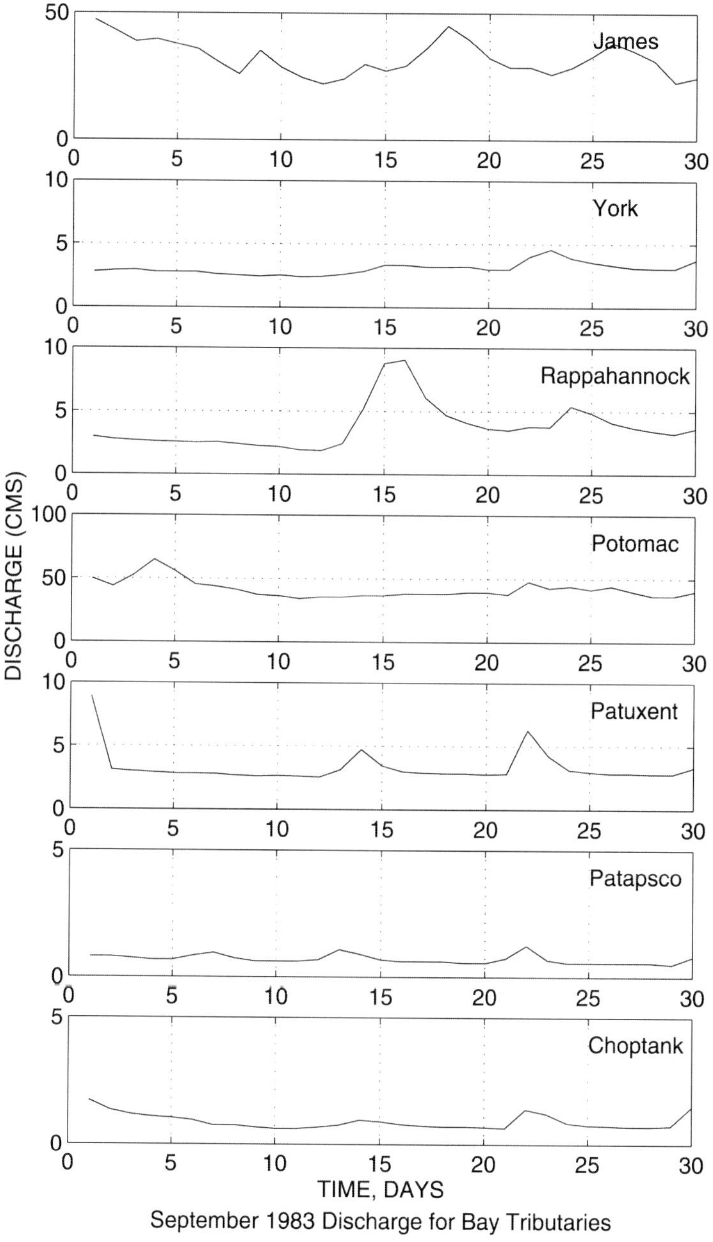

Figure 6.3.2 Chesapeake Bay Test Case: Freshwater inflows from tributaries

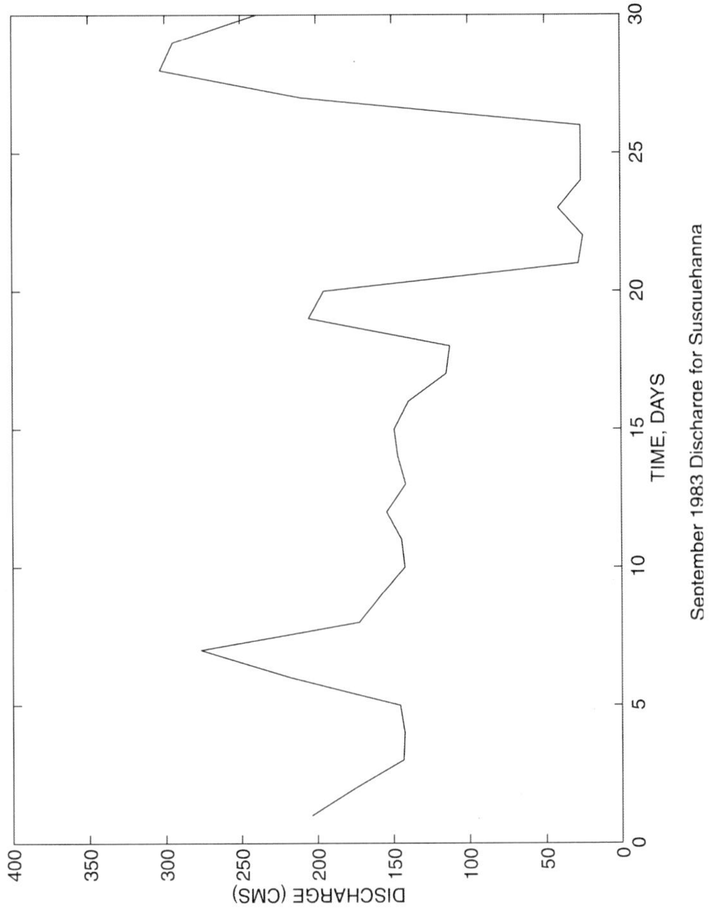

Figure 6.3.3 Chesapeake Bay Test Case: Freshwater inflow from the Susquehanna River

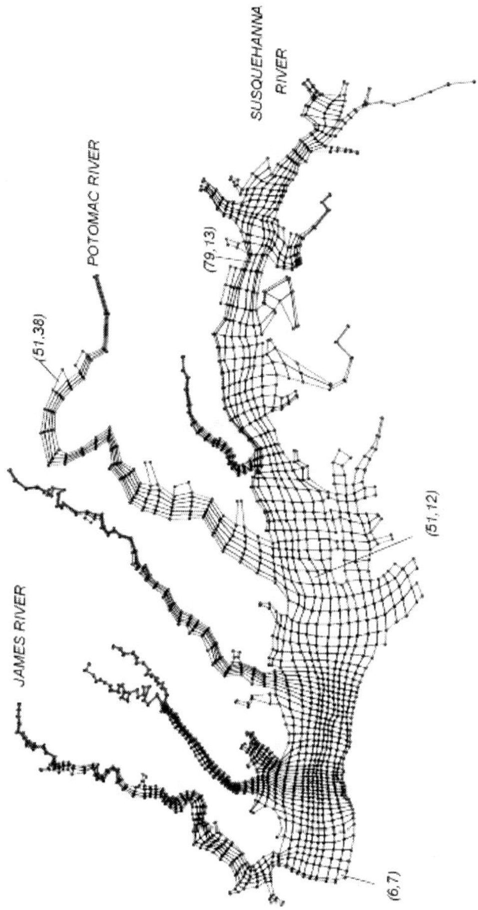

Figure 6.3.4 Chesapeake Bay Test Case: Boundary-fitted planform grid

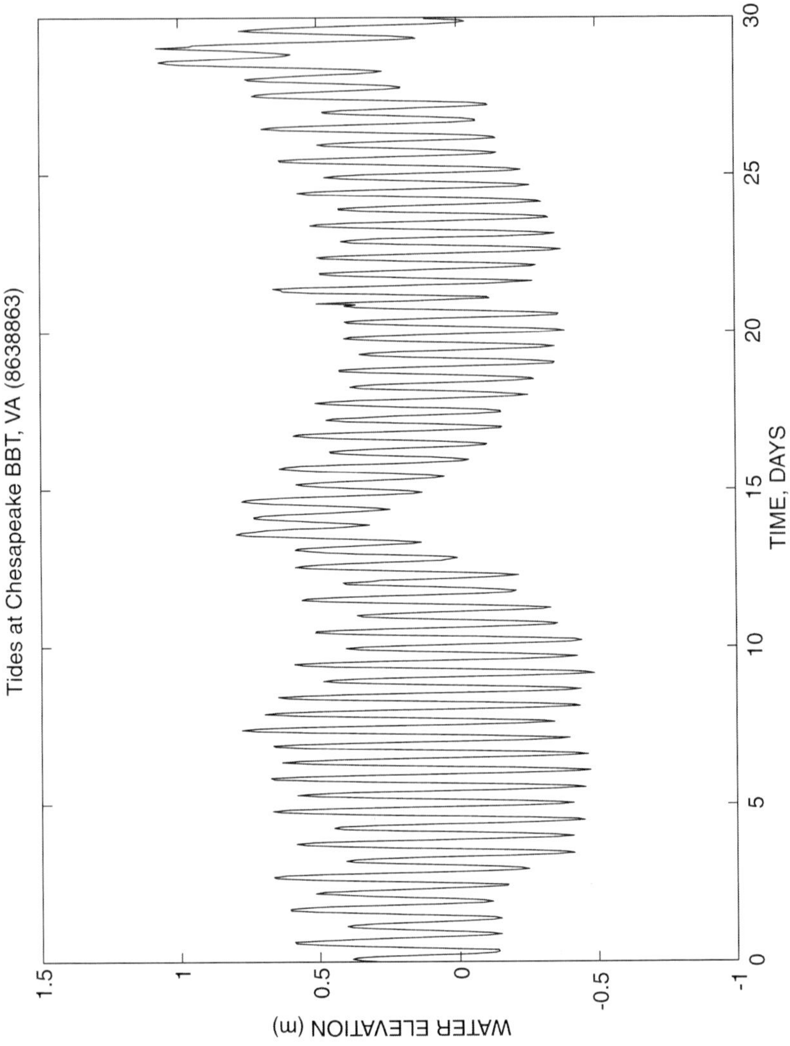

Figure 6.3.5 Chesapeake Bay Test Case: Water surface elevations at bay mouth

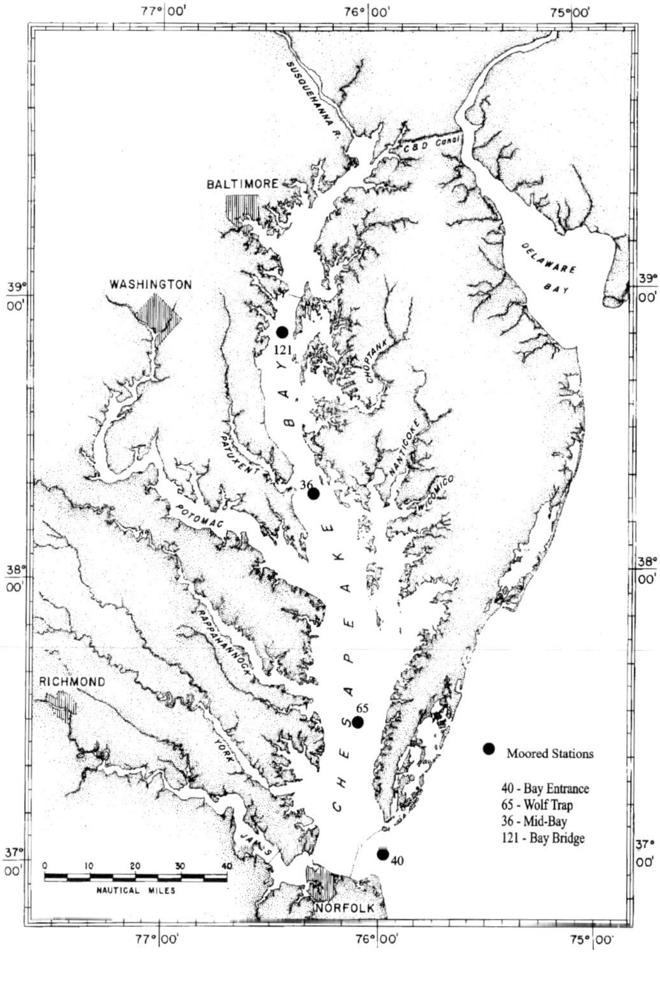

Figure 6.3.6 Chesapeake Bay Test Case: Locations of data stations

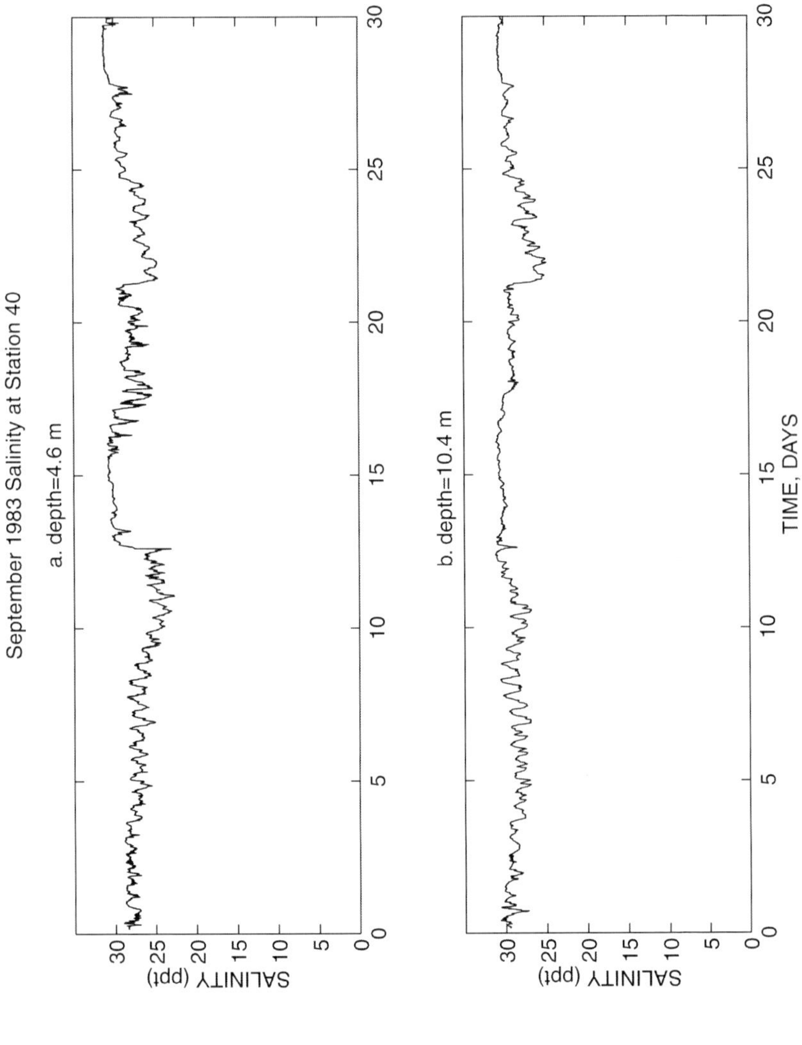

Figure 6.3.7 Chesapeake Bay Test Case: Salinity at bay mouth (Station 40)

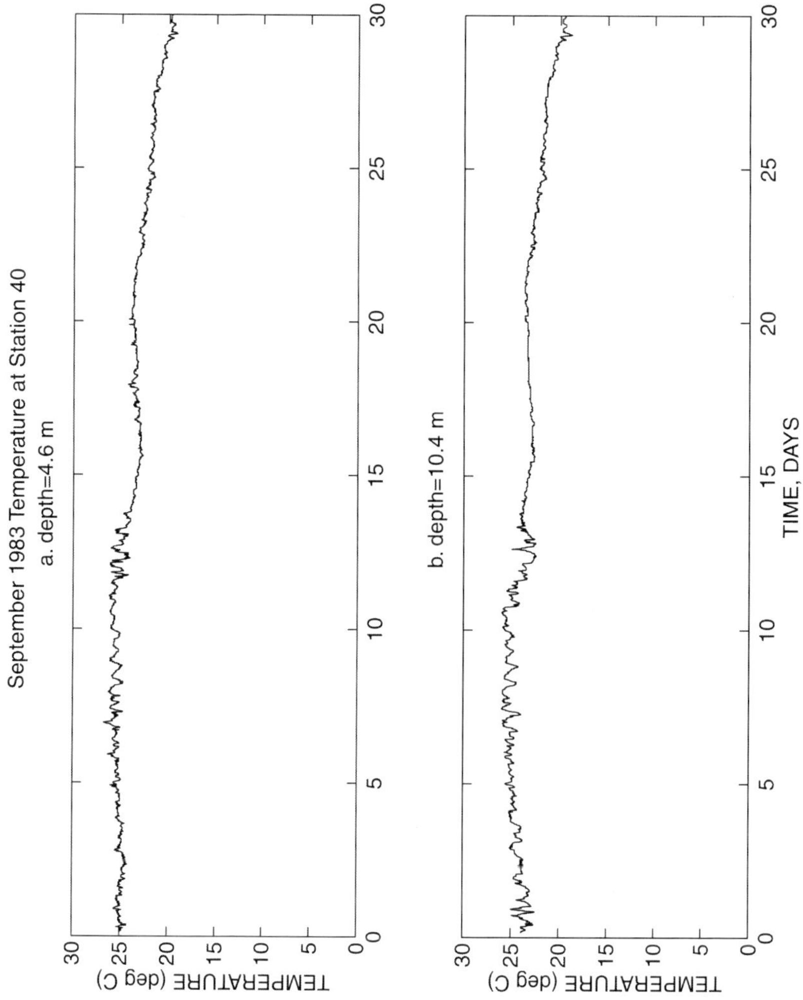

Figure 6.3.8 Chesapeake Bay Test Case: Temperature at bay mouth (Station 40)

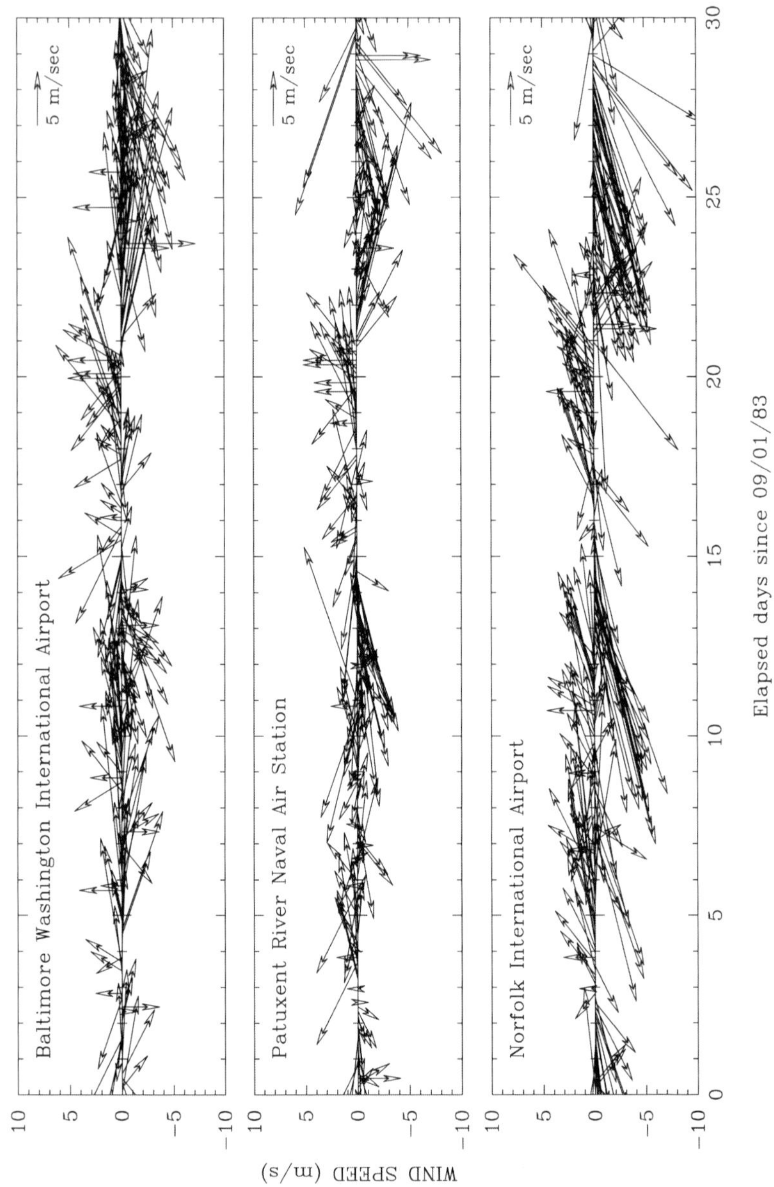

Figure 6.3.9 Chesapeake Bay Test Case: Wind data

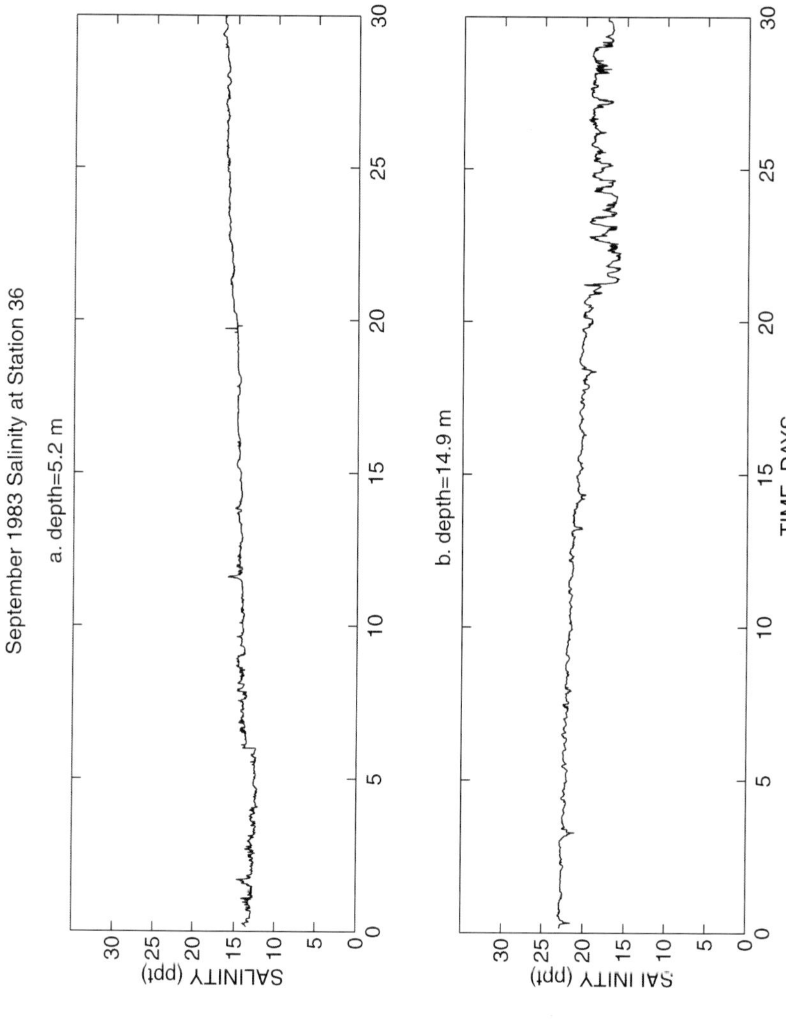

Figure 6.3.10 Chesapeake Bay Test Case: Salinity data at Station 36

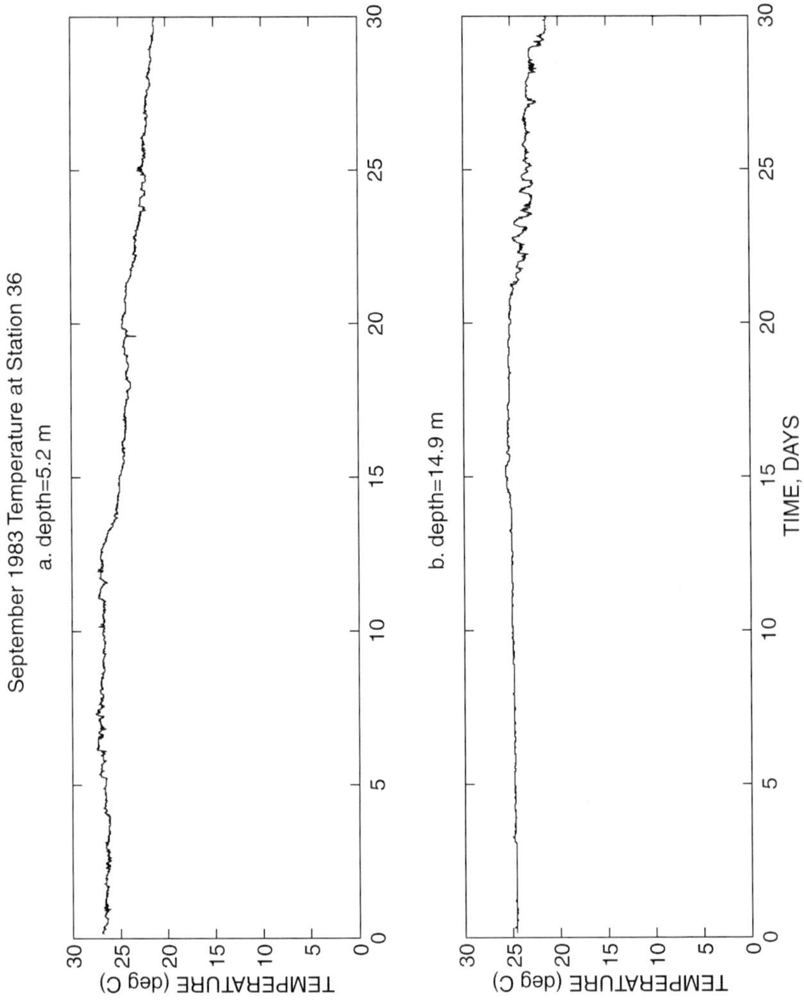

Figure 6.3.11 Chesapeake Bay Test Case: Temperature data at Station 36

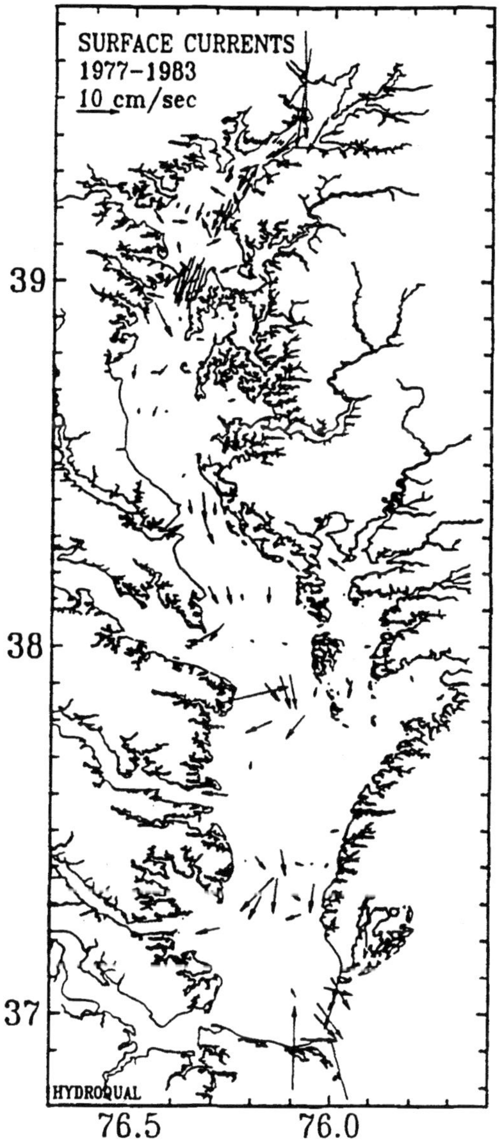

a. Near-surface

Figure 6.3.12 Chesapeake Bay Test Case: Residual currents computed from observed data (Sheet 1 of 2)

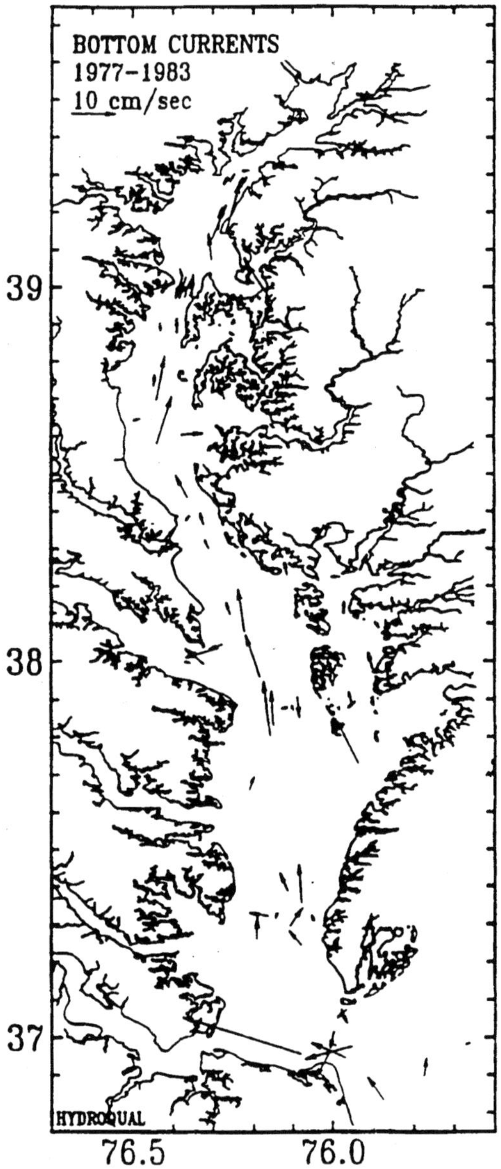

b. Near-bottom

Figure 6.3.12 Chesapeake Bay Test Case: Residual currents computed from observed data (Sheet 2 of 2)

6.4 SAN FRANCISCO BAY TEST CASE

Contributor: Peter E. Smith

6.4.1 Background

Three-dimensional (3D) hydrodynamic models of San Francisco Bay were developed by the U.S. Geological Survey (USGS) beginning in the late 1980's (Smith and Cheng, 1990). The code by Sheng et al. (1986) was the first to be applied by USGS. Later a special (semi-implicit) version of the Estuarine and Coastal Ocean Model (ECOM-si), developed under contract for the USGS by Blumberg (1991), was tested and evaluated. Recent applications are being done using the USGS Semi-implicit 3D (Si3D) model (Smith, et al., 2005; Smith, 2006). Because most of the USGS model applications have been used to increase understanding of how hydrodynamics affects aspects of the ecology of the bay, studying the mechanisms affecting long-term transport and the tidally-averaged (residual) circulation is emphasized in this test case. Owing to the complex bathymetry in the bay, a 3D model is needed that is both stable and accurate when the grid resolution is not highly refined. Even with the speed of modern-day computers and the recent advances in the efficiency of hydrodynamic model algorithms, it is not possible to use grids with the highest desirable level of refinement for long-term (seasonal or longer) simulations of the bay without requiring excessively large amounts of computer resources.

The San Francisco Bay test case described here is especially challenging because it includes a period during the winter of 1998 when the freshwater inflows into the bay were unusually high causing strong density-driven (gravitational) circulation and vertical stratification of salinity. The test case is a challenging one, both numerically and in terms of the physical processes that must be simulated, but was designed so that sophisticated gridding or wetting-and-drying schemes are not required. By avoiding these requirements, models without these schemes will be able to use the test case.

6.4.2 Objective

The objective is to assess the ability of a 3-D baroclinic model to predict circulation in a stratified estuary with strong tidal forcing and steep bathymetric gradients. A particular emphasis is on predicting the tidally averaged, vertical profile of density-driven circulation as it is affected by freshwater inflow to the estuary and the spring-neap tidal cycle. A model used on the test case should be computationally efficient enough so that the proposed seven-month simulation can be done without requiring more than one or two days of computer run time on a fast workstation.

6.4.3 Approach

The computational domain for the test case simulation includes all of San Francisco Bay and part of the coastal ocean (Figure 6.3.1). Bathymetric data are provided for a rectangular grid of uniform, square cells in the horizontal plane (Figure 6.3.2). Forcing of the model is through three open boundaries (one on the west, two on the east) where sea level (or flow) and salinity are specified from a measured time series. The effects of temperature gradients and surface wind stress are neglected in the simulation, although the effects of salinity gradients are not. In addition to bathymetry data, input for the simulation includes data for initial conditions and boundary conditions. Vertical variations in the salinity initial condition are provided at a fixed grid interval of 2.0 meters using points located at the center of each three-dimensional grid cell. The time period of the test simulation is seven months during the winter and spring of water year 1998 beginning on December 1, 1997 and extending through June 30, 1998. In addition to this being a period for which significant measured data were collected, it also includes a large freshwater inflow event into San Francisco Bay that caused strong density-driven (gravitational) circulation and vertical salinity stratification in the lower parts of the estuary. To validate a model, time series data are provided for water levels at 10 stations, salinities at 7 stations, and velocity profiles at 2 Acoustic Doppler Current Profiler (ADCP) stations.

6.4.4 The Physical Domain

San Francisco Bay is the largest estuary along the West Coast of the United States. The waters of the bay (including mudflats) have a surface area of 1,240 square kilometers (Conomos, et al., 1985) and consist of four major subembayments: Suisun Bay, San Pablo Bay, Central Bay, and South Bay (Figure 6.4.1). The 10-kilometer long Carquinez Strait connects Suisun and San Pablo Bays. The bottom topography of the bay strongly affects circulation and can be characterized as a drowned river floodplain incised by narrow channels that are 10 to 20 meters deep. Most of the bay is quite shallow, having an average depth (excluding mudflats) of only six meters below a datum of mean lower low water (MLLW) (Conomos, et al., 1985). At the deepest point near the Golden Gate, water depths exceed 100 meters.

The tides in San Francisco Bay are mixed, predominantly semidiurnal but influenced by a significant fortnightly, spring-neap cycle. An annual cycle consists of winter and summer periods when tides vary most (strong spring tides and weak neap tides) and spring and fall periods when tides vary least (weak spring tides and strong neap tides) (Smith, et al., 1995). The tidal range at the Golden Gate during a strong spring tide can reach 2.5 meters; the range of a subsequent neap tide can be one-half that of the spring tide. During spring tidal periods, currents at the Golden Gate can reach maximum speeds in excess of 2 meters per second. The configuration of the bay leads to tides that are a mixture of progressive and standing waves (Walters, et al., 1985). The enclosed embayment of South Bay oscillates mostly as a standing wave; resonance causes the amplitude of the tide height to increase toward the southern end of the bay. The tides in the northern reach propagate more as a progressive wave with amplitudes being diminished through Carquinez Strait and Suisun Bay.

Ninety percent of the annual volume of freshwater inflow to the bay enters from the delta of the Sacramento and San Joaquin Rivers (Figure 6.4.1). The tidally averaged freshwater flow from the delta into the bay is referred to as delta outflow. Because of the significant seasonal cycle of precipitation in California, delta outflow is naturally highest during the wet months of winter and lowest during the dry months of summer and early fall. Annual delta outflow varies widely from year-to-year because of annual fluctuations in precipitation.

Within San Francisco Bay, mixing the freshwater inflow from the delta with sea water from the Pacific Ocean results in longitudinal and vertical density gradients, primarily due to salinity gradients. In the northern reach of the bay, the longitudinal density gradient is sufficiently strong throughout most of the year to maintain a tidally averaged, two-layer gravitational circulation in the deep-water channel downstream of Suisun Bay. This density-driven circulation is characterized by tidally averaged currents that are landward in the lower part of the water column and seaward in the upper part (Hansen and Rattray, 1965; Officer, 1976; Smith, et al., 1991). As a general rule, the strength of density currents increases with increasing delta outflow as the longitudinal salinity difference between the ocean and the freshwater interface is established over a shorter reach of the estuary. Landward flowing density currents in the channels downstream of Suisun Bay have magnitudes on the order of 10 to 20 centimeters per second with even higher currents at some locations (Smith, et al., 1995). In addition to the longitudinal salinity gradient, the strength of density currents depend on water depth, the intensity of vertical mixing, and tidal energy (Walters, et al., 1985). Other factors being equal, density currents will be higher where the depth of water is greater. For those parts of the northern reach of the bay where channel depths are at a minimum—approximately 11 meters below low water at Pinole Shoal in San Pablo Bay and at the western end of Suisun Bay—the density currents are usually weak or absent. In the much deeper channels of Central Bay the magnitude of density currents can be large (~40 centimeters per second) (Smith, et al., 1995). The spring-neap cycle in the tides can cause significant variation in the magnitude of density currents at a particular location. As the amplitude (and energy) of the tide wave varies over a 14-day spring-neap tidal cycle, the amount of vertical mixing and the strength of gravitational circulation also will vary. During neap tides when tidal energy and vertical mixing are at a minimum, density currents are greatest. During spring tides, when tidal energy and vertical mixing are at a maximum, density currents are least.

During the winter season of most years, one or more large delta outflows to San Francisco Bay occur, creating conditions of greatly reduced salinity and increased salinity stratification in the northern reach of the bay that persist until outflow recedes and mixing returns the estuary to antecedent conditions. The degree of stratification is influenced by the water buoyancy from delta outflow and by the amount of vertical mixing derived from the energy in the tides. Walters, et al. (1985) and Smith, et al. (1995) discussed the effect of the spring-neap cycle in tidal energy on vertical stratification. Stratification generally is greater during neap tides than during spring tides because of less vertical mixing.

Water year 1998, the subject of this test case, was a strong El Niño year along the Pacific coast and was characterized by above average precipitation from January through June throughout the San Francisco Estuary watershed (Knowles, et al., 1998). Delta outflows were especially high during most of the winter/spring period in 1998, causing strong density currents and significant variations in the longitudinal and vertical salinity gradients in the lower estuary. A peak discharge of 9,450 cubic meters per second occurred on February 9th. During this period the freshwater interface was pushed down the estuary into San Pablo Bay. Flows were high enough to cause significant intrusion of low salinity water into South Bay (Schemel, 1998).

6.4.5 The Numerical Grid and Bathymetry

Bathymetry for the test simulation is defined for the numerical grid shown in Figure 6.4.2. The horizontal grid is composed of 151 x 169 square cells, each 500 meters per side. This is considered a course resolution grid for Carquinez Strait and Suisun Bay, but is adequate (for this test case) for representing the lower part of the estuary (downestuary of Carquinez Strait) where the estuarine geometric features are somewhat larger. The 500-meter grid was chosen over a finer grid so that the test case could be run economically using relatively modest computer resources. Grid cells shown in Figure 6.4.2 outside the numerical model boundary of San Francisco Bay represent land cells that are considered dry. Depth is provided at the corners of each cell measured in tenths of a meter downward from a local datum of mean lower low water. The depths in the file are written in decimeters. The depth at any dry corner of a cell is representated in the data file by a value of -90. An entire cell is defined as dry (a land cell) if any one or more corners are dry, or if the cell is specifically identified as a single land cell (an individual land cell, usually in the interior of a water body, that cannot be identified from the bathymetry file and must be identified separately) . A zoomed-in view of the numerical grid for Suisun Bay is shown in Figure 6.4.3 to illustrate the convention used to define land cells, model boundary land cells (land cells that are adjacent to the model water boundary), and single land cells. Also shown are thin-wall barriers which are impermeable barriers to flow along a cell face and have no thickness. As part of the data set, a special file is included that defines the (i,j) nodal coordinates for each of the thin-wall barriers and single land (dry) cells.

The raw data used for defining the bathymetric grid came from tens of thousands of soundings made by the National Oceanic and Atmospheric Administration (NOAA) during 1990–91. The raw bathymetry data are available on the World Wide Web (NOAA, National Ocean Service, Estuarine Bathymetry, accessed March 13, 2007). These data were entered into a Geographic Information System (GIS) and used to interpolate the depths at the corners of a 100-meter square grid using the GRID module of ArcInfo (ESRI, accessed March 13, 2007). The 500-meter grid was then developed from the 100-meter resolution bathymetric grid (U.S. Geological Survey, Access USGS, accessed March 13, 2007).

The bathymetric data should be corrected to a standard vertical datum such as the National Geodetic Vertical Datum of 1929 (NGVD29) or North American Vertical Datum of 1988 (NAVD1988). For this test case, NGVD29 was used as the reference datum for all the boundary condition data for water surface elevations. The bathymetry, therefore, should be corrected to NGVD29 before using it in a numerical model. An approximate correction can be made by adding one meter to all the depth values in the bathymetry file.

For the test simulation, the relatively small area of intertidal mudflats in the bay have been converted to permanent land or water areas to prevent any wetting and drying of grid cells. In particular, water depths along the northern shoreline of San Pablo Bay and along the southernmost shorelines of South Bay have been modified. For a z-coordinate (level plane) model, a choice of Δz= 2.0 meters will prevent wetting and drying of surface-layer cells for the largest tides that occur in San Francisco Bay. For smaller choices of Δ a model that allows for wetting and drying of surface-layer cells is required.

6.4.6 Initial Conditions

The hydrodynamic initial conditions in the bay at $t = 0$ are quiescence (zero velocity); the free surface elevation can be assigned a value of zero relative to mean sea level (NGVD29). The water temperature can be held constant at 17 degrees Celsius for the entire simulation. The salinity initial condition was estimated by using 3-D interpolations between all data available at the start of the simulation on December 1, 1997. The initial condition is supplied in a file with values defined at a vertical grid interval of 2.0 meters and at the center of each horizontal grid cell shown in Figure 6.4.2. Forty-five vertical layers are defined to accommodate the maximum cell depth at the Golden Gate. Although the maximum depth near the Golden Gate actually exceeds 100 meters, the bathymetry for this test case has been smoothed slightly so the maximum depth in the model grid is only 90 meters. To simplify reading the salinity initial condition file, dry points are not eliminated from the file but are flagged with a salinity of zero. There are no wet points where the initial salinity equals exactly zero. The salinity initial condition is the result of interpolations among data stations and was not derived from a prior "spin-up" model simulation to ensure consistency with hydrodynamic calculations. It is expected that at least several weeks of simulation will be needed to obtain realistic and consistent calculations of the hydrodynamic and salinity fields.

6.4.7 Boundary Conditions

Along the open ocean boundary, time series of water surface elevation and salinity must be defined. To simplify the test simulation, a single time series for water surface elevation is applied along the entire western boundary; it was derived from a weighted average of measured coastal sea level data collected by the NOAA at Point Reyes and Monterey Bay, California. The original data for these two stations and a map showing the station locations are available on the World Wide Web (NOAA, National Ocean Service, National Water Level Observation Network, accessed March 13, 2007). The open ocean boundary along the south was closed-off for this test simulation. Experiments with different boundary conditions for the coastal ocean have shown that results inside San Francisco Bay are not sensitive to small phase and amplitude variations along the ocean boundary. Salinity along the ocean boundary can be maintained at ocean values of 34.[1]

Boundary conditions for water level and salinity, or flow and salinity, must be specified at the Sacramento and San Joaquin River boundaries on the east side of the model domain. Files containing boundary-condition data for water level, flow, and salinity are available in Appendix A of this report. The modeler can choose a boundary condition of water level or flow, and use the other data for model validation. The data stations used in deriving the boundary conditions are shown in Figure 6.4.4. At the four flow stations ultrasonic velocity meters (UVMs) or Sideward-looking Acoustic Doppler Current Profilers (SL-ADCPs) are used to continuously measure tidal flows. These flows were tidally averaged using a low-pass, digital filter and combined to estimate the values of delta outflow shown in Figure 6.4.5a. An extensive data set of flow measurements made in the fall of 1998 in the vicinity of Sherman Lake (Ruhl et al., 2002) were used to calibrate transfer functions between the tidal flows at the four flow-measuring stations and flows at the model boundary. Figures 6.4.5b and 6.4.5c show the tidal flows estimated at the model boundary for the Sacramento and San Joaquin Rivers. Figure 6.4.5a shows the tidally averaged flows for the two rivers which exactly sum-up to equal delta outflow. To properly model the salinity field in San Francisco Bay, the delta outflow to the model must be correct.

Data for the water levels near the cities of Collinsville and Antioch (Figure 6.4.4) were collected by the California Department of Water Resources (Interagency Ecological Program, accessed March 13, 2007). The salinity data near these cities were collected by the USGS. A small shift was applied to the Collinsville data to account for the distance between the station and the model boundary. Because of the high delta outflows that occurred during the test simulation, salinity on the boundary was often zero.

All data values used for boundary conditions are provided at 15-minute time steps. For model applications, a smaller time step most likely will be required for boundary inputs; intermediate values can be linearly or parabolically interpolated.

6.4.8 Validation Data

Sources of the data to be used for validation include 10 stations for water level, 7 stations for salinity, and 2 stations for velocity profiles (measured with ADCPs). The location of the data stations are shown in Figure 6.4.6. A sample of the salinity data

---

[1] Salinity is expressed according to the Practical Salinity Scale, 1978 (Unesco, 1979). The salinity of freshwater is zero and of coastal ocean water near San Francisco Bay is approximately 34.

illustrating the very low salinities that occurred during the high delta outflows in February (days 32–59) is shown in Figure 6.4.7. Salinity was computed directly from specific conductance at each of the stations. Water temperatures also are available at all the stations where salinity was measured. At five of the salinity monitoring stations—San Mateo Bridge, Pier 24 (Bay Bridge), Point San Pablo, Selby (also called Wickland Oil Terminal), and Martinez—specific conductance was recorded at two depths, one near the surface and another at approximately mid-cross-section depth. At the Presidio station, salinity was available at only one depth that was about 1.3 meters below MLLW. The three stations in the northern reach that monitored salinity at two levels show clearly that the stratification varied with the spring-neap tidal cycle. The stratification was greater during the neap tidal periods than during the spring tidal periods as is shown in Figure 6.4.8 for the salinity measured at Point San Pablo. This effect was due mostly to a spring-neap variation in the vertical turbulent exchange of salt between the surface and lower layers and to a spring-neap variation in the gravitational circulation.

The two ADCPs (Figure 6.4.6) were deployed facing upward from the bottom of the water column to measure a long-term time series of velocity profiles. The vertical region measured by each ADCP begins at about 2 meters above the bed and extends to about 2.5 meters below the free surface. Water velocities are available at depth intervals (called bins). The ADCP located just south of the Richmond Bridge was deployed by the USGS and Stanford University for the three month period between April and June of 1998. The bin height of this instrument was set to 0.5 meter and the instrument was deployed in water of depth approximately 15 meters below MLLW. The ADCP located in the outbound channel near the Golden Gate is part of the NOAA's PORTS[2] for San Francisco Bay (U.S. Geological Survey, SFPORTS, accessed March 13, 2007). The bin height of this instrument was set to 2.0 meters and the instrument was deployed in water of depth approximately 50 meters below MLLW. Because of numerous gaps in the PORTS ADCP data, only data for January 1998 are available to validate the model.

Graphs of low-pass filtered (tidally averaged) currents for five of the ADCP bins measured during April-May 1998 near the Richmond Bridge are shown in Figure 6.4.9. Similarly, graphs of low-pass filtered currents for five of the ADCP bins measured during January 1998 near the Golden Gate are shown in Figure 6.4.10. A graph of the tidal current speed in the near-surface bin is included in both figures to help identify the spring and neap tides. A relation between the density-current speed and the tidal energy is apparent at both ADCP locations. At the Richmond Bridge location, during neap tides when tidal forcing was weak and vertical mixing was relatively small, the landward flow along the bottom was greatest; maximum current speeds were about 25 centimeters per second. During spring tides when tidal forcing was strong and vertical mixing was relatively large, the landward bottom currents were small (<10 centimeters per second) and the overall vertical shear in the profile was less than that during neap tides. During the spring tides, the near surface currents were strongest in the downestuary direction. During the period of time depicted in

[2] Physical Oceanographic Real-Time System.

Figure 6.4.9, the variability in delta outflow to San Francisco Bay was small, so changes in the magnitude of the gravitational circulation are not attributable to outflow variations. At the Golden Gate location, the tidally averaged currents were mostly directed upestuary for the entire January period. There were some variations in delta outflow during January of 1998, so it is possible that part of the variability in the nontidal currents could be explained by outflow variations. Figures 6.4.11 and 6.4.12 display in another way the differences in the tidally averaged velocity profiles measured during spring and neap tidal conditions at both ADCP measurement locations.

6.4.9 Remarks

The emphasis in this test case is to have a 3-D model reproduce the tidally averaged velocity profiles measured using the two ADCPs. Reproducing the horizontal and vertical changes in the salt field during the hydrograph of delta outflow and for the several months afterward is a second objective. The data set is useful for evaluating the performance of various turbulence submodels used in 3-D models for simulating turbulent vertical mixing.

6.4.10 References

Blumberg, A.F., 1991, A primer for ECOM-si [Estuarine, Coastal, and Ocean Model -- semi-implicit]: Hydroqual, Inc., Mahwah, N.J., Report prepared for the U. S. Geological Survey, 68 pp.

Conomos, T.J., Smith, R.E., and Gartner, J.W., 1985, Environmental Setting of San Francisco Bay: Hydrobiologia, vol. 129, pp. 1-12.

ESRI, ArcInfo: accessed March 13, 2007 at URL *http://www.esri.com/software/arcgis/arcinfo* .

Hansen, D.V., and Rattray, M., 1965, Gravitational circulation in straits and estuaries: Journal of Marine Research, no. 23, pp. 104-122.

Interagency Ecological Program, IEP HEC-DSS Time-Series Databases: accessed March 13, 2007 at URL *http://wwwiep.water.ca.gov/data.html* .

Knowles, Noah, Cayan, D.R., Uncles, R.J., Peterson, D.H., 1998, Simulated effects of delta outflow on the bay—1998 compared to other years: IEP Newsletter, Interagency Ecological Program for the Sacramento-San Joaquin Estuary, vol. 11, no. 4, Fall 1998, p. 29-32 (accessed March 13, 2007 at the URL *http://www.iep.ca.gov/report/newsletter/*).

National Oceanic and Atmospheric Administration, National Ocean Service, Estuarine Bathymetry, Pacific Coast Estuaries: accessed March 13, 2007 at URL *http://estuarinebathymetry.noaa.gov/pacific.html* .

National Oceanic and Atmospheric Administration, National Ocean Service, National Water Level Observation Network: accessed March 13, 2007 at URL *http://tidesandcurrents.noaa.gov/.*

Officer, C.B., 1976, Physical oceanography of estuaries (and associated coastal waters): New York, Wiley-Interscience, 465 pp.

Ruhl, C.A., Burau, J.R., and Oltmann, R.N., 2002, Hydrodynamic field study of a shallow estuarine subembayment, Sherman Lake, California, Proceedings of the Conference on Hydraulic Measurements and Experimental Methods 2002 on CD-ROM, American Society of Civil Engineers, Environmental and Water Resources Institute, Stanley Hotel, Estes Park, CO, July 28-August 1, 2002, 10 pp.

Schemel, Lawrence, 1998, Effects of delta outflow and local streamflow on salinity in South San Francisco Bay—1995-1998: IEP Newsletter, Interagency Ecological Program for the Sacramento-San Joaquin Estuary, vol. 11, no. 4, Fall 1998, pp. 32-33 (accessed March 13, 2007 at the URL *http://www.iep.ca.gov/report/newsletter/*).

Sheng, Y.P., Parker, S.F., Henn, D.S., 1986, A three-dimensional estuarine hydrodynamic software model (EHSM3D): Report prepared for the U.S. Geological Survey under Contract no. 14-08-001-21730, 162 pp.

Smith, P.E., 2006, A semi-implicit, three-dimensional model for estuarine circulation: U.S. Geological Survey Open-File Report 2006-1004, 176 pp. (accessed March 13, 2007 at the URL http://pubs.usgs.gov/of/2006/1004/).

Smith, P.E., and Cheng, R.T., 1990, Recent progress on hydrodynamic modeling of San Francisco Bay, California, Proceedings of the Conference on Estuarine and Coastal Circulation and Pollutant Transport Modeling: Newport, R.I., November 15-17, 1989, pp. 502-510.

Smith, P.E., Cheng, R.T., Burau, J.R., and Simpson, M.R., 1991, Gravitational circulation in a tidal strait, Proceedings of the 1991 National Conference on Hydraulic Engineering: American Society of Civil Engineers, Nashville, Tenn., July 29-August 2, 1991, pp. 429- 434.

Smith, P.E., Donovan, J.M., and Wong, H.F.N., 2005, Applications of 3D Hydrodynamic and Particle Tracking Models in the San Francisco Bay-Delta Estuary, Proceedings of the 2005 World Water and Environmental Resources Congress on CD-ROM, American Society of Civil Engineers, Environmental and Water Resources Institute, Anchorage, Alaska, 8 pp.

Smith, P.E., Oltmann, R.N., and Smith, L.H., 1995, Summary report on the Interagency hydrodynamic study of San Francisco-Bay-Delta Estuary, California:

Interagency Ecological Program for the Sacramento-San Joaquin Estuary, Technical Report no. HYDRO-IATR/95-45, California Department of Water Resources, Sacramento, Calif., 72 pp.

Unesco, 1979, Ninth report of the joint panel on oceanographic tables and standards: Unesco Technical Papers in Marine Science, vol. 36, no. 24.

U.S. Geological Survey, Access USGS—San Francisco Bay and Delta, San Francisco Bay Bathymetry Web Site: accessed March 13, 2007 at URL *http://sfbay.wr.usgs.gov/sediment/sfbay/* .

U.S. Geological Survey, USGS Research Site for SFPORTS, Current profiles: accessed March 13, 2007 at URL *http://sfports.wr.usgs.gov/sfports.html* .

Walters, R.A., Cheng, R.T., and Conomos, T.J., 1985, Time scales of circulation and mixing processes of San Francisco Bay waters: Hydrobiologia, vol. 129, pp. 13-36.

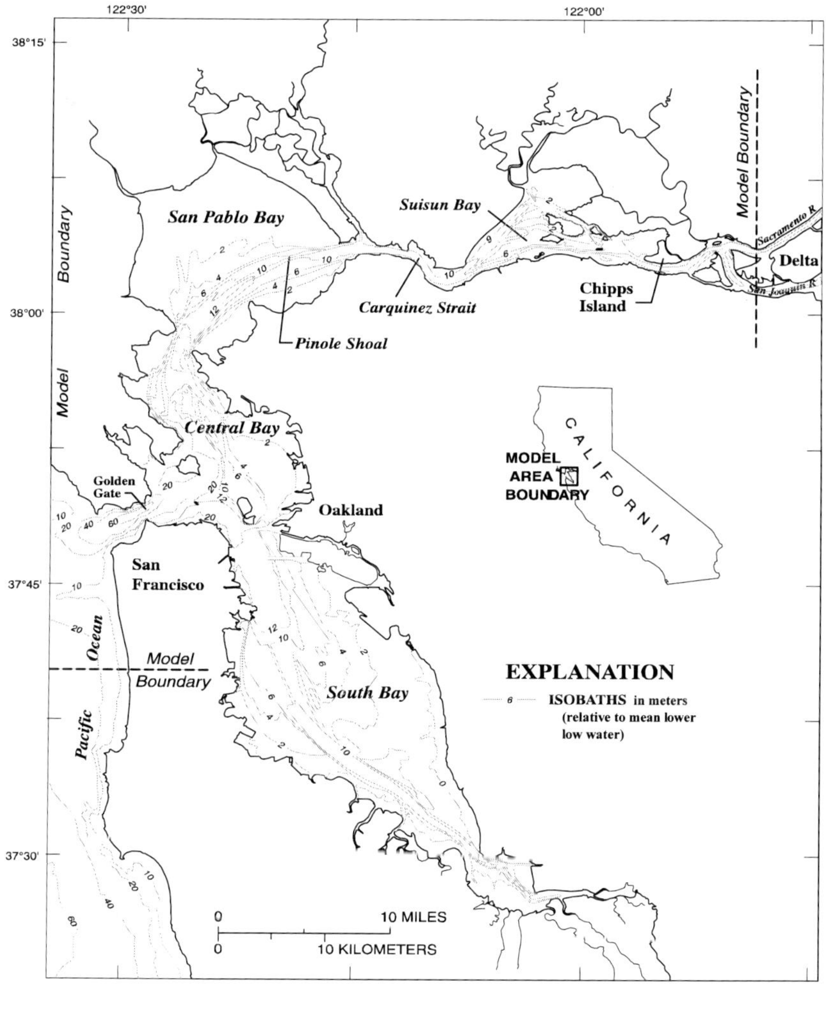

Figure 6.4.1 San Francisco Bay Test Case: Base map

San Francisco Bay is that part of the estuary seaward of Chipps Island including South Bay, Central Bay, San Pablo Bay, Carquinez Strait, and Suisun Bay. The northern reach of the bay is the region north of the Golden Gate. The Sacramento-San Joaquin River Delta lies landward of Chipps Island and is formed near the confluence of the Sacramento and San Joaquin Rivers.

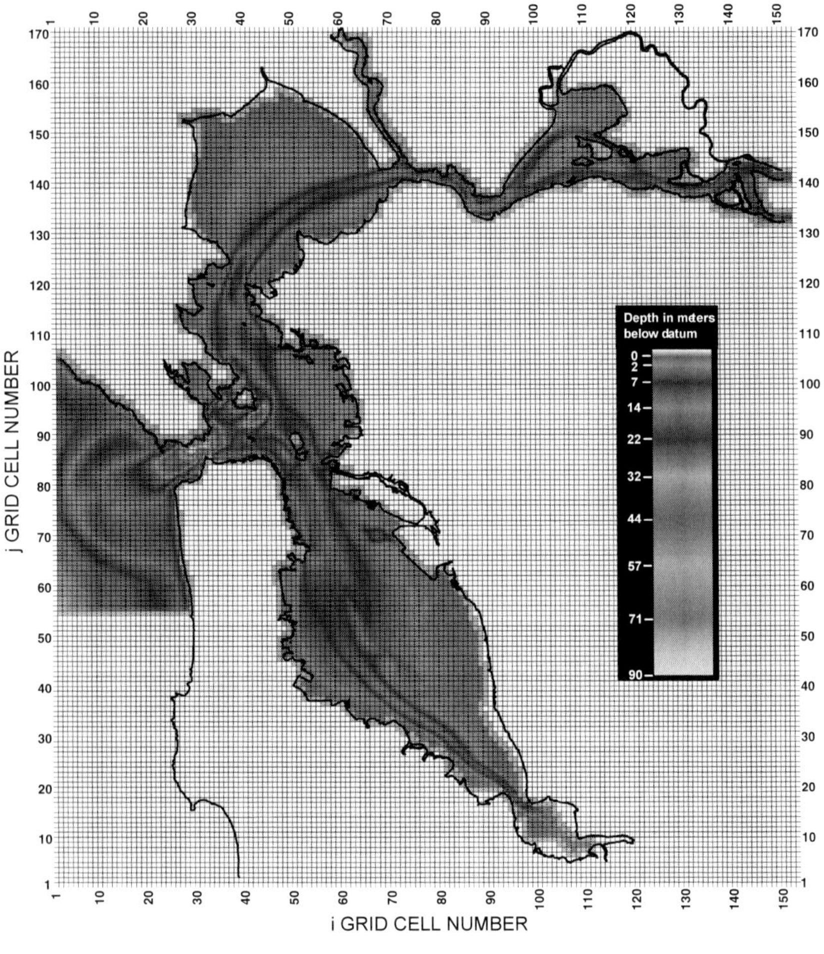

Figure 6.4.2 San Francisco Bay Test Case: Numerical grid and bathymetry
Each grid cell is a 500-meter by 500-meter square. The datum used for displaying bathymetry is mean lower low water. The black line defines the actual shoreline boundary of the bay.

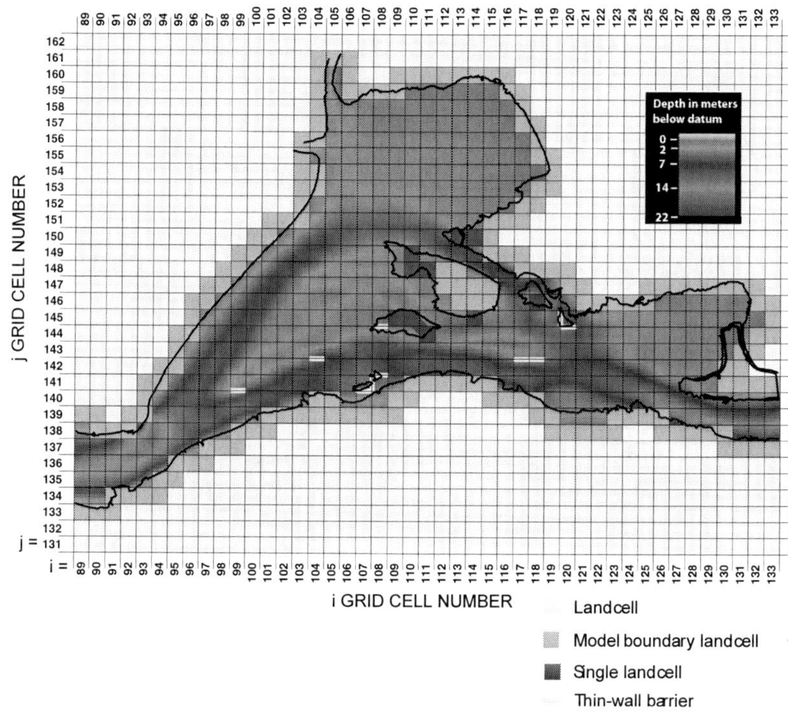

Figure 6.4.3 San Francisco Bay Test Case: Numerical grid of Suisun Bay
The grid illustrates the various types of land cells and shows locations of thin-wall barriers. Thin-wall barriers have no thickness but are impermeable to flow. The datum used for displaying bathymetry is mean lower low water. The black line is used to show the actual shoreline boundary of Suisun Bay.

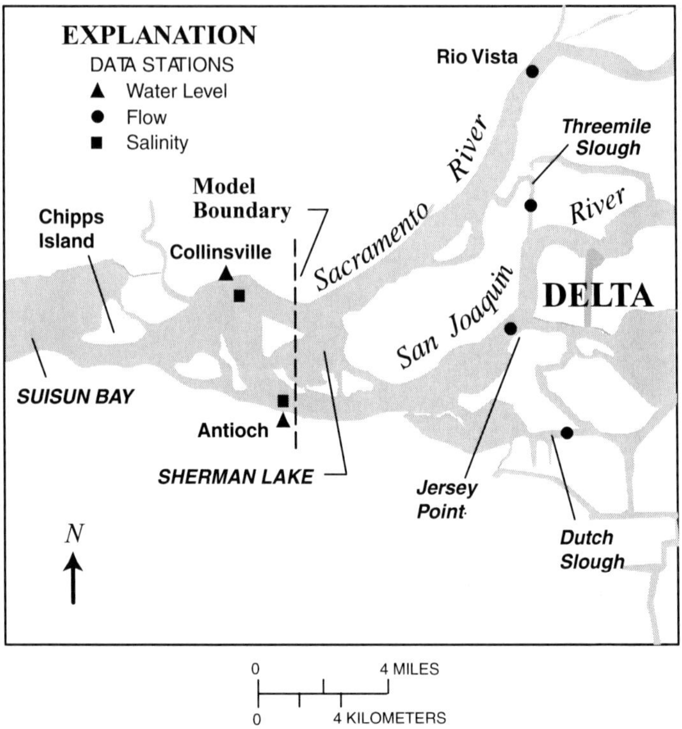

Figure 6.4.4 San Francisco Bay Test Case: Locations of monitoring stations used in defining boundary condition data for the Sacramento and San Joaquin Rivers

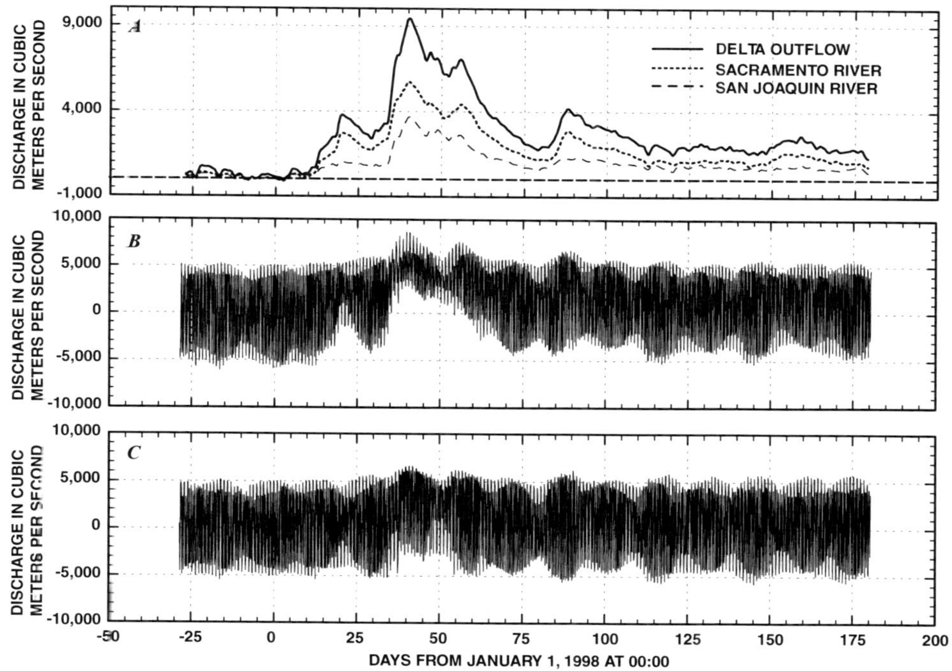

Figure 6.4.5 San Francisco Bay Test Case: River inflows

(*A*) Tidally averaged delta outflow into San Francisco Bay, (***B***) Sacramento River tidal flow just downstream of Sherman Lake, and (***C***) San Joaquin River tidal flow just downstream of Sherman Lake.

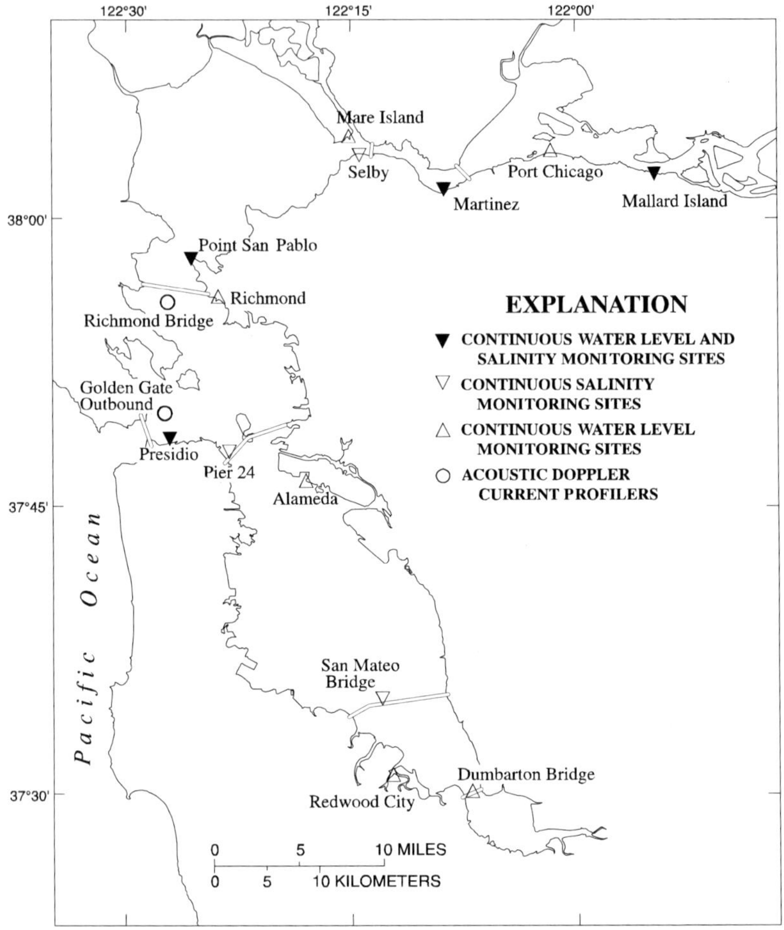

Figure 6.4.6 San Francisco Bay Test Case: Locations of stations for model validation

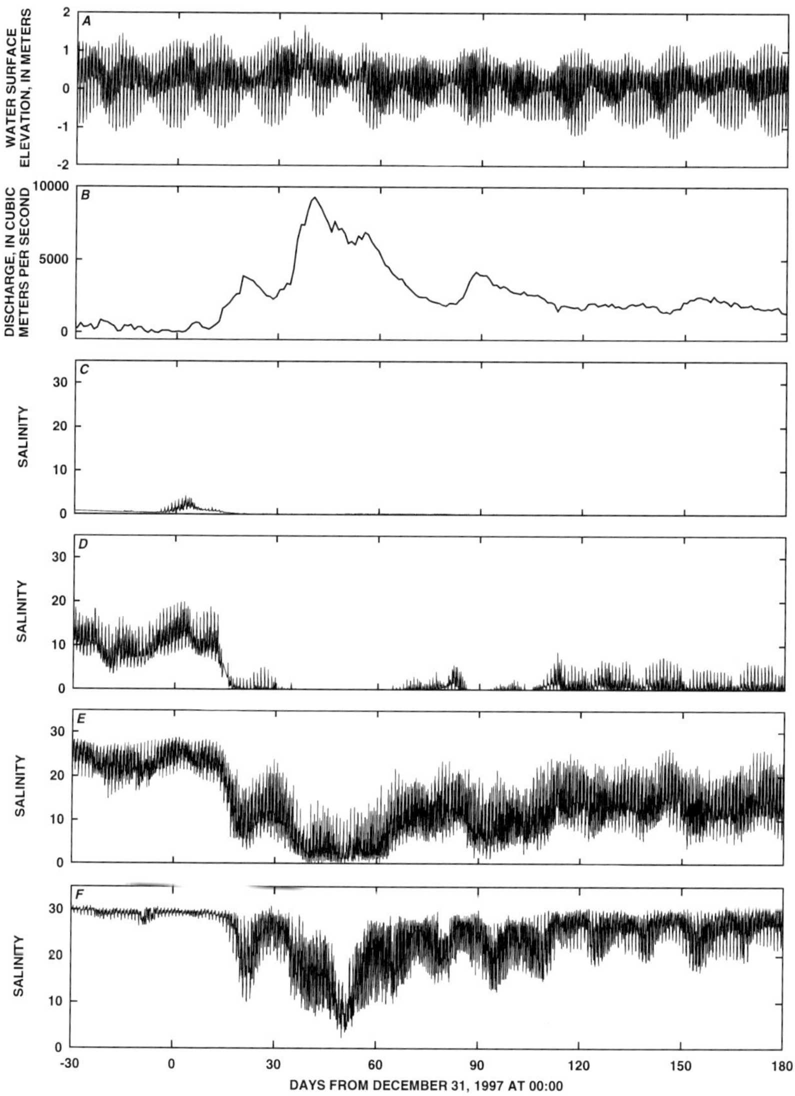

Figure 6.4.7 San Francisco Bay Test Case: Measured data for water surface elevation, delta outflow, and salinity

(*A*) Water surface elevation at Presidio measured from the National Geodetic Vertical Datum of 1929, (***B***) Delta outflow, (***C***) Bottom salinity at Antioch, in Practical Salinity Units (PSU), (***D***) Surface salinity at Martinez, in PSU, (***E***) Surface salinity at Point San Pablo, in PSU, and (***F***) Surface salinity at Presidio, in PSU. The salinity data for Antioch are used for the boundary condition on the San Joaquin River.

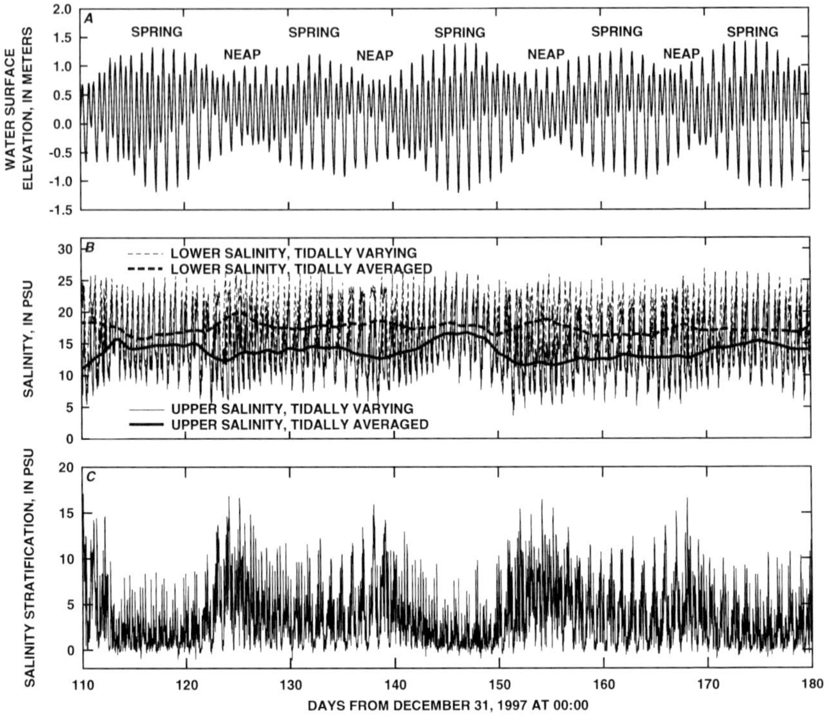

Figure 6.4.8 San Francisco Bay Test Case: Measured data showing the effect of the spring-neap tidal cycle on the salinity stratification at the Point San Pablo monitoring station

(***A***) Tidally varying water surface elevation at Presidio measured from the National Geodetic Vertical Datum of 1929, (***B***) Tidally varying and tidally averaged salinity in Practical Salinity Units (PSU) for the upper and lower measuring sensors at Point San Pablo, and (***C***) Tidally varying salinity stratification in PSU at Point San Pablo.

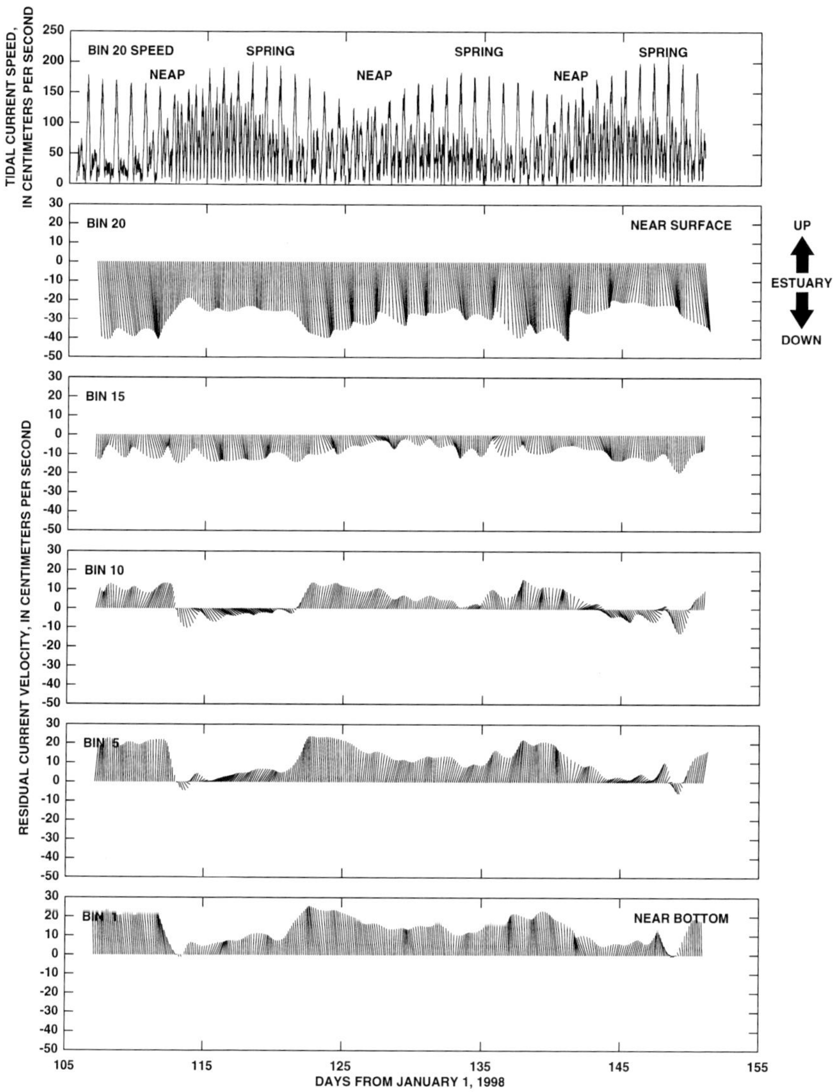

Figure 6.4.9 San Francisco Bay Test Case: Low-pass filtered currents from the Richmond Bridge Acoustic Doppler Current Profiler

Tidal current speed is plotted in the top graph to identify spring and neap tidal conditions. The water velocities were measured at depth intervals called bins and then low-pass filtered. The bin height of the profiler was set to 0.5 meter and was deployed in water of depth approximately 15 meters below mean lower low water. The length and the angle of the sticks shown in the graphs indicate the speed and direction of the water velocity vector. A stick pointing vertically in the positive direction indicates true north.

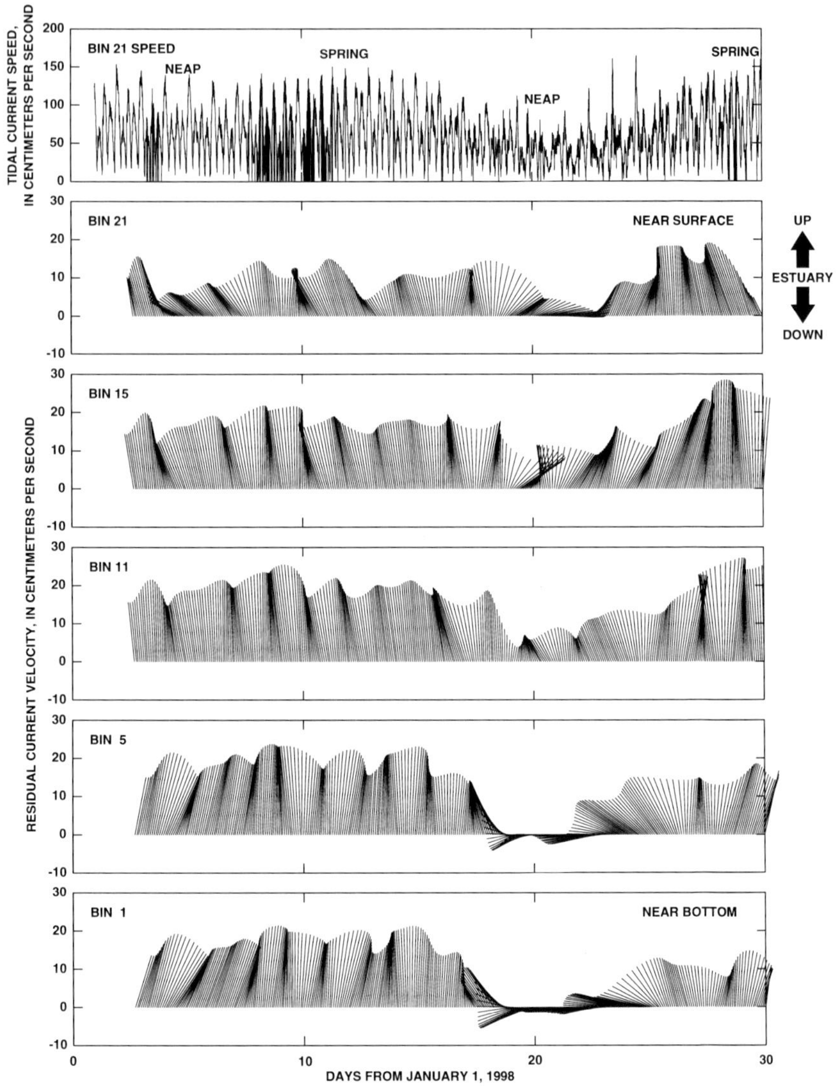

Figure 6.4.10 San Francisco Bay Test Case: Low-pass filtered currents from the Golden Gate Acoustic Doppler Current Profiler

Tidal current speed is plotted in the top graph to identify spring and neap tidal conditions. The tidal water velocities were measured at depth intervals called bins and then low-pass filtered. The bin height of the profiler was set to 2.0 meters and was deployed in water of depth approximately 50 meters below mean lower low water. The length and the angle of the sticks shown in the graphs indicate the speed and direction of the water velocity vector. The currents in the figure have been rotated so that a stick pointing vertically in the positive direction indicates a current flowing northeast at an angle measured 60 degrees clockwise from true north.

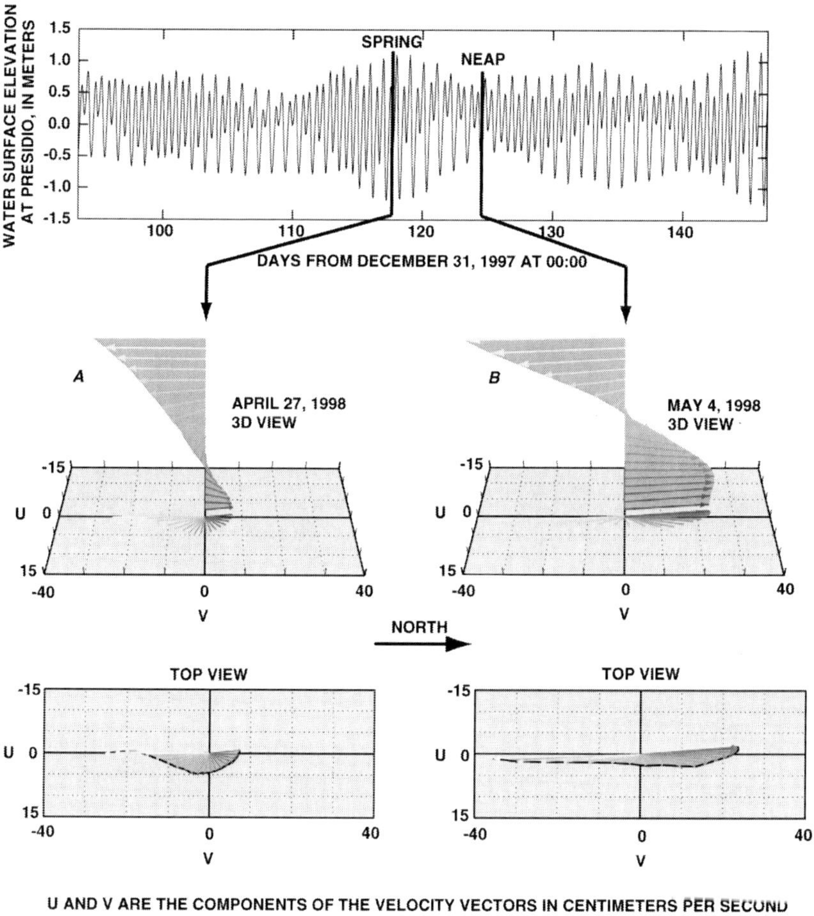

Figure 6.4.11 San Francisco Bay Test Case: Typical tidally averaged velocity profiles for (***A***) spring and (***B***) neap tidal conditions as measured by an Acoustic Doppler Current Profiler near the Richmond Bridge

The graph at the top shows water surface elevation plotted relative to National Geodetic Vertical Datum of 1929 at the Presidio with spring and neap tidal conditions identified. The velocity profiles are defined with the u-component positive to the east; the v-component is positive to the north.

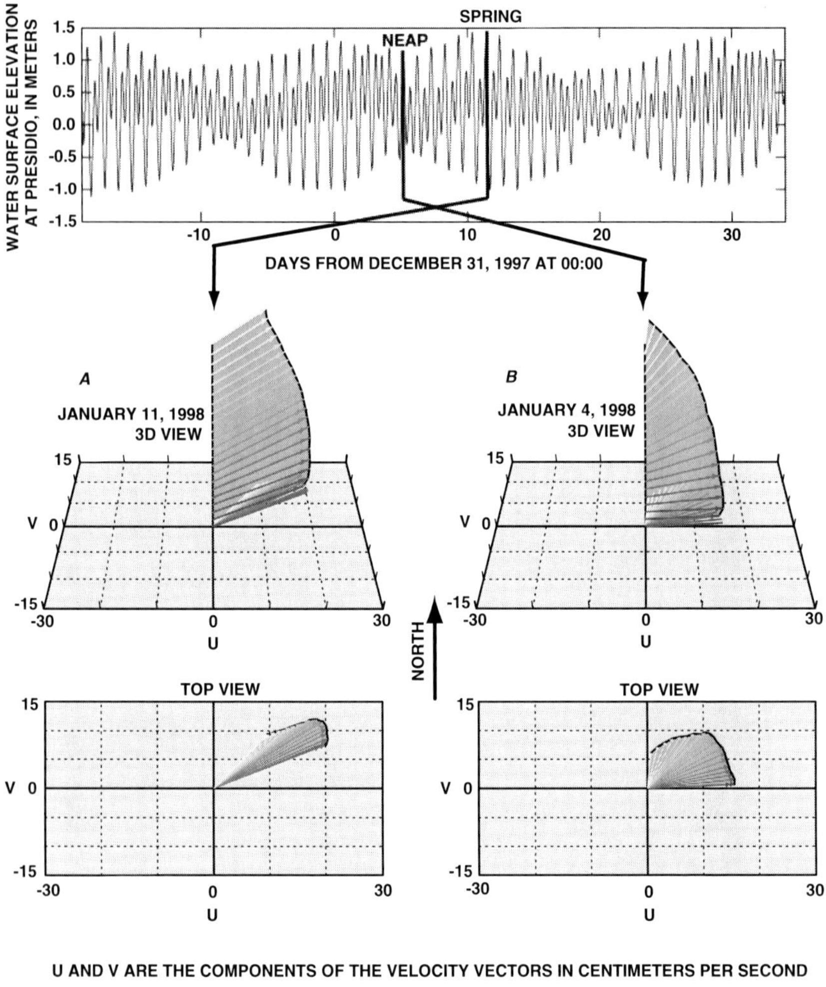

Figure 6.4.12 San Francisco Bay Test Case: Typical tidally averaged velocity profiles for (***A***) spring and (***B***) neap tidal conditions as measured by an Acoustic Doppler Current Profiler near the Golden Gate

The graph at the top shows water surface elevation plotted relative to National Geodetic Vertical Datum of 1929 at the Presidio with spring and neap tidal conditions identified. The velocity profiles are defined with the u-component positive to the east; the v-component is positive to the north.

6.5 APALACHICOLA BAY TEST CASE

Contributors: Wenrui Huang, Tien-Shuenn Wu, William K. Jones, Gilmar Rodriguez, and Steven C. McCutcheon

6.5.1 Background

Apalachicola Bay is a national resource noted for excellent oyster production and other special characteristics. The bay has been designated as a National Estuarine Research Reserve, Outstanding Florida Water, and State Aquatic Preserve. Hydrodynamic modeling of Apalachicola Bay has been conducted to support water management activities to preserve the ecosystem of the bay. A one-month observational data set (Raney, et al. 1985) was used by Wu and Jones (1991) and Wu et al. (1997) in a preliminary circulation modeling study. Field observations by Jones and Mozo (1993, 1994) and Jones and Rodriguez (1995) provide a complete data set for the calibration and validation of a three-dimensional hydrodynamic model of Apalachicola Bay (Huang and Jones, 1997). Model simulations have been used to investigate the circulation and salinity dynamics in the bay (Huang et al. 2000; Huang and Jones, 2001). Model simulations have also been used to examine the effects of water management scenarios on the estuarine salinity and oyster mortality (Huang, 2000; Livingston et al., 2000). In these model simulations, approximately fifteen-hours of CPU time were required for each one-year simulation using a 16-processor SGI Origin 2000 computer at Florida State University. Salinity is an important factor affecting the oyster and shellfishing industry in this highly productive estuary. Therefore, the model's ability to simulate the salinity dynamics is the focus of the model test case in Apalachicola Bay.

6.5.2 Objectives

The objective is to present a model test case for validating the ability of a 3-D hydrodynamic model in the prediction of circulation and salinity dynamics under the forcing of tides, winds, and freshwater inflow. The emphasis is on the comparison of model predictions and field observations of salinity in the bay, especially at the oyster reefs. In order to apply the model as a cost-effective tool to predict the long term impacts of water management scenarios on the estuary, it is desired that the CPU time required for a yearlong model simulation should be less than one day.

6.5.3 Approach

The computational domain for this test case includes Apalachicola Bay, St. Vincent Sound, East Bay, and St. George Sound (Figure 6.5.1). Bathymetric data are provided at the centers of the orthogonal curvilinear grid system in the horizontal plane (Figures 6.5.2 and 6.5.3). Horizontal variations in the salinity initial condition are provided by using linear interpolation and extrapolation over the model domain based on the available field data. A complete data set for a seven-month period between May 1 and November 30, 1993 is provided to define model boundary conditions and validate the model performance. The data in May is used in model "spin-up" to obtain realistic circulation conditions at the beginning of June, and the data during the period between June 1 and November 31 are used for model test and validation. Boundary condition data consist of water surface elevations, salinity, and temperature at five tidal inlets; fresh water inputs from Apalachicola River and the tributaries; rainfall runoff from the coastal watershed; and wind speeds and directions on the water surface. For model validation purposes, time series data are provided at several stations within the interior of the bay. These consist of water levels at two stations and salinity at six stations.

6.5.4 The Physical Domain

Apalachicola Bay is a multiple-inlet, bar-built estuary on the Florida Panhandle. There are four natural openings (Indian Pass, West Pass, East Pass, and Lanrard Reef) and one man-made opening (Sikes Cut) that connect the bay to the Gulf of Mexico (Figure 6.5.1). The bay is a highly productive estuarine system that supports a diverse and abundant commercial, non-commercial, and recreational fishery. The bay has the third largest catch of shrimp statewide and accounts for 90 percent of Florida oysters and 10 percent of the nationwide harvest. The bay is a shallow water system with an approximate 3.0 m depth, which is affected by wind, tide and buoyancy forces. Major tidal constituents in the bay are diurnal and semi-diurnal components (Jones, et al. 1994) that cause periodic changes of water level and salinity in the bay. Tidal forcing from multiple inlets causes complex circulation in the bay (Huang, et al., 2000). The bay axis is approximately in east-west direction, which is perpendicular to the southward river discharge. Due to the long wind fetch along the east-west bay axis and the large inlets located in the east and west ends of the estuary, surface winds play a significant role in estuarine circulation and salinity transport. The majority of freshwater is discharged into the bay from the Apalachicola River. Historically, higher river flows occur in the late winter and early spring and lower flows occur in the late summer and early fall. Between seasons, the system can experience an order-of-magnitude change in flows with average annual low flows of 270 m^3/s and average annual high flows of 2130 m^3/s. The annual mean flow is 725 m^3/s at Chattahoochee, Florida. Buoyancy force in the bay is due to the existence of salinity gradients. Salinity stratification varies from strongly stratified near the river mouth to the well mixed near the open ocean boundaries.

6.5.5 The Numerical Grid and Bathymetry

Bottom topography in the bay was determined by the US Army Corps of Engineers in 1986 (Figure 6.5.2). Bathymetry for the test simulation is defined at the grid centers for the numerical grid shown in Figure 6.5.3. The horizontal grid (Huang and Jones, 1997) is composed of 941 orthogonal curvilinear cells. The largest grid cell (1.65 km x 1.0 km) is midway between West Pass and Sikes Cut along the bayside of little St. George Island. St. George Sound grid cells are generally about 1.0 km^2. The smallest grid cell is located in the Sikes Cut (175m x 496m). The latitudes, longitudes and depth at the center of each grid cell is given in the file *depthgridcenter.txt*.

Based on these data, the user may either construct the horizontal grid shown in Figure 6.5.3 or interpolate the depth field on other horizontal grids. In the vertical, five uniform sigma levels have been used, but the modeler is free to choose other vertical coordinate systems. Validation data are at the top (near surface) and at the bottom (0.5m above the bottom).

6.5.6 Initial Conditions

The initial salinity is linearly interpolated and extrapolated in the horizontal plane throughout the model domain for salinity and temperature based on the available field data. In the vertical direction, the initial salinity is specified as uniform distributions. Initial temperature is held constant at 20 degree Celsius. The initial hydrodynamic conditions in the bay are specified as quiescence with free surface elevation taken as zero relative to mean sea level. Model simulations by Huang and Jones (1997) indicate that realistic circulation and salinity can be calculated by the end of a 31-day period by using the initial conditions and the boundary conditions in May 1993. The initial salinity and temperature at the center of each cell are given in the file *initial_s.txt*.

6.5.7 Boundary Conditions

Boundary conditions for model validation consist of field observations of river inflows, winds, tides, temperatures and salinity. Ocean boundary conditions are specified for Indian Pass, West Pass, Sikes Cut, East Pass and Lanard Reef. Flows in the Apalachicola River and the river tributaries were obtained by simulations from a one-dimensional river model. Rainfall runoff from local coastal watershed was determined from hydrologic model simulations. Wind forcing was obtained from a weather station on the St. George Island causeway. All data values used for boundary conditions are provided in the time series format. The details of the boundary conditions are described as follows.

Ocean boundary conditions include water levels, salinity, and temperature at five tidal inlets (Indian Pass, West Pass, Sikes Cut, East Pass, and Lanard Reef) as shown in Figure 6.5.1. Hourly time series data of water levels, salinity, and temperature were specified using the field observations (Figures 6.5.4 and 6.5.5). Surface and bottom salinity measurements were provided to address vertical stratification. Temperature of 20 degree Celsius was used at open ocean boundaries. At the Skies Cut, West Pass, and Indian Pass boundaries where a single grid cell is used, surface elevation and salinity were directly specified using field observations. At East Pass and the opening between Dog Island and Lanard Reef where several grid cells are used, surface elevation was uniformly distributed while salinity was linearly interpolated in both horizontal and vertical directions between data from measurement stations. The hourly water levels at open ocean boundaries are given in file *ocean_ele.txt*. The surface and bottom salinity data at half-hour interval are given in the file *ocean_s.txt*.

Surface wind conditions are specified using data obtained from a meteorological station established on the St. George Island causeway near mid bay (Jones et al., 1994). Measurements of average and peak wind speeds along with direction were taken at thirty-minute intervals from a tower 10 m above the water surface. The wind data were smoothed using a three-hour low-pass filter. Wind speeds and directions (Figure 6.5.6) are homogeneously applied to each surface grid in the computational domain. The hourly wind speeds and directions are given in the file *winds.txt.*

Flow records from the USGS (1995) Sumatra gage are used as freshwater input from the Apalachicola River (Figure 6.5.7). The Sumatra gage is about 33.2 km (river mile 20.6) upstream from the river mouth. Below Sumatra, the Apalachicola River divides into a number of distributaries or rivers that feed off the main stem of the river. Estimates of the distribution of freshwater inflow to Apalachicola Bay were made using a one-dimensional hydrodynamic model DYNHYD (Ambrose and Barnwell, 1989) adapted to the river distributary system. Time dependent river discharge rates at the Sumatra gage were used as the upstream river boundary condition, while tidal data were used as the downstream open boundaries. The hourly freshwater inputs to the bay from the river model are given file *river_dis.txt.*

In addition to Apalachicola River flow, rainfall runoff from the local watershed (Whiskey George, Cash Creek, and Carrabelle River basins) also contributes freshwater input to the bay. Local rainfall is generally highest in the summer and fall. Upper East Bay is influenced by the rainfall runoff from the Whiskey George and Cash Creek watersheds, which have a combined watershed of 64,000 acres. To the east of the bay is the Carrabelle River system consisting of the New and Crooked rivers. The drainage basin is substantially larger at 259,000 acres, but has similar discharge patterns compared to the Whiskey George and Cash Creek watersheds. The hydrological model HSPF (USEPA, 1984) was used to estimate surface runoff from these local watersheds. Input information for the HSPF model includes soil types, land cover and slopes, rainfall, and evaporation. Rainfall and evaporation were input into the model in time series form. The rainfall data used were from three rainfall gages operated by the Florida State Climate Center, in the vicinity of the study area: St. James, Carrabelle, and Tates Hell. Evaporation data was obtained from the NOAA station at Milton and Santa Rosa County. Rainfall runoffs from the HSPF model were used as the freshwater input boundary conditions in the hydrodynamic model (Figure 6.5.7). The hourly rainfall runoff is also included in the file *river_dis.txt.*

6.5.8 Validation Data

The data for model validation in Table 6.5.1 are provided for two stations measuring water levels, one station for currents, and six stations measuring salinity. The locations of the data stations are shown in Figure 6.5.1. The water levels (Figure 6.5.8) at station S397 near mid bay, and station S399 at upper East Bay, can be used to examine the model performance in reproducing the water levels under the tidal forcing from five tidal inlets. Since model predictions of salinity variation in the bay are important to the water management of Apalachicola Bay, hourly time series of salinity are provided at six stations across the bay (Figure 6.5.9). Stations S394 and S414 are located at two major oyster reefs (Cat Point and Dry Bar). S399 is located in the northern end of the bay near the river distributaries. S392 and S387 are located in the eastern part of the bay. With the time series of hourly salinity at these stations, the model performance can be validated in the prediction of both the spatial and temporal salinity variations in response to the forcing of winds, tides, and river inflows. If a model can be satisfactorily validated, it can be used to simulate the impact of upstream freshwater diversion on the salinity change, and ultimately on the aquatic ecosystem in the estuary. The hourly water levels for model validation are given in the files *Els397_validation.txt* and *Els399_validation.txt*. The hourly salinity data are given in the file *S_validation.txt*.

Table 6.5.1 Depth of instruments from bottom

Station ID	S414	S399	S398	S394	S392	S387
Depth (ft) from bottom	1	2	2	2.4	2	2

6.5.9 Remarks

A complete data set for a model test case in Apalachicola Bay during the seven-month period between May 1 and November 30 is provided in Appendix A. These data files include:

- Readme.txt: Description of the format for each data files.
- Initial_s.txt: Initial salinity field.
- Depthgridcenter.txt: Bathymetry data given at grid center.
- Ocean_ele.txt: Water levels at open ocean boundaries.
- Ocean_s.txt: Salinity and temperature at open ocean boundaries.
- River_dis.txt: Freshwater inputs from Apalachicola River and distributaries, and rainfall runoff from coastal watersheds.
- Winds.txt: Wind speeds and directions.
- Els397_validation.txt: Water level at station s397 for model validation.
- Els399_validation.txt: Water level at station s399 for model validation.
- V_validation.txt: Currents at station s400 for model validation.
- S_validation.txt: Salinity data at stations in the bay for model validation.

6.5.10 References

Ambrose, R. B., Jr., T. O. Barnwell, 1989. Environmental Software at the U. S. Environmental Protection Agency's Center for Exposure Assessment Modeling, Environmental Software, 4, No. 2, pp, 76-93.

Huang, W., and W.K. Jones, 2001. Characteristics of long-term transport in Apalachicola Bay. Journal of American Water Resource Association, Vol. 37, No.3, pp. 605-615.

Huang, W., 2000. Long-term simulations of flow scenario effects on salinity in Apalachicola Bay. Report of Civil Engineering Department of Florida State University, submitted to Northwest Florida Water Management District.

Huang, W., K. Jones, and T.Wu, 2000. Modeling tidal circulation in a barrier-island estuary – Apalachicola Bay. Proceedings of the 6th Conference in Coastal and Estuarine Modeling (ASCE) , New Orleans. November 3-5, pg. 1216-1233.

Huang, W. R. and W.K. Jones, 1997. "Three-Dimensional Modeling of Circulation and Salinity for the Low River Flow Season in Apalachicola Bay", Florida, Northwest Florida Water Management District, Water Resources Special Report 97-1.

Jones, W.K. and M.R. Mozo, 1993. "Apalachicola Bay freshwater needs assessment program, Geophysical data collection program". Northwest Florida Water Management District, Water Resources Special Reports, 93-5 (Volume 1).

Jones, W.K. and M.R. Mozo, 1994. "Apalachicola Bay freshwater needs assessment program, Geophysical data collection program". Northwest Florida Water Management District, Water Resources Special Reports, 94-1,94-6,94-7.(Vol.2-4).

Jones, W.K. and G. Rodriguez, 1995. "Apalachicola Bay freshwater needs assessment program, geophysical data collection program". Volumes 5-6, Northwest Florida Water Management District, Water Resources Special Reports, 95-1,95-2.

Jones, W.K., B. Galperin, T.S. Wu, and R.H. Weisberg, 1994. "Preliminary circulation simulations in Apalachicola Bay, Florida", Northwest Florida Water Management District, Water Resources Special Report 94-2.

Livingston, R., Lewis, G, Woodsum, G., Niu, X., B. Galperin, B. Galperin, W. Huang, 2000. Modeling oyster response to variation in freshwater input. Journal of Estuarine, Coastal and Shelf Science, 50, page 655-672.

Raney, D. C., I. Huang, and H. Urgun. 1985. "A hydrodynamic and salinity model for Apalachicola Bay, Florida". BER Report No. 339-183, for the U.S. Army Corps of Engineers, Mobile District.

U. S. Environmental Protection Agency, 1984. "Application guide for hydrological simulation program - FORTRAN (HSPF)". Prepared by Anderson-Nichols and Co., Palo Alto, CA., Contract No. 68-01-6207.

U. S. Geologic Survey, 1995. "Water Resources Data, Florida, Water year 1995", Volume 4, Northwest Florida, Data Report FL-95-4.

Wu, T. S., and W. K. Jones, 1991. "Preliminary Circulation Simulations in Apalachicola Bay", The 2nd International Estuarine and Coastal Modeling, November 13-15, 1991, Tampa, Florida, pp. 344-356.

Wu, T. S., J. M. Hamrick, S. C. McCutcheon, and R. B. Ambrose Jr., 1997. "Benchmarking the EFDC/HEM3D Surface Water Circulation and Eutrophication Models", Next Generation Environmental Models and Computational Methods, Edited by G. Delic and M. F. Wheler, Printed by Society for Industrial and Applied Mathematics (SIAM), Library of Congress Catalog Card Number: 96-71103, ISBN 0-89871-378-1. pp. 157-161.

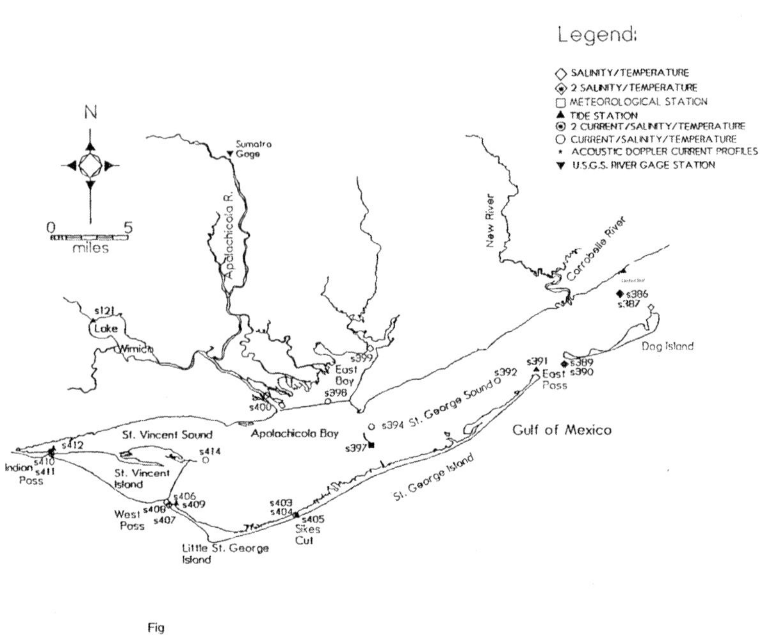

Figure 6.5.1 Locations of Current, Salinity, Tide, and Meteorological Data Collection Stations for the Apalachicola Bay

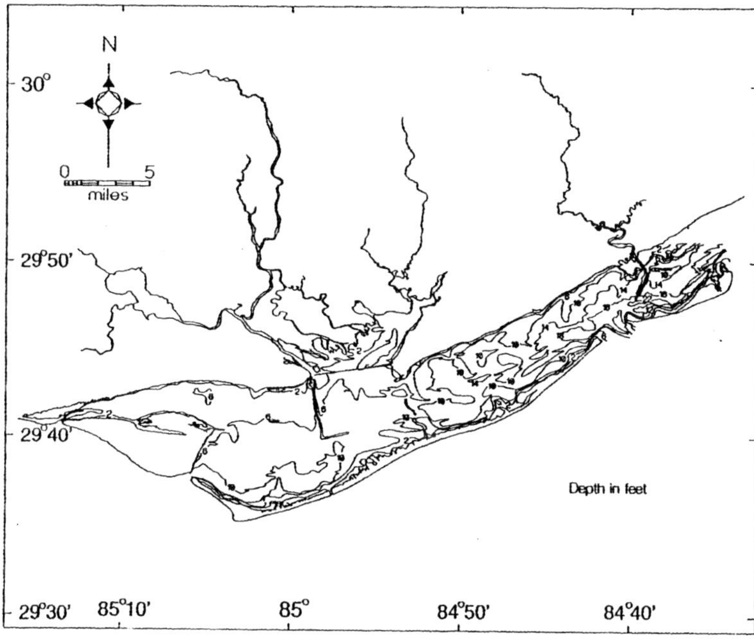

Figure 6.5.2 Apalachicola Bay Test Case: Bathymetry

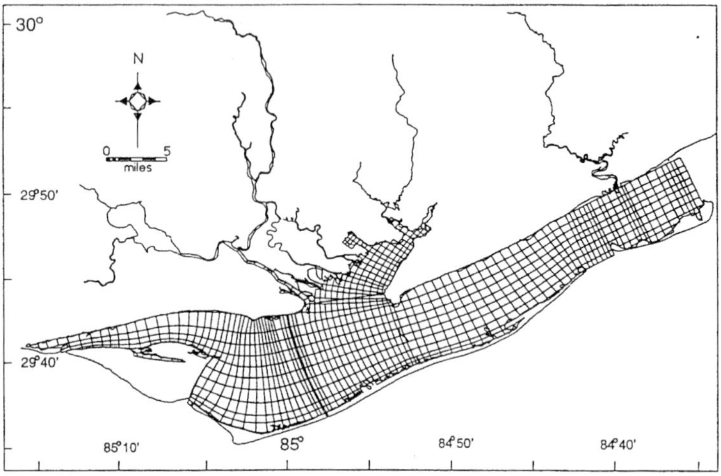

Figure 6.5.3 Apalachicola Bay Test Case: Horizontal Grid System

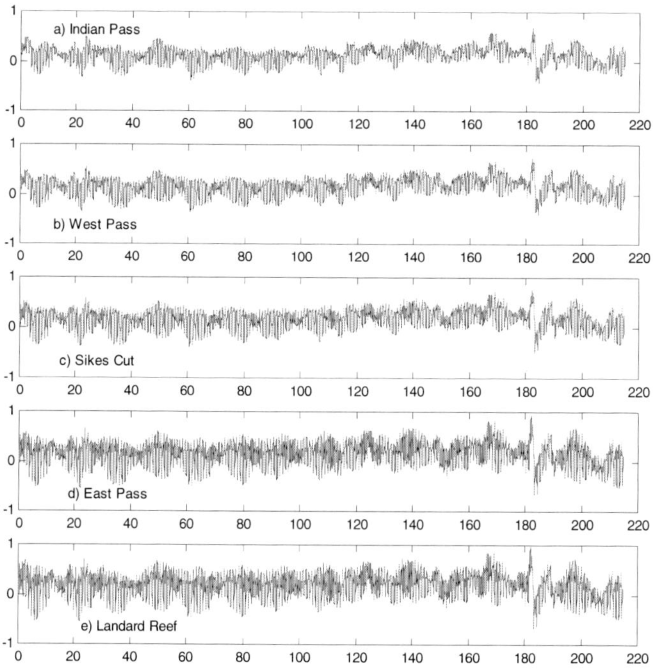

Figure 6.5.4 Apalachicola Bay Test Case: Water levels at open ocean boundaries

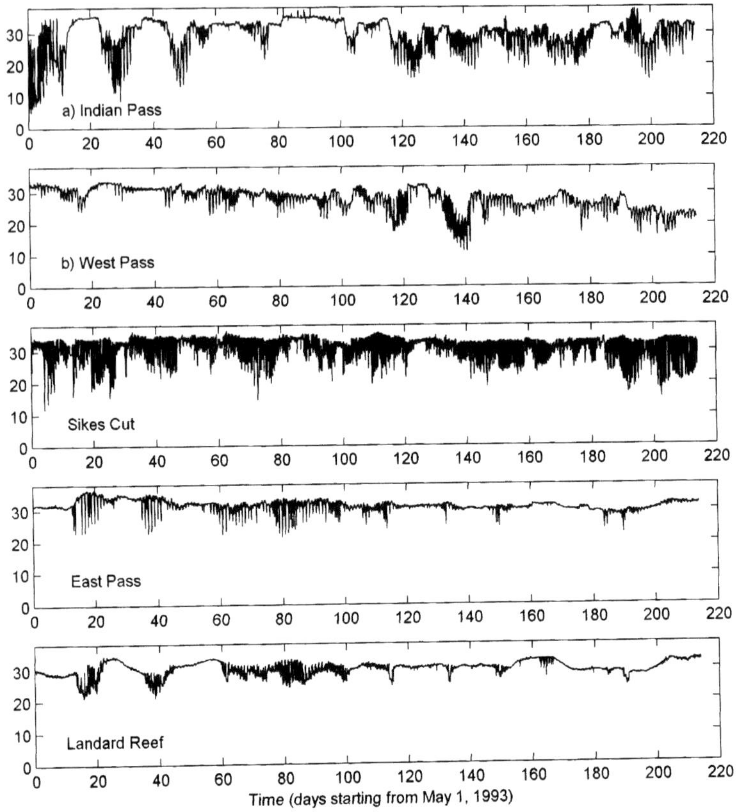

Figure 6.5.5a Apalachicola Bay Test Case: Surface salinity (ppt) at open ocean boundaries.

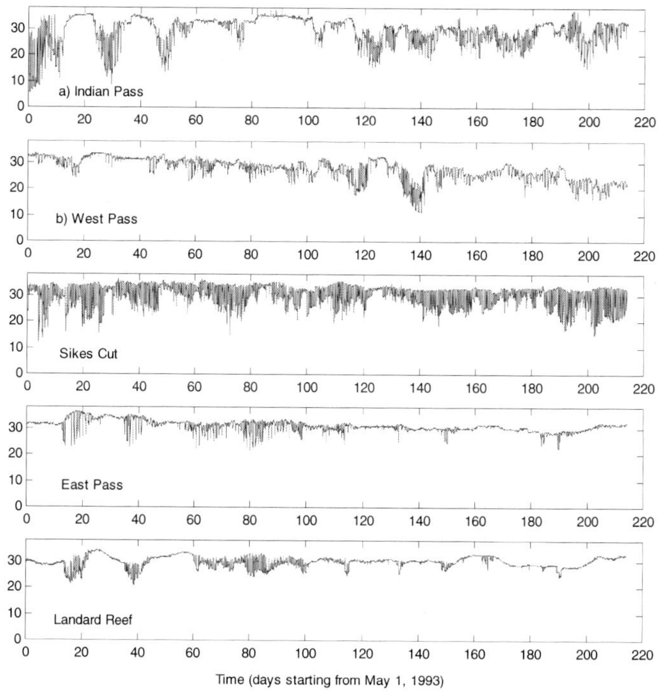

Figure 6.5.5b Apalachicola Bay Test Case: Bottom salinity (ppt) at open ocean boundaries.

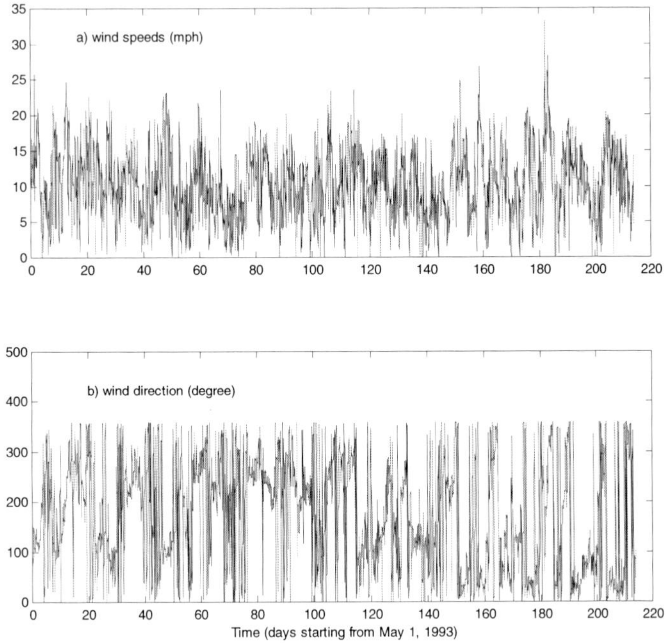

Figure 6.5.6 Apalachicola Bay Test Case: Wind speeds and directions.

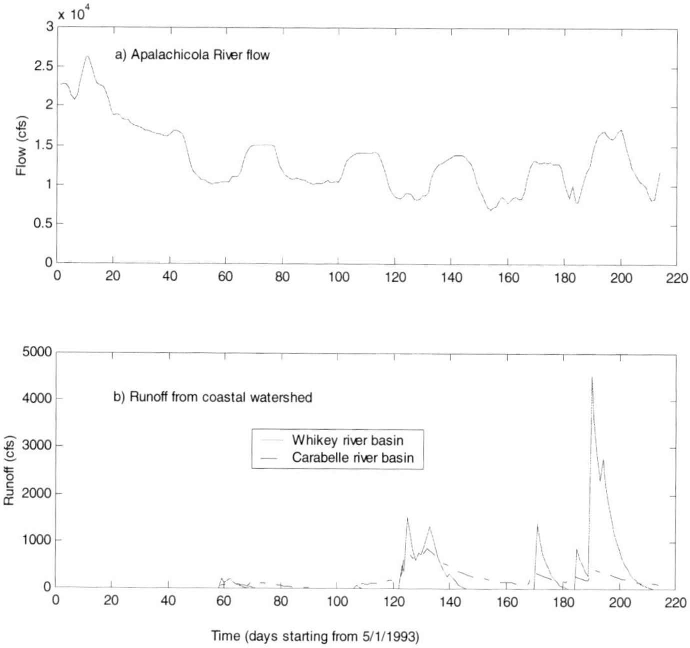

Figure 6.5.7 Apalachicola Bay Test Case: Freshwater Inflows (Whiskey River Basin)

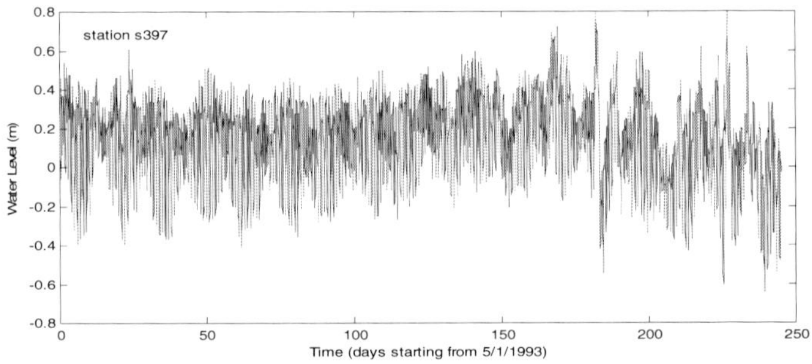

Figure 6.5.8 Apalachicola Bay Test Case: Water levels in the bay for model validation

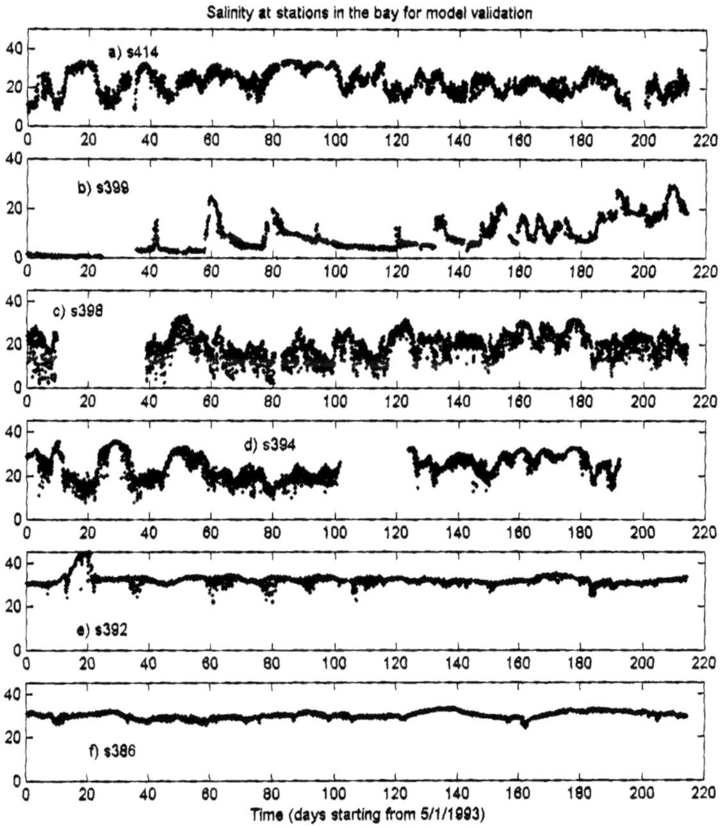

Figure 6.5.9 Apalachicola Bay Test Case: Salinity (ppt) at stations in the bay for model validations

6.6 MELDORF BAY TEST CASE

Contributors: Roberto Mayerle and Christian Winter

6.6.1 Background

Hydrodynamic modeling of the Meldorf Bay and adjacent tidal areas has been conducted in conjunction with the research project *Predictions of Medium-Scale Morphodynamics – PROMORPH* funded by the German Ministry of Education and Research from 2000 to 2002 (Mayerle and Zielke, 2005, Zielke et. al 2000, Palacio et al., 2001, Wilkens et al., 2001, and Mayerle and Palacio, 2002). The modelling study was driven by the need to predict morphological changes in the bay over periods of several years. Such modeling efforts require accurate and long-term two-dimensional depth-integrated (2DH) and three-dimensional (3D) simulations of water levels and flow covering long periods. Furthermore, the computational time required to carry out such simulations must be reasonable.

6.6.2 Objectives

In a domain with intertidal flats and channels with boundaries open to the sea, the objective is to determine a 3D free-surface numerical hydrodynamic model's

- ability to simulate water levels,
- ability to simulate 2DH and 3D fields of velocity over tidal cycles,
- ability to handle domains that fall dry during a tidal cycle, and
- computational efficiency

6.6.3 Approach

A curvilinear numerical grid is provided. Data for two test conditions is made available. The first involves a one month astronomic tide simulation covering the month of May 1990, while the second test condition is for a five day complete meteorologically forced simulation over 15-21 May, 1999. Water level condition data are given along the open sea boundaries of the model domain. Astronomical constituents based on long term water level measurements are given. Water level time series of gauge stations at several locations are provided. For the second test condition current velocity data over the vertical for a cross-section covering a full tidal cycle are also provided. Observed data is presented in both graphical and digital form for comparison with model results.

6.6.4 The Physical Domain

The domain of investigation is located on the North Sea coast of Germany between the Elbe and Eider estuaries covering an area of 392 km^2 (Figure 6.6.1). The bathymetry of the domain with intertidal flats (about 50% of the area) and channels is controlled by the combined effect of tides, wind-driven currents and wind waves. Water depths in the tidal channels vary from 5 to 20 m. The tidal range is around

3.5m. Maximum wave heights of up to 3-3.5m have been observed at storm conditions. Sediments are mainly sandy and partly muddy.

6.6.5 The Numerical Grid and Bathymetry

The plan form numerical grid with the bathymetry and location of the data stations and transect is shown in Figure 6.6.2. There are about 30,000 surface cells (about 7,100 inactive) with grid spacing ranging from 60m to 180m. For test condition one the model bathymetry was generated using measurement data from around 1990. The grid and bathymetrical data are given in file MELDORF90.GRD. For test condition two the model bathymetry was generated using echo sounding measurement data from around 1998. In higher areas along the tidal flats, the bathymetry was obtained from bathymetrical maps from 1990. The grid and bathymetrical data are given in the file MELDORF99.GRD.

Grid nodes are listed in xyz format. The coordinates are given in meters (Gauss-Krueger System -third meridian). All depths are given in meters, referenced to NN (Normal Null) which is approximately equal to the mean sea level. Depths are defined positive downwards. Nodes listed with depths equal to –999 are inactive. Further instructions for reading the data are given in the header of the file.

6.6.6 Initial Conditions

The initial water surface elevation should be set to zero elevation relative to the mean sea level. A sufficient (about one day) spin up period should be applied for both test simulations. Meteorological data are given for each test condition as follows:

Test Condition 1:
For the land based station Buesum(see Figure 6.6.3) wind-speed and -direction data is presented in the files BUESUM90.METEO. All times are referenced to UTC[3].

Test Condition 2:
For the land based station Buesum (see Figure 6.6.3) meteorological data is presented in the file BUE99.METEO. The following data is provided amongst others: wind speed, wind direction, air temperature, air pressure and solar radiation. Note that the time is referenced to MESZ[4] = UTC + 2hrs. Further instructions for reading the data are given in the header of the file.

6.6.7 Boundary Conditions

Test Condition 1

Boundary forcing data can be obtained by specifying astronomical tides determined by tidal analysis of water level observations. Measurement data of two years (January

[3] UTC: Universal Time Coordinated (former GMT)
[4] MESZ : Central European Summer Time

1989 until December 1990) has been analyzed with the GETIJSYS package (DELFT HYDRAULICS, 1992). The analyzed period enabled the determination of 40 tidal constituents for three stations located along the western open boundary. Results of the tidal analysis are given by the amplitudes and phases (the tidal constants) of the analyzed constituents (Table 6.6.1). Details of the analysis can be found in Mayerle and Palacio (2002). Amplitudes and phases of the tidal components are stored in the file TIDES90.AST. All phases are referenced to UTC. The water elevation is defined positive upwards, referenced to NN (=Mean Sea Level). Read instructions can be found in the header of the file.

Table 6.6.1 Amplitudes and phases for 40 astronomical components

AMPLITUDES [m]					PHASES [DEG]			
	Blauort	Tertius	Trischen			Blauort	Tertius	Trischen
M2	1.5164	1.5603	1.5349		**M2**	326.9	327.6	323.1
S2	0.3506	0.3698	0.3432		**S2**	37.2	38.3	35.1
N2	0.2615	0.2695	0.2668		**N2**	295.8	295.3	294.8
2MN2	0.1851	0.1856	0.1902		**2MN2**	173.8	176.7	165.2
SSA	0.1704	0.1364	0.0906		**SSA**	181.4	191	237.9
M4	0.1092	0.1082	0.1118		**M4**	148.5	148.5	153.1
K2	0.0996	0.1050	0.0975		**K2**	37.2	38.3	35.1
O1	0.0946	0.0992	0.1018		**O1**	241.4	234.5	234.4
MU2	0.0856	0.1013	0.0888		**MU2**	59.4	57.5	49.1
K1	0.0766	0.0820	0.0832		**K1**	22.2	24.3	21.6
MF	0.0736	0.0746	0.0741		**MF**	174.2	172.9	171.3
NU2	0.0507	0.0523	0.0518		**NU2**	295.8	295.3	294.8
MN4	0.0432	0.0410	0.0422		**MN4**	104.6	106	118.1
MNS2	0.0398	0.0440	0.0411		**MNS2**	46.9	41	25.6
3MN4	0.0391	0.0345	0.0354		**3MN4**	334.9	332.7	334.9
M6	0.0333	0.0417	0.0568		**M6**	22.9	44.2	19.6
MSN2	0.0292	0.0253	0.0332		**MSN2**	234.2	230.1	203.6
2SM6	0.0282	0.0253	0.0305		**2SM6**	247.6	217.7	226.1
MK4	0.0277	0.0284	0.0289		**MK4**	347.2	338	344.1
P1	0.0251	0.0269	0.0273		**P1**	22.2	24.3	21.6
3MS2	0.0231	0.0127	0.0184		**3MS2**	178.7	228	207.9
MM	0.0220	0.0204	0.0203		**MM**	229.6	248	231.8
2MS6	0.0214	0.0229	0.0359		**2MS6**	60.5	85.3	64.9
Q1	0.0213	0.0181	0.0202		**Q1**	171.9	174.7	180
2MN6	0.0211	0.0280	0.0382		**2MN6**	344	7	343.8
M8	0.0205	0.0198	0.0174		**M8**	280	282.8	265.3
3MS8	0.0172	0.0153	0.0115		**3MS8**	343.1	345	314.6
MK3	0.0165	0.0053	0.0117		**MK3**	136.9	181.8	152.5
2MSN8	0.0164	0.0158	0.0203		**2MSN8**	17.2	12.5	10.8
3MS4	0.0129	0.0159	0.0141		**3MS4**	223.9	223	211.5
2MNS4	0.0104	0.0092	0.0090		**2MNS4**	213.7	202.4	204
4MS10	0.0100	0.0104	0.0071		**4MS10**	161.4	202.5	184.4
M3	0.0095	0.0060	0.0074		**M3**	148.6	109	127.2
S4	0.0055	0.0051	0.0050		**S4**	18.1	346	23.9
4MS6	0.0044	0.0015	0.0038		**4MS6**	281	278.7	357.3
3MNS6	0.0037	0.0074	0.0054		**3MNS6**	181	170.3	110.8
2SM2	0.0030	0.0019	0.0037		**2SM2**	66.5	64.1	18.6
2(MS)8	0.0030	0.0017	0.0019		**2(MS)8**	333.2	320.7	336.9
MSN6	0.0022	0.0032	0.0085		**MSN6**	157.1	234.7	165.4
2SMN4	0.0010	0.0029	0.0026		**2SMN4**	146.7	65.7	98.4

Test Condition 2:

The tidal wave of the North Sea approaches the model area uniformly from the west. It can be observed that there is nearly no time lag between the Northern and Southern boundary water elevations. Water levels from the gauge station 'Trischen' which is located at the western model open boundary can therefore be used for open boundary forcing. Water level time series are given for a one-week period (15th to 21st of May 1999) in the file TRISCHEN99.WL. All times are referenced to UTC. The water elevation is defined positive upwards, referenced to NN (=mean sea level). Read instructions can be found in the header of the file.

6.6.8 Validation Data

Test Condition 1:

Water surface elevations at stations Blauort, Steertloch, Trischen, Flackstrom, and Tertius are available from the month of May 1990. The data is listed in the files >NAME<90.WL in which >NAME< refers to the name of the station. Format and read instructions are included in the headers of the files. The time series for the station Trischen is presented graphically in Figure 6.6.4. All times are referenced to UTC. The water elevation is defined positive upwards, referenced to NN (=mean sea level). Read instructions can be found in the header of the file. Water temperature and conductivity in the model domain does not change significantly in time and space. Typical water temperature and salinity values of 13.5°C and 25.5 ppt can be considered in the simulations. See Figures 6.6.7 and 6.6.8 for profile measurements.

Test Condition 2:

Water surface elevations at stations Blauort, Steertloch, Flackstrom, Buesum and Tertius are available from the 15th to the 21st of May 1999. The data is listed in the files >NAME<99.WL in which >NAME< refers to the name of the station. Format and read instructions are included in the headers of the files. The time series for the station Tertius is presented graphically in Figure 6.6.5. Note that the data can be deficient during low water level for the station Buesum due to device failure. All times are referenced to UTC. The water elevation is defined positive upwards, referenced to NN (=mean sea level). Read instructions can be found in the header of the file.

Three-dimensional water velocity data from ADCP (Acoustic Doppler Current Profiler) measurements along the transect indicated in Figure 6.6.3 is available from 7:45 to 19:45 hrs [UTC] of the 20th of May 1999. The period of water velocity measurements is indicated in Figure 6.6.5. The transect is about 3 km long and comprises two channels. Measurements were carried out using a ship mounted ADCP. A total of 28 transects are available for the period in question. Beginning times of transects are given in Table 6.6.2. An example velocity transect is presented graphically in Figure 6.6.6. All times are referenced to UTC. Instructions to read these files are given in the file READ_ADCP.TXT.

Observed water temperature and conductivity in the model domain does not change significantly in time and space. No stratification could be observed. Figures 6.6.7 and 6.6.8 show plots of the vertical distribution for different positions and times over a tidal cycle. Typical water temperature and salinity values of 13.5 °C and 25.5 ppt respectively can be considered. Bathymetric data were obtained from the Office for Rural Development in Husum (ALR) and by the Federal Maritime and Hydrographic Agency in Hamburg (BSH). Water surface elevations were obtained from the Office for Rural Development in Husum (ALR). Current velocities across the transect were measured by the Research and Technology Centre Westcoast of the University of Kiel.

Table 6.6.2 Starting times of velocity transect measurements 20th of May 1999

Filename	Time (UTC)
P3A002T.dat	6:36:12
P3A003T.dat	7:00:24
P3A004T.dat	7:27:17
P3A005T.dat	7:47:54
P3A006T.dat	8:07:55
P3A007T.dat	8:29:23
P3A008T.dat	8:58:17
P3A009T.dat	9:17:59
P3A010T.dat	9:41:16
P3A011T.dat	10:01:52
P3A012T.dat	10:23:49
P3A013T.dat	10:40:53
P3A014T.dat	11:11:44
P3A015T.dat	11:42:59
P3A016T.dat	12:12:54
P3A017T.dat	12:44:25
P3A018T.dat	13:15:18
P3A019T.dat	13:44:19
P3A020T.dat	14:10:53
P3A021T.dat	14:44:49
P3A022T.dat	15:17:18
P3A023T.dat	15:40:01
P3A024T.dat	16:05:53
P3A025T.dat	16:30:12
P3A026T.dat	16:54:35
P3A027T.dat	17:22:43
P3A028T.dat	17:46:16
P3A029T.dat	18:05:07

6.6.9 Remarks

Test Condition 1:

Various concerns can result in the need to compute 2D and 3D flow fields in tidal areas such as Meldorf Bay and its adjacent channels. The adequate handling of open sea boundary conditions is vital for proper simulation of water levels in the investigation area. Furthermore, the gradients of the free surface elevation should be properly modeled in order to get the velocity distributions and thus the sediment transport in good agreement with observed results. Since it is intended to apply the model for simulating long term periods (month to years), it should be able to handle neap and spring tide conditions.

Test Condition 2:

In addition to accurately computing water levels and velocities and thus providing flow fields to conduct 3D modelling of sediment transport and morphodynamics over several years, such models should also be able to handle domains that fall dry during a tidal cycle. Furthermore the computational efficiency should be as high as possible in order to handle periods of several years.

The Data files listed below are given in Appendix A of this report.

Data files Test Case 1	Data files Test Case 2
MELDORF90.GRD	BUE99.METEO
BUESUM90.METEO	MELDORF99.GRD
TIDES90.AST	TRISCHEN99.WL
TRISCHEN90.WL	BLAUORT99.WL
TERTIUS90.WL	BUESUM99.WL
BLAUORT90,WL	STEERTLOCH99.WL
STEERTLOCH90.WL	FLACKSTROM99.WL
FLACKSTROM90.WL	TERTIUS99.WL
	READ_ADCP.TXT
	P3AXXX.DAT

6.6.10 References

DELFT HYDRAULICS, 1992. GETIJSYS, Analysis and Prediction of Tides, User Manual, May 1992, Delft.

Mayerle, R. and Zielke, W., 2005, Predictions of meso-scale morphodynamics – PROMORPH. Die Küste 69 (2005).

Mayerle, R. and Palacio, C., 2002, Assessment of Open Sea Boundary Condition Approaches for Coastal Models, in 13th Congress of the Asia and Pacific Division of the International Association for Hydraulic Engineering and Research, Asia and Pacific Division of IAHR, Singapore.

Palacio, C., Winter, C. and Mayerle, R. 2001, Set-up of a hydrodynamic model for the Meldorf Bight, presented at EWRI 2001,World Water and Environmental Resources Congress, Orlando, Florida.

Wilkens, J., Asp, N., Ricklefs, K. and Mayerle, R. 2001, Medium-scale morphodynamic modelling in the Meldorf Bight, ASCE, Proceedings (CD-ROM) of World Water & Environmental Resources Congress, May, 2001, Orlando, FL.

Zielke, W., Gross, G., Hoyme, H, Mayerle, R. Ricklefs, K, Winter, C, Eppel, D. and Witte, G. 2000, Predictions of Medium-Scale Morphodynamics – Promorph-poster presented at the International Conference on Coastal Engineering, ICCE, July 2000, Sydney, Australia.

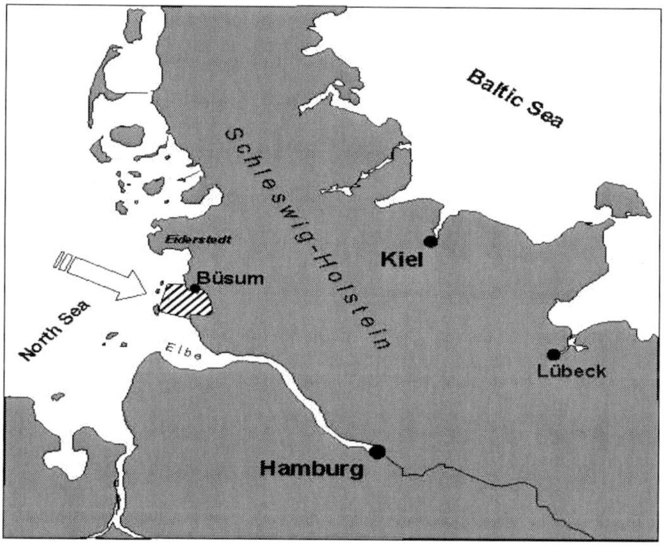

Figure 6.6.1 Meldorf Bay Test Case: Base Map

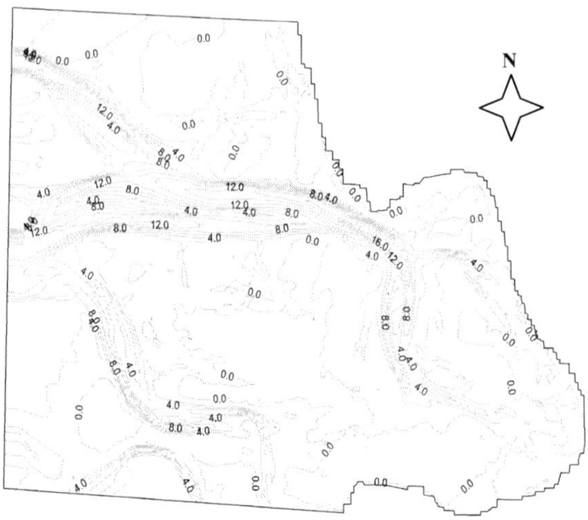

Figure 6.6.2 Meldorf Bay Test Case: Bathymetry

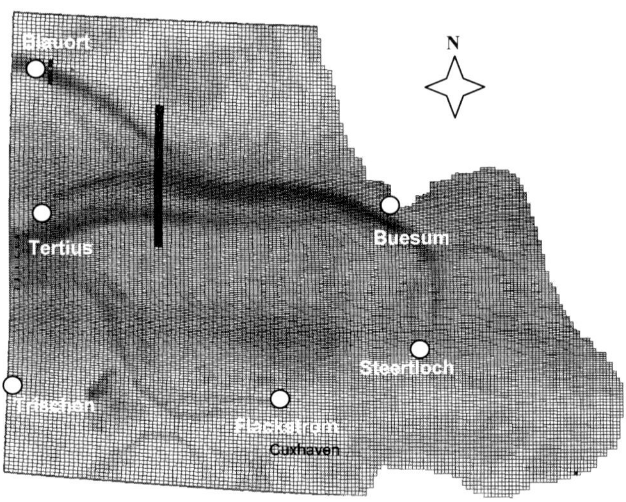

Figure 6.6.3 Meldorf Bay Test Case: Model grid and location of water level gauges

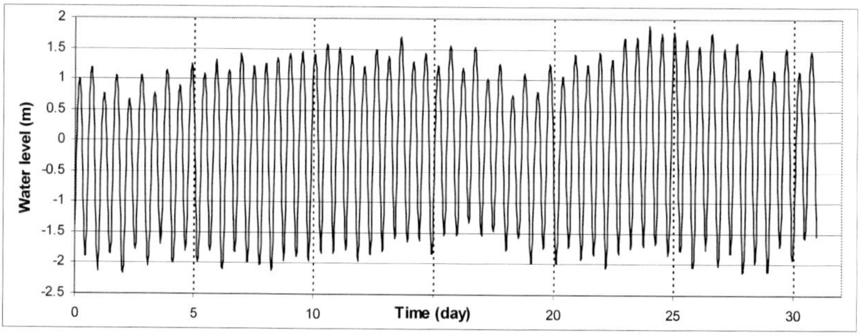

Figure 6.6.4 Meldorf Bay Test Case: Water level elevation at station Trischen, May 1990

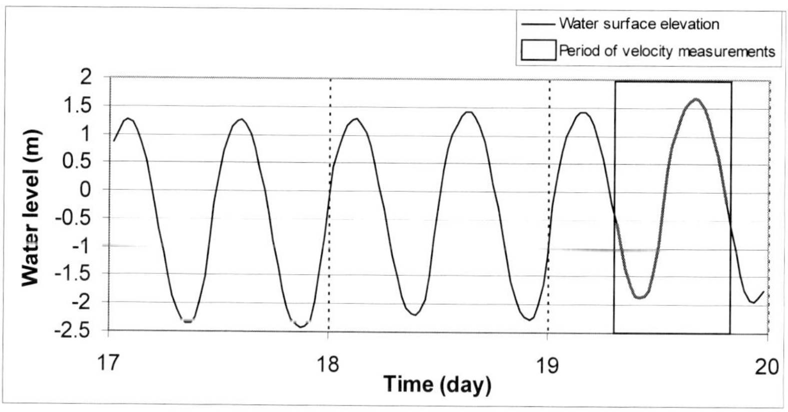

Figure 6.6.5 Meldorf Bay Test Case: Water level elevation and period of velocity measurements at station Tertius

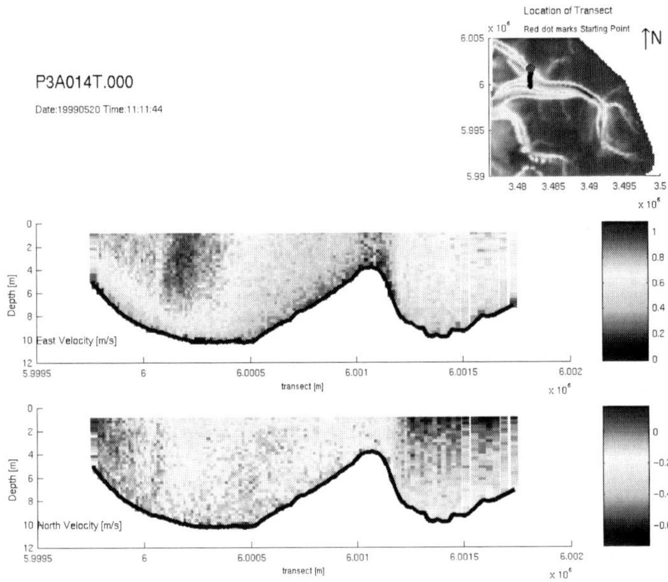

Figure 6.6.6 Meldorf Bay Test Case: Measured water velocities (May 20th 1999, 11:11)

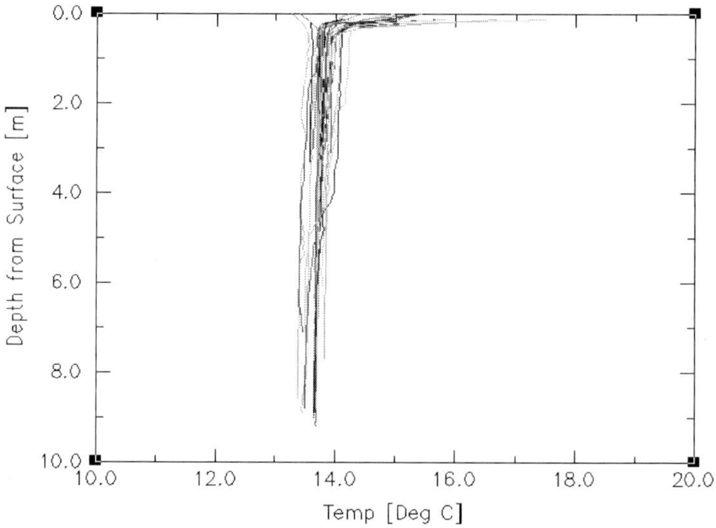

Figure 6.6.7 Meldorf Bay Test Case: Measured water temperature profiles for May 20th 1999

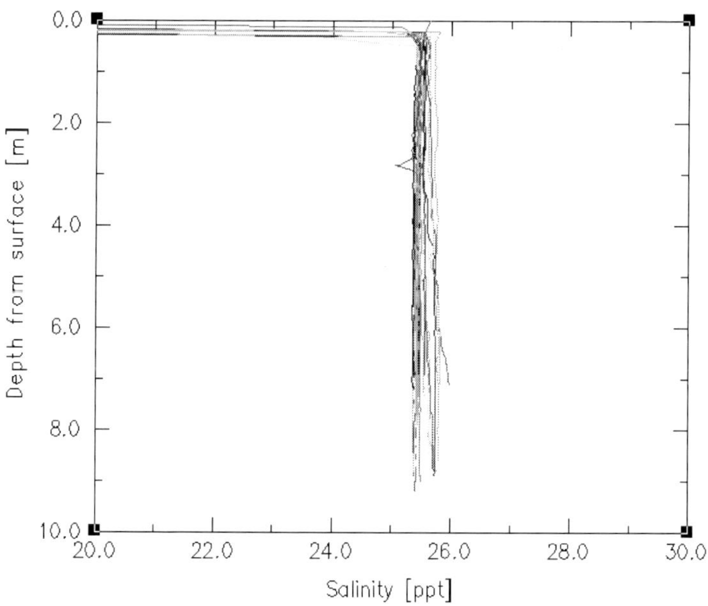

Figure 6.6.8 Meldorf Bay Test Case: Measured water salinity profiles for May 20th 1999

6.7 TOKYO BAY TEST CASE

Contributors: Mutsuto Kawahara, Toshio Kodama, and Yan Ding

6.7.1 Background

The simulation and prediction of three-dimensional flow in coastal areas are needed for planning navigation channel construction, designing harbor and offshore structures, assessing the environmental and ecological impacts of coastal engineering projects. This test case provides a set of comprehensive 3D tidal current measurements in Tokyo Bay taken during a period from 25 August to 25 October 1983. The database consists of the bathymetry of the bay, tidal levels at eight stations and currents measured at several locations as well as different water depths. In addition, it suggests the values of physical constants and boundary forcing conditions including incident tidal waves, river inflow velocities, and wind velocities above the water surface. Kodama et al. (1996) have conducted the hydrodynamic modeling of the 3D tidal current by utilizing this database. A multiple-level finite elemental model was validated in their studies by comparing the numerical results with the observed tidal current data. A series of numerical experiments were carried out to carefully examine the tidal circulations affected by the forcing factors of Coriolis force, river inflows and wind shear stresses, either individually or combined (Kodama et al. 1996).

The correct specification of tidal wave boundary conditions is of vital importance in modeling tidal flows in harbors, estuaries, and tidal rivers. The persistent difficulties in specifying the tidal waves at open boundary are mainly caused by two reasons. First, there may not be any measurements of water elevations at the specified open boundary. However, as long as water elevations and/or currents have been measured at some locations in the considered water region, the tidal waves at the specified open boundary can be estimated by means of optimal theories to minimize the discrepancies between simulations and observations in the region. In fact, parameter identification techniques have been widely used in the simulation of flood waves, ocean circulations, and tidal flows. Lardner (1993) presented a procedure for optimal control of open boundary conditions in a numerical tidal model. Gunson and Malanotte-Rizzoli (1996) then provided an optimal theory for identifying open boundary conditions and initial conditions in an open-ocean flow model. In the case of Tokyo Bay, using the linearized shallow water equations, Kodama and Kawahara (1992) proposed a procedure by means of an optimal theory to estimate the tidal waves at the open boundary. Then, Kodama et al. (1991) used the nonlinear shallow water equations to improve the estimated amplitudes of the incident tidal waves.

Second, because the incident wave trains will reflect on solid walls in the computational domain, if the open boundary does not permit the reflective waves to freely go out, the outgoing waves will reflect on the boundary toward the internal domain to generate spurious reflective waves (Kodama et al. 1991). Generally, the incident waves are continuous wave trains propagated from the deep-water region of sea; the reflective waves are generated from the internal domain including the open boundary during wave reflections. The spurious reflective waves always destabilize the computation of the initial transient flow on the open boundary when the tidal flow starts from the static state ("cold start"), and predict the initial tidal flow with seriously non-physical oscillations, because the reflective waves introduce the modes of free oscillations in the computational domain (Blumberg and Kantha, 1985; Kodama and Kawahara, 1992). Directly specifying the water elevations cannot avoid this non-physical reflection at the boundary, in principle. The outgoing waves should freely transmit through the open boundary without the spurious reflections on the boundary. Therefore, it is necessary to distinguish between the incident waves and the reflective waves at the open boundary. The latter should be allowed to go out freely, in order to eliminate the spurious reflective waves (Kodama et al. 1991). This treatment, which is called "non-reflective open boundary condition", is identical to the absorbing wavemaker in laboratory experiments of wave generations (Schaffer et al. 1994; Lin and Liu, 1999).

6.7.2 Objectives

This test case provides a database of field data consisting of tidal levels and tidal currents at a large number of measuring stations in and around simulation domain. It can be used to determine the 3D hydrodynamic model's capability of predicting tidal processes. The following tidal processes generated by hydrodynamic model can be validated:

- How realistic are the predicted tidal levels during the observed period?
- Are the 3D currents and residual current structure simulated under incident tidal waves in reasonable agreement with those measured at several observation stations over a tidal cycle?

6.7.3 Approach

In general, before running a 3D tidal current model, users have to use the bathymetric data provided in the database to generate a 3D computational mesh. Then users need to define the boundaries in the computational domain, and impose the incident tidal waves and river inflows on the corresponding boundaries. Users also need to set up some of physical parameters to include other external forces in the model, e.g., the Coriolis force and wind shear stresses. All of the data about the bathymetry of the bay, incident tidal waves, and discharges of river inflows have been provided in this test case.

As an example for explaining the validation approach, a multiple-level finite element model proposed by Kawahara et al. (1983) and further modified by Kodama et al. (1996), is introduced as follows. Assuming the vertical acceleration of a large water body is negligible, and the hydrostatic pressure distribution is adopted, the three-dimensional computational domain can be idealized as a multiple-level domain. To do so, only a two-dimensional computation at each level is necessary. This model is general enough to include the effects of Coriolis force, river inflows and wind shear stresses. The non-reflective wave condition has been applied on the open boundary to filter off spurious reflection waves. The density of water is treated as a constant in each level, but it may be different from other levels. To simulate 3D tidal currents, the following data have been already provided:

- Finite element mesh, nodal connections in vertical direction, different set of nodal points on boundaries for specifying incident tidal wave, river inflows, and coastal lines and/or river banks;
- Incident tidal wave conditions: Four major tidal constituents generated the incident tidal waves;
- Physical constants;
- Resistance coefficients of wind and bottom friction;
- River inflow velocities;
- Initial condition: cold start;
- Time increment: 15s for the multiple-leveled model;
- Selected nodal points and variables for outputs of numerical results: tidal levels at eight tide elevation gauges, and tidal currents at ten observation stations;
- Calculation of residual current: the computational period is six days in the multiple-leveled model; the tidal currents at final tidal period of M_2 were used to calculate the residual currents.

6.7.4 Physical Domain

Tokyo Bay shown in Figure 6.7.1 is about 48km long and 37km wide, located off the southeast coast of Honshu Island, Japan, connected to the West Pacific Ocean. It provides a spacious harbor area for several Japanese cities, including Tokyo and Yokohama. The three-dimensional tidal currents in the bay were measured during the period from 25 August to 25 October 1983. The measurements of the tidal currents were made to establish a fundamental database on flow profiles in the bay. Meanwhile, oceanographic and meteorological data such as tidal levels, wind, precipitation, atmospheric pressure, etc., have been collected at some observation points shown in Figure 6.7.1. During the observation period, there were totally ten observation stations for monitoring 3D tidal current, eight tide gauges for recording surface elevation, and others around the bay for measuring wind velocity, atmospheric pressure, river discharge, and precipitation (Yokohama, 1983). Mainly, four rivers (i.e. Edogawa River, Arakawa River, Tama River, and Tsurumi River) discharge freshwater into the bay. The water depth contours in Tokyo Bay are shown in Figure 6.7.2 (Maritime Safety Agency, 1984). The locations of the tide gauges and the observation stations of tidal currents in a triangular mesh of surface level in the multiple-level finite elemental model are indicated in Figure 6.7.3.

6.7.5 Numerical Grid and Bathymetry

Kodama et al. (1996) have applied a multiple-level finite element model to simulate the tidal currents during the observation period in Tokyo Bay. The triangular finite element meshes of Tokyo Bay shown in Figure 6.7.3 (surface mesh only) and Figure

6.7.4 (meshes in five different levels) has been employed. In the surface mesh system, the total numbers of nodes and elements at the surface level are 685 and 1216 respectively. The numerical meshes for the multiple-level finite element model at other levels are the projections of the surface triangular mesh. The water depth in the first layer between the first level and second level is 5m; the water depth of the other layers is 10m. Users can directly utilize this mesh system as their computational mesh, of which the digital data can be found in the file MESH.DAT in Appendix A of this report. The data showing the connection of nodal points and triangular elements between two adjacent levels are stored in the files NOD_CONK.DAT and ELE_CONK.DAT. The other information about the total nodal number and triangular element number can be found in the file LEVEL.DAT. All of these data files are given in Appendix A. In case of testing a fully three-dimensional model for tidal flows, users are advised to utilize the mesh data to generate their own three-dimensional mesh. In addition, users can use the surface mesh data in the file SURF_MESH.DAT that consists of water depth measured from the mean water level and wind velocities at nodal points of the surface triangular mesh.

6.7.6 Initial Conditions

The physical constants are listed in Table 6.7.1, including the air density, the Coriolis parameter, the horizontal eddy viscosity coefficient, the friction coefficients at the water bottom and the interfaces between two levels, and the profile of water density along vertical levels (The water density is assumed to be constant at each level). The static state of water body is used as the initial conditions (cold start), i.e., the initial water elevation is the static horizontal plane relative to the mean sea level.

Table 6.7.1 Physical constants in the multiple-level model

Wind drag coefficient	0.00015
Air density (kg/m^3)	1.29
Friction coefficient at interface	0.001
Friction coefficient at bottom	0.0026
Horizontal eddy viscosity coefficient (m^2/s)	10.0
Coriolis parameter	0.000084
Water density profile (from surface layer) (kg/m^3)	1010.0, 1015.0, 1020.0, 1025.0, 1030.0

6.7.7 Boundary Conditions

Incident Tidal Wave

In the test case, the incident tidal waves are imposed on the entrance (open boundary) of the bay (Figure 6.7.3). They consist of the four major tidal constituents, i.e., M_2, S_2, K_1, and O_1. The amplitudes of the four tidal constituents have been estimated by means of an identification method, by which a non-reflective boundary condition was imposed on the open boundary (Kodama et al. 1991). Using these incident tidal waves, the multiple-level tidal model has generated the optimal tidal levels in excellent agreement with the observed tidal levels. The superposition of the incident tidal waves is represented as follows:

$$\eta(t)=\sum_{m=1}^{4} a_m \sin\left[\frac{2\pi}{T_m}(t-t_0)-\phi_m\right], \tag{6.7.1}$$

where $\eta(t)$ means the observed tidal level at the time t; t_0 is an initial time, equal to 0.0hr; for each tidal constituent, a_m, T_m, and ϕ_m are its tidal amplitude, tidal period, and phase delay, respectively. The amplitudes, periods, and phase delays of the four tides are listed in Table 6.7.2. Users can find the data in the file TABLE78.XLS. A set of nodal points relative to the location of the open boundary has been retrieved from the suggested mesh, and they can be found in the file NOD_OPNBC.DAT. If users choose the finite element mesh provided in the case as their computational mesh, they might directly impose the tidal wave on the open boundary.

Table 6.7.2 Tidal constituents of incident waves

Constituents	Amplitude a_m (m)	Period T_m (hour)	Phase delay ϕ_m (Rad)
M_2	0.21	12.42	-5.8398
S_2	0.15	12.00	-5.6018
K_1	0.14	23.93	-1.3290
O_1	0.10	25.82	-1.3211

For specification of tidal waves in the open boundary (Figure 6.7.3), users may choose one of the following two approaches. To avoid the complicated treatment of non-reflective boundary condition as proposed by Kodama et al. (1996), the simple approach is to directly specify the water elevations on the open boundary. The time series data of water elevations during 12 days at the open boundary can be found in the file OUTLET_WL.DAT, which are the computed water elevation at the middle point of the entrance of the bay by the multiple-level model (Kodama et al. 1996). The water elevations are the tidal levels on the open boundary considering both incident tidal waves and reflective waves. In order to eliminate the spurious reflective wave on the open boundary, the user may use a modified form of Sommerfeld radiation boundary condition, and refer to Blumberg and Kantha (1985) for details.

The second approach for specification of the open boundary condition is to impose a non-reflective boundary condition. Note that the incident waves have been identified from optimal estimation of predictive errors through taking account of the reflective waves generated from solid boundaries in the computational domain (Kodama et al. 1991; Kodama and Kawahara, 1992); the sum of the four major constituents is only the part of incident waves at the open boundary, not the total water elevation in the boundary. Therefore, since the water elevation at the open boundary consists of the incident wave and the reflective wave, the reflective wave in the boundary should be calculated by considering the reflection on the internal boundaries due to the attack of incident waves. The calculation of reflective waves can be carried out in the simulation process. Detailed treatment about the reflective wave can be found in Kodama et al. (1996).

Due to the complexity in the approach of Kodama et al. (1996), we introduce a relatively simple way to estimate the reflective wave on the open boundary. It is assumed that water elevation on open boundary can be expressed as a linear combination of the incident and reflective wave:

$$\eta(n_0,t) = \eta^I(n_0,t) + \eta^R(n_0,t), \tag{6.7.2}$$

where $\eta(n_0,t)$ denotes the water elevation on the open boundary; n_0 is the normal direction at the boundary inward to the domain, η^I and η^R are the incident wave and the reflective wave components, respectively. The reflective wave at the adjacent node $n_0+\Delta n$ in the internal domain can be calculated as follows:

$$\eta^R(n_0+\Delta n,t) = \eta(n_0+\Delta n,t) - \eta^*(n_0+\Delta n,t), \tag{6.7.3}$$

where $\eta(n_0+\Delta n,t)$ is the computed water elevation at the adjacent node including reflective waves; $\eta^*(n_0+\Delta n,t)$ denotes water elevation at the adjacent node in the case where the incident wave goes through the node without reflective wave. For linear wave, it can be calculated as follows:

$$\eta^*(n_0+\Delta n,t) = \sum_{m=1}^{M} a_m \sin\left(\frac{2\pi}{T_m}t - k_m\Delta n + \phi_m\right), \tag{6.7.4}$$

where M denotes the total number of tidal constituents, ϕ_m is the phase lag, k_m represents the wave number, i.e.,

$$k_m = \frac{2\pi}{T_m C}, \tag{6.7.5}$$

where $C=\sqrt{gh}$ is the wave velocity, and h is water depth at the open boundary. Then, the reflective wave component $\eta^R(n_0,t)$ at the open boundary can be computed by using the radiation condition,

$$\frac{\partial \eta^R}{\partial t} - C_r\frac{\partial \eta^R}{\partial n} = 0, \tag{6.7.6}$$

where C_r is the wave velocity of reflective wave, equal to C under the assumption of shallow water. This equation can be solved simply by means of a finite difference formulation in a subdomain near the open boundary by considering the reflective wave $\eta^R(n_0+\Delta n,t)$ as the known boundary condition. For the case of using the finite element mesh in Tokyo Bay, a complete set of nodal points in the sub-domain can be found in the file NOD_OPNBC.DAT, and the subdomain for computing the reflective wave is located at the entrance of the bay. Users may also define a sub-domain that consists of the open boundary and the adjacent grid line according to their own mesh.

Wind Velocity

The steady wind field and the wind speed at each numerical node have been linearly interpolated from the averaged values measured at the available wind gauges around Tokyo Bay (Figure 6.7.1). The wind speeds at the gauges were measured at standard 10m elevation above ground. A maximum wind speed of 10.0m/s was observed during the period of observation of the tidal currents. The interpolated wind velocity profile is shown in Figure 6.7.5. For the digital data of the wind field based on the surface mesh, users may refer to the file WIND_VEL.DAT in Appendix A.

River Inflow Discharges

Four river inflow discharges considered in the case during the observation period are given in Table 6.7.3. The four rivers flow into the west of the bay (Figure 6.7.1). The velocities of the river inflows are imposed on the corresponding boundary points. For simplicity, only the averaged velocities are considered during the computational period. The nodal points related to the positions of river inlets have been already extracted from the finite element mesh, which can be found in the file NOD_RIVER.DAT.

Table 6.7.3 River inflow velocities into Tokyo Bay

River	Velocity (m/s)
Edogawa River	1.0
Arakawa River	0.5
Tama River	0.5
Tsurumi River	0.5

6.7.8 Validation Data

Harmonic Constants of Tidal Levels

The tide-generating force can be expressed as a series of harmonic constituents. Each tidal constituent has its individual period and amplitude. On the basis of the facts, the observed tidal levels at each tide gauge station during the observation period are represented as a Fourier series of the four major astronomical constituents, i.e., M_2, S_2, K_1, and O_1. Then, the observed tidal levels in the observation period can be calculated from the superposition of the four major constituents by means of Equation (6.7.1). However, the initial time t_0 for validation data is equal to 8.8hr for each tidal constituent. Note that the tidal levels were measured from the mean water elevation. The values of each amplitude and phase delay for the corresponding tidal constituent at eight tide gauge stations are listed in Table 6.7.4. Therefore, model testers can use these harmonic constants to regenerate the observed tidal levels at each tide gauge station.

Table 6.7.4 Tidal constituents of tidal levels at eight tide gauge stations

Tide Gauge Stations	Amplitudes of tidal constituents (m) a_m				Phase delay of tidal constituents (rad) ϕ_m			
	M_2	K_1	S_2	O_1	M_2	K_1	S_2	O_1
Funabashi	0.4842	0.2645	0.2270	0.1935	2.6957	3.1367	3.1142	2.7981
Samugawa	0.5220	0.2540	0.2680	0.1900	2.6721	3.1206	3.0508	2.7471
Haneda	0.5110	0.2640	0.2480	0.1960	2.8990	3.0543	2.9112	2.9147
Anesaki	0.5600	0.2700	0.2900	0.2300	2.7227	3.0631	3.1206	2.9845
Yokohama	0.4725	0.2490	0.2293	0.1960	2.6803	3.1187	3.0938	2.7997
Kimitsu	0.4409	0.2442	0.2148	0.1902	2.6300	3.0997	3.1301	2.7887
Futtsu	0.4500	0.2500	0.2100	0.1800	2.6005	3.1067	3.1416	2.7576
Yokosuka	0.4124	0.2405	0.1986	0.1875	2.6663	3.1161	3.1168	2.7997

Harmonic Constants of 3D Tidal Currents

The tidal currents in Tokyo Bay were observed at 10 velocimeter stations as well as different water depths, of which the locations in the bay are shown in Figure 6.7.3. The effective tidal current data have been collected in five stations (i.e., St.1, 2, 5, 9, and 10) for 15 days, other four stations (i.e., St.3, 4, 6, and 8) for 30 days, and only St. 7 for 60 days. By means of the similar description of tidal level data, the tidal current data at different stations and water depths are also expressed as a combination of harmonic constituents. The selected harmonic constituents are totally classified as ten classes, from K_1 tide to MS_4 tide. The two tidal current components northward and eastward during the observation period can be represented as the following superposition of the ten tidal constituents:

$$U_N = \sum_{m=1}^{10} U_m \sin(\frac{2\pi}{T_m}t + \frac{\pi\alpha_m}{180}), \tag{6.7.7}$$

$$U_E = \sum_{m=1}^{10} V_m \sin(\frac{2\pi}{T_m}t + \frac{\pi\beta_m}{180}), \tag{6.7.8}$$

where U_N and U_E are the northward and eastward current components respectively; U_m and V_m are the amplitudes of the two components in the corresponding tidal constituents; α_m and β_m are their phase delay with the unit of degree. The tidal period of the ten tidal constituents can be found in Table 6.7.2 and Table 6.7.5. These harmonic constants of tidal constituents in the upper layer (near water surface), middle layer, and lower layer (near bed surface) at each observation station are sequentially listed in Table 6.7.6 –Table 6.7.8. The locations (levels) of the three layers at ten different velocimeter stations are show in Table 6.7.9. Therefore, model testers can use the harmonic constants to replicate the observed tidal current data, and then compare the time series of computed tidal currents (or tidal current ellipses) with the observed.

Table 6.7.5 Tidal periods of tidal constituents

Tidal Constituents	P_1	Q_1	K_2	N_2	M_4	MS_4
Periods T_m (hour)	24.07	26.87	11.97	12.66	6.21	6.10

3D Residual Current data

Residual current is defined as a residual water movement when the tidal currents are averaged over several tidal cycles. Tidal movement, wind stress, bottom friction stress, and density gradients may drive residual circulation. Residual currents are generally one or two orders of magnitude less than the tidal currents themselves, but they play a crucial role in water qualities in coastal waters. This database provides the residual currents on upper, middle, and bottom layers, obtained from the tidal currents during the observation period in Tokyo Bay. The values of two current components northward and eastward at the ten observation stations and the three layers are listed in Table 6.7.9. The distribution of residual circulation on the upper layer is shown in Figure 6.7.6.

The observation data described above are archived in Appendix A of this report. Read instructions can be found in the file README.TXT. Users can find the digitized data in some EXCEL files including the harmonic constants of the tidal levels and the tidal currents.

As a test example, the comparisons of tidal levels and currents between the observations and simulations by the finite element model (Kodama et al. 1996) are shown as follows. Figure 6.7.7 shows the comparisons of the time variations of tidal levels at eight tide gauges. Figure 6.7.8 provides an example of the comparisons between the computed tidal ellipses driven by the four major tidal constituents listed in Table 6.7.2 and the observed tidal ellipses generated only by the M_2 tide of which the current data are shown in Table 6.7.6–Table 6.7.8. The comparisons have been done in three different layers, i.e., upper (near water surface), middle, and lower (near bed surface) layers. In Figure 6.7.9, the computed residual currents at the three different layers are compared with the observed steady currents at the ten observation stations listed in Table 6.7.9.

6.7.9 Remarks

This test case provides a set of rather comprehensive database of 3D tidal currents in Tokyo Bay observed during a period from 25 August to 25 October 1983. The database consists of the bathymetry of the bay, tidal levels at eight stations and currents measured at several locations as well as different water depths. In addition, it suggests the values of physical constants and boundary forcing conditions including incident tidal waves, river inflow velocities, and wind velocities above the water surface. This database is adequate for users to validation of any 3D hydrodynamic model developed for tidal current simulations in coastal waters.

6.7.10 Reference

Blumberg, A. F., and Kantha, L. H. (1985), Open boundary condition for circulation models, J. Hydr. Engrg., ASCE, 111(2), 237-255.

Gunson, J. R., and Malanotte-Rizzoli, P. (1996), Assimilation studies of open-ocean flows, 1, Estimation of initial and boundary conditions, J. Geophys. Res., Vol. 101 , No. C12 , p. 28,457-28472.

Lardner, R. W. (1993), Optimal control of open boundary conditions for a numerical tidal model, Computer Methods in Applied Mechanics and Engineering, 102, 3, pp367-387.

Kawahara, M., M. Kobayashi, and K. Nakata (1983), Multiple level finite element analysis and its applications to tidal current flow in Tokyo Bay, Appl. Math. Modelling, Vol.7, 197-211.

Kodama, T., and Kawahara, M. (1992). Multiple level finite element analysis for tidal current flow with non-reflective open boundary condition, Structural Eng./Earthquake Eng., JSCE, Vol.9, No.1, 77s-87s.

Kodama, T., T. Kawasaki, and M. Kawahara (1991), A finite element method for shallow water equation including open boundary condition, Int. J. Numer. Methods Fluids, Vol.13, 939-953.

Kodama, T., S. S. Y. Wang, and M. Kawahara (1996), Model Verification on 3D Tidal Current Analysis in Tokyo Bay, Int. J. Numer. Methods Fluids, Vol.22, 43-66.

Lin, P., and Liu, P. L.-F. (1999), Internal wave-maker for Navier-Stokes equations models, J. Wtrwy., Port., Coast., and Oc. Engrg., ASCE, 125, 4, 207-215.

Maritime Safety Agency (1984), Maritime Chart of Tokyo Datum, p90.

Schaffer, H. A., Stolborg, T., and Hyllested, P. (1994). Simultaneous generation and active absorption of waves in flumes, Int. Symp. Waves-Phys. And Numer. Modelling, IAHR, Delft, The Netherlands, 90-99.

Yokohama Branch, Second Port Construction Bureau, Ministry of Transport, Government of Japan (1983), Report of Measurements of Tidal Current in Tokyo Bay, 1983 (in Japanese).

Table 6.7.6 Harmonic constants of northward and eastward velocity components (Upper Layer)

St.			M_2	S_2	K_1	O_1	P_1	Q_1	K_2	N_2	M_4	MS_4
1	North U_N	U_m	13.2	10.4	5.9	3.2	2.0	3.6	2.8	2.6	2.4	3.1
		α_m	51.9	94.5	113.8	81.7	113.8	83.6	94.5	98.5	240.2	304.3
	East U_E	V_m	4.2	4.0	0.8	1.3	0.3	0.7	1.1	2.7	2.0	2.3
		β_m	64.1	149.2	342.1	2.8	342.1	316.9	149.2	301.9	37.6	63.9
2	North U_N	U_m	12.2	8.8	4.1	0.2	1.4	1.8	2.4	2.6	1.3	1.4
		α_m	27.7	69.4	35.8	114.7	35.8	64.4	69.4	68.0	149.4	122.5
	East U_E	V_m	19.6	9.2	8.1	4.3	2.7	1.2	2.5	2.0	1.3	1.7
		β_m	59.3	100.9	44.1	29.9	44.1	355.4	100.9	64.3	173.5	212.7
3	North U_N	U_m	11.0	8.8	2.2	2.7	0.7	1.0	2.4	0.4	2.1	0.8
		α_m	58.2	86.7	113.3	98.9	113.3	169.9	86.7	1.0	112.9	205.1
	East U_E	V_m	11.2	7.5	1.3	0.3	0.4	1.3	2.1	2.1	1.0	0.8
		β_m	84.1	111.8	131.8	266.8	131.8	129.0	111.8	19.2	177.7	259.0
4	North U_N	U_m	9.2	6.1	2.0	2.5	0.7	1.5	1.7	2.0	1.3	0.9
		α_m	72.9	113.5	123.6	88.4	123.6	272.2	113.5	61.1	161.9	220.6
	East U_E	V_m	7.6	3.9	1.1	2.0	0.4	1.6	1.1	2.5	0.3	0.8
		β_m	104.3	125.2	83.8	105.2	83.8	49.6	125.2	78.9	198.1	318.4
5	North U_N	U_m	2.2	1.8	3.2	0.8	1.1	1.5	0.5	1.9	1.2	0.5
		α_m	121.6	52.0	44.2	84.7	44.2	36.0	52.0	112.5	239.9	334.5
	East U_E	V_m	7.9	5.4	4.7	1.3	1.6	1.9	1.5	1.5	1.2	0.7
		β_m	37.3	77.7	168.6	93.7	168.6	27.0	77.7	265.7	78.2	76.8
6	North U_N	U_m	7.8	3.3	2.4	3.5	0.8	1.5	0.9	2.6	0.7	0.7
		α_m	70.4	85.5	148.6	98.7	148.6	205.2	85.5	41.6	282.7	274.9
	East U_E	V_m	4.2	3.4	2.9	1.2	1.0	1.0	0.9	1.2	0.3	0.7
		β_m	72.0	94.3	101.0	104.9	101.0	327.4	94.3	85.1	65.1	42.9
7	North U_N	U_m	5.3	3.1	0.2	0.9	0.1	0.8	0.9	1.1	0.9	0.2
		α_m	52.0	78.8	17.6	220.1	17.6	152.6	78.8	56.1	152.8	43.7
	East U_E	V_m	7.1	6.4	2.8	1.0	0.9	0.4	1.7	1.2	0.5	0.5
		β_m	52.6	91.6	101.5	36.0	101.5	16.5	91.6	58.2	157.5	196.8
8	North U_N	U_m	3.1	3.4	1.6	1.4	0.5	1.1	0.9	1.5	1.2	0.3
		α_m	36.9	74.7	67.7	99.2	67.7	102.6	74.7	282.7	123.9	305.3
	East U_E	V_m	4.4	2.2	1.6	1.8	0.5	0.5	0.6	2.5	0.5	0.5
		β_m	95.0	136.8	121.8	24.7	121.8	212.3	136.8	44.2	130.2	66.1
9	North U_N	U_m	7.4	3.5	2.8	1.8	0.9	1.8	0.9	2.6	0.5	0.7
		α_m	78.2	90.5	40.8	50.4	40.8	260.7	90.5	70.5	292.7	258.9
	East U_E	V_m	2.9	3.4	4.3	2.7	1.4	4.2	0.9	2.0	0.4	1.1
		β_m	115.6	131.4	51.9	91.8	51.9	317.3	131.4	33.5	152.1	237.5
10	North U_N	U_m	7.4	4.8	1.2	1.3	0.4	0.4	1.3	0.6	0.2	0.7
		α_m	73.3	80.0	359.6	115.6	359.6	4.4	80.0	144.4	270.0	181.0
	East U_E	V_m	3.6	2.2	2.2	2.1	0.7	2.8	0.6	1.7	0.7	0.0
		β_m	87.3	136.5	80.4	36.1	80.4	270.8	136.5	119.4	160.4	15.7

*Note: The unit of U_m and V_m is cm/s; the unit of α_m and β_m is degree.

Table 6.7.7 Harmonic constants of northward and eastward velocity components (Middle Layer)

St.			M_2	S_2	K_1	O_1	P_1	Q_1	K_2	N_2	M_4	MS_4
1	North U_N	U_m	13.8	8.0	5.7	5.9	1.9	3.0	2.2	6.1	3.3	2.6
		α_m	93.4	127.9	112.8	81.2	112.8	48.5	127.9	28.9	309.1	346.4
	East U_E	V_m	2.5	3.4	0.8	1.3	0.3	3.5	0.9	1.7	2.4	3.4
		β_m	285.6	294.0	156.8	248.3	156.8	148.0	294.0	29.0	55.1	113.8
2	North U_N	U_m	12.5	5.5	4.3	1.0	1.4	0.5	1.5	2.2	0.7	0.5
		α_m	41.3	69.9	41.6	353.9	41.6	98.1	69.9	315.3	129.9	87.7
	East U_E	V_m	18.4	10.2	5.1	4.1	1.7	1.8	2.5	4.1	2.0	1.9
		β_m	58.6	95.7	45.2	51.2	45.2	20.5	95.7	12.4	139.3	201.9
3	North U_N	U_m	12.8	7.2	1.7	1.5	0.6	1.7	2.0	0.7	0.1	0.9
		α_m	67.9	109.0	127.9	47.7	127.9	112.7	109.0	74.7	32.9	85.1
	East U_E	V_m	7.8	1.3	3.4	4.2	1.1	2.0	0.4	1.6	0.8	0.8
		β_m	88.3	77.3	126.5	100.0	126.5	328.8	77.3	100.3	28.2	97.2
4	North U_N	U_m	9.8	6.4	2.8	3.0	0.9	0.8	1.7	1.3	0.2	0.6
		α_m	63.6	98.5	73.6	45.2	73.6	7.7	98.5	82.4	111.9	301.9
	East U_E	V_m	10.1	4.9	1.8	1.7	0.6	0.9	1.3	2.6	0.5	0.5
		β_m	73.1	108.1	86.2	84.0	86.2	7.1	108.1	19.4	159.4	156.5
5	North U_N	U_m	4.3	1.6	2.1	1.7	0.7	1.5	0.4	3.4	0.4	0.7
		α_m	95.1	141.4	162.3	115.7	162.3	68.2	141.4	17.7	201.8	258.8
	East U_E	V_m	8.7	4.3	0.9	0.3	0.3	2.3	1.2	6.4	1.0	1.3
		β_m	99.0	116.7	247.1	74.2	247.1	155.7	116.7	56.3	212.4	242.7
6	North U_N	U_m	8.7	5.4	2.3	1.8	0.8	0.4	1.5	1.6	0.5	0.6
		α_m	72.2	109.9	61.0	54.4	61.0	61.5	109.9	52.0	179.7	226.3
	East U_E	V_m	3.9	1.6	2.0	1.0	0.7	0.9	0.4	0.7	0.1	0.3
		β_m	74.2	122.8	87.6	76.4	87.6	15.4	122.8	56.5	137.2	216.9
7	North U_N	U_m	5.3	3.3	1.2	0.7	0.4	0.4	0.9	1.2	0.3	0.2
		α_m	69.5	119.1	35.5	358.0	35.5	59.7	119.1	13.7	181.8	227.0
	East U_E	V_m	5.4	2.3	2.2	2.0	0.7	0.3	0.6	1.3	0.1	0.2
		β_m	88.3	133.6	66.9	48.4	66.9	115.6	133.6	66.7	135.7	205.4
8	North U_N	U_m	2.2	0.9	1.0	0.8	0.3	0.3	0.2	0.9	0.2	0.2
		α_m	112.5	119.1	41.6	20.2	41.6	330.4	119.1	108.4	215.1	47.9
	East U_E	V_m	1.1	1.8	0.6	0.8	0.2	0.2	0.5	0.4	0.3	0.5
		β_m	59.4	92.7	116.2	53.0	116.2	255.4	92.7	210.1	144.6	201.3
9	North U_N	U_m	4.6	3.2	1.5	1.5	0.5	0.7	0.9	0.7	1.0	1.0
		α_m	65.3	117.7	105.9	49.4	105.9	305.4	117.7	66.3	263.5	296.8
	East U_E	V_m	2.2	1.3	2.1	1.0	0.7	0.7	0.3	0.6	0.4	0.1
		β_m	38.2	45.2	82.0	1.7	82.0	105.7	45.2	286.0	51.9	320.3
10	North U_N	U_m	2.9	2.1	1.0	1.3	0.3	0.6	0.6	0.3	0.6	0.0
		α_m	75.0	122.3	65.5	45.6	65.5	113.9	122.3	290.7	172.4	102.0
	East U_E	V_m	1.9	1.5	1.4	1.8	0.5	1.0	0.4	0.8	0.2	0.1
		β_m	35.2	128.2	343.6	31.9	343.6	79.9	128.2	71.3	91.3	151.3

*Note: The unit of U_m and V_m is cm/s; the unit of α_m and β_m is degree.

Table 6.7.8 Harmonic constants of northward and eastward velocity components (Lower Layer)

St.			M_2	S_2	K_1	O_1	P_1	Q_1	K_2	N_2	M_4	MS_4
	North	U_m	14.8	7.7	6.2	3.7	2.1	1.0	2.1	3.4	1.1	0.6
1	U_N	α_m	74.2	108.0	105.0	34.1	105.0	58.9	108.0	33.2	313.0	340.0
	East	V_m	0.6	1.0	0.6	0.6	0.2	0.4	0.3	1.3	0.6	1.0
	U_E	β_m	8.1	316.3	13.5	13.5	13.5	160.2	316.3	197.4	211.2	281.1
	North	U_m	N/A	N/A	N/A	N/A	N/A	N/A	N/A	N/A	N/A	N/A
2	U_N	α_m	N/A	N/A	N/A	N/A	N/A	N/A	N/A	N/A	N/A	N/A
	East	V_m	N/A	N/A	N/A	N/A	N/A	N/A	N/A	N/A	N/A	N/A
	U_E	β_m	N/A	N/A	N/A	N/A	N/A	N/A	N/A	N/A	N/A	N/A
	North	U_m	5.9	4.9	1.5	1.6	0.5	1.2	1.3	1.0	0.4	0.1
3	U_N	α_m	44.0	67.5	110.6	90.9	110.6	3.4	67.5	76.3	113.5	131.3
	East	V_m	7.5	5.1	3.4	3.0	1.1	1.2	1.4	2.9	0.3	0.3
	U_E	β_m	85.4	89.7	113.6	85.9	113.6	45.6	89.7	92.5	224.4	270.6
	North	U_m	8.6	4.9	2.0	1.7	0.7	0.3	1.3	1.9	0.8	0.5
4	U_N	α_m	58.2	92.6	73.9	41.2	73.9	47.5	92.6	61.0	181.9	221.6
	East	V_m	10.4	5.7	1.9	1.6	0.6	0.8	1.5	1.2	0.5	0.7
	U_E	β_m	66.7	103.3	53.4	42.3	53.4	354.2	103.3	56.3	144.0	194.1
	North	U_m	N/A	N/A	N/A	N/A	N/A	N/A	N/A	N/A	N/A	N/A
5	U_N	α_m	N/A	N/A	N/A	N/A	N/A	N/A	N/A	N/A	N/A	N/A
	East	V_m	N/A	N/A	N/A	N/A	N/A	N/A	N/A	N/A	N/A	N/A
	U_E	β_m	N/A	N/A	N/A	N/A	N/A	N/A	N/A	N/A	N/A	N/A
	North	U_m	8.6	5.6	3.2	1.7	1.1	0.5	1.5	1.1	0.6	0.3
6	U_N	α_m	70.1	104.3	76.1	45.3	76.1	91.8	104.3	33.2	141.4	234.3
	East	V_m	4.1	2.4	1.5	1.1	0.5	0.2	0.6	0.7	0.5	0.4
	U_E	β_m	77.4	113.3	100.2	53.1	100.2	103.4	133.3	18.4	154.6	219.2
	North	U_m	N/A	N/A	N/A	N/A	N/A	N/A	N/A	N/A	N/A	N/A
7	U_N	α_m	N/A	N/A	N/A	N/A	N/A	N/A	N/A	N/A	N/A	N/A
	East	V_m	N/A	N/A	N/A	N/A	N/A	N/A	N/A	N/A	N/A	N/A
	U_E	β_m	N/A	N/A	N/A	N/A	N/A	N/A	N/A	N/A	N/A	N/A
	North	U_m	N/A	N/A	N/A	N/A	N/A	N/A	N/A	N/A	N/A	N/A
8	U_N	α_m	N/A	N/A	N/A	N/A	N/A	N/A	N/A	N/A	N/A	N/A
	East	V_m	N/A	N/A	N/A	N/A	N/A	N/A	N/A	N/A	N/A	N/A
	U_E	β_m	N/A	N/A	N/A	N/A	N/A	N/A	N/A	N/A	N/A	N/A
	North	U_m	N/A	N/A	N/A	N/A	N/A	N/A	N/A	N/A	N/A	N/A
9	U_N	α_m	N/A	N/A	N/A	N/A	N/A	N/A	N/A	N/A	N/A	N/A
	East	V_m	N/A	N/A	N/A	N/A	N/A	N/A	N/A	N/A	N/A	N/A
	U_E	β_m	N/A	N/A	N/A	N/A	N/A	N/A	N/A	N/A	N/A	N/A
	North	U_m	N/A	N/A	N/A	N/A	N/A	N/A	N/A	N/A	N/A	N/A
10	U_N	α_m	N/A	N/A	N/A	N/A	N/A	N/A	N/A	N/A	N/A	N/A
	East	V_m	N/A	N/A	N/A	N/A	N/A	N/A	N/A	N/A	N/A	N/A
	U_E	β_m	N/A	N/A	N/A	N/A	N/A	N/A	N/A	N/A	N/A	N/A

*Note: The unit of U_m and V_m is cm/s; the unit of α_m and β_m is degree.

Table 6.7.9 Measured steady flow at 10 stations in Tokyo Bay during the period from August 25 to October 25, 1983.

Station Number	Layer	Vertical Location	Northward Velocity (cm/s)	Eastward Velocity (cm/s)
1	Upper	3m below surface	1.5	-2.8
	Middle	15m above bottom	15.8	-1.8
	Lower	5m above bottom	5.1	1.3
2	Upper	3m below surface	-12.3	-7.0
	Middle	4m above bottom	-5.8	-6.8
	Lower	N/A	N/A	N/A
3	Upper	3m below surface	-5.4	-5.7
	Middle	13m above bottom	7.0	10.0
	Lower	5m above bottom	3.2	5.3
4	Upper	3m below surface	-3.2	-8.1
	Middle	14m above bottom	1.9	2.4
	Lower	5m above bottom	0.5	0.4
5	Upper	3m below surface	-4.5	-11.9
	Middle	10m above bottom	0.6	4.0
	Lower	N/A	N/A	N/A
6	Upper	3m below surface	10.1	0.3
	Middle	10m above bottom	1.9	-1.1
	Lower	5m above bottom	0.0	-2.1
7	Upper	3m below surface	-3.8	0.2
	Middle	6m above bottom	0.8	2.5
	Lower	N/A	N/A	N/A
8	Upper	3m below surface	-3.7	-0.2
	Middle	4m above bottom	2.2	1.3
	Lower	N/A	N/A	N/A
9	Upper	3m below surface	8.5	4.2
	Middle	4m above bottom	0.9	0.3
	Lower	N/A	N/A	N/A
10	Upper	3m below surface	-2.2	-1.3
	Middle	3m above bottom	-0.4	-3.5
	Lower	N/A	N/A	N/A

Figure 6.7.1 Observation points in Tokyo Bay

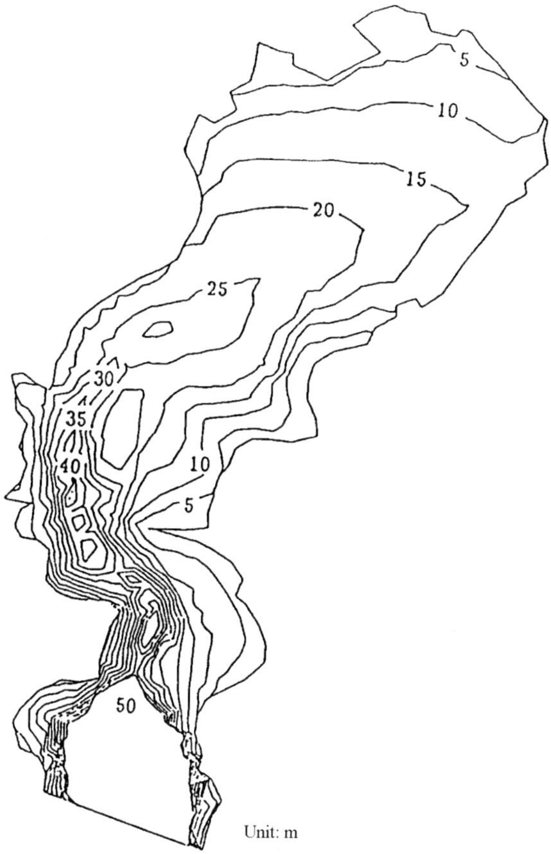

Figure 6.7.2 Water depth contours in Tokyo Bay (unit: m)

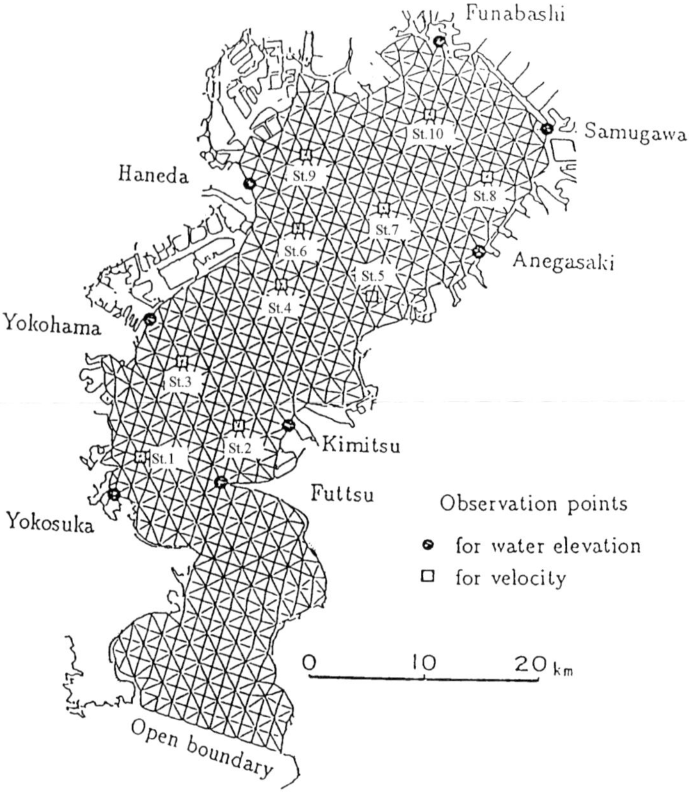

Figure 6.7.3 Tide gauges and observation stations of tidal current (marked as St.) in a surface finite element mesh

Figure 6.7.4 Finite element meshes at five layers from surface to water bottom

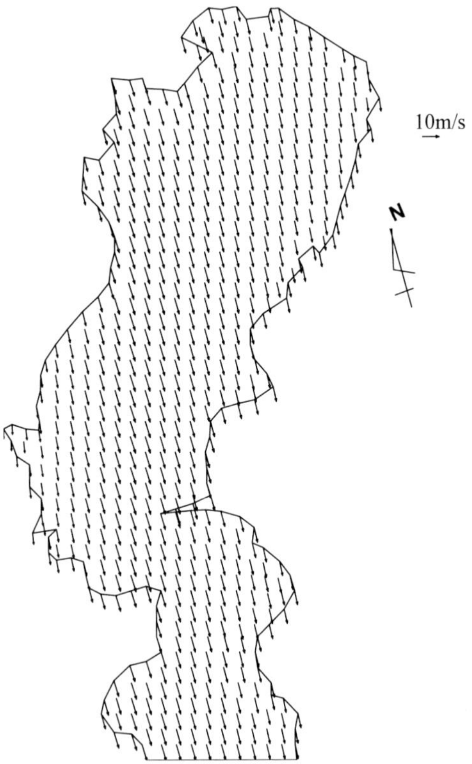

Figure 6.7.5 Steady wind velocity field

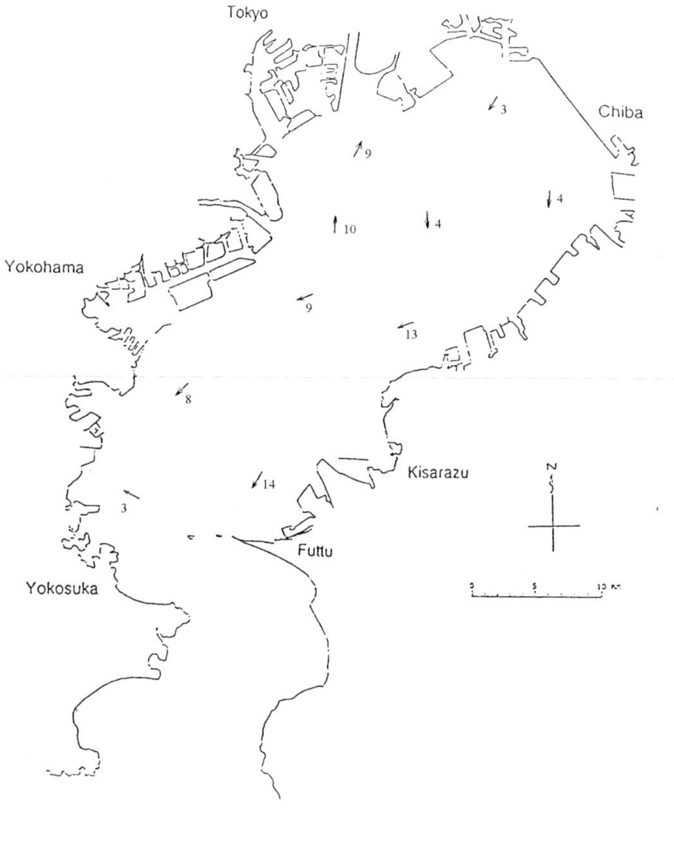

Figure 6.7.6 Measured steady flow in upper layer (unit: cm/s)

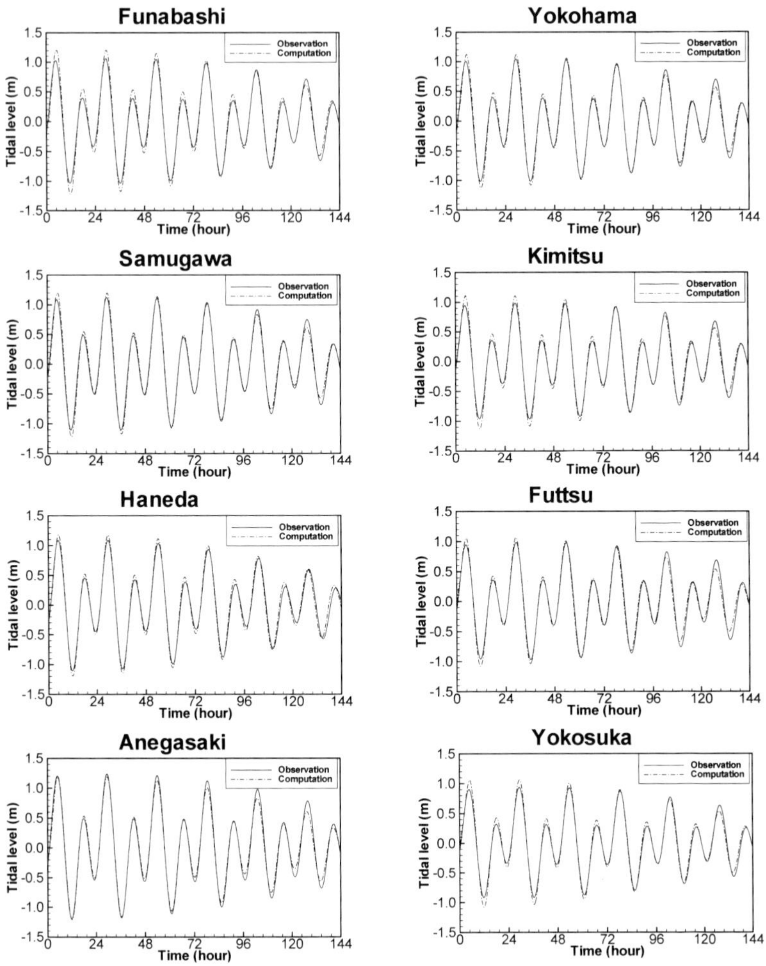

Figure 6.7.7 Comparisons of time variations of tidal levels at the eight tide gauges

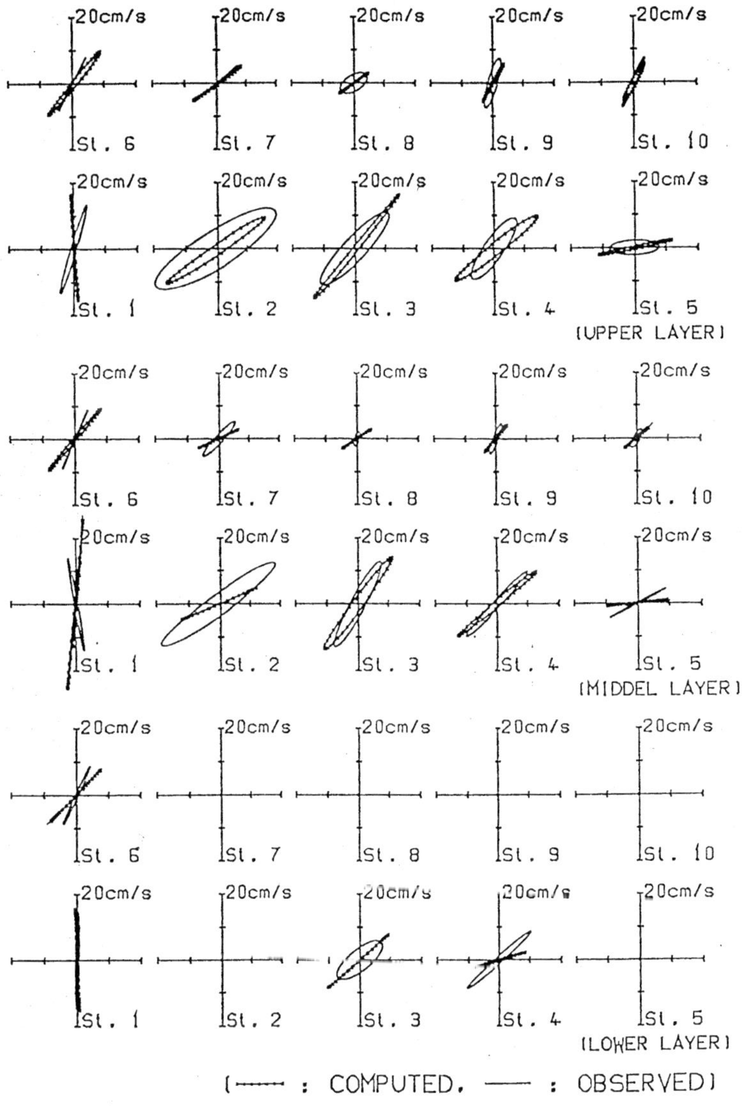

Figure 6.7.8 Comparisons of M_2 tidal ellipses at ten observation stations. The horizontal and vertical directions represent the eastward velocity and the northward velocity, respectively.

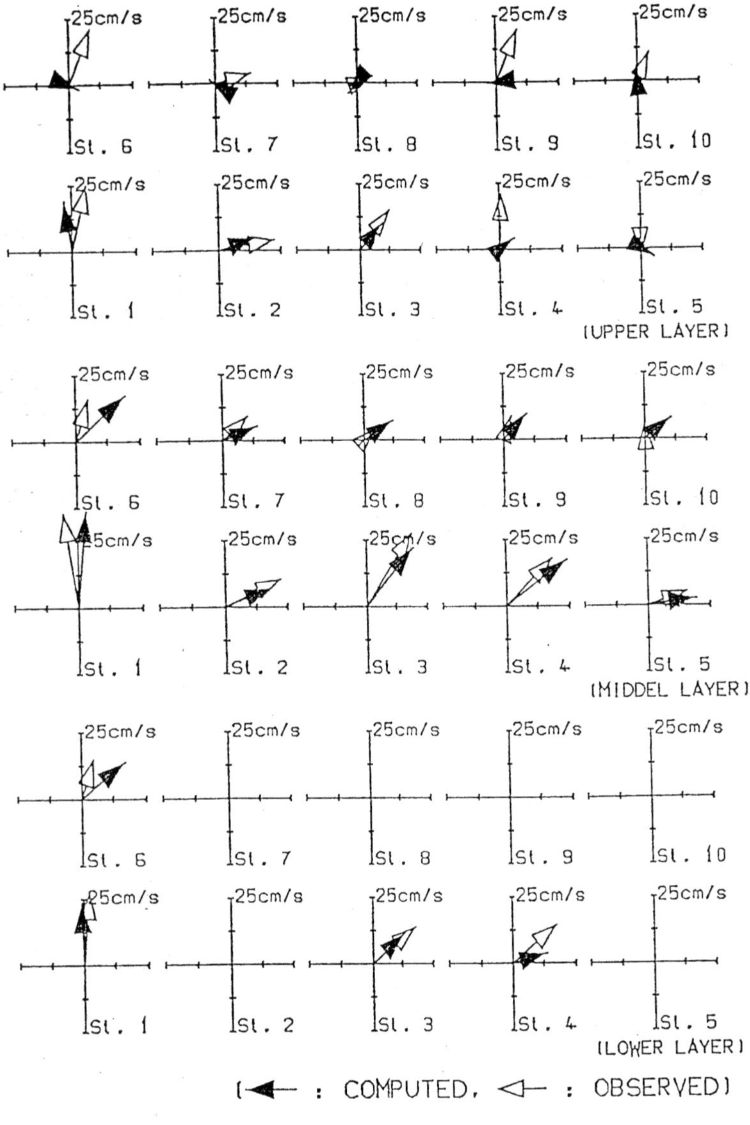

Figure 6.7.9 Comparisons of residual currents at different layers. The horizontal and vertical directions represent the eastward velocity and the northward velocity, respectively.

6.8 VICTORIA BENDWAY, MISSISSIPPI RIVER, TEST CASE

Contributors: Yafei Jia and Sam S.Y. Wang

6.8.1 Background

River engineering projects for navigation safety, flood control, bank protection, channel stabilization, bridge safety, stream ecological system restoration and enhancement, etc. have increasing needs of computational simulations to improve the cost-effectiveness of engineering analysis, design, short- and long-term environmental impact assessment. Both computational simulation model developers and users may be benefited by using a test case based on a set of field data collected from an inland waterway to conduct a model validation.

The test case presented here is based on the data set provided by the U.S. Army Corps of Engineers for the investigation of the effectiveness of submerged weir field installed in the Victoria Bendway of the Mississippi River for enhancing channel navigability. The test case was developed by the National Center for Computational Hydroscience and Engineering at The University of Mississippi.

The data set was collected in a reach of the Mississippi River, the Victoria Bendway, for field study to investigate the effectiveness of hydraulic structures, including three spur dikes on the point bar of the inner (convex) bank and six submerged weirs near the outer (concave) bank for the enhancement of navigability in this highly curved bendway. Three-dimensional velocity components were measured at 34 cross-sections in the study reach. In each cross-section, velocity components were measured along a large number of vertical lines with many measuring points along each vertical line from near the free surface to near the channel bed. The fine resolution of the measured data provides numerical modelers a good opportunity to validate their three-dimensional hydrodynamic models' capability and accuracy in simulating a flow field affected by a series of in-stream structures in a natural river.

6.8.2 Objectives

The objective of this test case is to provide both numerical modelers and users a means to determine the capabilities of three-dimensional free surface flow models for predicting realistic flow fields in a natural environment with highly complex boundary geometry and bathymetry. The complexities of this case include varying shapes of channel cross-sections, width and bathymetry, channel curvature, exposed spur dikes and submerged weirs. As a result, the flow field is expected to be truly three-dimensional with various vortices and secondary currents.

6.8.3 Approach

A short reach, the Victoria Bendway, including the hydraulic structures (dikes and weirs) may be used for the validation test of a three-dimensional model. Due to the complexity of the bathymetry and structures, a refined numerical grid is suggested for this computational domain. To keep the computational effort to a minimum, users of this test case are advised to simulate a longer reach of the Mississippi River first using a two-dimensional depth-averaged shallow water flow model. Because the velocity and free surface elevation distributions along the boundaries of the short reach are unknown, this preliminary step will allow the users to select two channel cross-sections sufficiently far away from the short reach for prescribing approximate boundary conditions. The 2D solution at the boundaries of the short channel reach can then be used for prescribing more desirable and realistic inlet and outlet boundary conditions of the three-dimensional computational domain. A two-dimensional model can also be used to calibrate the site-specific parameters such as Manning's n or bed roughness with adequate accuracy for the three-dimensional model validation test.

6.8.4 The Physical Domain

The Victoria Bendway of the lower Mississippi River is located near the confluence with the White River. It is seen from Figure 6.8.1 that both rivers are meandering in this area with one loop of the White River's meander segment being cut off by the Mississippi River. As a result, the discharge in the Mississippi River upstream the Victoria Bendway varies due to the confluence which poses some difficulty for determining the boundary conditions needed for simulating the flow field. The White River from Arkansas State joins in and immediately branches out of the Mississippi River at the upstream of the Victoria Bendway, and it returns to the Mississippi River again at the downstream portion of the point bar.

Victoria Bendway is a highly curved reach. The ratio of the radius of curvature to the channel width varies from 1 to 3 approximately, depending on the water stage. It has a 108° heading change and a radius of 1280 meter. It is expected that the secondary current would be quite strong.

Six submerged weirs were constructed across the thalweg near the concave side of the Victoria Bendway in 1995, oriented towards to the upstream with an angle from 69 to 76 degree between the weir and the bend longitudinal line (Figure 6.8.2). Post-construction survey shows deposition at the upstream reach of weirs and scouring appears within and the downstream channel of the weir system. The average water depth over the weirs is 11.3 meter (37 ft) with a range from 6.7 m to 20.4 m (22 ft to 67 ft), according to Waterway Simulation Technology, Inc., 1999.

There are three long spur dikes built on the flood plain or point bar of the Victoria Bendway, for converging the flow to the main channel; the large area of the point bar is thus protected from erosion. The discharge in the main channel would be stronger when additional flow is diverted from the point bar during high flows into it due to these dikes. There are many other spur dikes built along the banks of the channel both up and downstream of the Victoria Bendway, but they are located outside the 3D simulation domain and thus need not to be considered.

The extensive bed elevation survey data available for this study were taken in 1998. The bathymetry data available covers the channel from the upstream to the downstream of the Victoria Bendway. The data resolution in the bendway is extremely fine compared with those for other parts of the studied channel. The bathymetry data (Figure 6.8.2) were measured by acoustic devices positioned with a Cartesian coordinate system. Bed elevation data measured in other time periods (1994, 1996) are also available but they are either incomplete in covering the entire study reach or have much lower resolution in the Victoria Bendway than those measured in 1998.

For investigating performance of these submerged weirs, velocity data were taken in using ADCP instrumentation along 34 cross-sections. These data were measured on June 11 and June 12, 1998, about two years after the submerged weirs had been installed in the channel. The flow discharge and water stage were almost constant during the survey.

6.8.5 Initial Conditions

If one plans to validate a steady flow model, the initial condition is simply a guessed solution, which can be arbitrary or assumed at rest (cold start). An educated guess would be better, because it will take less computing time for the iterative solution to converge to the steady state flow condition. The same guessed solution can also be used for unsteady models with steady boundary conditions. Since the data was surveyed at a flow condition close to steady state, the validation is actually conducted under assumed steady state condition. However the test case can be used for validation of both steady state and unsteady state models.

6.8.6 Boundary Conditions

Ideally the boundary conditions at the upstream inlet (velocity or discharge distributions) and downstream outlet (stage) should be prescribed based on the measured data. Because they are not available, one may want to set the locations of the inlet and outlet in a relatively straight reach which is sufficiently far away from the modeling domain. By doing so, the unnecessary complications affecting the accuracy of the prescribed boundary conditions can be minimized, including the reduction of influence of the instream structures on the flow field at the inlet and outlet, and the difficulties of realistically distributing the total discharge to specific discharge or velocities at numerical nodes over the cross-section areas of the inlet and outlet.

Due to the fact that the detailed velocity measurements are not available at desirable inlet and outlet cross-sections, one may apply a depth-averaged model to simulate the flow with a longer reach (for computing efficiency) and convert the two-dimensional velocity distributions at the inlet and outlet cross-sections into 3D velocity fields as the prescribed boundary conditions for validating the 3D free-surface flow simulation models. For this reason, the validation test example given in this test case has applied this approach. Of course, one should not be restricted by this approach if one has an alternative to prescribe the inlet discharge (or velocity) distribution and outlet stage boundary conditions.

To prescribe velocity boundary conditions on the bed, bank and structural surfaces one may choose log-law of the wall, partial-slip or total slip condition. Manning's coefficient, n, is needed and can be obtained by a site specific parametric calibration process. A calibrated value of 0.047 was found to be acceptable for this test case. The Stricker's relation,

$$n = \frac{d^{1/6}}{A} \tag{6.8.1}$$

may be applied to obtain its equivalent wall roughness height, d, where A is an empirical constant. A = 19 was adopted (Chien and Wan, 1999). The flow discharge for this test case is 14,000 m^3/s (494,400 cfs).

6.8.7 Validation Data

Velocity data measured with ADCP instrumentation (600Hz) on June 11-12, 1998, were provided for this model validation case. Figure 6.8.3 shows the survey paths for measuring the velocity field in the Victoria Bendway. A depth-averaged velocity vector of the measured velocities is shown at each survey point where many velocity vectors were measured along a vertical line from free surface to channel bed. The velocity data measured on June 11, 1998 contain 17 paths with a total of 2210 vertical profiles, while the data taken on June 12, 1998 includes 17 paths with a total of 2494 vertical profiles. Each vertical profile is a line along which 3D velocity components (u, v, w) were measured; number of points along a vertical line varies from less than 10 to over 100, depending on the water depth. The average length of a survey path is about 1000 m, and the average time for each survey path is about 9.4 min. As shown in the figure, most paths are not in a straight line due to drifting of the survey vessel in the strong current in the channel. The first weir is located near the 2nd path line. Its front side is hard to recognize from the bathymetry due to sediment deposition in the channel upstream of this weir and/or displacement of rocks from the top of the weir. Scour holes have developed between weirs and downstream of the last weir.

Measured vectors at several levels are shown in Figure 6.8.4 (6.7.4-a: surface level, 6.8.4-b: $0.4h$ level, and 6.8.4-c: $0.05h$ level, where h is flow depth). For clarity, not all of the points along these survey paths are shown. The fluctuation of velocities, especially their directions, can be observed, and the degree of fluctuations is less near the free surface than near the bed. The u, v and w velocity components are defined in the Cartesian coordinate system, aligned with the x (east), y (north) and z (vertical) directions, respectively.

Figure 6.8.5 shows the measured total velocity distribution in several cross sections. One sees clearly the total velocity is high in the central part of the main channel and the main flux of the flow represented by high velocities gradually shifts closer to the concave bank while flowing downstream. It can also be seen there are strong fluctuations and irregularity and the velocity over the point bar is approximately of the same magnitude as that of the main channel. This is due to the fact that the flow in the river is highly turbulent with a wide range of fluctuations; the velocity measurement was conducted in a relatively short period of time which is insufficient for catching the mean flow velocity accurately. Since these sections are arranged in the order from upstream to downstream of the Bendway, one can tell that the shape of the cross-sections changes dramatically: the lower part of the bend is deeper than the upper part with a more distinct main channel; the change of the flood plain is just opposite. The flood plain widens the bendway channel abruptly, and it diverts a large proportion of discharge from the main channel at the upstream of the bend and returns the flow back in the lower part of the bend. Because the flood plain is very wide, the flow flux from the flood plain to the main channel is not negligible.

Figure 6.8.6 shows the measured velocity vector components (*u, v,* and *w*) and magnitude in two sections (Sec. 1 and 19) of the main channel portion.

To illustrate the secondary current in the channel bend, the measured velocities were projected to the transverse planes perpendicular to the longitudinal direction. Because the measured flow direction fluctuates and the curvature of the channel is not constant, the longitudinal and transverse directions were not easy to define. For this plot, they were determined according to the shape of the main channel. Figure 6.8.7 shows the resultant transverse velocity profiles in several cross sections. It is not surprising that due to the highly irregular plane form and the bathymetry of the Bendway, a classical helical motion seen from curved channel experiments is not clearly displayed; rather, numerous vortices are shown at various locations. The secondary motion near the intersection of the main channel and the flood plain, driven by the flow returning from the flood plain to the main channel, can be seen quite clearly for most of the sections (14, 19, 30, 31). Figure 6.8.8 shows a mesh in the main channel with longitudinal and transversal line along these directions and normal to each other. The secondary current in Figure 6.8.7 is obtained by projecting the measured flow velocity to cross-sections of this mesh. In fact, the secondary current of the simulated velocity can be computed using this grid as supplied in the file (newmesh.dat).

To simulate the flow field in the Victoria Bendway, appropriate boundary conditions and a computational mesh are necessary. The measured velocity data are provided in a file (measure.dat) which includes the measuring locations in the Cartesian coordinate system and the vertical distance from the free surface to each velocity measuring point. As seen from the field data measured, the velocity field is indeed complex, especially around the submerged weirs. The complex flow field contains the detailed hydrodynamic mechanisms which can enhance our understanding of the hydrodynamics for explaining not only why the flow field is improved for navigation, but also why the local scours are formed. In order to validate a three-dimensional model's capability in simulating these detailed near field flows, the users of this test case should design the numerical grid with sufficient resolution.

Due to the fact that the mesh generation is a manpower-intensive and time-consuming task, if the user would like to take the advantage of the mesh generated for performing a test similar to the example below, he/she may use the structured quadrilateral mesh provided for users' convenience as shown in Figure 6.8.9. The data file (geo.dat) also includes total number of I and J lines of mesh (I_{max} , J_{max}), nodal coordinates (x, y), initial free surface elevation, bed elevation, roughness height, and nodal type. The vertical nodal location and distribution for 3D simulation are up to users to decide. Two of the three spur dikes on the point bar are quite high, while the other is low compared to the water stage. The two higher ones are configured as emerging dikes, and the lower one is treated as submerged. All the submerged weirs are represented as humps on the bed. The computational domain and bed bathymetry are shown in Figure 6.8.3.

The inlet boundary conditions involving distributions of flow discharge and direction across the inlet are provided in a file (bnd.dat). The water stage at the outlet section could be set as constant which is given in a file (flowb.dat). The flow discharge for this test case is 14,000 m^3/s (494,400 cfs). This discharge could be specified as the boundary condition and distributed along the inlet cross-section according to a function of flow depth. The discharge distribution provided could save time and effort to conduct the 3D test.

6.8.8 Remarks

A three dimensional velocity data set in the Victoria Bendway of the Mississippi River is introduced for conducting a validation test of 3D free surface numerical models. The geomorphology, hydrology, hydraulic structures, and flow field characteristics in this channel reach are described. The velocity data set is quite comprehensive, even though some degree of uncertainty is unavoidable. The data illustrate complex three-dimensional secondary helical current structures, which provide a powerful validation test case for checking the applicability of the turbulence closure and the three-dimensional model flow formulation. Files for the datasets (in Appendix A) are as summarized below:

- geo.dat: mesh, initial water surface elevation, bed elevation, nodal type, and roughness height
- bnd.dat: unit flow discharge components (q_x, q_y) along the inlet section.
- flowb.dat: the water stage at the downstream section
- newmesh.dat: mesh to project longitudinal and transverse velocity components.
- measure.dat: measured velocity data.

6.8.9 A Validation Test Example

A validation test result (Jia and Wang, 2000, Jia et al. 2001) of a 3D free surface flow model, CCHE3D, is presented as an example to demonstrate the workability of the test case.

It should be noted that this model has already been tested by the authors previously using several test cases based on analytic solutions, manufactured solution method, and laboratory measurements. In tests using analytic methods and manufactured solutions one can conclude that this model is free of errors of mathematical formulation, numerical procedure, and computer programming. When it was tested against the laboratory experiments, the model was proven to be capable of reproducing basic flow mechanisms observed. Finally, with the excellent agreement between the simulated and observed velocity field from real life flow in the complex natural river (Victoria Bendway of the Mississippi River), the validation of the model to predict the realistic flow field is also established.

The computational mesh shown in Figure 6.8.9 was used for this computation. The upstream and downstream boundary conditions for this computational domain were obtained from the solution of a depth-integrated two-dimensional flow model CCHE2D, with a computational domain covering a longer reach of the channel, extending to both upstream of the Victoria Bendway and downstream section before the return of the White River (Figure 6.8.1). Distributions of depth averaged velocity magnitudes and directions along the inlet section and water surface elevation along the outlet section of the 3D domain were obtained. In addition, the Manning's coefficient n was calibrated with this two-dimensional flow simulation and the available surface elevation data measured in the same day when the velocity data were taken.

Figures 6.8.10a and 6.8.10b show samples of the velocity comparisons at selected points in the main channel. At each measuring point along a vertical line from free surface to the bed, velocity components u, v, w and total velocity are compared separately. In general, the agreements of the simulation and the measured data are very realistic. As noted earlier, the measured velocity fluctuates strongly, one should pay more attention to the general trend of the mean flow rather than the random details. It was observed that discrepancies are relatively high behind the weirs (Figure 6.8.10c) or where the bed elevation changes abruptly. In these areas, the velocity fluctuations are very strong particularly near the upper and lower part of the flow along a vertical line. Generally, at locations where bed elevations have larger variation, the turbulence fluctuations are greater, and the discrepancy between the measured velocities and models results are higher.

Almost all measured data show fluctuation and variation along vertical lines; the simulation results are smooth curves. This is because the numerical model simulates mean turbulent flow field while the measured velocities are taken in highly turbulent natural conditions in a short period of time. Based on the experiences of physical modelers as well as the knowledge of turbulent flows, the time period sampling for obtaining a mean turbulent flow velocity has a significant effect on the magnitude and direction of the time mean velocity at a measuring point.

6.8.10 Acknowledgement

The data set was provided by Dr. Steve Scott, Hydraulic Engineer of Coastal and Hydraulic Engineering Laboratory of ERDC of the U.S. Army Corps of Engineers in Vicksburg, Mississippi.

This work is a result of research sponsored by the USDA Agriculture Research Service under Specific Research Agreement No. 58-6408-2-0062 (monitored by the USDA-ARS National Sedimentation Laboratory).

6.8.11 References

Chien, N. and Wan, Z. 1999, Mechanics of Sediment Transport, ASCE press, ASCE, 1801 Alexander Bell Drive, Reston, Virginia 20191-4400.

Jia, Y., Scott, S., and Wang, S.S.Y., 2001, "3D Numerical Model Validation Using Field Data and Simulation of Flow in Mississippi River", ASCE, Proceedings (CD-ROM) of World Water & Environmental Resources Congress, May, 2001, Orlando, FL.

Jia, Y., and Wang, S.S.Y., 2000, "Numerical simulations of the channel flow with submerged weirs in Victoria Bendway, Mississippi River", Technical Report No. NCCHE-TR-2000-3, National Center for Computational Hydroscience and Engineering, The University of Mississippi.

Waterway Simulation Technology, Inc. 1999, "A physical model test plan for bendway weir design criteria", Vicksburg, MS.

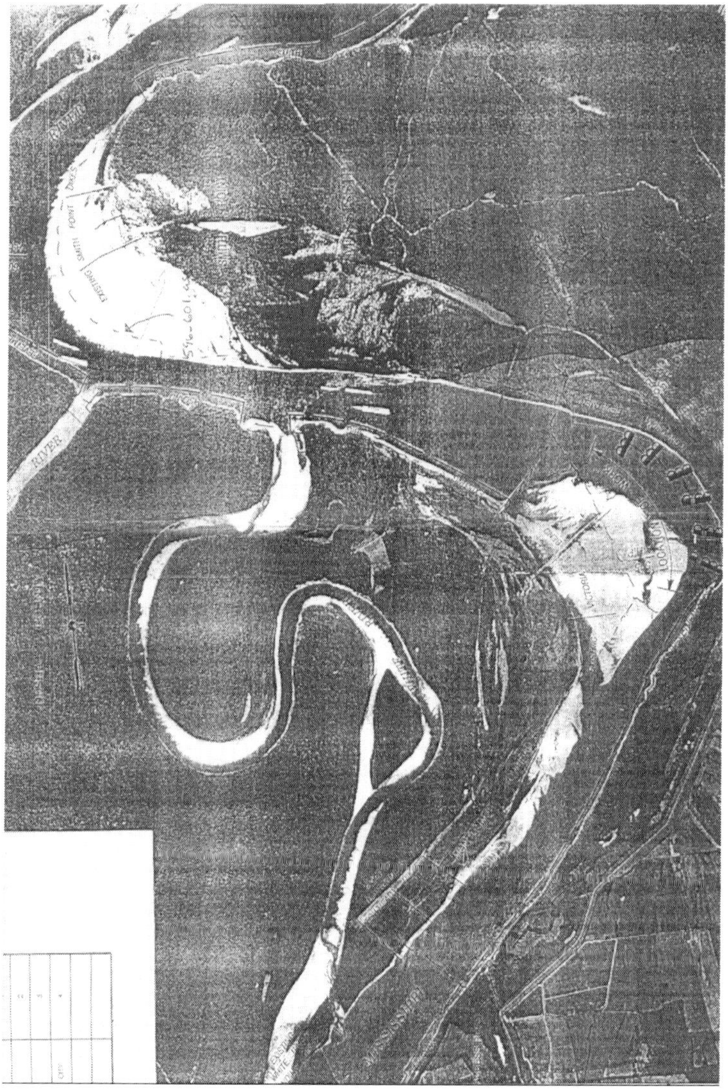

Figure 6.8.1 Victoria Bendway, Mississippi River Test Case: Base Map. The confluence of the White River made it difficult to determine the upstream boundary flow condition for the bendway simulation.

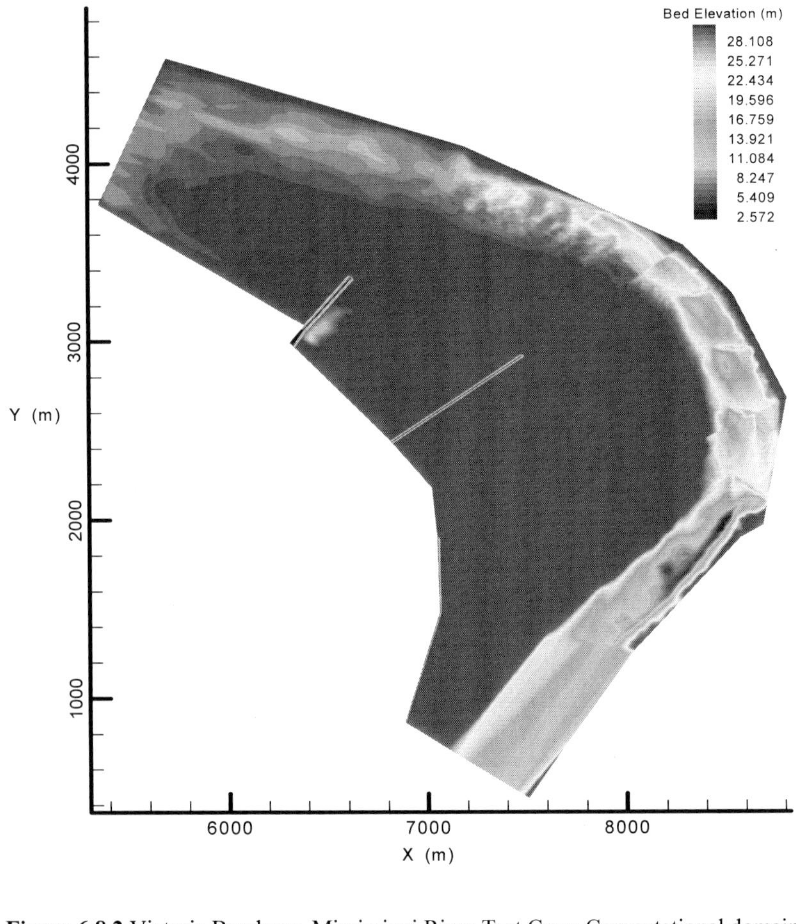

Figure 6.8.2 Victoria Bendway, Mississippi River Test Case: Computational domain and bed topography for the 3D simulation.

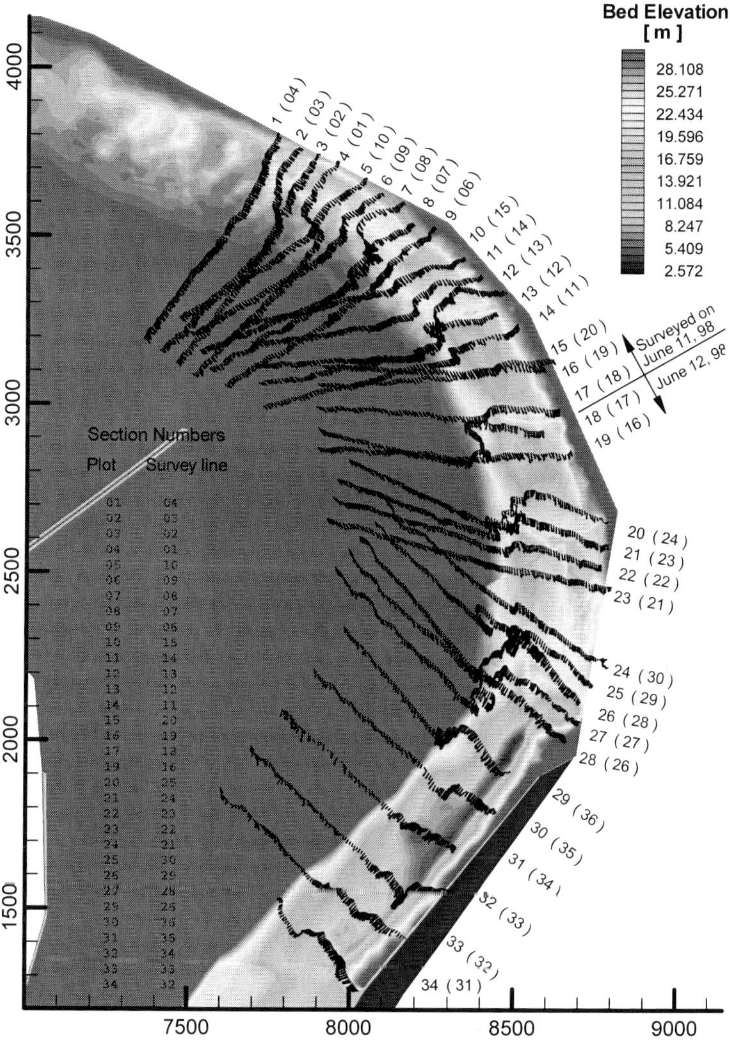

Figure 6.8.3 Victoria Bendway, Mississippi River Test Case: Survey lines and depth-averaged velocities

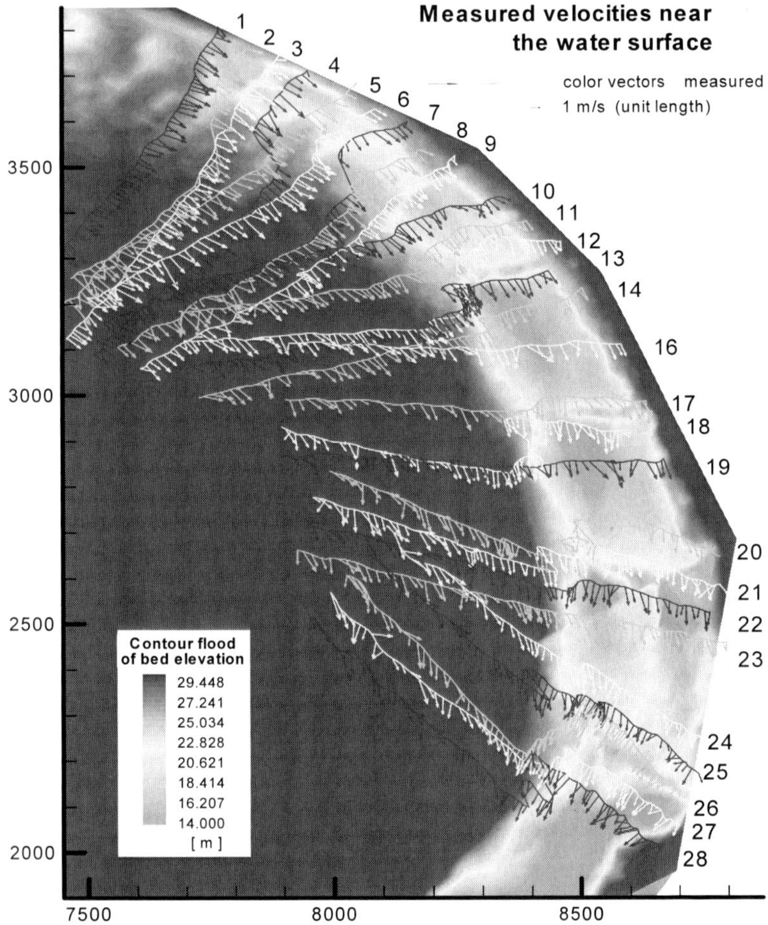

Figure 6.8.4-a Victoria Bendway, Mississippi River Test Case: Measured velocity field at water surface level.

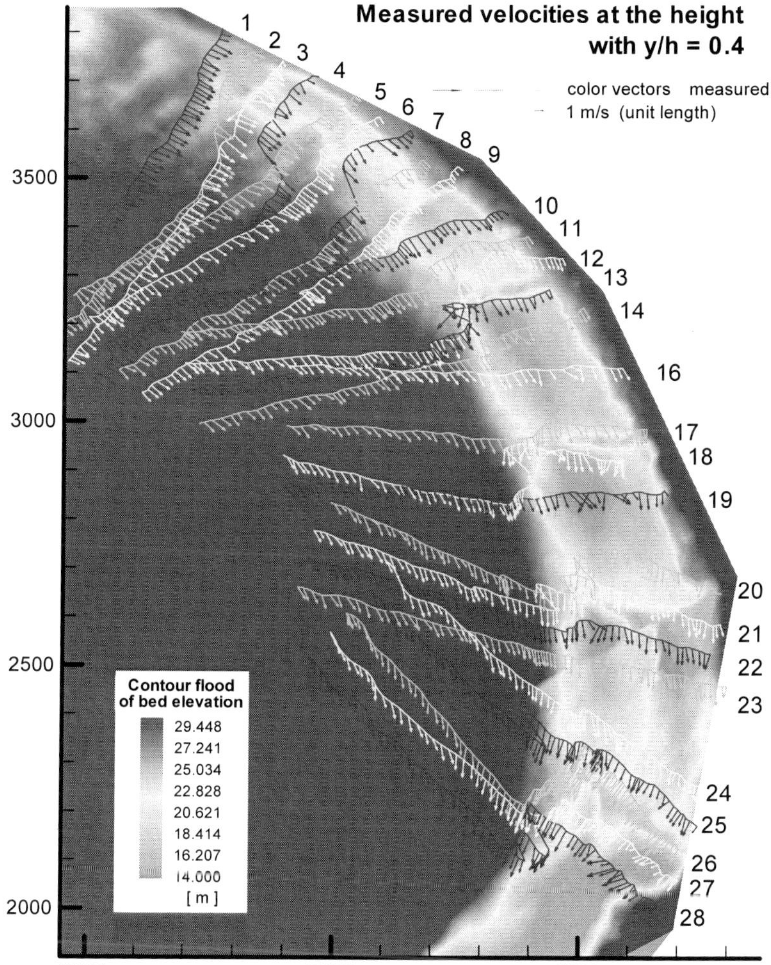

Figure 6.8.4-b Victoria Bendway, Mississippi River Test Case: Measured velocity field at 0.4h level.

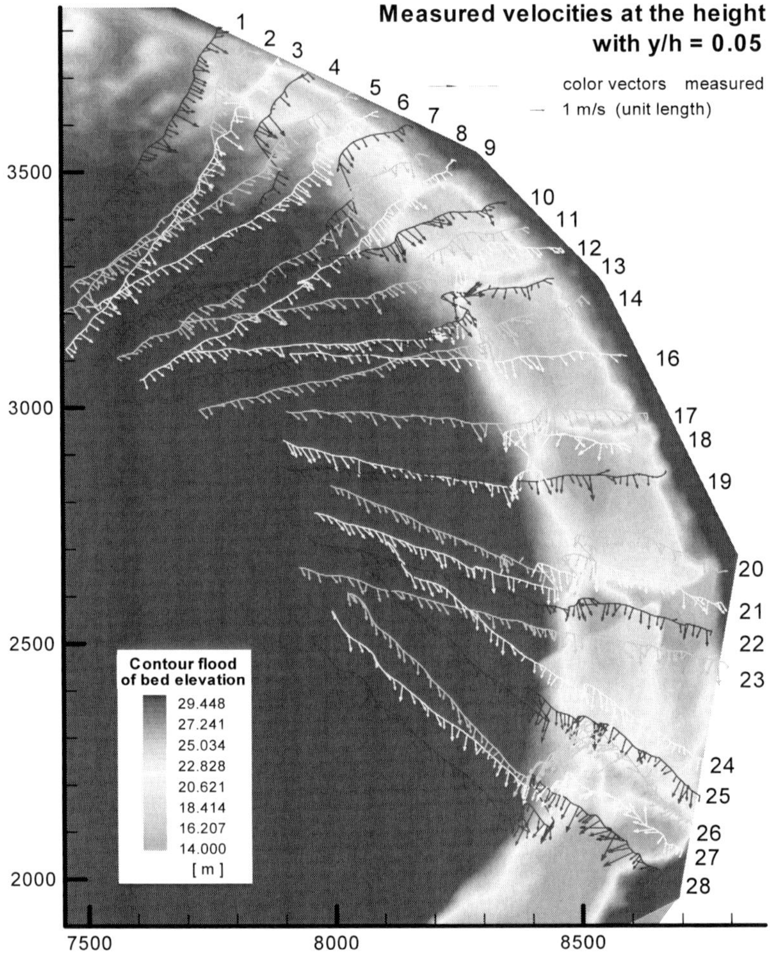

Figure 6.8.4-c Victoria Bendway, Mississippi River Test Case: Measured velocity field at *0.05h* level.

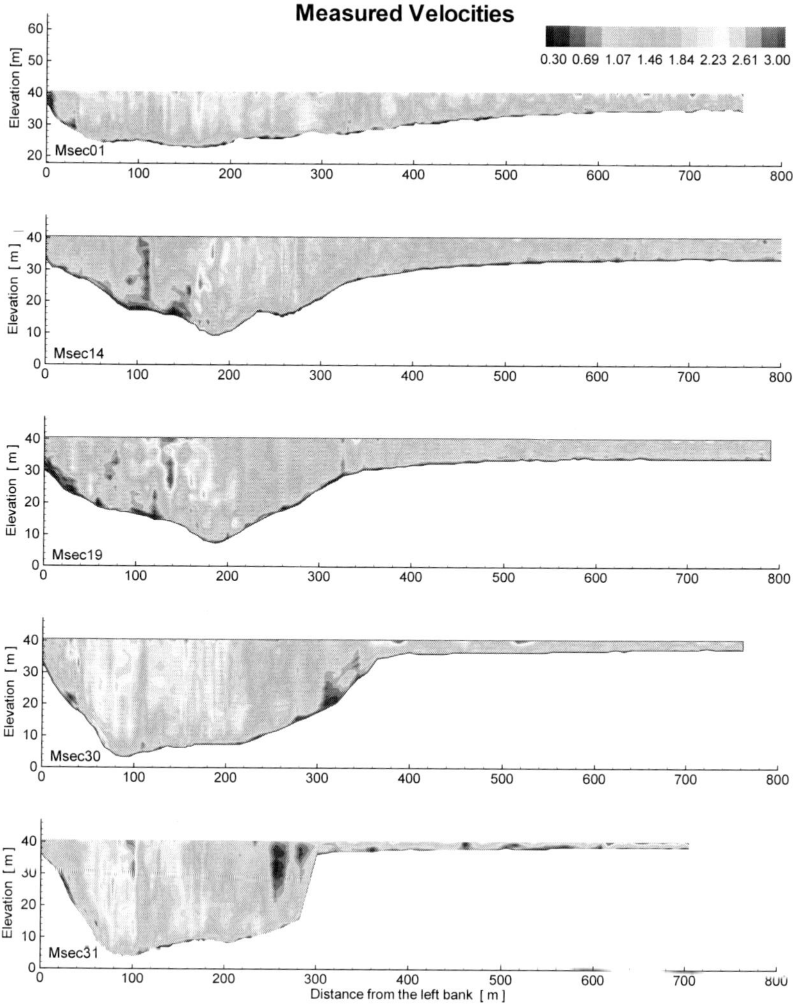

Figure 6.8.5 Victoria Bendway, Mississippi River Test Case: Measured total velocity in several sections along the survey path lines.

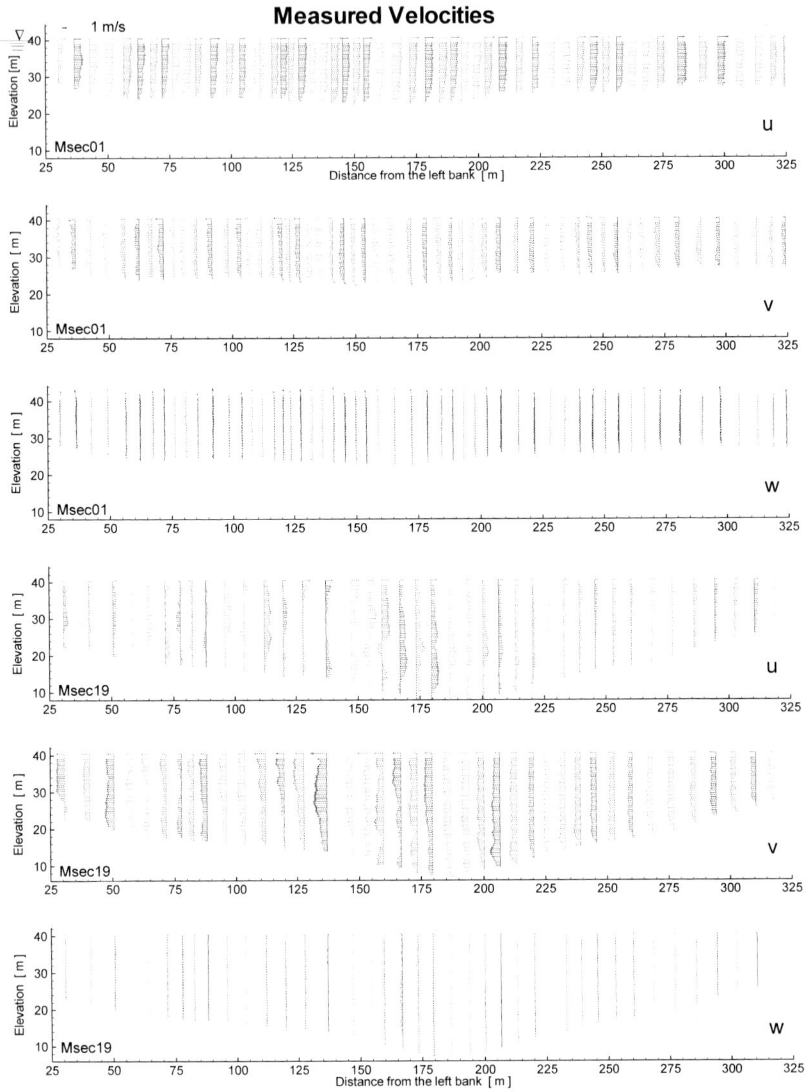

Figure 6.8.6 Victoria Bendway, Mississippi River Test Case: Measured three dimensional velocity components in section 1 and 19.

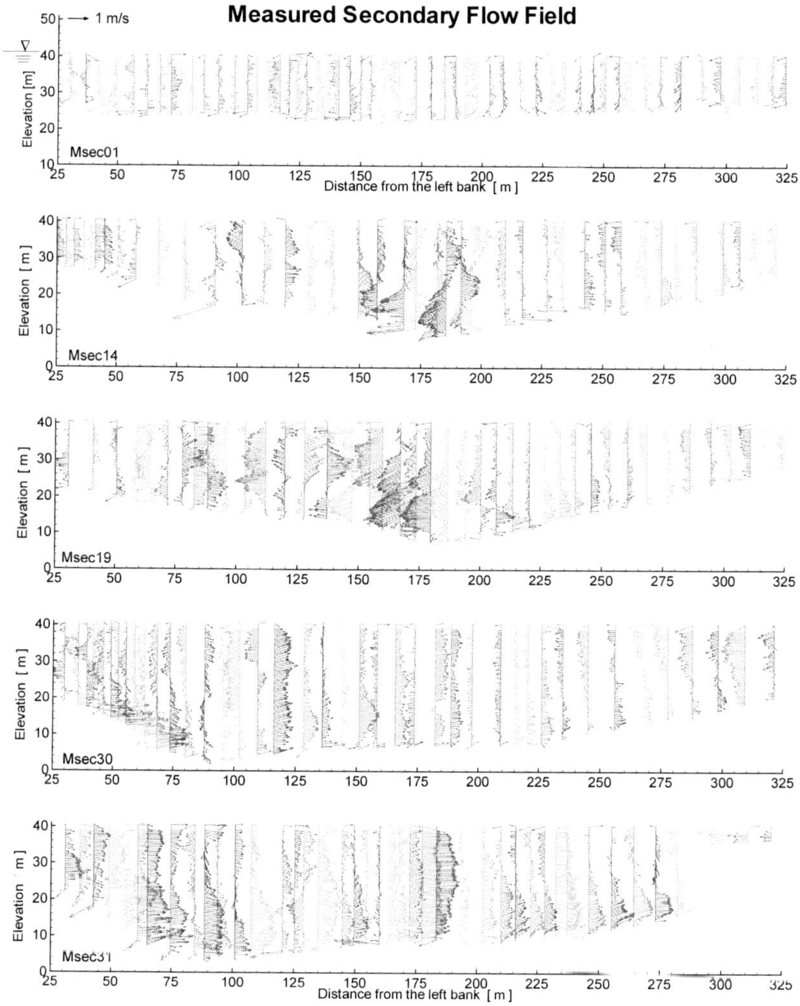

Figure 6.8.7 Victoria Bendway, Mississippi River Test Case: Secondary flow pattern. Pattern obtained by projecting the measured velocity to the section normal to the main flow direction.

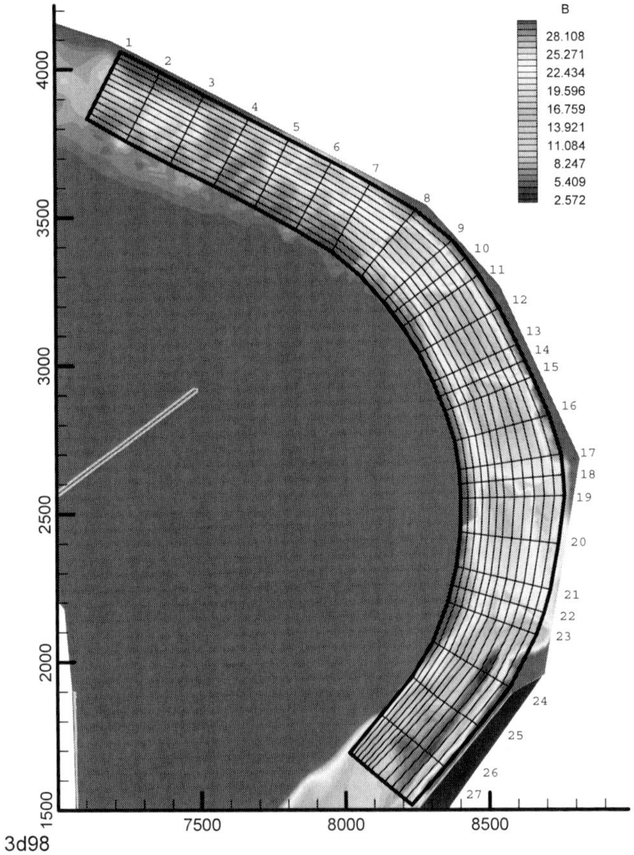

Figure 6.8.8 Orthogonal mesh along the main channel of Victoria Bendway

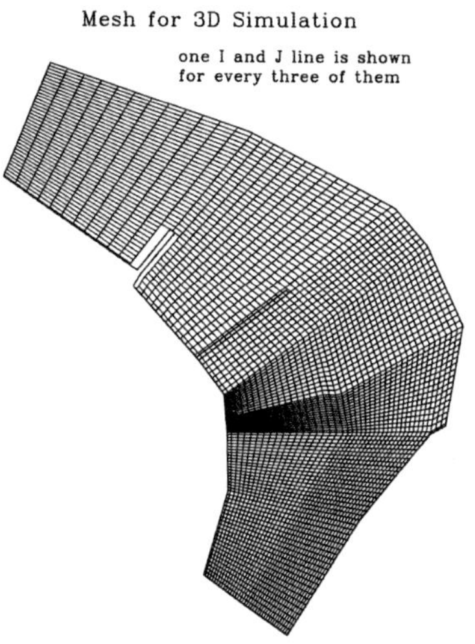

Figure 6.8.9 Victoria Bendway, Mississippi River Test Case: Quadrilateral structured mesh for the Victoria Bendway simulation

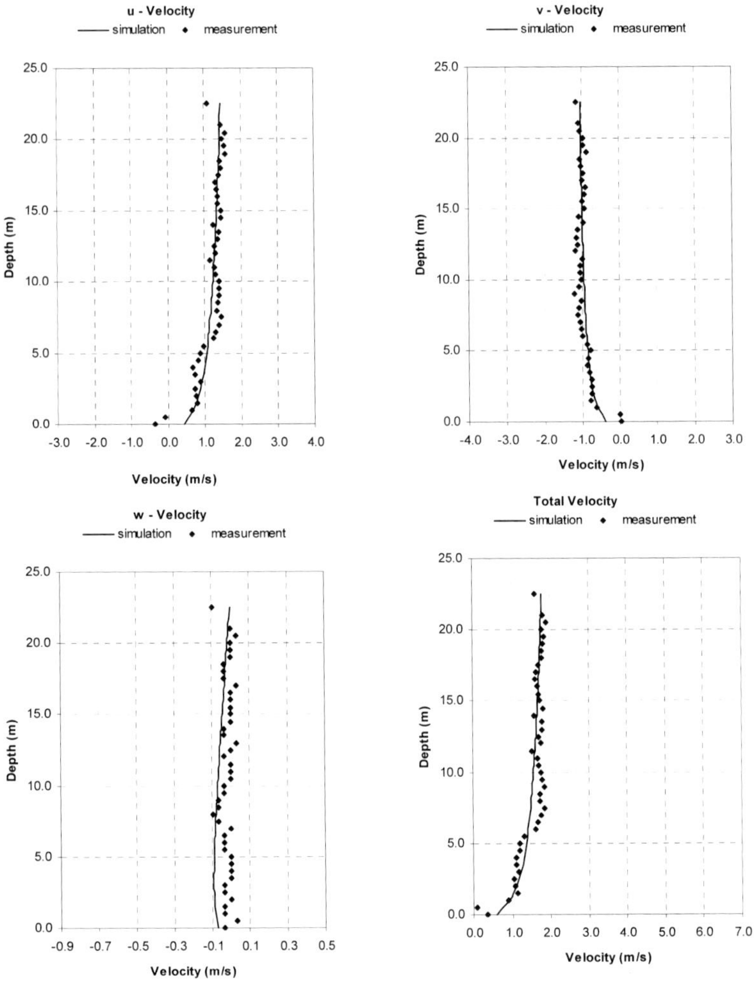

Figure 6.8.10a Comparison of computed and measured velocity components at several points: Section 9, point 39

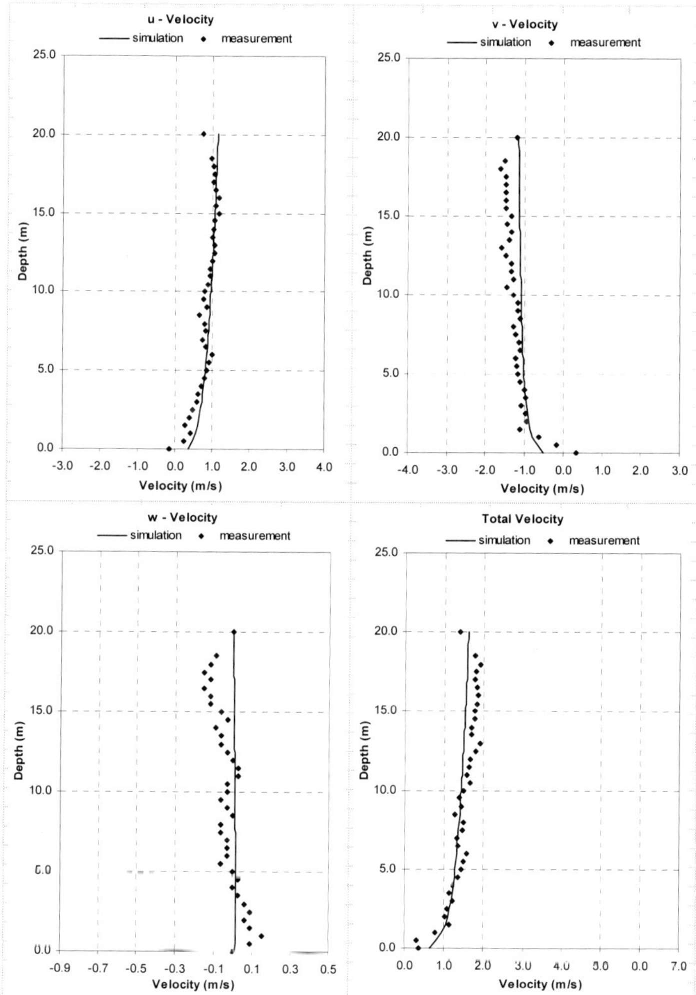

Figure 6.8.10b Comparison of computed and measured velocity components at several points: Section 12 point 9.

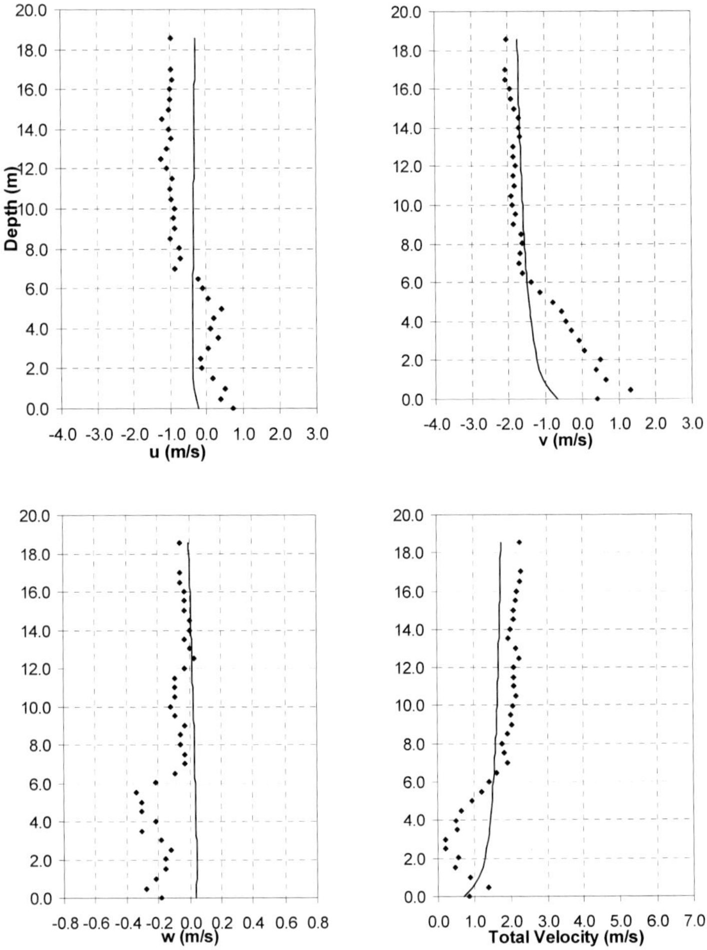

Figure 6.8.10c Comparison of computed and measured velocity components at several points: Section 26 point 21.

CHAPTER 7

A SYSTEMATIC MODEL VERIFICATION AND VALIDATION PROCEDURE

Sam S.Y. Wang

7.1 INTRODUCTION

In the preceding Chapters 3, 4, 5 and 6, twenty two (22) test cases developed by the Task Committee Members are selected for inclusion in this report, each of which serves specific purpose(s) of verification and validation of three-dimensional free surface flow models. Prospective users should not assume, however, that they can apply any one of the provided test cases to carry out the entire verification and validation process of a numerical model. To conduct a complete verification and validation process, one needs to select a series of well-designed tests and carry them out systemically in order to insure that a model can satisfy the following requirements:

- It is consistent and convergent to the solution of the boundary value problem defined by governing differential equations and prescribed boundary/initial conditions, which it intends to solve;
- Its solution can achieve the required order of accuracy by refining the resolution of numerical grids;
- It is capable of simulating most of the basic physical processes relevant or important to the physical system to be modeled; and
- It can predict the over-all behaviors of the real-life problem realistically with reasonable accuracy in both field properties and their trends of spatial and temporal variations.

The series of verification and validation tests should begin with a Mathematical (Code) Verification by using one of the tests provided or self-designed based on analytic, prescribed or manufactured solution for linear or nonlinear cases respectively. In addition, a grid convergence analysis to determine the order of convergence is also needed. Once the model is proven mathematically correct, and capable of achieving the required order of convergence or accuracy quantitatively, it should go through a Physical Process Validation, which is to test the model by using one or more test cases based on laboratory experimental results. This validation step is to determine whether the model is capable of reproducing basic physical processes, at least those relevant to the physical problem to be investigated. Before completing the Physical Process Validation, a Calculation Verification is highly recommended for estimating the error due to calculations using different grid resolution or size distributions.

These two steps, namely the Mathematical (Code) Verification and the Physical Process Validation, are usually conducted once for each model or each version of a model whenever it is developed or significantly upgraded.

When a numerical model has passed the first step, the model is certified to be mathematically correct. This step needs to be repeated only when a new numerical scheme is adopted or the code has gone through a major upgrade and/or update. Passing the second step indicates the model satisfies the basic physical principles and is capable of reproducing most of the basic physical processes essential to the physical system to be studied. The second step needs to be repeated whenever the model is applied to the study of a different physical problem, because it may have different physical processes of importance. If so, the series of test cases for Physical Process Validation should be reselected or designed. Then, the model needs to be re-tested.

Before a numerical model is to be applied to an investigation of a real-world problem, one more step, the Application Site Validation, is required. This step is to be carried out in two sub-steps: Calibration and Validation, in this order. An appropriate portion of the field data collected at the study site is used for the calibration of the specific values of the model parameters to include the specific characteristics of the study site. Due to the fact that these model parameters are usually not measurable directly and are unique for each study site representing site characteristics, one must use the calibrated model to predict flow field variables and their variation trends in space and time under prescribed forcing and boundary conditions. Only by doing so, the results produced by the model can be meaningfully compared with the field data measured at the same site under the same boundary and forcing conditions. One is reminded that the field data used for validation should be the ones that have not been used in performing the calibration. If a reasonable agreement between the model simulations and the field measurements is obtained without tuning the model parameters again, the numerical model is then validated for applications to the study of this site-specific problem. This success does not, however, imply that the model is validated for applications to the study of similar problems at a different site, nor for a significantly different phenomenon or event at the same site, especially at an extreme event or at a different time (e.g., months or years later) This is due to the fact that the site conditions are drastically different from those used during calibration or might have been changed by hydrologic event(s) during the lapse of time. Therefore, the most reliable Application Site Validation should be the one to be conducted once for each application site on a case by case basis.

As presented in Chapter 2, Terminology and Basic Methodology, it is recommended for the testers to perform a Grid Convergence or Calculation Verification. A member of the Committee has attempted to work out an example of the Calculation Verification of a numerical model simulating a real-life field application and hoping to have it included in the report. Unfortunately, it was quite involved and took much more time than that he can spare for this voluntary work. Therefore, model testers are suggested to follow the technique and simple examples given in Chapter 2 to estimate the order of error of the numerical solutions and/or Grid Convergence Index.

7.2 A MODEL VERIFICATION AND VALIDATION PROCEDURE

The proposed procedure includes a series of test cases in the following three steps.

(1) Mathematical (Code) Verifications based on analytic solutions of linearized governing differential equations (Chapter 3) and/or the prescribed or manufactured solutions of both linear and nonlinear governing differential equations (Chapter 4);

(2) Physical Process Validations based on laboratory experimental measurements of basic physical processes relevant to the physical problem to be studied (Chapter 5); including a Calculation Verification (Chapter 2); and

(3) Application Site Validation based on field data collected at a specific problem site (Chapter 6), including a Grid Convergence or Calculation Verification (Chapter 2).

The series of tests are designed to diagnose the model's defects and to measure the model's quality in terms of convergence, consistency, stability and accuracy as well as its ability to predict the correct physics of physical system, and realistic behaviors of the real life problem.

A prospective user of a model for conducting a particular application may select appropriate test cases from this report or other published literature, or design his/her own test cases to construct a series of rigorous and systematic verification and validation tests before the model is adopted to simulate the problem he/she intends to investigate. The user-designed test cases are very important, because the users know what capabilities and the level of accuracy are required to conduct the application study successfully, so that they can design the special tests to serve the purposes. Prospective model users are reminded that it is inadequate to test a model by using only one of the test cases from this report or in the open literature to verify and validate a model.

The purpose of this chapter is to present a proposed procedure for conducting an appropriate, rigorous and systematic model verification and validation test. It is presented with the assistance of examples whenever appropriate, to demonstrate the detailed processes.

Let's assume that a research engineer is assigned to study the hydrodynamic effects of a submerged weir system installed near the outer bank on the bed of a channel bendway as shown in Figure 7.2.1. The purpose of this investigation is to understand why it improves the navigability of large barge-tow systems traveling through the river bendways. With better understanding of the hydrodynamics produces by the weir system, a scientific tool can be developed to design the optimal bendway weir systems in the future, rather than continuously relying on the experience of a few expert practitioners. Due to the constraints of time and funds available, the engineer finds that the application of an appropriate free-surface flow numerical model is a feasible alternative to achieve the objectives of this investigation. Before commencing the actual research, he/she is advised to follow a systematic model verification and validation procedure. This procedure includes the following major steps:

Figure 7.2.1 Victoria Bendway, Mississippi River Test Case

7.2.1 The First Step: Mathematical (Code) Verification

In this step, one is to make sure that the numerical model being considered is mathematically correct, or free of mathematical mistakes. Without this assurance, the mathematical mistakes will be the sources of problems or inaccuracies in the model results, which are hard to find and difficult to explain by physical principles. This step can be performed by selecting a test case from either Chapter 3, to test a linear or linearized nonlinear problem, or Chapter 4, to test a fully nonlinear problem. To conduct a linear test, one need to first apply the same idealization and simplification assumptions, which were adopted for obtaining analytic solution of the test case, and to linearize numerical model by eliminating all nonlinear terms in the model equations. Then, input the same boundary and initial conditions, geometric and physical parameters of the test case into the numerical model to compute the results.

Finally by comparing the simulated results with the values of the analytic solution calculated at the same locations and time, if agreement is acceptable, one can conclude that the linearized numerical model is verified. Sometimes disagreements may be found, which indicate that the numerical model must have some mathematical or coding related mistakes. These latter findings are equally valuable, which can lead to corrections of mistakes or improvements of the model's capabilities. For the prospective user, this finding may also lead to his/her consideration adopting an alternative model for the investigation. The process to conduct a nonlinear test is similar. The difference is for the user to select a nonlinear analytic solution (prescribed or manufactured, Chapter 4, or self-designed by the user). Obviously, the verification of nonlinear models is more rigorous, because all or most nonlinear terms of the governing differential equations can be included rather than omitted in the verification test.

The detailed process of conducting the First Step, Mathematical (Code) Verification, is described below using a hypothetic example. If model ABC is considered for adoption to the investigation of a problem, it should be first tested by selecting an analytic, prescribed or manufactured solution, from an idealized set of governing equations with prescribed forcing, boundary and initial conditions of a known solution domain. Then, the equations of the model have to be modified, the forcing, boundary and initial conditions, as well as the geometrical domain and physical parameters have to be set to those values used in obtaining the analytic solution. And then, the results of model simulation are compared to those evaluated from the analytic solutions at the same locations and time. Finally, this process is to be repeated once or more for a few refined grid(s) to quantitatively determine the convergence and order of errors due to calculations.

The reader is reminded that this step is most likely a pure mathematical exercise. A good agreement achieved from this test, can only certify that the model is free of mathematical mistakes, coding errors, and its solutions are convergent, consistent and having the desired or designed order of accuracy. If, on the other hand, the agreement is not acceptable, the user may ask the model developer to improve the model or to consider an alternative model for adoption.

As one should aware that the linear test cannot assure that a physical system's nonlinear characteristics have been modeled correctly because all nonlinear terms have been temporarily omitted from the governing equations, and usually these nonlinear terms cause more problems than the linear terms. Therefore, if the prediction of nonlinear behaviors is important to the physical system to be studied, additional verification is needed. One may use one of the two test procedures presented in Chapter 4. This test actually is simpler to perform, because it has an advantage of not requiring the simplification of the governing equation system to deactivate the nonlinear terms in the numerical model. Due to the fact that all the terms in the differential equations including nonlinear and linear terms can all be tested together easily, one may wish to use this test along to conduct Mathematical (Code) Verification.

In practice, by using one of the test cases in Chapter 4, one needs to only modify the numerical model by adding a source or forcing term in each of the numerical equations. At each node and a particular time, the source or forcing term is simply a numerical value obtained from one of the analytic forcing functions. Details are given in Chapter 4. Again, one needs to input to the numerical model the geometric and physical parameters and boundary/initial conditions exactly the same as those used in the test case and obtain the numerical solutions by executing the slightly modified numerical model. And finally, the numerical solutions are compared to the corresponding values of the analytical solutions either prescribed or manufactured.

To complete the Mathematical (Code) Verification of a nonlinear numerical model, a Grid Convergence Test or Calculation Verification is needed. This test shall provide the user with an estimate of discretization error or order of accuracy in the numerical solution. Therefore, one may conclude that the numerical model has been verified to have the same order of accuracy as the one planned when the model was developed.

If a numerical model fails a Mathematical (Code) Verification test, there is a strong possibility that some mathematical mistakes in derivations, or solutions, or some coding errors have been committed. By inclusion or exclusion some terms or groups of terms and repeating the tests, the modeler may have an alternative way to identify the sources of the mistakes, which may not be found by using other debugging techniques.

It is very important to re-emphasize that the Mathematical (Code) Verifications, using either linear or the prescribed or manufactured nonlinear analytic functions, could be purely exercises of mathematics. In principle, the selected, prescribed or manufactured analytic functions are solutions of a set of modified differential equations, which do not have to have any physical meaning. Any functions, as long as they satisfy the boundary conditions and have continuous, definite and non-trivial values of system variables and their derivatives in the solution domain, may be used for the prescribed solution forcing approach.

Calculation Verification using the GCI (Contributed by Pat Roache)

Since the Calculation Verification has been mentioned several times previously in this report, it may be a good idea to give a brief description of performing procedure. The minimal procedure for formal Verification of Calculations (either for Physical Process calculations or Application Site calculations) reports the estimate of ordered numerical uncertainty using the Grid Convergence Index (GCI). The analyst obtains solutions on two geometrically similar grids using identical model boundary conditions, initial conditions, coefficients, and basin geometry. In a one-dimensional case added by editor the analyst chooses some quantity f of interest, e.g. integrated functionals of the solutions such as flow rate, boundary shear stress, sediment load, etc. The f values used may be fractional, percentage, or absolute. Then ε is defined as

$$\varepsilon = \frac{f_2 - f_1}{f_1}$$

where f_1 and f_2 are the f values in the fine and coarse grids respectively. Then the GCI for the fine grid solution is calculated by

$$\text{GCI}\left[\text{fine grid}\right] = F_s \frac{|\varepsilon|}{r^p - 1} \quad , \quad F_s = 3$$

where r is the grid refinement ratio ($r = \Delta_2/\Delta_1$ and Δ is some representative size of the grid elements) and p is the rate of convergence, e.g. $p = 2$ for a second-order method. For best results, it is important that the two grids be close to geometrically similar. This is often easily obtained by using a grid halving to obtain the second (coarse) grid, giving $r = 2$, but this is not necessary. If the analyst has convincingly Verified that the code actually attains the theoretical order of accuracy (e.g., $p = 2$ for a second-order method), at least on a "nearby" problem, then the conservative value of $F_s = 3$ can be replaced by the more optimistic value $F_s = 1.25$. See Chapter 2 and references therein for details, example calculations, and further discussion.

7.2.2 The Second Step: Physical Process Validation

After the model has successfully passed the Mathematical (Code) Verification, the next step is to confirm that the model has the capability to reproduce key physical processes relevant to the physical system to be studied. To achieve this goal, one may design a series of tests based on laboratory experiments, in which the properties of concerned physical processes have been measured quantitatively. This test is referred to as the Physical Process Validation as presented in Chapter 5. Prospective model users may select some of the test cases given in Chapter 5 to construct this series of tests. They are also encouraged to select other cases from published literature and/or to design additional cases by themselves, especially if some physical process(es) of key importance to the physical system are not covered by test cases in Chapter 5 nor in other published literature. After the series of tests has been selected for the physical process validation, one is to run the numerical model by simulating one of these cases at a time using the same geometric and physical parameters, and boundary and initial conditions.

It is recommended that this step includes Calculation Verification (Chapter 2). If the identified physical processes having been measured in the laboratory are reproduced by the verified computational model simulation, and the agreements are within reasonable error bounds, then the model is validated to have the capability of reproducing the key physical processes of the physical system.

The prospective model users are reminded, however, that there are differences between the physical and mathematical models, and inaccuracies in the measurements of the physical experimentation as well as in the calculations (discretization, round-off errors) of the numerical simulation. Therefore, it is advisable to run more than one simulation (if data is available) for making additional comparisons between the numerical simulations and laboratory measurements. If reasonable agreements are obtained both in values of field variables and may be more important in trends of variations in space and time of the field variables, the physical processes validation has succeeded.

To demonstrate the detailed validation process, a numerical model CCHE3D is chosen to be tested by the Physical Process Validation using two test cases selected from Chapter 5.

(A) *Test Case of Flow in a U-Shaped Channel*

To perform this test, one simply provides the given input data set from Appendix A at the end of the report to the numerical model, executes the model, and compares the computed results to the experimental measurements. The input data given for this test case includes: the geometric configuration (Figure 7.2.2) of the modeling domain, the boundary conditions at inlet, outlet, and solid walls, and the roughness height of the bed, etc. Also given in this data file are the flow field properties at measuring stations.

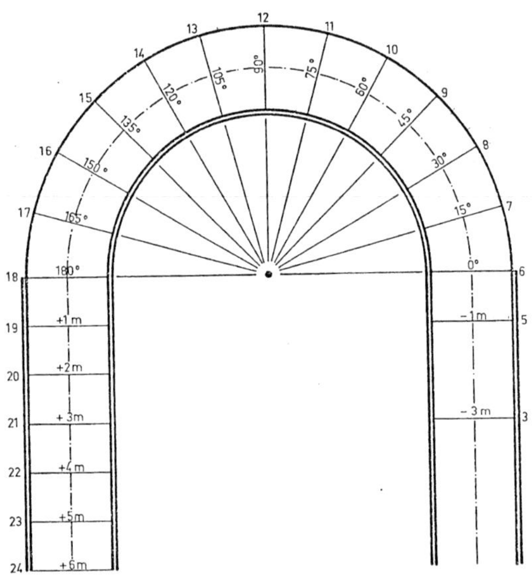

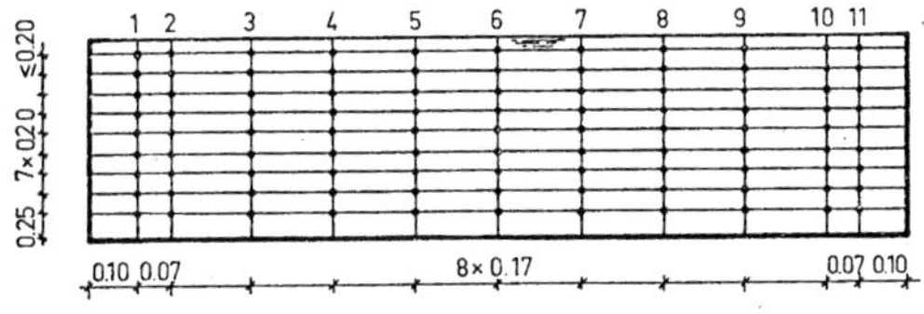

Figure 7.2.2 Configurations of the U-shaped channel and locations of the cross sections. Nodal points in the cross section are the velocity measuring points (de Vriend, 1979).

The comparisons of various flow field properties and their spatial variations between calculated results using CCHE3D (Jia and Wang, 1992) and measured data by de Vriend (1979) are shown in Figure 7.2.3 and Figure 7.2.4. From these figures, one sees good agreements (although not perfect) in both values and the trends of spatial variations of free surface elevation, the longitudinal flow velocity and transverse velocity. Additional evidence has been presented in Chapter 5. These findings have validated the fact that the model (CCHE3D) tested does have the capability to simulate the secondary motions and water surface super elevations of the flow field in a curved channel. The resulting vector field at (90°) cross-section simulated using a nonlinear k-ε closure included in the CCHE3D model indicates that the model is able to capture the smaller vortices near the outer bank (Figure 7.2.5) which could not have been reproduced by the linear k-ε models, or zero equation models. This example demonstrates that the Physical Process Validation can contribute to the detection of a model's shortcomings, which could lead to significant improvements of the model's capabilities.

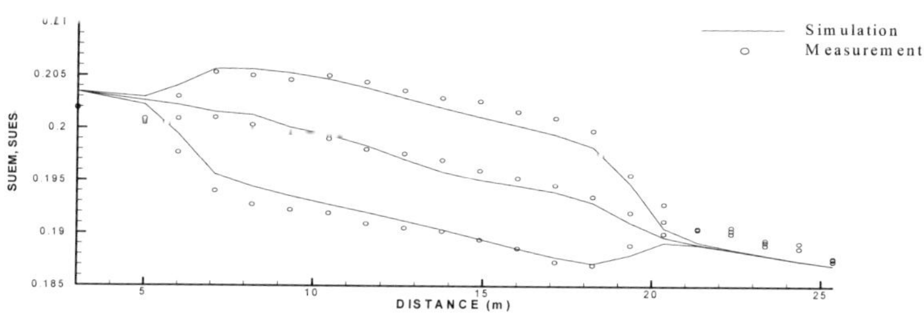

Figure 7.2.3 Comparisons of free surface profiles along the U-shaped channel. The solid lines are simulated results, the symbol ○ represents measured data.

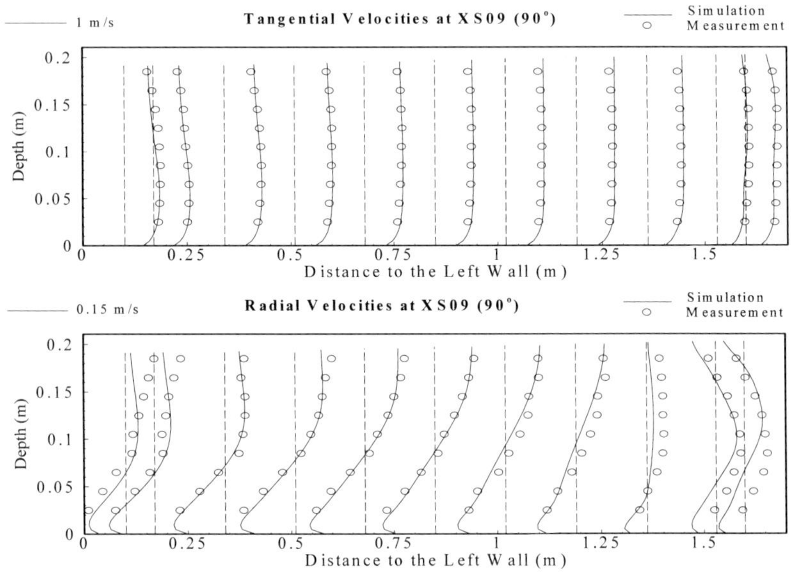

Figure 7.2.4 Comparisons of computed and measured longitudinal and transverse velocity profiles in the cross-section of $\alpha=90^0$

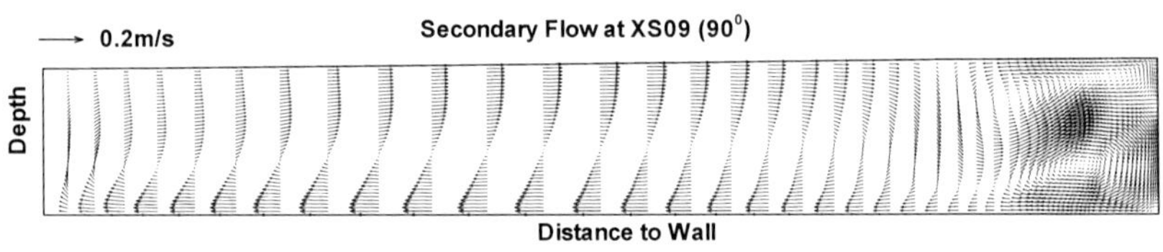

Figure 7.2.5 Vector field of the transverse velocity in the cross-section of $\alpha=90^0$

(B) Test Case of Flow around a Submerged Trapezoidal Spur Dike:

Flowfield around a submerged spur dike or weir is quite complex and three-dimensional (Figure 7.2.6). Various vortices are generated by this structure, which are expected to introduce significant effects on the hydrodynamics and local scours around the dike. The test case is selected to validate whether the numerical model can adequately reproduce the complex flowfield and hydrodynamics around such a structure. This case is chosen, because each weir of the bendway weir system has a shape similar to this.

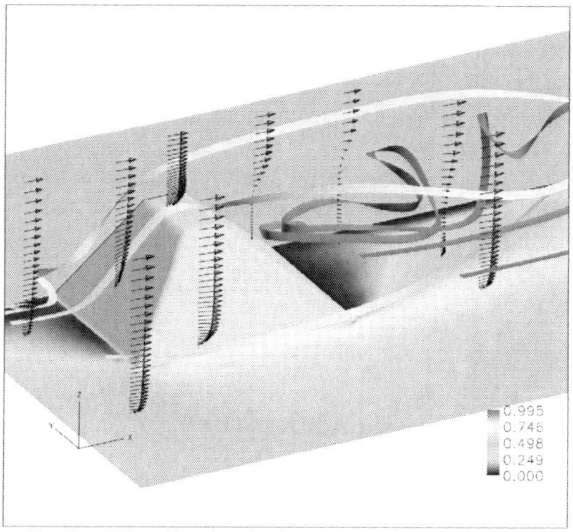

Figure 7.2.6 Qualitative Visualization in Stream-Ribbons of Flow pattern around the submerged trapezoidal dike

A sketch of the laboratory model tested at the USDA National Sedimentation Laboratory in Oxford, Mississippi (Kuhnle, et al, 1997, 1999) is shown in Figure 7.2.7. The detailed description of this laboratory experiment is given in Section 5.6.

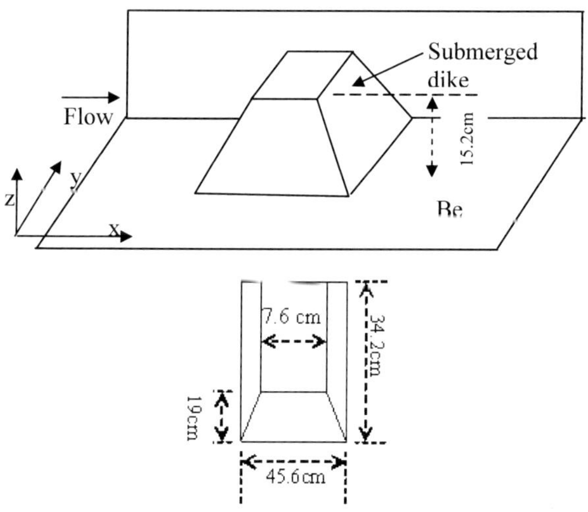

Figure 7.2.7 Sketch of trapezoidal shaped dike model (not in scale)

Again, this validation test can be conducted by reading in the input data set from Appendix A at the end of this report, and executing the numerical model (CCHE3D) (Jia and Wang, 1996, 2000). The comparisons between the simulated values and the laboratory measurements are shown in Figures 7.2.8 – 7.2.10. From the comparison of the results of numerical simulations and the experimental measurements, it was confirmed that the numerical model, CCHE3D, is capable of reproducing the essential physical processes in the flow around a submerged dike or weir. It is believed that the assumption of non-hydrostatic pressure distribution and the adoption of the standard k-εturbulence closure scheme have resulted in good agreements.

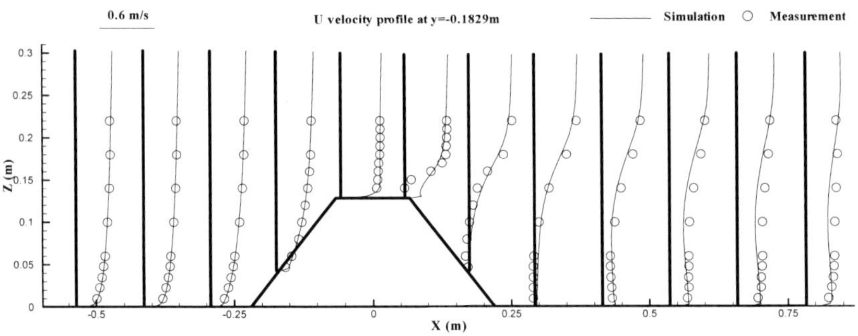

Figure 7.2.8 Comparison of simulation and measured longitudinal velocity in a longitudinal cross-section.

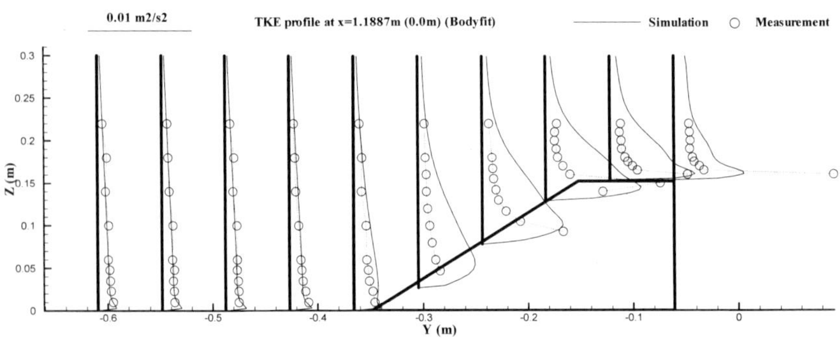

Figure 7.2.9 Comparison of simulation and measured turbulent kinetic energy (k) in a transverse cross-section. (The small scale is adopted to exaggerate the discrepancies near the dike surfaces for clear visualization)

Due to the availability of a large number of data collected at well-designed collection locations with high resolution in the key areas around the structure, an over-all agreement was examined with results shown in Figure 7.2.10. The root mean square error of velocity magnitudes is approximately 0.03 m/s. All results indicate that the tested model is validated to have the capability of reproducing the essential physical processes and both the values and variation trends of the flow properties around a submerged weir fairly accurately. During a discussion about source of disagreements between the physical and numerical modelers at a seminar held jointly by the USDA National Sedimentation Laboratory and the NCCHE of the University of Mississippi, the participating research scientist from both groups agreed that both physical measurements and numerical solutions may have some errors especially at the surfaces and corners of the dike. More detailed results were presented in Section 5.5 of this report.

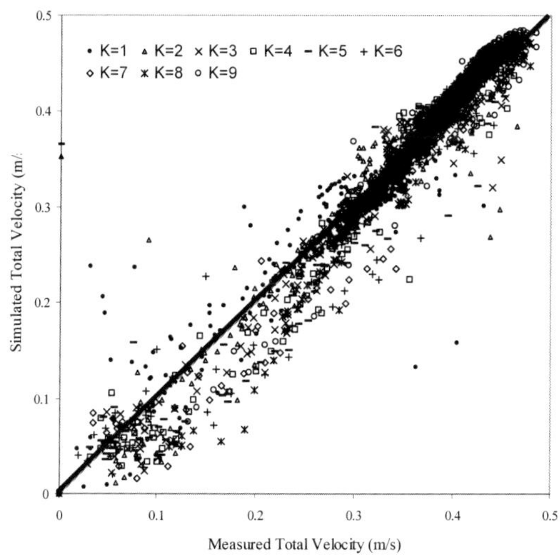

Figure 7.2.10 Overall agreement of simulated and measured total velocity magnitude.

7.2.3 The Third Step: Application Site Validation

To some model users, the Application Site Validation is the most important step, because it is the ultimate test to determine whether a numerical model can predict the behavior of a flowfield realistically in a natural or man-made physical system. In the past, the majority of numerical modelers have performed their model validations by using this step only and quite a few of them are still doing so even at the present time. What they have been doing, however, was to simulate the site specific problem with their numerical model without calibration. When the agreement between the model simulations and the field data is not obtained, they just simply fine tune the model parameters until the results are in good agreement. They then claim that the model is validated. What they had done was actually only the calibration of the model parameters. There was no effort in conducting validation of the numerical model or code.

There are two problems with their approach:

- The field data collected were usually too few and inadequate for conducting a justifiable parametric calibration, let alone the validation,
- the agreements were usually achieved by tuning the adjustable model parameters until the simulation matches the measurement. This can hardly be called validation.

Consequently, this traditional silver bullet (calibration only) approach has not been accepted by professional societies as evidenced by their journal publication policy (See Chapter 1), as well as the professionals in the field.

To conduct an application site validation recommended by this task committee, one must:

(i) have an adequate amount of high quality field data collected (at well-selected locations and time) of the study site;
(ii) apply only a portion of the data collected to the calibration of the site specific model parameters, and
(iii) use the remaining portion (having not been used in calibration) of the collected field data solely for model validation.

Let us continue to use the Victoria Bendway Flow Test Case (Section 6.8) as an example (Jia et al. 2001). First, by using a portion of the discharge measurements at the upstream end of the study reach, and the water surface elevation at the downstream boundary of the study reach, the equivalent river bed roughness, or the Manning's n, is calibrated. Secondly, by applying the same geometrical and physical parameters including the calibrated Manning's n value (fixed), and boundary conditions, the simulated velocity field is obtained. Thirdly from the results of the comparison between the numerical simulations and field measurements (Figure 7.2.11), one can conclude whether or not the numerical model (CCHE3D) is capable of reproducing the flow field realistically within an acceptable error bound. Due to the differences in field characteristics between the real-life field problem and the idealized numerical model, one should be prepared to accept an allowable error bound between the simulated results and the measured data over the entire study domain. But, one should require that the numerically predicted trends of variations of field variables, both spatial and temporal, are in reasonable agreement to those measured in the field.

In addition, the prospective users of the numerical models are recommended to conduct a Grid Convergence Test or Calculation Verification for estimating the calculation errors or the uncertainty of the accuracy of the calculated results.

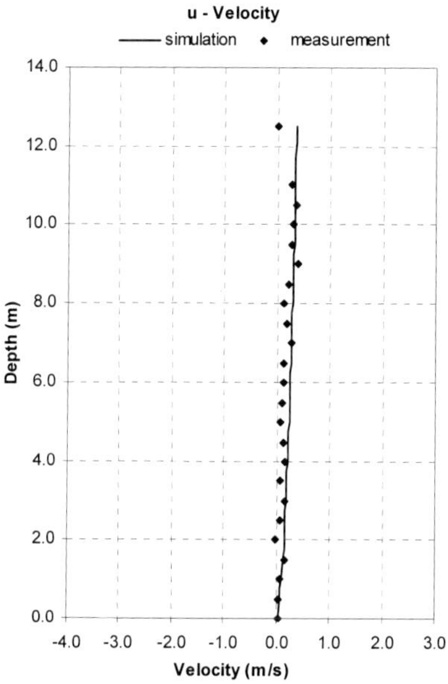
u - Velocity
simulation
measurement
Depth (m)
Velocity (m/s)

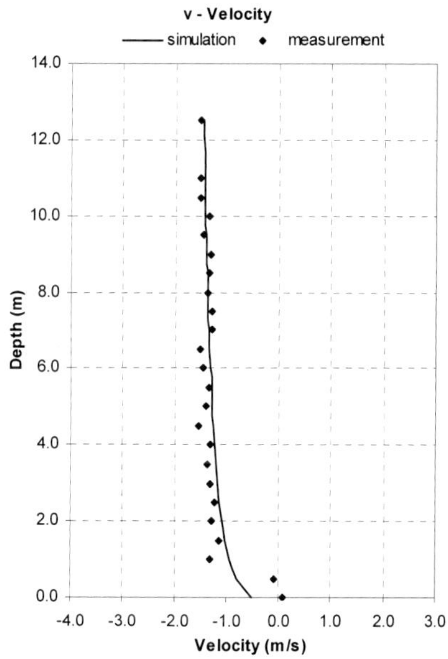
v - Velocity
simulation
measurement
Depth (m)
Velocity (m/s)

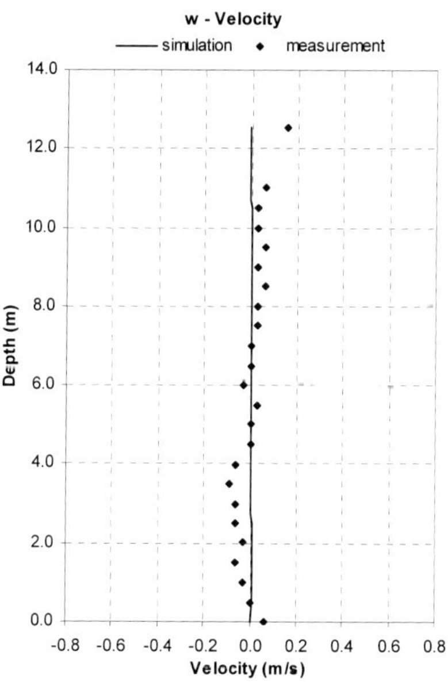
w - Velocity
simulation
measurement
Depth (m)
Velocity (m/s)

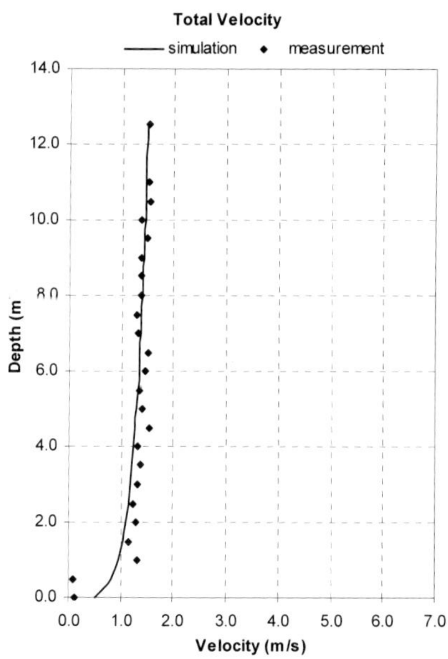
Total Velocity
simulation
measurement
Depth (m)
Velocity (m/s)

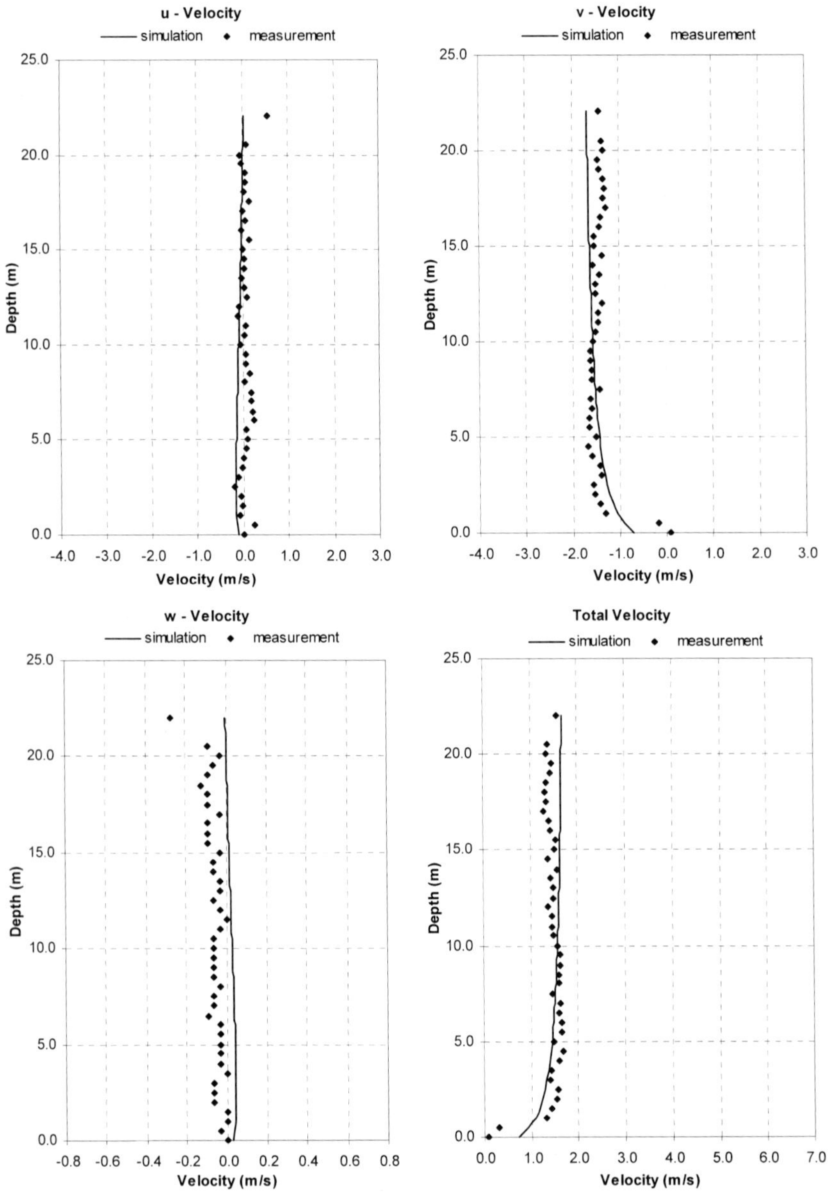

Figure 7.2.11 Comparison of simulated and measured velocity profiles in the Victoria Bendway, Mississippi River

7.3 CONCLUDING REMARKS

It has been demonstrated by examples, that the newly developed standard verification and validation procedure can be applied to test a model systematically in three major steps: (1) Mathematical (Code) Verification, (2) Physical Process Validation, and (3) Application Site Validation. In addition, the model testers are recommended to perform a Calculation Verification in all three steps.

Rather than unnecessary or even trivial, as regarded by some professionals in the past, the Mathematical (Code) Verification has been proven to be both a very important and a fundamental test step. It is the only step when the analytic and numerical solutions are obtained from exactly the same set of differential equations having the same basic assumptions and simplification, the same values of physical parameters, the same boundary and initial conditions, and the same solution domain. More importantly, there is no need to tune any numerical model parameters. If there is any disagreement in solutions, the numerical model must have some inaccuracy and/or mistakes requiring corrections and improvements. In addition, this test is the only one which can be used to determine the accuracy of numerical model results quantitatively. It can also be used to estimate the model's order of accuracy and to compare the numerical accuracies of different numerical grid resolutions. Consequently, when a numerical model has passed this mathematical code verification test, the model users are assured that it does not have mathematical formulation mistakes, major numerical solution errors, computer coding errors, etc.

After a numerical model has passed this test step, the next task is to show that it is capable of simulating the basic physical processes relevant to the physical system to be studied. For example, to study the flowfield around a submerged weir system in the Victoria Bendway, one would want to make sure that the numerical model chosen has the capability to simulate the characteristics of a flowfield around a submerged structure, as well as to reproduce a flowfield in a curved channel. Without this step, if one is to apply the model directly to the complex real-life problem with multiple interacting forcings and influences, one may have difficulties finding out just what are the causes of the problems, if found. The advantages of using laboratory experiments to validate the numerical model's capability in reproducing the basic physical processes are (1) that the laboratory experiments are conducted in a more controllable and repeatable environment by eliminating as many non-essential forcings as possible; and (2) that the numerical model's capability in predicting the responses to each primary forcing can be validated separately.

The Application Site Validation is intended to answer the final question: how realistic are the predictions of the numerical model, when it is applied to the investigation of a site-specific real-life problem. Due to the unavoidable differences between the idealized system being simulated by a numerical model and the highly complex real-life system in nature, it is expected that there are discrepancies between the numerical model's predictions and the field measurements. The model users should be prepared to accept a reasonable level of accuracy in the predicted field variables, but to demand that the numerically predicted trends of spatial and temporal variations of the field properties to be physically correct. With continuing advancement in physical theories, mathematical methodologies, field measurement and instrumentation technologies, it is likely that the numerical modeling capability and accuracy in reproducing realistic field behavior will continue to improve. As a result, more and more complex and realistic systems can be simulated with higher and higher accuracy.

7.4 REFERENCES

De Vriend, D.J. (1979). Flow measurements in a Curved Rectangular Channel, Laboratory of Fluid Mechanics, Department of Civil Engineering, Delft University of Technology, Internal Report No. 9-79.

Jia, Y., and Wang, S.S.Y., 1992, "Computational model verification test case using flume data". Hydraulic Engineering, pp436-441, ASCE.

Jia, Y., and Wang, S.S.Y., 1996, “A modeling approach to predict local scour around spur dike-like structures”, Proceedings of the Sixth Federal Interagency Sedimentation Conference, p. II-90-97.

Jia, Y., and Wang, S.S.Y., 2000, “Numerical Study of Turbulent Flow around Submerged Spur Dikes”, 4th International Conference for Hydroscience and Engineering, 2000, Seoul, Korea.

Jia, Y., Scott, S., and Wang, S.S.Y., 2001, “3D Numerical Model Validation Using Field Data and Simulation of Flow in Mississippi River”, ASCE, Proceedings (CD-ROM) of World Water & Environmental Resources Congress, May, 2001, Orlando, FL.

Kuhnle, R., Alonso, C.V., and Shields, F.D., 1997, “Geometry of scour holes around spur dikes, and experimental study”, Wang, S.S.Y., Langendoen, E.J., and Shields, F.D., (ed.) Proceedings of the Conference on Management of Landscapes Disturbed by Channel Incision, Center for Computational Hydroscience and Engineering, School of Engineering, The University of Mississippi, p. 283-287.

Kuhnle, R., Alonso, C.V., and Shields, F.D., 1999,”Geometry of scour holes associated with 90° spur dikes”, Journal of Hydraulic Engineering, ASCE, 125(9), p.972-978.

CHAPTER 8

SYSTEMS ANALYSIS CONSIDERATIONS

Richard A. Schmalz, Jr. and Bernard B. Hsieh

8.1 INTRODUCTION

Contributor: Richard A. Schmalz, Jr. and Bernard B. Hsieh

With the increase in computational resources and the advancement of multidimensional modeling techniques, the ability to simulate complex three-dimensional free surface flows has greatly improved. The increased interest in model validation has led to the desire to employ a more systematic approach. Toward this end, we refer the reader to the works of Lynch and Davies (1995) in the ocean sciences and Wilkes (1995) in the atmospheric sciences. Practical aspects of skill assessment and acceptance criteria have been addressed by Hess and Bosley (1992), Bosley and Hess (1998), and Hess and Gross (1999). Another related area of rapid development comprises data assimilation, or the use of real time observations to adjust the model prediction during the simulation. These methods are briefly introduced in Section 8.2 via Kalman filtering. The reader is referred to the inverse methods of Bennett (1992), data analysis methods of Daley (1991), and topics in models, estimation, and control by Maybeck (1979). These references are by no means exhaustive but will serve as starting points for further investigation.

Here we consider a framework for improved flow model validation using system analysis as shown in Figure 8.1.1. Section 8.2 addresses the activities in the left half of the figure, while in Section 8.3, activities in the right half are considered as discussed below.

In Section 8.2 an optimal design of field data collection is proposed. The use of modern computing techniques including computational intelligence and numerical models and their integration has become commonplace in managing water resource projects. Almost every management decision, particularly for a complex and dynamic system, requires the construction of a mathematical flow model and its validation through field data collection. The field data collection program in most cases consumes a major portion of the modeling budget. Therefore a systematic design to achieve the maximum data collection is a highly desirable goal. The required computational tools include a numerical flow model, a stochastic estimator, sensitivity analysis, nonlinear frequency domain based analysis, and artificial neural networks (ANNs). The design procedures consist of: 1) preliminary numerical testing, 2) the deterministic-stochastic approach, and 3) the moving installation monitoring.

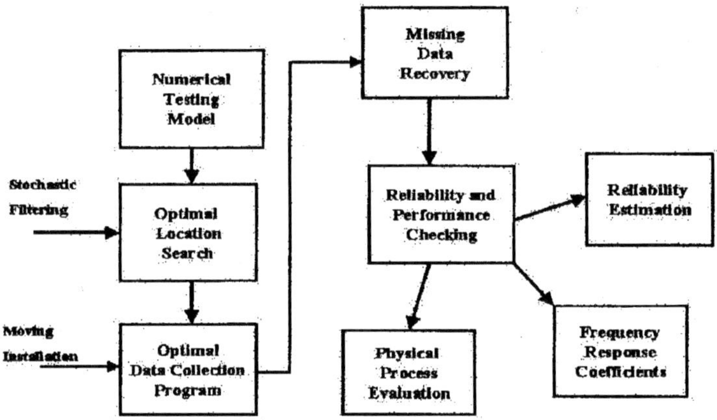

Figure 8.1.1 A framework for improved model validation using system analysis

Section 8.3 deals with the reliability of the flow models. The missing data recovery system can be used to extend the limited measurement set to maximize the validation horizon after data collection has been completed. Validation procedures for measuring model reliability have typically consisted of simple statistical comparisons between model predictions and field observations; here system analysis techniques are used to determine a performance measure with an associated reliability index and weighted coherence function to augment the standard statistical procedures. The model validation procedure presented considers not only the performance for general trends but also during extreme events, when the forcing functions are at extrema.

With the advent of real time measurement systems and operational weather and river flow forecasts, the ability now exists to perform nowcasts (present state) and forecasts (future state) of water levels and currents in bays and coastal shelves. The National Oceanic and Atmospheric Administration (NOAA), National Ocean Service (NOS), is developing nowcast/forecast systems to supplement its Physical Oceanographic Real Time Systems (PORTS) in Tampa Bay, Galveston Bay, New York/New Jersey Harbor, San Francisco Bay, and Chesapeake Bay with several new systems in the planning stages. Two and three dimensional hydrodynamic models have been designed by NOS and are run in experimental mode prior to operational implementation. Such systems have been run in Chesapeake Bay (Bosley and Hess, 1998), New York/New Jersey Harbor (Wei and Sun, 1998; Wei et al., 1998), and in Galveston Bay (Schmalz, 1999). A description of evolving coastal ocean nowcast/forecast systems for the Atlantic Ocean continental shelf and Gulf of Mexico and their relationship to the above PORTS nowcast/forecast systems is given by

Aikman et al. (1996) and Kelley et al. (1997). Detailed aspects of NOS nowcast/forecast system objectives and the potential for provision of real-time vessel to bottom clearance estimates are presented by Parker (1998) and in Parker and Huff (1998), respectively. In Section 8.4, the focus is on the development of the experimental nowcast/forecast system in Galveston Bay. Definition of terms and the process which was followed to arrive at the present nowcast/forecast system are first discussed. Next, the design and improvement of the nowcast /forecast system is briefly outlined. A one year validation of the nowcast/forecast system is then performed using the NOS (1999) acceptance criteria as targets. Next a physical interpretation of the statistical evaluation is made. Future directions in conjunction with several techniques developed in the Sections 8.2 and 8.3 are presented in the concluding summary section.

8.1.1 References

Aikman, F. III, G. L. Mellor, T. Ezer, D. Sheinin, P. Chen, L. Breaker, K. Bosley, and D. B. Rao, 1996. Towards an operational nowcast/forecast system for the U.S. East Coast, In: *Modern Approaches to Data Assimilation in Ocean Modeling*, P. Malanotte-Rizzoli (ed.), Elsevier Oceanography Series, 61, 347-376.

Bennett, A. F., 1992: *Inverse methods in physical oceanography*, Cambridge University Press, Cambridge, UK.

Bosley K. T. and K. W. Hess, 1998: Development of a experimental nowcast-forecast system for Chesapeake Bay, *Proceedings of the 5th Estuarine and Coastal Modeling Conference*, ASCE, 413-426.

Daley, R., 1991: *Atmospheric data analysis*, Cambridge University Press, Cambridge, UK.

Hess, K. W. and K. T. Bosley, 1992: Methodology for validation of a Tampa Bay circulation model, Proceedings *of the 2nd Estuarine and Coastal Modeling Conference*, ASCE, 83-94.

Hess, K. W. and T. Gross, 1999: User-based skill assessment methods for NOS oceanographic forecast systems, *Proceedings of the 3rd Conference on Coastal Atmospheric and Oceanic Prediction and Process*, AMS, 215-219.

Kelley, J. G. W., F. Aikman, L. C. Breaker, and G. L. Mellor, 1997: Coastal ocean forecasts, Sea Technology, 38(5), 10 - 17.

Lynch, D. R. and A. M. Davies (eds), 1995: Quantitative skill assessment for coastal ocean models, American Geophysical Union, Washington, DC.

Maybeck, P. S., 1979: *Stochastic models, estimation, and control*, Volume 141-1 of Mathematics in Science and Engineering, Academic Press, New York.

NOS, 1999: NOS procedures for developing and implementing operational nowcast and forecast systems for PORTS, NOAA Technical Report NOS CO-OPS 0020, Silver Spring, Maryland.

Parker, B. B., 1998. Development of model based regional nowcasting/forecasting system. In: Spaulding, M.L. and A.F. Blumberg (eds), *Proceedings of the 5th International Estuarine and Coastal Modeling Conference*, ASCE, 355-373.

Parker, B. B. and L. C. Huff, 1998. Modern under-keel clearance management, *International Hydrographic Review*, Monaco LXXV(2), 143-165.

Schmalz, R. A., 1999: Demonstration of a Nowcast/Forecast System for Galveston Bay, *Proceedings of the 3rd Conference on Coastal Atmospheric and Oceanic Prediction and Process*, AMS, 213-214.

Wei, E. J. and C. L. Sun, 1998: Development of a New York/New Jersey Harbor water level and current nowcast/forecast System, *Proceedings of the 2nd Conference on Coastal Atmospheric and Oceanic Prediction and Process*, AMS, 15-22.

Wei, E. J., M. Chen and A. Zhang, 1998: Development of a New York/New Jersey Harbor water level and current nowcast system, *Proceedings of the Ocean Community Conference*, MTS, 422-425.

Wilkes, D. S., 1995: *Statistical methods in the atmospheric sciences: An Introduction*, Academic Press, San Diego, CA.

8.2 OPTIMAL DESIGN OF FIELD DATA COLLECTION

Contributor: Bernard B. Hsieh

8.2.1 Introduction

A design strategy using a stochastic algorithm to select the number and locations of measurement developed by Hsieh (1997) combined deterministic and stochastic models by embedding a numerical approximation within a stochastic filter, such as Kalman filtering. The method can identify the regions of maximum variance, which are the most significant sampling locations. One of the alternatives to search for the most significant sampling locations is to perform a sensitivity analysis on the model output variation due to the input variables. Hsieh (1997) proposed using a nonlinear response system computation and artificial neural network algorithms for designing a moving installation-monitoring program. The direct benefit of this approach is to use fewer instruments and to get more sampling locations. For the second stage, a method using response techniques to maximize the simulation reliability was explored (Hsieh, 1998). This is demonstrated by the missing data recovery system. The goal is to construct a larger, more reliable data system using a neighboring sampling parameter in which the developed recovery system also can be used to simulate the approximation of "real data" for the second location while using the information from the first location with field measurement and numerical simulation and a neighboring second point with numerical simulation. This is very useful when the standard validation process has reached its limit and the modeler would like to estimate possible measurement time series for the second location without performing the actual monitoring. However, the results are estimates of the measurements with a total error comprised of measurement and estimation error.

8.2.2 Overview of Required Computational Tools

The design framework requires three different computational tools, namely, a numerical flow model, a stochastic estimator, and a system simulation model. The numerical flow model is first outlined, followed by two approaches to perform the stochastic estimation. For the system simulation model, both the traditional nonlinear frequency response techniques as well as Artificial Neural Network (ANN) frequency based approaches are considered.

8.2.2.1 Numerical Flow Model

For most scientific and engineering analysis of fluid dynamic phenomena, the numerical flow model usually replaces the derivatives in the governing equations with algebraic terms, resulting in a system of algebraic equations to be solved. Physical processes impacting the circulation and vertical mixing that are modeled include tides, wind, density effects, freshwater inflows, turbulence, and the effect of the earth's rotation. The basic inputs for the salt transport hydrodynamic model are surface elevations and salinity distributions over the tributary, freshwater inflows, and wind forcing. The basic outputs are surface elevations, tidal currents, and salinity concentrations over the entire computational domain. This is the simulation mode of the system. The basic equations (Johnson et al., 1991) for an incompressible fluid in a right-handed Cartesian coordinate system (*x,y,z*) are:

$$\partial u/\partial x+\partial v/\partial y+\partial w/\partial z=0 \tag{8.2.1}$$

$$\begin{aligned}&\partial u/\partial t+\partial u^2/\partial x+\partial uv/\partial y+\partial uw/\partial z=fv-1/\rho_0\partial p/\partial x+\partial/\partial x(A_H\partial u/\partial x)\\&+\partial/\partial y(A_H\partial u/\partial y)+\partial/\partial z(A_v\partial u/\partial z)\end{aligned} \tag{8.2.2}$$

$$\begin{aligned}&\partial v/\partial t+\partial uv/\partial x+\partial v^2/\partial y+\partial vw/\partial z=-fu-1/\rho_0\partial p/\partial y+\partial/\partial x(A_H\partial v/\partial x)\\&+\partial/\partial y(A_H\partial v/\partial y)+\partial/\partial z(A_v\partial v/\partial z)\end{aligned} \tag{8.2.3}$$

$$\partial p/\partial z=-\rho g \tag{8.2.4}$$

$$\begin{aligned}&\partial T/\partial t+\partial uT/\partial x+\partial vT/\partial y+\partial wT/\partial z=\partial/\partial x(K_H\partial T/\partial x)\\&+\partial/\partial y(K_H\partial T/\partial y)+\partial/\partial z(K_v\partial T/\partial z)\end{aligned} \tag{8.2.5}$$

$$\begin{aligned}&\partial S/\partial t+\partial uS/\partial x+\partial vS/\partial y+\partial wS/\partial z=\partial/\partial x(K_H\partial S/\partial x)\\&+\partial/\partial y(K_H\partial S/\partial y)+\partial/\partial z(K_v\partial S/\partial z)\end{aligned} \tag{8.2.6}$$

$$\rho=\rho(T,S) \tag{8.2.7}$$

where (*u,v,w*) are velocities in the (*x,y,z*) directions, *f* is the Coriolis parameter defined as $2\Omega\sin\phi$, where Ω is the rotational speed of the earth and is the latitude ϕ, $\rho$is water density (ρ_0 is a reference water density) , *p* is pressure, *T* is temperature, *S* is salinity, and (A_H,K_H) are turbulent eddy coefficients. Equation (8.2.4) implies that vertical accelerations are negligible and thus the pressure is hydrostatic. These hydrodynamic equations are usually solved by numerical methods using finite-element and/or finite-difference approximations.

8.2.2.2 Stochastic Estimator

Estimation of the internal states of a stochastic dynamical system is a topic with important applications, especially in the realization of modern control strategies where the states are not accessible for measurement (Borrie, 1992). The Kalman filter, which applies to linear and some non-linear systems, is the most promising estimator presently available.

A linear time-variant system S at time *t* with update intervalΔt can be expressed as a sum of process and measurement noise corruption as follows.

$$x(t+\Delta t) = A(t)x(t) + B(t)u(t) + H(t)w'(t) \tag{8.2.8}$$

$$y(t) = C(t)x(t) + D(t)u(t) + v'(t) \tag{8.2.9}$$

where

x(t) represents the (*m*) system states,
u(t) represents the (*q*) systems inputs,
y(t) represents the (*r*) system outputs,
w'(t) represents an *s*-vector of Gaussian extended white noise, strength *Q(t)*,
v'(t) represents an *r*-vector of Gaussian extended white noise, strength *R(t)*, uncorrelated with *w'(t)*, and
A(t),B(t),C(t),D(t),H(t) are time-dependent matrices.

The system S has no direct knowledge of the noise signals, *w(t)* and *v(t)*, but its estimator, S, is required to take into account the presence of *w(t)*, and to attenuate the effects of *v(t)*. The quality of the result depends on the choice of the estimator gain matrix. If the choice is optimal in a particular sense, the estimator is a Kalman-Bucy filter. The extended Kalman filter (Bras, 1985) has been developed using Taylor series expansion to solve the nonlinear problem. Verlaan and Heemink (1988) developed an approximate Kalman filter (RRSQRT). This efficient approximate algorithm can reduce both computation time and the required memory by several orders of magnitude.

Although numerical hydrodynamic models have often provided reliable results, it should be noted that these models are approximations to the actual processes in the system. Errors in bathymetric features, flow field, transport coefficients, boundary conditions, and the numerical discretization of the partial differential equation introduce uncertainty or noise into the modeling process. This modeling error, or system noise, is propagated through time and space by the deterministic model (Budgell, 1981). According to that consideration, if time series observations of surface elevation, tidal current or salt concentration are available from the estuarine system, the modeling error at each time step can be estimated and the model results can be corrected. By updating the computed concentrations using observations, less error is propagated through the model.

A major difficulty associated with this procedure is that the observations will also contain a certain degree of measurement noise. Therefore, the actual corrections to be applied to the computed variables will not be the difference between the observed and computed values, but rather some function of that difference. A means of computing the optimal corrections to be applied to the computed values at each time-step is through the use of the Kalman filter, an optimal control estimator of the system.

8.2.2.3 Sensitivity Analysis

As an alternative to the stochastic estimator, a method is proposed for assessing the sensitivity of the outputs from physically based models to the input parameters. It is based on a classical perturbation analysis of the flow equations, except that perturbations affect parameters instead of the output variables. The equations that describe the sensitivity of model outputs to the parameters are classical partial differential equations (PDEs) of hydrodynamic behavior. Solving these PDEs allows the determination of those parts of the model where it is worth carrying out additional measurements to refine knowledge of the parameters. In this case, the boundary uncertainties, such as tidal elevations and tidal currents, are generated to represent the sensitivity response coefficients. The variation of residual flow also could be used to identify the most sensitive areas, which might need additional sampling. The computation procedures of the sensitivity analysis are much simpler then those of the stochastic estimator described above.

8.2.2.4 Nonlinear Frequency Domain Analysis

With the numerical solutions (or field measurements) over time, output functions such as surface elevation, tidal velocity, and salinity level for every interior point of the modeling domain can be expressed as a function of the historical variation of boundary forcings and sources and sinks. To incorporate the system model within the numerical model input/output structure, a moving response function with fixed inputs is proposed as shown in Figure 8.2.2. Each system output, such as the salinity level, corresponds to one computational cell/node from the numerical simulation and the input functions of the system will be the boundary conditions of the numerical model. Under this assumption, the relative response of any two locations can be represented by the ratio of the multiple frequency response function. A multiple linear system frequency domain approach transfers input/output series from the time to the frequency domain using Fourier transforms. The frequency response function (FRF) or MFRF (Multiple FRF) is obtained from each frequency band through the transformation. Mathematically, if N-dimensional (N inputs) transfer function vectors A(f) are defined using the N-dimensional cross-power spectrum vector of the output with inputs $G_{xy}(f)$ an (N x N) matrix of the power and cross spectrum of all the inputs $G_{xx}(f)$, then the multiple linear system can be written in matrix notation as

$$G_{xy}(f) = G_{xx}(f)A(f) \tag{8.2.10}$$

where

$$A(f) = [A_1(f), \ldots\ldots, A_N(f)]^T \tag{8.2.11}$$

$$G_{xy}(f) = [G_{1x}(f), \ldots\ldots, G_{Nx}(f)]^T \tag{8.2.12}$$

$$G_{xx}(f) = \begin{bmatrix} G_{11}(f) & G_{12}(f) & \ldots\ldots & G_{1N}(f) \\ G_{21}(f) & G_{22}(f) & \ldots\ldots & G_{2N}(f) \\ \cdot & \cdot & \cdot & \cdot \\ G_{N1}(f) & G_{N2}(f) & \ldots\ldots & G_{NN}(f) \end{bmatrix} \tag{8.2.13}$$

The MRFA A(f) is solved by

$$A(f) = G_{xx}^{-1}(f)G_{xy}(f) \quad (8.2.14)$$

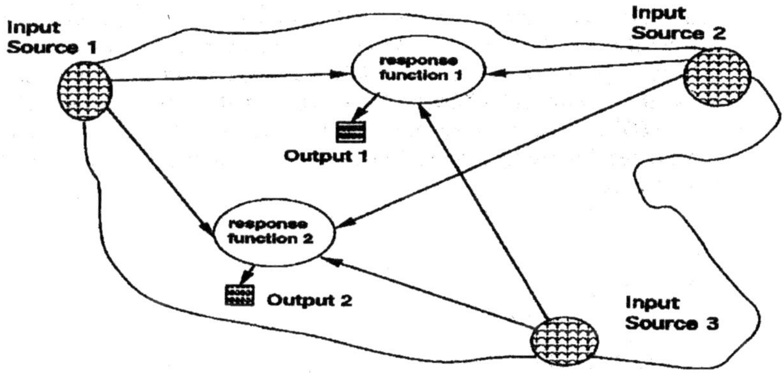

Figure 8.2.2 Incorporating the numerical model into a moving response fixed input system model design

The basic approach for solving the nonlinear system considered here is based on nonlinear decomposition and time shifting procedures to correct the group time delay and amplitude distortion problems. Hsieh and McAnally (1998) developed these decomposition and composition procedures to solve this problem. The major procedures are: 1) the identification of nonlinearity, 2) system decomposition, 3) shifting procedures for individual frequency time delay, 4) solving a series of equivalent linear systems, and 5) composition procedures.

8.2.2.5 Artificial Neural Networks Approach

Artificial neural networks (ANNs) are a relatively new system analysis modeling technique for directly addressing both nonlinearity and time-delay problems. Hsieh (2000) examines the advantages of using this tool instead of traditional nonlinear frequency domain based analysis. ANNs are able to solve problems in a way that resembles human intelligence (Khonker and Klinting, 1998). In the sense that observations provide knowledge, they are able to capture the knowledge within a data set. Unlike traditional artificial intelligence and statistical solution approaches, ANNs are able to solve problems without any a priori assumptions. As long as enough data are available, a neural network will extract any regularities or patterns that may exist and use them to form a relationship between input and output. ANNs have probably become the most efficient tools for generalization problems. The technique is also able to provide a map from one multivariable space to another through training, even when given a set of data with noise. These properties make ANN well suited to problems of estimation and prediction of flow phenomena. Usually, the data set is divided into training, cross-validation, and testing portions. The training part is used to identify the optimal weights to bridge the input/output series, while the cross-validation is used to monitor the training process to avoid over-training. The testing part is used to examine the performance of the ANN and it is not used in the training process.

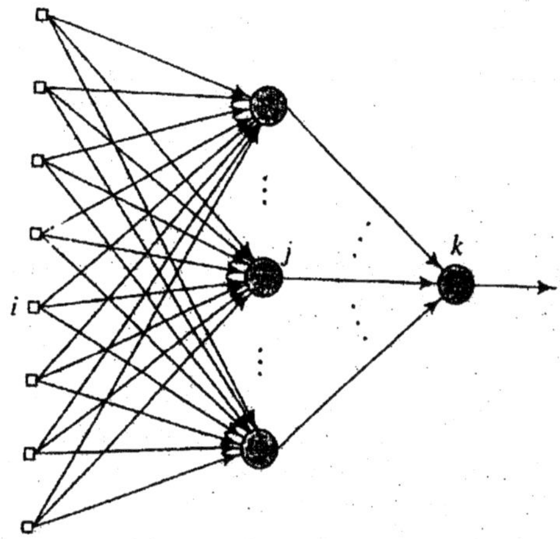

Figure 8.2.3 Fully connected feed-forward network with one hidden layer (3 nodes) and output layer

The most popular ANN algorithm is the classical multi-layer perceptron (MLP) model. MLPs (Figure 8.2.3) are feed-forward neural networks trained with a standard back-propagation algorithm. They are supervised networks, so they must be trained for the desired response. They can learn how to transform the input data into the desired response, so they are widely used for pattern classification. With one or two hidden layers an MLP can approximate the performance of optimal statistical classifiers in difficult problems. Two other algorithms, namely time-lagged neural networks (TDNN) and recurrent neural networks (RNN) are more powerful algorithms to solve time-series forecasting and prediction problems requiring the capability of addressing time-delay problems.

The nonlinear frequency domain based system analysis described above must be performed many times to solve nonlinear and time-delay problems. The new ANN approaches can solve this problem directly. To incorporate ANN into a frequency domain basis, the proper transformation/inverse transformation procedures have to be made since several very narrow frequency bands dominate the tidal system. This approach has been developed by Hsieh (2000) and applied to some hydrodynamic modeling applications. It could replace the traditional nonlinear frequency domain approach for some problems in the near future. The development is summarized as follows.

Step 1. Fourier transform and filtering processes

The first step is to convert the input/output series from the time to the frequency domain through a Fourier transform. Since the most significant frequencies are related to the tidal constituents, the non-tidal frequencies should be filtered out.

For each input $x_a(f)$ $(a=1,.....,l)$, the Fourier transform for each effective frequency band is:

$$X_a(f)=\sum x_a(n)e^{-i2\pi fn}\Delta n \tag{8.2.15}$$

Equation 8.2.15 can be decomposed into real and imaginary parts as

$$X_{ar}(f)=\sum x_a(n)\cos(2\pi fn)\Delta n \tag{8.2.16}$$

$$X_{ai}(f)=\sum x_a(n)\sin(2\pi fn)\Delta n \tag{8.2.17}$$

Let $m=2l$ (l input becomes m input), then

$$X_1(f)=x_{1r}(f) \qquad X_2(f)=x_{1i}(f) \tag{8.2.18}$$

....

$$X_{m-1}(f)=x_{lr}(f) \qquad X_m(f)=x_{li}(f) \tag{8.2.19}$$

Similarly for each output y_b $(b=1,...q)$ $(o=2q)$

$$Y_{br}(f)=\sum y_b(n)\cos(2\pi fn)\Delta n \tag{8.2.20}$$

$$Y_{bi}(f)=\sum y_b(n)\sin(2\pi fn)\Delta n \tag{8.2.21}$$

$$Y_1(f)=y_{1r}(f) \qquad Y_2(f)=y_{1i}(f) \tag{8.2.22}$$

....

$$Y_{o-1}(f)=y_{qr}(f) \qquad Y_o(f)=y_{qi}(f) \tag{8.2.23}$$

<u>Step 2. ANN model development</u>

Since a input and b output system have been converted to m input and o output series with a frequency domain basis, the development of the ANN model learning process is no longer based in the time domain. A process for determining the number of subsets needs to be performed. The determination is based on both the length of the original record and the pattern feature included. More importantly, the phase angles for each frequency band have to be shifted to a reference subset. This process must also be reversed (shifting back) when the ANN model has completed the learning process.

Step 3. Inverse Fourier transform process

The final procedure for this approach is to convert the simulated frequency series back to a time domain again (inverse Fourier transform). It is noted that the inverse process may only require the testing portion of the output function.

Each simulated b output can be represented as

$$Y'_{br}(f) = Y'_{2j-1}(f) \quad and \quad Y'_{bi}(f) = Y'_{2j}(f) \tag{8.2.24}$$

$$Y'_b(n) = (Y'_{br}(f) - iY'_{bi}(f))e^{i2\pi fn}\Delta f \tag{8.2.25}$$

The final form of this output is:

$$y'_b(n) = \sum [Y'_{br}(f)\sin(2\pi fn) + Y'_{bi}(f)\cos(2\pi fn)]\Delta f \tag{8.2.26}$$

The design architecture of this approach with four-input/two-output MPL system is as shown in Figure 8.2.4.

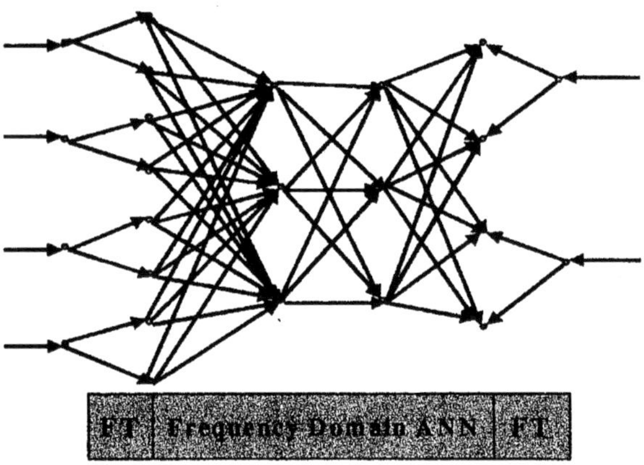

Figure 8.2.4 A hybrid approach combines ANN and Fourier transform

8.2.3 Optimal Design Methodology

Traditionally, the validation of numerical hydrodynamic models uses field measurements from predetermined sampling locations to compare with the numerical simulation results. The design of the monitoring system is based on the bathymetry, model boundary locations, and concentration gradients by experienced modelers and field personnel. However, the most significant locations, in terms of variations of parameters may not be known in advance. To help select the proper locations, a numerical model with a screening-level grid (mesh) system for testing sensitivity should be constructed and run first. It is suggested that short-term measurements, obtained during a week of intensive surveys, can be used to test the model for this preliminary study.

For any multidimensional hydrodynamic model, the set of difference equations can be written in vector notation.

$$A_{1n}H_n = A_{2n}H_{n-1} + A_{3n}U_n \qquad (8.2.27)$$

where $A_{1n,2n,2n}$ matrices whose elements depend on the hydrodynamic parameters;
H_n vector in which the elements are the surface elevations or concentration in grids at time step n;
U_n vector whose elements are all inputs.

8.2.3.1 Development of a Deterministic-Stochastic Approach

Kalman filtering combined with a deterministic flow and transport model can be used to provide the variance of the estimation error. The mathematical procedures are summarized as follows.

$$A_n = \{A_{1n}\}^{-1} A_{2n} \qquad B_n = \{A_{1n}\}^{-1} A_{3n} \qquad (8.2.28)$$

The Kalman filtering algorithm is applicable to the hydrodynamic flow system whose behavior can be described by a state equation of the form

$$H_n = A_n H_{n-1} + B_n U_n + W_n \qquad (8.2.29)$$

and a measurement process, which can be described by a measurement equation of the form

$$Y_n = C_n H_n + V_n \qquad (8.2.30)$$

where A_n, B_n model matrices at time step n;
W_n system noise at time step n;
Y_n vector containing all surface elevation or concentration measurements at time step n;
C_n measurement matrix at time step n;
V_n measurement error matrix at time step n.

A simple form of the Kalman filter algorithm is used with initial conditions H_0 and P_0 is given as:

$$\overline{\mathrm{H}}_n = A_n \mathrm{H}_{n-1} + B_n U_n \tag{8.2.31}$$

$$P_n = A_n \mathrm{P}_{n-1} A_n^T + Q_n \tag{8.2.32}$$

$$K_n = P_n C_n^T \{C_n P_n C_n^T + R_n\}^{-1} \tag{8.2.33}$$

$$\mathrm{H}_n = H_n K_n \{Y_n - C_n \overline{\mathrm{H}}_n\} \tag{8.2.34}$$

$$\mathrm{P}_n = P_n - K_n C_n P_n \tag{8.2.35}$$

where $\overline{\mathrm{H}}_n$ measurement update, which is the optimal linear estimate of the state vector at time n using measurements up to time step n;

H_n time update, which is the optimal linear estimate at time n using measurements up to the previous measurement time;

$\mathbf{P}_{n-1}$ covariance matrix of the measurement update error;

P_n covariance matrix of the time update error;

Q_n covariance matrix of the system error;

K_n Kalman gain matrix;

R_n covariance matrix of the measurement.

Equations 8.2.29 through 8.2.35 constitute the discrete form of the Kalman-Bucy filter.

The Kalman filtering is embedded within a deterministic flow and transport model. This deterministic-stochastic algorithm can be used over the complete computational domain to obtain the estimation error. The vector of the variance of estimation errors can then be used to design the optimal monitoring network. The optimal design strategy for the network design, is obtained by minimizing the cost under the budget constraint of given threshold values for variances of estimation error. For example, the threshold value for the standard deviation in the whole study area can be chosen to be 5 cm in water surface elevation, in accordance with the aims of the monitoring network and management policy. The candidates of potential monitoring locations can be determined by this stage.

8.2.3.2 Design of a Moving Installation Monitoring System

After the monitoring network has been defined using the above two steps, improving design reliability and extending the data usage within the available budget are usually considered. What is the monitoring strategy when it requires more sampling locations than the instruments can provide? The ANN approach can be used to design a moving installation data collection plan.

For example, the goal is to monitor three current meter time series for one year but only two meters are available. The strategy using the system response approach is to record the first two locations for the first half year, then remove the meter from the second gauge. Put the second meter at the third location for recording the time series for the second half year. With these data, two ANN models can be constructed and the same technique can be used to fill out the information for location 2 (second half-year) and location 3 (first half-year). In these cases, the filled portion is equivalent to simulating the testing process from known series to unknown series. This idea can extend to a more complex design. Figure 8.2.5 show five instruments are available to collect the information in a cross-section with nine different locations. The square symbol represents the period data is collected while the circle symbol shown indicates no data collection activities. The station in the center of the cross-section collects the information for the entire period. After the first half period (left part of the figure) has ended, four instruments are (clockwise) switched to the next station to complete the second half-monitoring (right part of the figure) activities. Since more neighboring stations are nearby, the reliability for filling missing information is higher (multiple input/single output case). One requirement for applying this technique is that the common recording period needs to be long enough to cover the significant variations occurring in the variable. For tidal systems the measurement period should extend at least 29 days to allow for a determination of the major tidal constituents. While the results are estimates of the measurements with a total error comprised of measurement and estimation error, they can be used to further build model confidence and as consistency checks.

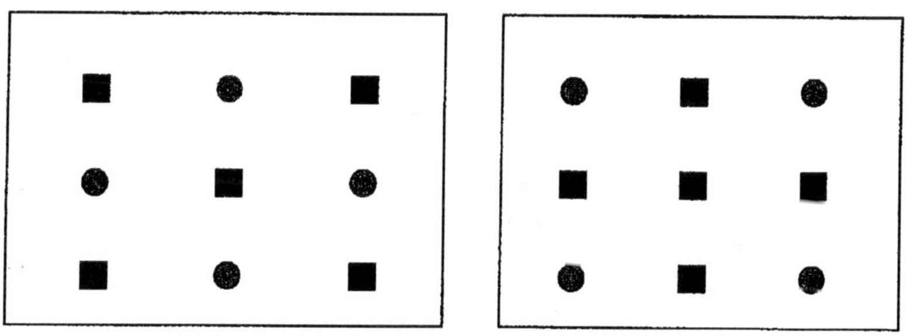

Figure 8.2.5 A moving installing monitoring system with five instruments and nine stations

8.2.4 Conclusions

A framework to improve model validation based on optimal design of field data collection has been presented. This section has reviewed the theoretical background from basic flow model, stochastic estimator, sensitivity analysis, and nonlinear frequency domain analysis to artificial neural networks for the design of the field data collection program. The artificial neural network techniques are proposed as an alternative to the traditional nonlinear system analysis. Due to the complex calculation procedures needed for performing the Kalman filtering, a good candidate for the complete integrated system is the numerical model, sensitivity analysis, and artificial neural networks.

8.2.5 Acknowledgements

The tests described and the resulting data presented herein, unless otherwise noted, were obtained from research conducted by the U.S. Army Engineer Research and Development Center, Vicksburg, MS. Permission was granted by the Chief of Engineers to publish this information.

8.2.6 References

Borrie, J. A., 1992: Stochastic Systems for Engineers, Prentice Hall, London, UK, 290p.

Bras, R. L., 1985: Random Functions and Hydrology, Addison Wesley, Reading, PA, USA, 559p.

Budgell, W.P. 1981: "A dynamic-stochastic approach for modeling advection-dispersion processes in open channels", Time Series Methods in Hydrosciences, Developments in Water Science, 17, 244-263.

Hsieh, B. B., 1997: "An optimal design of field data collection for improving numerical hydrodynamic model verification", Proceedings of the 27th Congress of the International Association for Hydraulic Research, 815-820.

Hsieh, B. B., 1998: "Maximum simulation reliability for three-dimensional flow model verification using multidimensional response techniques", Proceedings of ASCE Water Resources Conference, Memphis, TN, 1050-1055.

Hsieh, B. B., 2000: "A time domain and frequency domain based ANN for simulating tidal hydraulics systems", Proceedings of Hydroinformatics 2000, Cedar Rapids, Iowa, USA.

Hsieh, B. B and McAnally, W. H., 1998: "Evaluation of numerical model verification and reliability by system response techniques," Proceedings of 8th International Biennial Conference on Physics of Estuaries and Coastal Seas, The Hague, The Netherlands, 13-21.

Johnson, B. H., Kim, K. W., Heath, R. E., Hsieh, B. B., and Butler, H. L., 1991: Development and verification of a three-dimensional numerical hydrodynamic, salinity, temperature model of Chesapeake Bay, Volume 1, Technical Report HL-91-7, US Army Engineer Waterways Experiment Station, Vicksburg, MS.

Khondker, M.,-UI-H and Klinting, G. W. 1998: "Application of neural networks in real time flash flood forecasting", Proceedings of Hydroinformatics 98, Denmark, 777-781.

Verlaan, M. and Heemink, A. W., 1998: Kalman filtering in two and three-dimensional shallow water flow models, Proceedings of Hydroinformatics 98, Denmark, 1061-1068.

8.3 RELIABILITY ANALYSIS

Contributor: Bernard B. Hsieh

8.3.1 Introduction

One of the most significant elements of model development is using field measurements to validate the computational results. Due to budget constraints, it is almost impossible to collect enough information for validation. As we described in the last section, one approach is to design an optimal field data collection program using systems analysis. However, many projects begin after data collection has been completed, meaning that only very limited data are available to validate the model response over a large computational domain. Therefore, a method to increase the validation database, such as missing data recovery is needed. In addition, performance and reliability checking during the validation process should also be considered. Event analysis based on analysis of the physical relationship between input processes and output response should be performed. Each of these aspects is considered in turn below with a specific example included for a three-dimensional hydrodynamic model of Chesapeake Bay.

8.3.2 Missing Data Recovery System

The dynamic estimation method can be used to fill data at some stations with missing records. The procedure is similar to the moving installation method. Several different data recovery patterns, namely, self-recovery, neighboring station recovery, and multivariate parameter recover, can be performed by the Artificial Neural Network (ANN) technique. Hsieh and Pratt (2000) used this approach in Biscayne Bay, Florida. Their results indicated that the recovery reliability depends on the parameter characteristics. The partially recurrent network algorithm was found to be the most accurate data recovery system for this application.

Three types of data recovery system are defined as follows.

(1) Self-recovery: this type of recovery is based on a single time series. In this situation, no other series can be used as the reference to create the response bridge. The method is to break a long time series into two portions. The first part of the data is considered as the input, and the second part is regarded as the output function and contains the missing window. Having a high percentage of time series data, particularly if it contains significant patterns, is critical to the performance.

(2) Nearby neighboring station recovery: this is the most typical recovery case. Obviously, local recovery should have better performance than the remote recovery. If the involvement between input and output functions is a different parameter, this recovery is classified as the different parameter recovery. Otherwise, it is called the same parameter recovery.

(3) Multivariate parameter recovery: since the system response from input to output could involve more than one variable and exhibit a different time delays, this more complex system requires knowledge of the physical cause and effect to identify the system structure. For example, the salinity variation for a particular location could be caused by the tide, local wind, and nearby freshwater inflow in an estuarine system.

8.3.3 Frequency Response Coefficients

With input and output series, a system model can be constructed and validated. Extending the system structure for comparison, two multiple frequency response functions (MFRF) are established to state the numerical model performance as shown in Figure 8.3.1. The first system $MFRF_A$ represents a nonlinear system between input series and observed output. It indicates the measurement response, $MC_m(f)$. The computation response, $MC_c(f)$ which correlates the numerical model output and input series, is determined by the second system, $MFRF_B$. The ratio of measurement to computational response for each frequency band defines the frequency response coefficients, $FRC(f)$ as shown in Equation (8.3.1).

$$FRC(f) = MC_c(f) / MC_m(f) \tag{8.3.1}$$

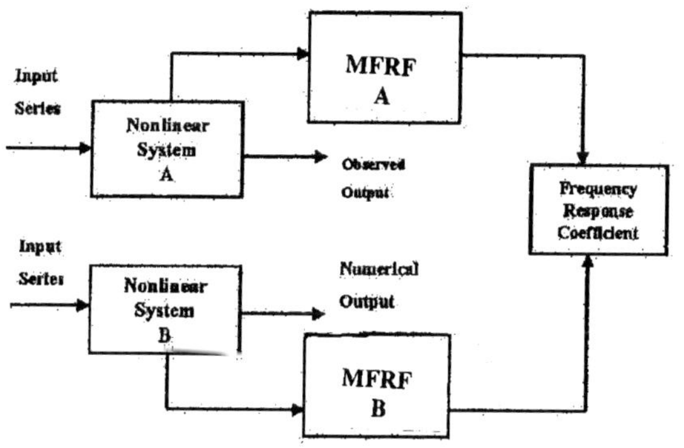

Figure 8.3.1 Frequency response coefficients for flow model validation

In a tidal system, the performance of model validation is primarily represented by significant tidal-forcing frequencies due to the tidal energy being concentrated in the semidiurnal and diurnal components. The advantage of this analysis is that it can provide information on tidal propagation.

8.3.4 Simulation Reliability Analysis

To improve simulation reliability, a computational procedure was developed by Hsieh and McAnally (1995) to examine model sensitivity and reduce uncertainty. The uncertainty can be estimated by the output function of a multiple linear/nonlinear system. Figure 8.3.2 shows a procedure for comparing the two simulation approaches. Simulation Process I uses known information to develop correlated relationships and to synthesize unknown input data. Simulation Process II uses all known information to obtain simulation output. If the model structure, estimated parameters, and boundary input are assumed correct for Process II, then its output can be assumed as "correct."

The amount of deviation between the two outputs depends on how much of the input data are unknown and what kind of estimation technique has been used to synthesize the unknown data. Three synthetic techniques, namely, nonlinear regression analysis, harmonic synthesis, and nonlinear frequency domain basis analysis, are examples.

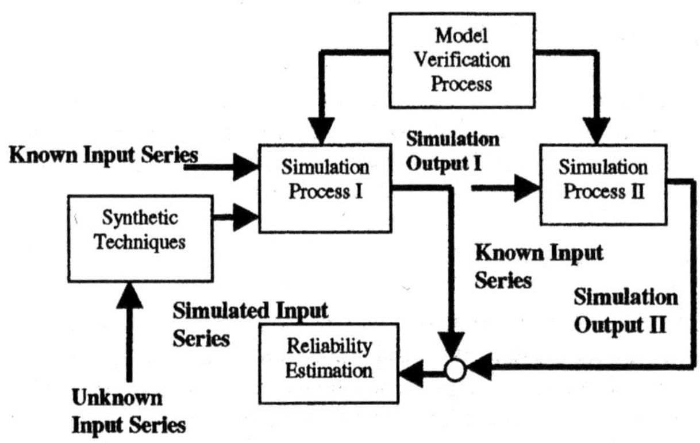

Figure 8.3.2 Reliability estimation of numerical model based on input uncertainties for two different simulation processes

Application of reliability and uncertainty analysis to examine simulation output is used by many researchers. The reliability of numerical simulation is related to the degree of output uncertainty. For a three-dimensional numerical simulation, the most important output parameters are flow patterns and the concentration of materials in transport. In general, the difference of predicted output and "true values" of output can be quantified in terms of a simulation reliability index.

Herein, two parameters are used to represent the reliability of the simulation. Initially, a design target is specified for the acceptable level of model vs observation deviation. A typical target criterion for the flow is 0.2 cm/s and for the salinity concentration is 0.2 PSU. The performance function can then be defined as the difference between the target values and the absolute value of error for each cell in the same layer. The reliability index is then calculated as the ratio of the mean and square root of the variance of the performance measures as given below.

$$PF_i = T - |obs_i - y_i| \quad , i = 1, m$$
$$RI = \mu_p / \sigma_p \tag{8.3.2}$$

where PF_i is the performance measure at time i;
T is the design target;
Obs_i is the observed value at time i;
y_i is the simulated value at time i;
RI is the reliability index;
μ_p is the mean of the performance measures;
σ_p is the standard deviation of the performance measures; and
m is the total number of observations.

Next a weighted coherence function, which varies between zero and one, is considered to address the model vs observation coherence over frequency. In an estuarine environment, the energy frequency spectrum is not equally distributed; therefore, a frequency weighted coherence function is useful in assessing the simulation reliability.

$$WC = \sum_{i=1}^{m} C_i V_i \tag{8.3.3}$$

where C_i is the coherence function between simulation and measurement for frequency i;
V_i is the percentage of total variance in the measurement series for frequency i; and m is the total number of frequency bands.

8.3.5 Event Analysis

Validating a three-dimensional numerical model requires a significant amount of time either for the adjustment of system parameters or for using statistical/mathematical methods to reduce the difference between the computed results and field measurements. However, the investigation of physical process based relationships between model response and the forcing functions are paid less attention to by researchers.

Traditionally, a good agreement of model validation is to observe the overall tendency against a desired criterion meeting certain statistical figures during the period of observations. No separate study of extreme events due to extreme values of the forcing functions is explicitly considered. However, through a physics-based investigation of model response under extreme values of the forcing functions, the validation can be performed for specific extreme events. Usually, the overall validation and reliability of the model can be enhanced by considering extreme events, separately.

8.3.6 A Three-dimensional Flow Model Example

Many three-dimensional hydrodynamic numerical model efforts have been applied by the Coastal and Hydraulics Laboratory at the US Army Engineer Research and Development Center (formerly the Waterways Experiment Station). The Chesapeake Bay, Delaware Bay, and Chesapeake and Delaware (C&D) canal system (Figure 8.3.3) is one example (Hsieh et al., 1993; Hsieh and Richards, 1996; Kim and Johnson, 1998). This section will demonstrate some approaches in testing performance and reliability for the flow models. It uses a 1993 data set to address newly developed frequency domain based artificial neural networks (FDANN) for missing data recovery (Hsieh, 2000). The experience gained from this modeling study (Hsieh and Richards, 1996) will demonstrate the frequency response coefficients, reliability estimation, and physical process evaluation.

The C&D Canal is a sea level, man-made canal joining two large estuarine systems. The main purposes for this investigation were to provide insight into the net transport through the canal for various forcing conditions over different averaging periods as well as to study the impact of deepening the navigation channel. Due to the more complicated geometry and the longer travel distance from the Chesapeake Bay side, the C&D canal receives stronger tidal signals from the Delaware Bay mouth. The Delaware estuary at it eastern boundary controls tidal flow in the canal with a mean tide range of 1.68 meters. Chesapeake Bay, at the western boundary of the canal has a mean tide range of 0.67 meters. There is a phase lag of 10 hours between tides on the western and eastern boundaries. This causes large tidal fluxes, often over 2500 cms, and maximum tidal currents of 1.07 m/s.

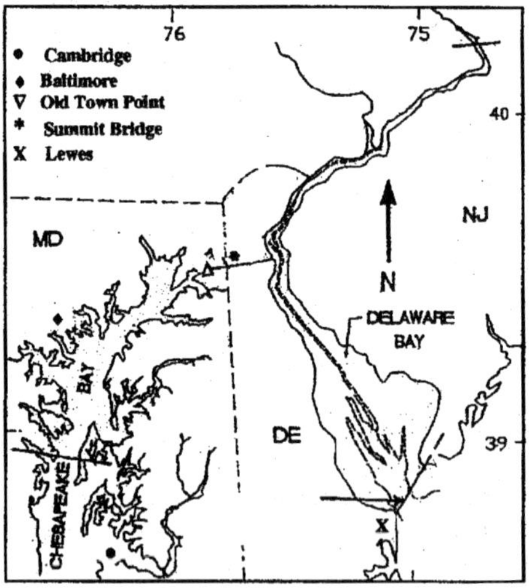

Figure 8.3.3 Upper Chesapeake Bay - C&D canal – Delaware Bay System

Here one seasonal model base, based on 1984 field measurements, was used to perform a series of simulation runs. This base covers the high-flow season (from April to June), which presents the most important fishery spawning period. The numerical model used was CH3D-WES; a three-dimensional finite difference formulation employing boundary fitted coordinates (Johnson et al., 1991). A grid was developed with 6210 computational cells and 1422 active horizontal grid cells. The model domain included the Upper Chesapeake Bay above the bay bridge, near Annapolis, through the C&D canal, and connecting the entire Delaware Bay from the bay mouth to the fall line at Trenton, NJ. Time-varying water surface elevations and salinity distributions were prescribed as the open boundary conditions. Nine major tributaries and two wind stations were addressed as other boundary forcings. To capture the important bathymetric features, the maximum number of vertical layers was set to 16.

Salinity time histories at both ocean ends of the model domain were not fully defined, so they had to be estimated. One alternative was to extend the boundary condition to the mouth of the Chesapeake Bay. In that case, the boundary uncertainty would have been minimum for addressing the C&D Canal problem. However, due to computational time considerations, an estimation technique was conducted to examine the dynamic change due to uncertainty in the boundary conditions. The solution was to use the result of a previous three-dimensional Chesapeake Bay numerical model (Johnson et al., 1991) at the closest computational point near the bay bridge to synthesize the boundary condition at the Chesapeake Bay side of the model limits. After these boundary conditions were estimated, the numerical model was validated by field measurements. Seasonal (Spring 1984) averaged flows and salinity was computed by taking the mean value of these two output variables over the total simulation period for each computational cell.

8.3.6.1 Missing Data Recovery

Numerical models are often used to simulate water resource systems and to manage their operations. However, the accurate usage of these models often requires extensive computational resources and validation using field measurements. The validation activities require boundary condition data that is often unavailable at suitable points along the computational domain. A means to generate this data is needed in most modeling situations and one method for generating the data is the use of ANN in either the time or frequency domains.

Two tidal stations (Cambridge and Summit Bridge), and one tidal current station(near Old Town Point) were used to demonstrate the tidal hydraulic processes in this area. Hourly data from April 19, 1993 to June 18, 1993 were used to construct the relationship for a set of input/output time series to describe tidal currents due to the slope of the water surface. The assumption is that the river channel component of current at Old Town Point depends on the relative surface elevations from Cambridge and Summit Bridge.

As a general principle, any ANN model will perform better if more patterns are trained. Accordingly, two-month's worth of data was divided into 24 subsets (60 hours for each). Each subset included at least two tidal cycles for the diurnal component. The number of subsets for training, cross-validation, and testing were 16, 2, and 6, respectively. The original data file was divided into 24 subsets to increase the number of training patterns as well as to capture the major tidal variation. With harmonic analysis, sixteen significant tidal constituents are extracted. These include ten semi-diurnal components (M2, S2, N2, K2, T2, L2, 2N2, ν2, λ2, and μ2) and six diurnal components (K1, O1, P1, Q1, M1, and J1). Thirty-two (including the imaginary part) frequency bands were used to perform the Fourier transform for every subset. The learning experience indicated that an important step was to adjust the corresponding phases. The shifting basis may select the first subset.

A two input/single output system (Cambridge tide can be decomposed into four-input/two-output system) is considered with 768 frequency response bands. The amplitude and phase angle were first computed to determine the quantity of shifting. A general recurrent network (GRN) model was tried to connect both amplitude and phase simultaneously. While the amplitudes show strong correlation, the variation of phase angles was found to be very difficult to correlate. The reason for this is due to the fact that the variation of phase angles follows a periodic function of 6.28 radians. The ANN cannot recognize the same corresponding trigonometric function, which could come from different radians. Secondly, the fact that the phases change continuity as the waves travel through each subset has to be considered. Therefore, the modification of phase angle with respect to the first subset is required. The real part and imaginary part of the corresponding function for corrected phases form the new input/output structure. Figures 8.3.4 and 8.3.5 show the original tidal signal and the real part after this shifting process for Summit Bridge, respectively.

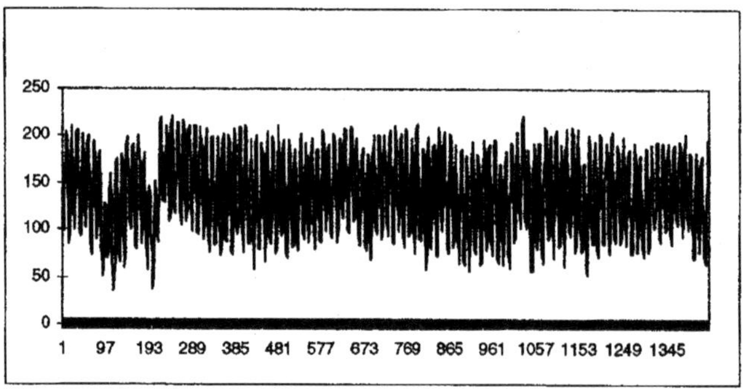

Figure 8.3.4 Tidal signals at Summit Bridge during the period between April 19 and June 18, 1993

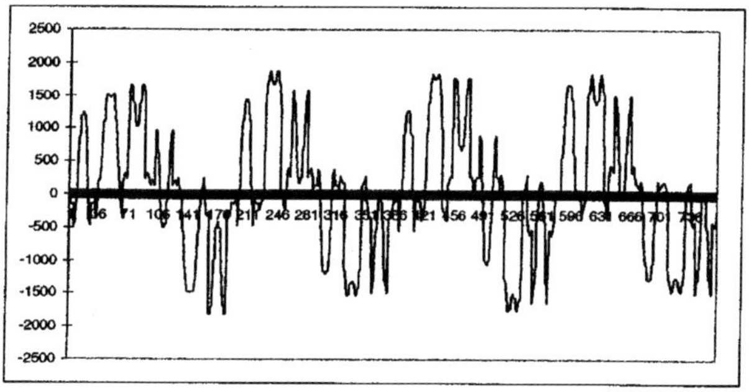

Figure 8.3.5 Real part of tidal signals from designed training set for Summit Bridge

The GRNs were again applied to this frequency series set with the four-input/two output system. Highly correlated results for the testing set have been found from both real and imaginary components (r=0.94 and r=0.96, respectively) from the Old Town Point current. Figures 8.3.6 and 8.3.7 represent the real and imaginary component for the test portion. The simulation of the test portion is equivalent to the missing data recovery problem. Using the existing field measurements the accuracy of the performance needs to be determined. For the real case, no data are available for the simulated testing portion (missing window).

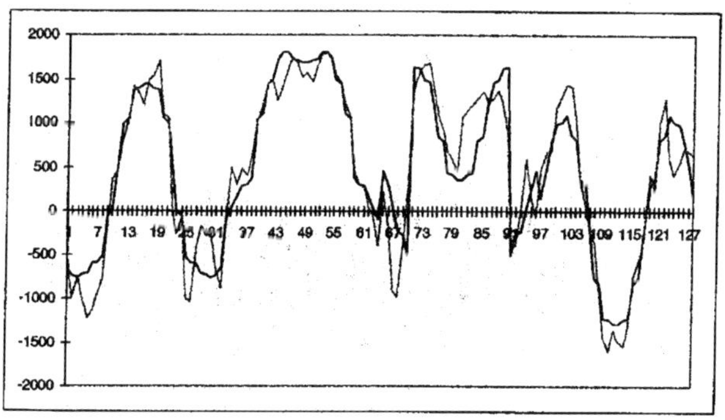

Figure 8.3.6 Real component of testing portion (missing window) for tidal current at Old Town Point (model values – light line)

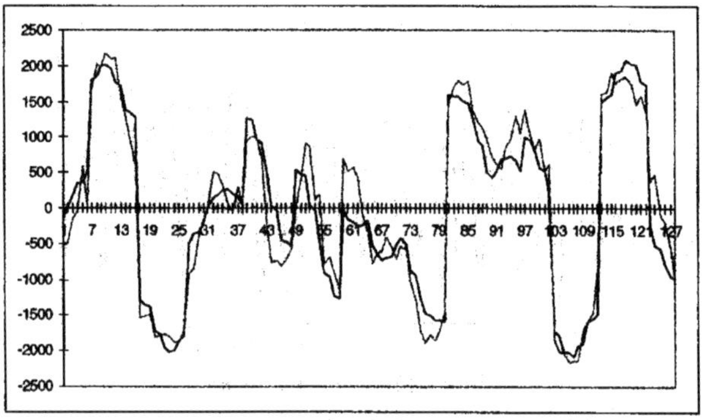

Figure 8.3.7 Imaginary component of testing portion (missing window) for tidal current at Old Town Point (model values – light line)

These simulated frequency series from testing portions of the ANN model are then converted back to the time domain again and compared to the field measurements. In Figure 8.3.8 the results of the time domain (TDANN) approach are compared with the field measurements while in Figure 8.3.9 the frequency domain (FDANN) approach is shown. Less accurate results (r=0.94) were obtained for the FDANN approach due to the fact that the frequency domain approach needs a much longer record to represent the variation for the entire frequency band. In addition, Hsieh (2000) indicated that the frequency band approach would tend to be less accurate than the time domain approach because the non-significant frequencies may contribute unnecessary amplitudes and phases when applying the Fourier transform and its inverse process.

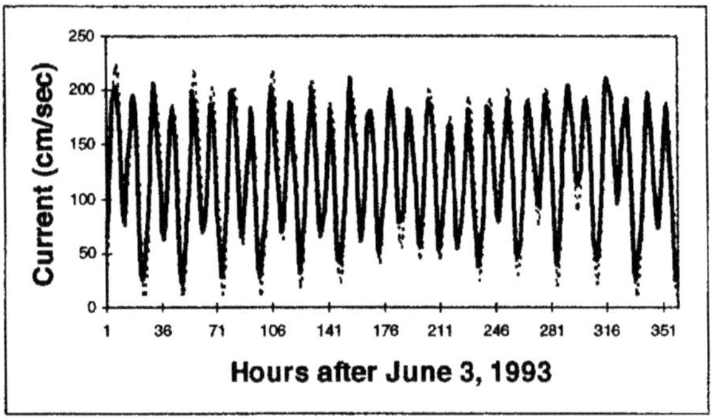

Figure 8.3.8 Missing window recovery for TDANN at Old Town Point (dashed line)

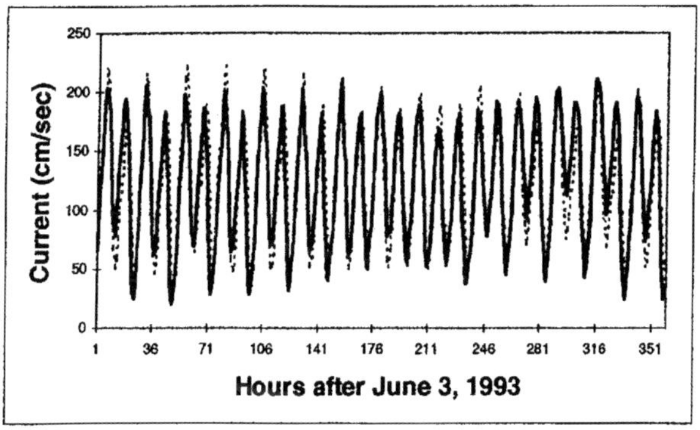

Figure 8.3.9 Missing window recovery for FDANN at Old Town Point (dashed line)

8.3.6.2 Frequency Response Coefficients Application

In the first phase of model validation, the prototype tidal current was compared with numerical solution for a month. Excellent agreement (3.5 cm/s standard deviation with average amplitude of 92 cm/s) for the near bottom layer at Summit Bridge, for example, was obtained. Using the system modeling approach to calculate the FRC, three forcings with five different sources (tides from Delaware Bay mouth and Upper Chesapeake Bay, riverflow discharges from the Susquehanna River and Delaware River, with wind stress from the gauge at Wilmington, Delaware) are considered as inputs, and the tidal current at Summit Bridge is regarded as output.

The analysis shows that the river flows and wind stress have weak correlation to tidal current for this station over the month. Five significant tidal forcing frequencies (S2, M2, N2 ,K1, and O1) from two boundaries with different phases are chosen as system inputs. The FRC for the validation point are presented in Table 8.3.1. Another useful parameter, phase difference between the two systems, is also calculated. The results show that this numerical model is undershooting the semidiurnal bands and is over shooting the diurnal bands. The field measurement leads the numerical result by about 24 degrees for the major component M2, for example. Since the Summit Bridge receives stronger tidal propagation signals from the Delaware Bay side, the physical causes and effects, such as the bathymetric refinement from the higher gradient area in the numerical model, need further study.

Table 8.3.1 FRC of tidal current validation at the Summit Bridge, C&D Canal

Significant Tidal Constituent	Frequency Response Coefficients	Phase Difference (radians)
S2	1.222	0.347
M2	1.208	0.414

Significant Tidal Constituent	Frequency Response Coefficients	Phase Difference (radians)
N2	1.209	0.003
K1	0.843	0.338
O1	0.856	0.239

8.3.6.3 Simulation Reliability Analysis

Here, the uncertainty of input parameter estimation is addressed only for the salinity boundary at the Chesapeake Bay side. The other input forcings for 1985, surface elevation, wind field and Delaware Bay salinity boundary were assumed to be known. The approach of this numerical experiment was to use several synthetic techniques to generate salinity boundary conditions with 1984 information and to compare the results with the boundary, which were the simulation results from the Chesapeake Bay numerical model. The latter boundary condition was considered to be a true value. More importantly, reliability estimation from simulated outputs with/without synthetic techniques was computed by two different indexes. The significance of this reliability estimate provides us with insight into the confidence we can have in various forms of estimation-how good a simulation we can expect if the boundary data are uncertain or no information exits.

Three synthetic techniques, namely, nonlinear regression analysis, harmonic synthesis, and nonlinear frequency domain analysis were used to create the salinity boundary for 1985. All techniques consider a multiple input/single output system, which means that the surface elevation at Annapolis, wind forcings at Baltimore-Washington International Airport, and the freshwater inflow from the Susquehanna River at Conowingo Dam flow gauge are the inputs and salinity at Annapolis is the output.

Under these estimation techniques, the surface and bottom salinity boundary conditions were estimated by constructing two sets of input/output structure. While the 1984 salinity boundary conditions showed significant changes with respect to the 1985 conditions, the method of nonlinear analysis resulted in the best estimation. The standard deviations for both surface and bottom boundary conditions for this method were 0.61 ppt and 0.79 ppt, respectively. It can be concluded that the uncertainties of input parameters, such as boundary conditions, can be highly reduced by a proper estimation technique.

A reliability index was computed with the target criterions for flow equal 0.2 cm/sec and for salinity equal 0.2 ppt. Layers 16, 13, and 10 (which corresponds to the top, middle, and near bottom layers of the canal) for salinity concentration, x-component and y-component of flow for seasonal average were selected for evaluation. As expected, the results (Table 8.3.2) show that the nonlinear frequency domain approach (NL FD) produced the highest reliability index for all testing parameters (flow and salinity) and boundary condition methods. The reason is that the nonlinear analysis is a much better estimation method to remove the degree of uncertainty.

Table 8.3.2 Reliability index for simulation results of flow and concentration

	Concentration			Flow (x-component)			Flow (y-component)		
Parameter	LR16	LR13	LR10	LR16	LR13	LR10	LR16	LR13	LR10
84 BC	-0.71	-0.78	-0.80	0.11	-0.02	-0.16	0.01	-0.08	-0.08
NL Regr	-0.01	-0.11	-0.17	3.74	3.55	3.77	2.07	2.00	2.47
Harm Sn	-0.13	-0.11	-0.08	3.09	3.24	2.64	1.32	1.59	2.12
NL FD	2.07	2.76	1.39	4.38	5.24	6.36	2.08	2.80	4.28

Four control points in the model, two in the Upper Chesapeake Bay, one in the C&D Canal, and one in the middle of Delaware Bay, were selected to compute the weighted coherence function for the salinity concentration for the time series generated by the 90 day simulation. In this case, the output function from each synthetic technique considers the system inputs and the 1985 true value calculation results are summarized as Table 8.3.3. Once again, the most highly correlated relationship was found for the nonlinear frequency domain (NL FD) method.

Table 8.3.3 Weighted coherence functions of salinity level for four selected control columns during the simulation period

Control Point	Chester River E.	Town Point	Reedy Point	60 Miles (Del.) Ch.
84 Boundary	0.582	0.553	0.440	0.424
NL Regr	0.793	0.798	0.796	0.797
Harm Sn	0.780	0.803	0.815	0.821
NL FD	0.936	0.950	0.914	0.948

8.3.6.4 Event Analysis

The model validation using mathematical/statistical methods sometimes has its limitation. The physical insight, particularly in the mutual relationship among the forcing functions, is often not considered during special extreme events. The model response can be studied, separately, for these events as discussed below to improve the model validation.

Initial Refinement

In the first phase of validation, a month-long comparison for surface elevation, tidal current, and salinity showed excellent agreement with field measurements. However, when the simulation period was extended to seasonal scale, the computed results indicated less response and produced the salt conservation problem. To seek improvement, two steps were used to perform the model refinements: uncertainties of boundary and initial conditions and bathymetry/channel refinement for grid modification. An example comparison after completion of this phase is shown in Figure 8.3.10.

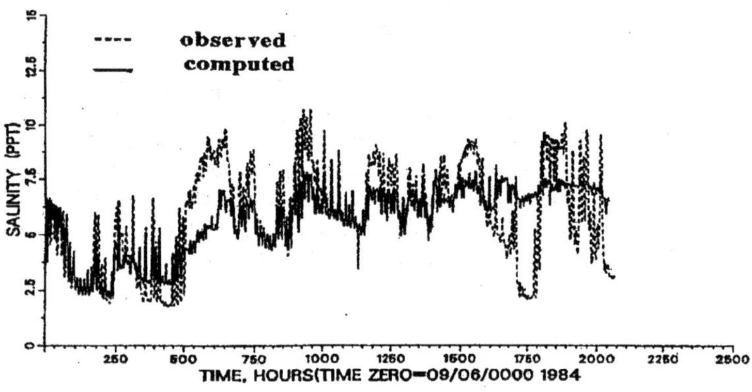

Figure 8.3.10 Model validation after completing the initial refinement process at Summit Brige, MD (4.5m from water surface)

Northeast Wind Effect

This initial refinement of the model generally improved validation; however, the spikes of the salinity distribution still could not be captured. For example, an extremely low salinity level occurred on November 16-18, 1984, at Summit Bridge. The salinity level suddenly dropped to lower than 2 ppt from about 7.5 ppt and then returned to 9 ppt after two days. It was found, after examining the forcing functions, that this was caused by a very strong eastward wind, small tidal ranges, and more freshwater inflow released from the Conowingo Dam. Therefore, less salinity entered the C&D Canal from the Delaware Bay boundary, and more water was brought by the wind from the Chesapeake Bay side. This strongly implies that the land-water conversion factor from the local wind field is sensitive to these events. It is suggested that this wind event could be related to the "Northeast" wind with long duration.

Event Based Model Evaluation

Instead of using system/statistical methods to identify the relationships, a simple data processing procedure was conducted. First, a low-pass filter to compare the subtidal components of these functions was performed. As the numerical model assumed, the salinity level at Summit Bridge is a function of source tides, local wind, and freshwater inflows. The significant components are the tidal forcing at the Delaware Bay mouth, east-west wind component from Wilmington, Delaware station, and the freshwater inflow from the Susquehanna River. The spring and neap tidal events strongly appear in the salinity time series. The strong weekly patterns in flow releases from the Conowingo Dam are explained by the minimum flow operation cycle during this period. While the fresh water contributes to the seasonal salinity level variation, the subtidal component of the tide dominates the salinity variation, particularly during the greater tidal range period. Because the canal crosses the two bays in a nearly east-west orientation, the east-west wind component actually dominated the local events. This is also particularly true during the small tidal range period.

From this simple analysis, it is concluded that the wind component contributes a significant portion of salinity variation. Usually, over-water winds were calculated using Hsu's (1986) formula, which suggests a simple dimensional relation between wind velocity over the water and over the land. In this study, no experimental data were available for specifying the coefficients of local wind (Wilmington, Delaware). After testing runs, the coefficients for the x-component were replaced by 2.00 (slope) and 1.35 (intercept). This modification significantly improved the model validation for these special events as shown in Figure 8.3.11.

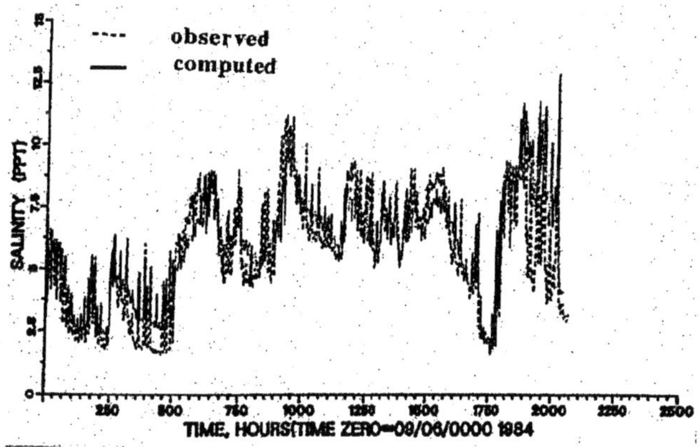

Figure 8.3.11 Final model validation at Summit Bridge, MD (4.5m from water surface)

8.3.7 Conclusions

This section identifies several important components of reliability analysis for flow model validation. While the missing data recovery system and physical process evaluation can improve model validation, the frequency response coefficient, reliability index, and weighted coherence function techniques can be used to further monitor the validation.

From the Upper Chesapeake Bay-C&D Canal-Delaware Bay Study, it is found that the flow model validation can be improved by an event analysis, which considers the physical processes among the forcing functions and by reducing the uncertainty in the boundary conditions. The developed computational procedures can provide guidance for designing an optimal data collection program and on how to assess model reliability.

With advancements in hydroinformatics, flow model validation can be incorporated within a system analysis framework. System analysis considerations can assist the model validation process and have the potential to enable more accurate and reliable results than based solely on traditional data vs model inter-comparison techniques.

8.3.8 Acknowledgements

The tests described and the resulting data presented herein, unless otherwise noted, were obtained from research conducted by the U.S. Army Engineer Research and Development Center, Vicksburg, MS. Permission was granted by the Chief of Engineers to publish this information.

8.3.9 References

Hsieh, B. B, Johnson, B.H., and Richards, D. R., 1993: A three-dimensional numerical model study for the Chesapeake and Delaware Canal and Adjacent Bays, *Technical Report HL-93-4*, US Army Engineer Waterways Experiment Station, Vicksburg, MS.

Hsieh, B. B and McAnally, W. H., 1995: "Improvement of simulation reliability for a 3D numerical hydrodynamic model," *Advances in Hydro-Science and Engineering, Volume 2*, Beijing, China, 1183-1188.

Hsieh, B. B, and Richards, D. R., 1996: Three-dimensional numerical simulation of seasonal flow and salt transport for the C&D Canal, *Technical Report HL-96-14*, US Army Engineer Waterways Experiment Station, Vicksburg, MS.

Hsieh, B. B., 2000: "A time domain and frequency domain based ANN for simulating tidal hydraulics systems", Proceedings of Hydroinformatics 2000, Cedar Rapids, Iowa, USA.

Hsieh, B. B. and Pratt, T. C., 2000: "Field data recovery in tidal systems usings ANNs", Proceedings of ANNIE, St. Louis, MO.

Hsu, S. A., 1986: "Correction of land based wind data for offshore applications: a further evaluation", *Journal of Physical Oceanography*, 16, 390-394.

Kim, K. W. and Johnson, B. H., 1998: A three-dimensional numerical model study to assess channel deepening in the Delaware River and Bay, *Technical Report CHL-98-20*, US Army Engineer Waterways Experiment Station, Vicksburg, MS.

Johnson, B. H., Kim, K. W., Heath, R. E., Hsieh, B. B., and Butler, H. L., 1991: Development and verification of a three-dimensional numerical hydrodynamic, salinity, temperature model of Chesapeake Bay, Volume 1, Technical Report HL-91-7, US Army Engineer Waterways Experiment Station, Vicksburg, MS.

8.4 GALVESTON BAY NOWCAST/FORECAST SYSTEM DEVELOPMENT AND VALIDATION

Contributor: Richard A. Schmalz, Jr.

8.4.1 Introduction

The National Oceanic and Atmospheric Administration's (NOAA), National Ocean Service (NOS), is developing nowcast/forecast systems in conjunction with its Physical Oceanographic Real Time Systems (PORTS) in Tampa Bay, Galveston Bay, New York/New Jersey Harbor, San Francisco Bay, and Chesapeake Bay with several new systems in the planning stages. To standardize development, an initial report (NOS, 1999) has been developed. We adopt the following definitions outlined therein: A *hindcast, nowcast, and forecast* are scientific predictions about the past, present, and future states, respectively, of water levels and/or currents (and possibly other relevant oceanographic variables such as salinity and temperature) in a bay or estuary made by a numerical, statistical, or hybrid model or method. These predictions rely on either observed or forecast data, not on hypothetical data. A *hindcast* incorporates past or historical observational data. A *nowcast* incorporates recent (and often near real-time) observed meteorological, oceanographic, and/or river flow rate data; covers the period of time from the recent past (up to a few days) to the present; and makes predictions for locations where observational data are not available. The present is the time at which the nowcast is made, and at which the most recent observations are from a few minutes to an hour old. A *forecast* incorporates meteorological, oceanographic, and/or river flow rate forecasts; makes predictions for locations where observational data will not be available; and is usually initialized by the results of a nowcast. Note under this definition, a nowcast and forecast may be run after the fact or in a hindcast mode. Indeed, this practice has been used to further refine and develop the above ongoing nowcast/forecast systems.

Herein, we focus on the nowcast/forecast system in Galveston Bay and the role of hindcasting in its development. The Galveston Bay PORTS is patterned after Bethem and Frey (1991) and was installed in June 1996 to monitor Galveston Bay as recommended by Williams et al. (1990). Water surface elevation, currents at prediction depth (4.7 m below MLLW), as well as near-surface and near-bottom temperature and salinity, and meteorological information are available at six-minute intervals at stations shown in Figure 8.4.1 in the format illustrated in Figure 8.4.2 (Appell et al., 1994). The nowcast/forecast system is based on the NOS Galveston Bay three-dimensional hydrodynamic model (GBM) and the National Weather Service (NWS) Aviation atmospheric model (AVN) (Kalnay et al., 1990). To simulate currents within the Houston Ship Channel (HSC), a finer resolution three dimensional Houston Ship Channel model (HSCM) has been developed. The GBM is used to provide bay wide water level and near entrance current forecasts as well as to directly provide water levels, density, and turbulence quantities to the HSCM for use in a one-way coupling. The coordinated model set comprises the hydrodynamic portion of the nowcast/forecast system.

We review the process which was followed to arrive at the present nowcast/forecast system and the role of hindcasting in the system development activities section. In the validation section, results of a one year validation of the nowcast/forecast system are discussed using the NOS (1999) acceptance criteria as targets. A water level event analysis is also presented. In conclusion, results are summarized and future directions toward operational nowcast/forecasting utilizing previous system analysis techniques are presented.

8.4.2 System Development Activities

The initial GBM and its calibration to May 1995 astronomical tide are first discussed. The results of the June 1995 hindcast are next presented and demonstrate the potential of the model to be further considered as a candidate for nowcast/forecasting in Galveston Bay. Necessary model extensions to enable nowcasting/forecasting over the range of conditions expected are then presented. Specifically, the inclusion of drying/wetting and salinity flux corrected transport (FCT) are considered. Hindcast results for the extended GBM are presented for the October 1994 flood to assess the FCT salinity transport and for the January 1995 "Northers" to assess the drying/wetting scheme. The development of the one-way coupled high resolution HSCM to the revised GBM is next discussed and assessed via the April 1996 hindcast. The coupled models were used to provide the hydrodynamics of the initial nowcast/forecast system (Schmalz, 1998a). The design of the nowcast /forecast system is briefly outlined and improvements are further developed in terms of several hindcast experiments.

8.4.2.1 Galveston Bay Model Astronomical Tide Calibration

A three-dimensional sigma coordinate Galveston Bay and near shelf (GBM) model has been developed (Schmalz, 1996) based on a version of the Blumberg and Mellor (1987) model extended to orthogonal curvilinear coordinates (see Mellor, 1996). The GBM computational grid in Figure 8.4.3 consists of 181x101 horizontal cells (dx = 254-2482 m, dy= 580-3502 m) with 5 sigma coordinate levels in the vertical. The model was spun up from rest over the first day, 1 May. The initial salinity and temperature fields were constructed from climatological considerations using a grid patch method developed for Long Island Sound (Schmalz, 1994). A sea surface temperature (SST) specification was used in lieu of heat flux. Climatological river inflows (Orlando et al., 1993) were included for the Trinity and San Jacinto Rivers and Buffalo Bayou. The salinity and temperature boundary conditions were determined based on National Marine Fisheries Service cruise data (Temple et al., 1977). Water surface elevations were specified at grid cells (3,2), (60,2), (120,2), (180,2), and (180,32) by modifying one year least square analysis (Schureman, 1958) harmonically derived tidal constituents at Freeport, Galveston Pleasure Picr, Galveston Pleasure Pier, Port Bolivar, and High Island, Texas, respectively. In general, at the offshore stations, phases were adjusted to account for up to a one hour time advance and a 1.02 amplitude adjustment was made. A linear spatial interpolation along the open boundary was used between the appropriate pair of boundary cells. Results of demeaned simulated water surface elevations versus harmonically reconstructed tidal signals are given in terms of an rms difference and a dimensionless (0 no shape error to 1 total shape error) relative error (Willmott et al., 1985) and ranged from 4 cm and 0.01-0.02, respectively, at Galveston Pleasure Pier and Pier 21 to 5 cm and 0.04, respectively, at Eagle and Morgans Points. To study the impact of long period constituents contained in the one-year least squares analysis, a 29-day harmonic analysis (Schureman, 1958) of simulated water levels was performed. Based on this analysis, a constituent amplitude weighted gain (model/observation) and phase difference in hours (model - observation) were computed and an rms error estimated (see Hess, 1994) as given in line 1 of Table 8.4.1. Rms errors were reduced to 3 cm, indicating that some error (order 2 cm) is contained in the longer period tidal response. Weighted gains were order 0.9 over the lower bay, indicating that the tidal response is 10 percent damped. Weighted phase errors are within one hour. Simulated level 3 current principal flood directions were computed and compared with March and April 1996 PORTS station observations given in Table 8.4.2. In general, May 1995 astronomic principal flood directions are within 30 degrees of PORTS station observations, which include meteorological effects.

Table 8.4.1 Galveston Bay Tide Station 29-Day Harmonic Water Level Analysis
Line (1,2,3) === (May 1995 GBM, April 1996 GBM, April 1996 HSCM)

Station	Gain (-)	Phase (hr)	RMS (cm)
Galveston	0.87	-0.04	3
Pier 21	0.85	0.13	3
677-1450	0.88	0.04	3
Eagle	0.97	0.32	3
Point	0.91	0.98	4
677-1013	0.90	0.79	4
Morgans	0.89	-0.68	3
Point	0.87	0.01	3
677-0613	0.87	0.08	3

Table 8.4.2 Galveston Bay PORTS Station Principal Flood Direction Analysis

	Model (deg True)			Observation (deg True)
Station	GBM (5/95)	GBM (4/96)	HSC (4/96)	(3-4/96)
Bolivar Roads	350	342	321	322
Redfish Bar	336	336	331	322
Morgans Point	317	313	318	341

Results of a 29-day harmonic analysis of principal direction level 3 currents are compared in line 1 of Table 8.4.3 with a 29-day harmonic analysis of March and April 1996 PORTS prediction depth currents. The weighted gain decreased significantly as one proceeds from Bolivar Roads in the entrance up to Morgans Point at the head of Galveston Bay with rms errors order 20 cm/s. Based on these results, the model tidal current as well as tide response is damped. A bottom roughness of z_0=2 mm was employed with a horizontal eddy viscosity coefficient of 0.005. Refer to Mellor (1996).

Table 8.4.3 Galveston Bay PORTS Current Station 29-day Harmonic Principal Direction Analysis at prediction depth (4.6m). Line (1,2,3) === (May 1995 GBM, April 1996 GBM, April 1996 HSCM)

Station	Gain (-)	Phase (hr)	RMS (cm/s)
Bolivar Roads	0.86	-0.50	17.8
	0.78	0.36	17.0
	0.63	-0.27	26.8
Redfish Bar	0.72	0.06	19.7
	0.70	-0.17	14.4
	0.70	-0.66	16.6
Morgans Point	0.32	-1.45	16.0
	0.42	-0.77	13.1
	0.68	-0.49	8.6

8.4.2.2 June 1995 Hindcast

Average daily flows were obtained from the USGS Houston Office, for Buffalo Bayou at Piney Point, Texas, Trinity River at Romayor, Texas and Lake Houston near Sheldon, Texas via a stage vs discharge relation. NDBC buoy 42020 (3m Discus) and 42035 (3m Discus) and C-MAN station S-2 Sabine and S-4 Port Aransas, Texas observations were obtained along with NWS surface weather observations at Houston IAH, Port Arthur, and WSO Galveston, Texas. Wind and pressure fields were developed over the model domain via inverse distance squared interpolation. Subtidal water levels at Galveston Pleasure Pier were used along the entire GBM open boundary. Texas Water Development Board (TWDB) salinity and temperature data were melded with climatological salinity and temperature data to form the initial density fields, surface temperature fields, and offshore boundary salinity and temperature conditions. Agreement between demeaned simulated and observed water levels was order 10 cm over the entire month using z_0=1 cm, and suggests that the further development of the GBM toward a nowcast/forecast system is warranted.

8.4.2.3 Galveston Bay Model Extensions

To further improve the Galveston Bay model, heat flux, drying/wetting, and flux-corrected salinity transport schemes were incorporated. The latter two are described herein, as it was found that a SST specification was more favorable than heat flux as discussed below.

8.4.2.4 Drying/Wetting Scheme

The scheme developed by Hess (1994) in Tampa Bay is modified for application in Galveston Bay to simulate winter time storms, during which northerly winds of up to 40 knots persist over the bay associated with cold front passages. An x-direction flow width reduction factor , $wx_{i,j}$, based on the upstream vertically integrated velocity, $u^n_{i,j}$, and ell depth, $d^n_{i,j}$, is computed at the beginning of each external mode time step and is specified using the relation:

$$
\begin{aligned}
&wx_{i,j} = \min(1, \max(0, (d^n_{i-1,j} - d_{u)} / d_T))) \quad u^n_{i,j} \geq 0 \\
&wx_{i,j} = \min(1, \max(0, (d^n_{i,j} - d_u / d_T))) \quad u^n_{i,j} < 0 \\
&\text{Note } wx_{i,j} \in (0,1) \text{ for } d^n_{*,j} \in (d_u, d_T), \quad \text{where } * = (i, i-1).
\end{aligned}
\tag{8.4.1}
$$

An analogous relationship is used to specify the y-direction flow width reduction. In the Galveston Bay model application, d_u= 0.25 m and d_T=0.5 m. Since the model is written in horizontal area format, one multiplies the x-direction flow width, $dx = 0.5(dx_{i,j} + dx_{i-1,j})$ by $wx_{i,j}$ and the analogous expression for dy by $wy_{i,j}$ to reduce the horizontal fluxes at each sigma level. The above linear cell depth relationship is used to reduce cell face flow widths when cell water depths drop below 0.75 m and fully eliminates flow paths when water depths drop below 0.25 m. For water depths greater than 0.75 m no reduction in flow width is made.

In addition to the flood width reduction, it is necessary to reduce the wind stress over cells with small water depths. The following approach was utilized for the above test case and in subsequent simulation.

$$
\begin{aligned}
&r_t = \min(1, \max(0, (\bar{d}_* - d_0) / d_1 - d_0))) \quad * \equiv (x, y) \\
&\text{where } \bar{d}_x = 0.5(d^n_{i,j} + d^n_{i-1,j}) \text{ and } \bar{d}_y = 0.5(d^n_{i,j} + d^n_{i,j-1}).
\end{aligned}
\tag{8.4.2}
$$

The factor r_t was applied to the surface wind stress terms with d_0=0.5 m and d_1=1.5 m, respectively. In general, the constants d_u and d_T in the cell width reduction formulas and d_0 and d_1 in the wind stress reduction relationship must be determined for each application and are a function of the wind event strength and of the estuarine tidal range, bathymetry, and morphology. The scheme fails if a negative water depth is computed in the external mode.

A five day test wind loading case, in which winds were ramped from 0 to 40 kts out of the north during day one, held constant at 40 kts out of the north for the next two days, then ramped to zero over the fourth day, and held at zero over the final day was used. May 1-5, 1995 astronomical tide conditions were specified with a -50 cm subtidal water level along the open boundaries ramped analogously to the wind. A large section of Trinity Bay dried and then reflooded after the wind and subtidal water level signal went to zero.

8.4.2.5 Flux Corrected Transport Scheme

A second order van Leer-type upstream-biased transport scheme (Lin et al., 1994) has been implemented to treat the very sharp horizontal salinity gradients in Galveston Bay. Consider the following parameter, $f^m_{i,j,k}$, to represent grid cell salinity at internal mode time level *m*. The scheme corrects the flux based on grid cell upstream velocity, $u^m_{i,j,k}$, x-direction cell width, $dx_{i,j}$, and internal mode time step length, ΔT, in the Equation (8.4.3) below. Analogous relationships hold for the y-direction, YFLUX, and sigma direction, ZFLUX. The XFLUX and YFLUX terms are multiplied by $\bar{d}_x = 0.5(d^m_{i,j} + d^m_{i+1,j})wx_{i,j}$ and. $\bar{d}_y = 0.5(d^m_{i,j} + d^m_{i,j+1})wx_{i,j}$. The XFLUX, YFLUX, and ZFLUX terms replace the original quantities used in Subroutine ADVT (refer to Mellor, 1996). Note the scheme employs a single increment from time level *m* to level *m+1* and hence, the diffusion terms are evaluated at time level *m* in the standard manner.

$$
\begin{aligned}
&\Delta\bar{F}^m_{i,j,k} = 0.5(f^m_{i,j,k} + f^m_{i-1,j,k})\\
&f\min = \min[f^m_{i-1,j,k}, f^m_{i,j,k}, f^m_{i+1,j,k}]\\
&f\max = \max[f^m_{i-1,j,k}, f^m_{i,j,k}, f^m_{i+1,j,k}]\\
&\Delta\bar{F}^m_{i,j,k} = sign(\Delta\bar{F}^m_{i,j,k}) * \min[|\Delta\bar{F}^m_{i,j,k}|, 2(f^m_{i,j,k} - f\min), 2(f\max - f^m_{i,j,k})]\\
&co = 0.5(1 - u^m_{i,j,k}\Delta T / dx_{i-1,j}) \quad u^m_{i,j,k} > 0\\
&co = 0.5(1 + u^m_{i,j,k}\Delta T / dx_{i-1,j}) \quad u^m_{i,j,k} \le 0\\
&XFLUX = u^m_{i,j,k}(f^m_{i,j,k} + co\Delta\bar{F}^m_{i-1,j,k}) \quad u^m_{i,j,k} > 0\\
&XFLUX = u^m_{i,j,k}(f^m_{i,j,k} + co\Delta\bar{F}^m_{i,j,k}) \quad u^m_{i,j,k} \le 0
\end{aligned}
\qquad (8.4.3)
$$

8.4.2.6 Hindcast Experiments

River inflows, water level residual forcings, wind and atmospheric pressure fields were included in all hindcasts in the same manner as previously described for the June 1995 hindcast. However, wind and pressure fields were developed over the model domain via two-step Barnes (1973) interpolation. During the October 1994 and January 1995 hindcast, the heat flux formulation developed by Martin (1985) was employed. In general, the heat flux formulation predicted temperatures 3-5 degrees cooler than the observations, and additional work is needed to further calibrate the flux parameters. For the April 1996 hindcast, a SST specification was employed. The mid-depth temperature comparisons were within 1 degree, and since SST information is available via PORTS, this approach has been adopted. Note the bottom roughness and eddy viscosity coefficient were set as in the astronomical tide calibration.

To test the flux-corrected transport scheme, a one-month simulation of October 1994 was performed. During 17-18 October, the flood of record occurred on both the San Jacinto and Trinity Rivers, with average daily flows on each river above 100,000 cfs.

For the next week, the water in Galveston Bay remained fresh. Demeaned water level comparisons are given in columns 1 and 3 in Table 8.4.4. Rms errors are order 10 cm with the relative errors order 0.05 indicating excellent agreement in shape characteristics. The water level response at Morgans Point shows the success of the model in capturing the major effects of the storm. The model over-prediction of water level may be due to the lack of a overland/marsh flooding algorithm. The present scheme will only allow an originally wet cell to dry and then subsequently wet. The flux-corrected salinity response at Port Bolivar is in excellent agreement with Texas Water Development Board observations. The simulated salinity response matches the observed abrupt decrease of order 28 PSU in salinity on 19 October. The excellent replication of the advection of the large horizontal salinity gradient over the last ten days of October is also to be noted. The flux corrected salinity scheme exhibits no under or overshooting and positivity.

To test the drying/wetting algorithm, the January 1995 period was considered, during which several storms occurred and portions of Trinity Bay near Round Point dried. Simulated water level responses at Round Point are in general agreement with the observations except that no clipping occurs and hence there is no loss of low water as cell width flow reduction factors remain greater than zero. Additional knowledge on the spatial extent of the wetting/drying region is needed to further verify the drying/wetting scheme. Demeaned simulated water levels throughout the bay are compared with observations for 1/95 in Table 8.4.4 and are less than 10 cm in rms and 0.05 in relative error.

To further resolve the currents within the Houston Ship Channel (HSC) in the Port of Houston, a refined channel model (HSCM) with order twice the horizontal resolution of the GBM was developed in three sections based on the Wilken (1988) elliptic grid generation program patterned after Ives and Zacharais (1987). Each grid section was linked to preserve orthogonality at the matching sections of the final composite channel grid (see inset solid region shown in Figure 8.4.3) consisting of 71 x 211 horizontal cells (dx=63-1007 m, dy=133-1268 m) with the same 5 sigma levels as in the GBM. In both models, bathymetry is based on historical hydrographic surveys (NGDC, 1988). However, the HSC bathymetry was incorporated into the HSCM grid based on Corps of Engineers channel survey data as given on nautical charts.

Table 8.4.4 Demeaned Water Level GBM vs Data Intercomparisions

	RMS (cm)		Relative Error (-)	
Station	(10/94)	(1/95)	(10/94)	(1/95)
Galveston Pleasure Pier	8	8	0.04	0.02
Galveston Pier 21	5	6	0.03	0.02
Morgans Point	9	7	0.04	0.02
Clear Lake	8	7	0.04	0.02
Eagle Point	8	8	0.07	0.03
Port Bolivar	11	10	0.11	0.06
Round Point	-	9	-	0.03

The two models were then nested in a one-way coupling scheme, wherein GBM water surface elevation, salinity, temperature, turbulent kinetic energy, and turbulent length scale time histories were saved at 6-minute intervals to provide boundary conditions to drive the HSCM. For salinity, temperature, turbulent kinetic energy, and turbulent length scale, a one-dimensional (normal to the boundary) advection equation is used. On inflow GBM values are advected into the HSCM domain, while on outflow HSCM internal values are advected through the boundary. Open lateral boundary coupling is accomplished by: 1) specifying open boundary cells on HSC, 2)locating the corresponding (nearest neighbor) open boundary cells on the Galveston Bay grid, 3) determining HSCM initial conditions via nearest neighbor on Galveston Bay grid and vertical sigma (GBM) - depth - sigma (HSCM) interpolation. Lateral flow boundary coupling is achieved by specifying river inflow cells on the HSCM grid and by using the corresponding flow and salinity and temperature boundary signals. Inflows and salinity and temperature boundary conditions are the same for Buffalo Bayou and San Jacinto Rivers, while the Trinity River is not included in HSCM. Surface boundary coupling is accomplished by placing the SST field on HSCM grid via nearest neighbor interpolation from the GBM grid. Wind and atmospheric pressure fields are directly determined on HSCM grid via 2-step Barnes (1973) interpolation.

During April 1996, PORTS current meters were in test operation and this period was used to evaluate both water level and current response. April 1996 demeaned water level comparisons are presented in Table 8.4.5 for the GBM and HSCM. Simulated water levels are nearly identical in each model, indicating that the coupling mechanics and grid topologies (grid structure and bathymetry) are compatible. Rms errors are order 5 cm and relative errors less than 0.05.

Table 8.4.5 Demeaned Water Level GBM and HSCM Model vs April 1996 Intercomparisions

	RMS (cm)		Relative Error (-)	
Station	(GBM)	(HSCM)	(GBM)	(HSCM)
Galveston Pleasure Pier	9	-	0.06	-
Galveston Pier 21	7	8	0.06	0.06
Morgans Point	5	5	0.02	0.02
Clear Lake	7	-	0.04	-
Eagle Point	5	5	0.02	0.02
Port Bolivar	7	7	0.06	0.06
Lynchburg Landing	6	6	0.02	0.02
Manchester Dock	7	8	0.02	0.03

Level 3 simulated principal direction currents are directly compared with observed currents in columns 1 and 3 and in 2 and 4 of Table 8.4.6 for the GBM and HSCM, respectively. Both models produce excellent agreement in shape with GBM currents closer in agreement with observations near the entrance and HSCM results improved over GBM results at Morgans Point. Due to data quality issues, Redfish Bar currents were not considered.

29-day harmonic and principal flood direction analyses analogous to those previously discussed for the May 1995 simulation were performed and are given in lines 2 and 3 in Tables 8.4.1-8.4.3 for the GBM and HSCM, respectively. Note the improvement in the current response at Morgans Point in the HSCM over that achieved in the GBM. Salinity and temperature rms errors were comparable in both models and ranged from 1-4 PSU and were order 1.5 °C, respectively.

Table 8.4.6 Principal Direction Current at Prediction Depth GBM and HSCM vs April 1996 Intercomparisions

	RMS (cm/s)		Relative Error (-)	
Station	(GBM)	(HSCM)	(GBM)	(HSCM)
Bolivar Roads	21	26	0.04	0.07
Morgans Point	17	13	0.34	0.11

Further details may be found in Schmalz (2000a; 2001).

8.4.2.7 Nowcast/Forecast System Description

The nowcast/forecast system consists of a data acquisition system in conjunction with a hydrodynamic model input preparation and output analysis system. The data delivery system consists of an Semi-Operational Data Acquisition and Archival System (ODAAS) maintained by the NOS Coast Survey Development Laboratory (CSDL) in which NWS AVN wind/pressure fields are automatically downloaded to CSDL machines. Additional scripts decode NWS Techniques Development Laboratory (TDL) storm surge water levels at Galveston Pleasure Pier. The NWS Western Gulf River Forecast Center (WGRFC) uploads to CSDL anonymous ftp, three day 6-hour interval forecasted river flow and stage for the Trinity River at Liberty, Texas and Lake Houston Dam near Sheldon, Texas, respectively. In addition, the previous day's hourly discharges at Liberty, Texas on the Trinity River and at Piney Point, Texas on Buffalo Bayou and stage for Lake Houston Dam near Sheldon, Texas are uploaded. A decode script accesses and decodes either the Houston/Galveston PORTS screen (see Figure 8.4.2) or PORTS Universal Flat File Format (PUFFF) files every 6 minutes and stores daily station files.

An initial design of a nowcasting/forecasting system using the above ODAAS has been completed. The design concept is modular such that refined hydrodynamic models can be readily substituted for the initial models. To this end, a separate nowcast/forecast program has been developed to establish hydrodynamic model forecast inputs. The program utilizes the following ten step procedure:

1) Setup 24 hour nowcast and 36 hour forecast time periods,
2) Predict astronomical tide,

3) Predict astronomical currents,
4) Read PORTS screen or PUFFF files and develop station time series,
5) Develop GBM subtidal water level signal,
6) Assimilate PORTS salinity and temperature data into GBM and HSCM initial conditions,
7) Establish GBM and HSCM salinity and temperature boundary conditions,
8) Establish GBM and HSCM SST forcings,
9) Establish USGS observed and NWS/WGRFC forecast freshwater inflows, and
10) Establish PORTS based and NWS/AVN wind and pressure fields.

Details of the above steps may be found in Schmalz (1998b). A one month demonstration using the PORTS Screen as input was performed over September 1997 in hindcast mode (see Schmalz, 1998c). Since the forecasts were performed after the fact, forecast results could be directly assessed against the observations. Due to transmission problems the PORTS screens were not reliably obtained on a six-minute basis, and an hourly sampling interval was used for the nowcasts. Water level comparisons were made by demeaning both model and observations and were not placed on a MLLW datum. Rms water level errors were order 10 cm during the nowcast and order 15 cm over the forecast periods. In an effort to improve reliability, the PUFFF files were accessed directly along with the instrument control files. Using this approach, the system data interval was decreased from one hour to six minutes. In addition, all water levels are reported on station MLLW datum during a second one-month demonstration over 9 April - 9 May, 1999 as described in Schmalz (2000b).

8.4.2.8 Nowcast/Forecast System Hindcast Experiment Improvements

An operational test bed was created by saving selected initial condition and restart files and by rerunning the nowcast/forecast system in hindcast mode. The initial condition files represent a cold start from climatological density initial conditions, which are adjusted for the given nowcast/forecast cycle and are used to reset the density structure.

The tidal dynamics were initially tested for the 14 December 1998 nowcast/forecast cycle and were rechecked for the 20 April nowcast/forecast cycle. The density structure was reset in each case. Forecast water levels at Galveston Pleasure Pier, while in excellent agreement for 14 December 1998 forecast cycle, were offset by approximately 13 cm from the 20 April forecast cycle predictions. The long period harmonic constituents Sa and Ssa were adjusted along the GBM open boundary to be consistent with Galveston Pleasure Pier values , the order 3 cm open boundary water level offsets were set to zero, and both test nowcast/forecast cycles were repeated. With these changes excellent agreement of both forecast cycle results against predictions is now achieved.

The restart file was saved for the 26 April nowcast/forecast cycle during which large oscillations in the water levels primarily during the nowcast cycle occurred. In addition, excessive stratification at Morgans Point was evidenced during the forecast cycle. It was also noted that the wind directions switched dramatically from nowcast to forecast. Several hindcast experiments were conducted to address these issues and resulted in the following modifications:

1) A revised procedure to eliminate excess stratification was used to reinitialize the initial temperature fields, which were then readjusted based on the PORTS measurements. The excessive temperature stratification at Morgans Point in the forecast no longer occurs.

2) A three hour moving average was used to smooth the nowcast and forecast water level residuals. The excessive water level oscillations at Galveston Pleasure Pier in the nowcast are removed.

3) NWS/AVN wind strengths were used directly without overwater correction.

4) Bottom roughness was reduced from z_0 =2 mm to z_0=1.5 mm.

5) NOS 1988 hydrographic survey results were used to update model bathymetries outside the navigation channels primarily within the Galveston Bay entrance via a general bilinear interpolation algorithm obtained from Sheinin (1999). Maximum changes in depth with respect to MLLW from the previous bathymetry were limited to 1.0 meter.

Improvements in water level are 0-1 cm, while prediction depth current improvements are modest at 2-3 cm/s. The bottom roughness, z_0=1.5 mm, corresponds to values of a sand bed (1 - 10mm) reported by Black (1987). Recent nowcast/forecast results over February 2000 are shown in Figures 8.4.4 and 8.4.5 for water levels and in Figures 8.4.6 and 8.4.7 for prediction depth currents at Bolivar Roads (just inside the Entrance) and at Morgans Point (Upper bay), respectively.

8.4.3 System Validation

Herein, we focus on the validation of the system over the one year period April 2000 through March 2001. Each 24 hour nowcast is concatenated, while the first 24 hours of the 36 hour forecast are concatenated to form the nowcast and forecast yearly time series, which are evaluated based on the NOS (1999) formal procedures. These are initially discussed in terms of three statistical measure sets followed by the results for water levels, principal component direction currents at prediction depth (4.7 m below MLLW), and near surface salinity and temperature. Next critical water level events are defined and evaluated. Critical principal component direction currents event assessment is presently under review. A physical interpretation of the statistical evaluation is discussed followed by conclusions and recommendations for improvement.

8.4.3.1 Statistical Measures

Three statistical measure sets are used in the evaluation with error equal to model prediction minus observation. While the formal NOS (1999) targets are emphasized, informal targets for several standard statistics not used in the formal procedures are also presented. Note at present one target is given for both nowcast and forecast. Each measure set is described in turn below.

Statistical Measure Set 1: The first set of measures involves the hydrodynamic quantities of water level, principal component direction currents at prediction depth, near surface temperature, and near surface salinity. Statistical error measures are based on a six-minute data interval.

CF=Central frequency associated with a given reference level (15 cm, 26 cm/s). *The NOS formal target is to exceed 0.9.*

POF=Positive outlier frequency associated with a given positive reference level (30 cm, 52 cm/s). *The NOS formal target is to be less than 0.01.*

NOF=Negative outlier frequency associated with a given negative reference level (-30 cm, -52 cm/s). *The NOS formal target is to be less than 0.01.*

RMSE = Root mean square error. *Informal NOS targets for water levels, prediction depth current strengths, surface salinity, and surface temperature are 10 cm, 20 cm/s, 2 PSU, and 2 ^{o}C, respectively.*

WILLMOTT RE= Willmott dimensionless relative average error in the range (0-1), where 0 corresponds to no error. Refer to Willmott et al. (1985). *The NOS informal target is to be under 0.05.*

BIAS=Intercept, b, of the linear regression of the model on the observed data, y=mx + b, where y corresponds to the observation and x the model series. *Informal NOS targets for water levels, prediction depth current strengths, surface salinity, and surface temperature are +/-5 cm, +/-10 cm/s, +/-1 PSU, and +/-1 ^{o}C, respectively.*

GAIN=Slope, m, of the linear regression of the model on the observed data, y=mx + b, where y corresponds to the observation and x the model series. *The NOS informal target is to be in the range of 0.9 to 1.1.*

STD ERROR=Standard error of the linear regression assuming a bivariate normal distribution for the observed data and model series. *Informal NOS targets for water levels, prediction depth current strengths, surface salinity, and surface temperature are 10 cm, 20 cm/s, 2 PSU, and 2 ^{o}C, respectively.*

CC=Linear correlation coefficient. *The NOS informal target is to exceed 0.9.*

Statistical Measure Set 2: The following outlier duration and frequency measures are used for water levels and principal component direction currents at prediction depth. Salinity and temperature are not included.

MDPO= Maximum Duration of Positive Outliers, equal to the number of consecutive hours during which the error (model - obs) exceeds a given positive reference level (30 cm, 52 cm/s). *The NOS formal target is to be less than 24.*

MDNO= Maximum Duration of Negative Outliers, equal to the number of consecutive hours during which the error (model - obs) is less than a given negative reference level (-30 cm, -52 cm/s). *The NOS formal target is to be less than 24.*

WOF=Worst Case Outlier Frequency associated with the occurrence of when the error magnitude exceeds a given reference level (30 cm, 52 cm/s) and that (1) the model prediction exceeds the tidal prediction and the observation is less than the tidal prediction or (2) the opposite circumstance to (1) holds. *The NOS formal target is to be less than 0.005.*

NFUF=Nowcast/Forecast Utility Frequency, conditioned on the observation differing from the tidal prediction by a given reference level (5 cm, 26 cm/s), of occurrence when the model prediction is closer to the observation than the tidal prediction. *The NOS informal target is to exceed 0.80.*

Statistical Measure Set 3: The third set of measures involves the difference in times between the observed and model predictions for zero crossing times, times of maxima, and times of minima associated with water levels and principal component direction currents at prediction depth (flood is considered positive). Salinity and temperature are not included. Outlier frequencies were not explicitly considered.

CF=Central frequency associated with a given reference time (here considered as 30 min for zero crossing and extrema for both water levels and principal component prediction depth currents). *The NOS formal target is to exceed 0.9.*

Associated with the times of maxima and minima are the levels of the maxima and minima themselves. The corresponding central frequency measure of the error, defined as model prediction minus observation, was considered as in statistical measure set 1 above.

8.4.3.2 Nowcast/Forecast Results

Water Level: Statistical measure set 1 results are given in Table 8.4.7. RMSEs increase from nowcast to forecast from 7.4 to 10.9 cm at Galveston Pleasure Pier in the GBM and from 8.9 to 12.6 cm at Morgans Point in the HSCM. All informal and formal NOS targets are met by the nowcast with slightly degraded results obtained for the forecast. Statistical measure set 2 results are shown in Table 8.4.8. All NOS formal targets are met at each water level station in both models. Forecast results are slightly degraded from those obtained during the nowcast. Statistical measure set 3 results are given in Table 8.4.9. Due to the flatness of the water level curves, an interval based method was used to compute the time of zero crossings and water level extrema. The zero crossing time targets are not met at Eagle Point. Extrema time criteria are not met at any of the stations. Forecast results are slightly degraded from nowcast results. Minimum and maximum levels nearly meet the criteria. Maximum levels are generally improved over minimum levels at most stations.

Table 8.4.7 Water Level (cm) Analysis April 2000 -March 2001: Statistical Measure Set 1. Note GPP = GBM results at Galveston Pleasure Pier, P21=GBM results at Galveston Pier 21, BR = GBM results at Bolivar Roads (Port Bolivar), EP=HSCM results at Eagle Point, and MP=HSCM results at Morgans Point. NOS informal targets in italics with formal targets non-italicized.

Statistical Measure	Nowcast					Forecast (1-24h)				
[NOS Target Value]	GPP	P21	BR	EP	MP	GPP	P21	BR	EP	MP
RMSE *[10cm]*	7.4	7.3	7.6	6.3	8.9	10.9	10.1	10.3	9.2	12.6
Willmott Relative Error *[<0.05]*	0.021	0.030	0.034	0.026	0.041	0.046	0.056	0.061	0.053	0.075
Bias *[+/-5 cm]*	-3.5	-3.7	-1.1	1.5	-5.5	-2.9	-3.3	-1.2	-0.4	-7.5
Gain *[0.9-1.1]*	0.978	1.01	1.005	0.995	1.056	0.892	0.913	0.908	0.862	0.924
Standard Error *[10cm]*	6.5	6.3	7.5	6.1	6.2	10.2	9.3	10.1	8.8	9.8
Correlation Coefficient *[>0.9]*	0.967	0.958	0.938	0.953	0.954	0.917	0.903	0.885	0.899	0.907
CF (15 cm) [>0.9]	0.966	0.966	0.973	0.975	0.945	0.888	0.914	0.919	0.919	0.832
NOF (30 cm) [<0.01]	0.002	0.000	0.002	0.001	0.001	0.002	0.002	0.003	0.001	0.001
POF (30 cm) [<0.01]	0.003	0.002	0.004	0.002	0.007	0.015	0.013	0.015	0.010	0.021

Table 8.4.8 Water Level (cm) Analysis April 2000 -March 2001: Statistical Measure Set 2.

Statistical Measure	Nowcast					Forecast (1-24h)				
[NOS Target Value]	GPP	P21	BR	EP	MP	GPP	P21	BR	EP	MP
MDPO (30 cm) [<24]	6	5	6	6	10	19	19	19	18	16
MDNO (30 cm) [<24]	3	1	3	2	3	15	7	7	2	4
WOF (30 cm) [<0.005]	0.002	0.001	0.004	0.001	0.004	0.014	0.011	0.011	0.006	0.012
NFUF (5 cm) *[>0.8]*	0.884	0.873	0.912	0.940	0.886	0.786	0.797	0.831	0.866	0.759

Table 8.4.9 Water Level (cm) Analysis April 2000 -March 2001: Statistical Measure Set 3.

Statistical Measure [NOS Target Value]	Nowcast GPP	P21	BR	EP	MP	Forecast (1-24h) GPP	P21	BR	EP	MP
Zero Time: CF (30 min) [>0.9]	0.856	0.834	0.768	0.939	0.565	0.532	0.542	0.555	0.655	0.556
Min Time: CF (30 min) [>0.9]	0.695	0.672	0.748	0.710	0.587	0.702	0.646	0.728	0.643	0.529
Min Level: CF (15 cm) [>0.9]	0.895	0.860	0.831	0.874	0.799	0.809	0.806	0.762	0.836	0.739
Max Time: CF (30 min) [>0.9]	0.801	0.724	0.757	0.716	0.787	0.792	0.701	0.681	0.676	0.708
Max Level: CF (15cm) [>0.9]	0.921	0.935	0.909	0.860	0.833	0.855	0.877	0.852	0.793	0.792

Principal Component Direction Current: Statistical measure set 1 results for principal component direction prediction depth currents at Bolivar Roads from the GBM and at Morgans Point from the HSCM are presented in Table 8.4.10. Note in the analysis, sigma level 1 model cell centered currents were used and the flood direction is positive. While the RMSE is order 20 cm/s at both locations, the peak current strengths decrease from order 100 cm/s at Bolivar Roads to order 50 cm/s at Morgans Point. NOS targets are met at Bolivar Roads and are not met at Morgans Point. Note the ADCP at Morgans Point was not in service during most of August 2000 due to a cabling problem. Observed current strengths after the redeployment remained at levels seen in May 2000, while model currents did not change even after increasing the channel depths to the new project design values (from 12.2 m to 13.7 m below MLLW). Efforts to adjust the nowcast and forecast currents at Morgans Point are presently under development. Statistical measure sets 2 and 3 are given in Tables 8.4.11 and 8.4.12, respectively. Results at Bolivar Roads are more accurate than those obtained at Morgans Point, where ebb current strengths on nowcast and forecast are reduced from observations

Surface Temperature: Near surface temperature nowcast/forecast results are presented in Table 8.4.13. Since a SST specification based on PORTS observations is used in both models, and persistence of the SST is made for the forecast, nowcast and forecast errors tend to be similar in both models. RMSE is order 1.5 °C at Bolivar Roads and 1.0 °C at Morgans Point. While no NOS criteria have been adopted, the CF(2 °C) > 0.9 informal criterion is met.

Surface Salinity: Near surface salinity nowcast/forecast results are presented in Table 8.4.14. Both nowcast and forecast results are similar with an RMSE of order 2.5 PSU at Bolivar Roads and at Morgans Point. No formal NOS criteria have been adopted for surface salinity. The CF (2 PSU) > 0.9 informal criterion is not met.

Table 8.4.10. Principal Component Direction (4.7m) Current (cm/s) Analysis April 2000 -March 2001: Statistical Measure Set 1.BR=GBM results at Bolivar Roads and MP=HSCM results at Morgans Point. NOS informal targets italicized while formal targets are non-italicized.

Statistical Measure [NOS Target Value]	Nowcast		Forecast (1-24h)		Persistence Forecast	
	BR	MP				MP
RMSE *[20 cm/s]*	20.3	27.4	22.3	24.6	33.1	24.4
Willmott Relative Error *[<0.05]*	0.053	0.320	0.064	0.285	0.124	0.268
Bias *[+/-10 cm/s*]	-6.8	-16.9	-6.4	-13.1	-1.7	-6.8
Gain *[0.9-1.1]*	1.085	0.942	1.043	0.986	0.747	0.676
Standard Error *[20 cm/s*]	18.9	21.2	21.4	20.8	30.6	22.5
Correlation Coefficient *[>0.9]*	0.920	0.626	0.896	0.641	0.772	0.561
CF (26 cm/s) [>0.9]	0.844	0.646	0.817	0.720	0.598	0.727
NOF (52cm/s)[<0.01]	0.005	0.000	0.004	0.001	0.052	0.009
POF (52 cm/s) [<0.01]	0.011	0.036	0.025	0.025	0.058	0.023

Table 8.4.11. Principal Component Direction (4.7m) Current (cm/s) Analysis April 2000 -March 2001: Statistical Measure Set 2.

Statistical Measure [NOS Target Value]	Nowcast		Forecast (1-24h)		Persistence Forecast	
	BR	MP	BR	MP	BR	MP
MDPO (104 cm/s) [<24]	4	6	7	0	6	0
MDNO (104 cm/s) [<24]	2	0	0	0	2	0
WOF (52 cm/s) [<0.005]	0.009	0.035	0.019	0.026	0.063	0.028
NFUF (26 cm/s) *[>0.8]*	0.919	0.299	0.904	0.390	0.729	0.626

Table 8.4.12. Principal Component Direction (4.7m) Current (cm/s) Analysis April 2000 -March 2001: Statistical Measure Set 3.
Note flood corresponds to maximum and ebb corresponds to minimum levels.

Statistical Measure [NOS Target Value]	Nowcast BR	Nowcast MP	Forecast (1-24h) BR	Forecast (1-24h) MP	Persistence Forecast BR	Persistence Forecast MP
Zero Crossing Time: CF (30 min) [>0.9]	0.528	0.590	0.533	0.586	0.565	0.940
Min Time: CF (30 min) [>0.9]	0.559	0.170	0.556	0.312	0.516	0.351
Min Level: CF (26 cm/s) [>0.9]	0.711	0.500	0.703	0.554	0.666	0.683
Max Time: CF (30 min) [>0.9]	0.852	0.547	0.845	0.609	0.653	0.426
Max Level: CF (26 cm/s) [>0.9]	0.902	0.919	0.864	0.857	0.662	0.814

Table 8.4.13. Surface Temperature (°C) Analysis
April 2000 -March 2001: Statistical Measure Set 1.
GPP=GBM results at Galveston Pleasure Pier, BR=GBM results at Bolivar Roads, EP=HSCM results at Eagle Point, and MP=HSCM results at Morgans Point. NOS informal targets italicized.

Statistical Measure [NOS Target Value]	Nowcast GPP	Nowcast BR	Nowcast EP	Nowcast MP	Forecast (1-24h) GPP	Forecast (1-24h) BR	Forecast (1-24h) EP	Forecast (1-24h) MP
RMSE *[+/-2 °C]*	1.14	1.32	1.66	0.967	1.25	1.25	1.52	1.02
Willmott Relative Error *[<0.05]*	0.007	0.040	0.078	0.025	0.008	0.036	0.072	0.028
Bias *[1 °C]*	-0.963	-0.523	0.664	1.19	-1.75	-0.235	0.164	-1.24
Gain *[0.9-1.1]*	1.009	0.980	0.909	1.024	1.042	0.973	0.941	1.028
Standard Error *[2 °C]*	0.858	0.787	1.00	0.82	0.94	0.80	1.05	0.92
Correlation Coefficient *[>0.9]*	0.993	0.972	0.937	0.966	0.991	0.970	0.928	0.958
CF (2 °C) [>0.9]	0.936	0.896	0.803	0.972	0.906	0.929	0.883	0.961

Table 8.4.14. Surface Salinity (PSU) Analysis
April 2000 -March 2001: Statistical Measure Set 1.
BR=GBM results at Bolivar Roads, EP=HSCM results at Eagle Point, and MP=HSCM results at Morgans Point. NOS informal targets italicized.

Statistical Measure [NOS Target Value]	Nowcast BR	Nowcast EP	Nowcast MP	Forecast (1-24h) BR	Forecast (1-24h) EP	Forecast (1-24h) MP
RMS *[+/-2 PSU]*	2.56	1.78	2.47	2.78	2.01	2.74
Willmott Relative Error *[<0.05]*	0.043	0.083	0.070	0.052	0.140	0.091
Bias *[1 PSU]*	0.07	1.21	-1.10	-0.683	1.62	-1.51
Gain *[0.9-1.1]*	0.966	0.826	1.052	0.985	0.765	1.102
Standard Error *[2 PSU]*	2.48	1.61	2.39	2.60	1.77	2.66
Correlation Coefficient *[>0.9]*	0.922	0.854	0.887	0.914	0.769	0.861
CF (2 PSU) [>0.9]	0.604	0.807	0.833	0.563	0.767	0.760

8.4.3.3 Forecast Age Assessment

Here we investigate the quality of the water level and principal component direction current forecasts by forecast hour. Three forecast types are considered: 1) model forecast, 2) astronomical tide plus persistence forecast, and 3) astronomical tide forecast. Forecast type 2 is obtained by using a persistence of the initial nontidal water level or nontidal principal component direction current over the entire 24 hour forecast. The first 24 hours of the 36 hour model forecast period are used for the evaluation.

Water Level: Statistics are given by forecast age in hours for water levels at Galveston Pleasure Pier for the GBM in Table 8.4.15 and at Morgans Point for the HSCM in Table 8.4.16, respectively. The error is equal to model prediction minus observation with the signal mean, SM, and RMS error, RMSE, given as informal statistics. A level of 15 cm is used to determine the central frequency, CF, with 30 cm levels used to determine the negative outlier frequency, NOF, and positive outlier frequency, POF. The number of consecutive forecast hours times 24 (forecast interval in hours) associated with each hour during which the model exceeded 30 cm, MDPO or was less than the observation by more than 30 cm, MDNO, are also given. The worst case outlier frequencies, WOF, associated with a reference level of 30 cm and the nowcast/forecast utility frequency using a 5 cm reference level, NFUF, are given as well. Note at all of the water level stations, the quality of the forecast does not degrade significantly over the 24 hour period. In addition, the model forecast is superior to the other two forecast types at all stations.

Table 8.4.15. Galveston Pleasure Pier Water Level Forecast
April 2000 -March 2001 Evaluation Statistics. Note the reference level is 15 cm.

GBM Forecast

HR	SM (M)	RMSE (M)	CF	NOF	POF	MDPO (HR)	MDNO (HR)	WOF	NFUF
1	0.04	0.09	0.94	0.00	0.01	19	0	0.01	0.85
6	0.05	0.12	0.85	0.00	0.02	12	0	0.01	0.75
12	0.03	0.11	0.92	0.00	0.01	24	24	0.02	0.81
18	0.00	0.11	0.87	0.00	0.01	24	24	0.01	0.75
23	0.02	0.11	0.85	0.00	0.01	22	22	0.02	0.76

Astronomical plus Persistence Forecast

HR	SM (M)	RMSE (M)	CF	NOF	POF	MDPO (HR)	MDNO (HR)	WOF	NFUF
1	0.01	0.12	0.87	0.01	0.03	19	19	0.03	0.80
6	0.00	0.14	0.85	0.02	0.03	10	10	0.03	0.80
12	0.01	0.10	0.91	0.00	0.02	24	0	0.00	0.86
18	0.00	0.11	0.86	0.01	0.02	24	24	0.01	0.79
23	0.02	0.11	0.87	0.01	0.02	22	19	0.02	0.82

Astronomical Forecast

HR	SM (M)	RMSE (M)	CF	NOF	POF	MDPO (HR)	MDNO (HR)	WOF	NFUF
1	0.04	0.15	0.76	0.01	0.05	48	19	-	-
6	0.05	0.16	0.73	0.02	0.04	34	10	-	-
12	0.03	0.15	0.76	0.00	0.03	53	24	-	-
18	0.04	0.14	0.77	0.01	0.03	48	24	-	-
23	0.03	0.15	0.74	0.01	0.04	62	22	-	-

Table 8.4.16. Morgans Point Water Level Forecast
April 2000 - March 2001 Evaluation Statistics. Note the reference level is 15 cm.

HSCM Forecast

HR	SM (M)	RMSE (M)	CF	NOF	POF	MDPO (HR)	MDNO (HR)	WOF	NFUF
1	0.03	0.08	0.95	0.00	0.00	19	19	0.00	0.90
6	0.08	0.12	0.89	0.00	0.01	22	0	0.01	0.80
12	0.10	0.14	0.76	0.00	0.02	24	19	0.02	0.69
18	0.09	0.14	0.79	0.00	0.03	24	0	0.02	0.70
23	0.04	0.12	0.83	0.00	0.02	22	0	0.02	0.77

Astronomical plus Persistence Forecast

HR	SM (M)	RMSE (M)	CF	NOF	POF	MDPO (HR)	MDNO (HR)	WOF	NFUF
1	0.00	0.13	0.85	0.02	0.02	19	19	0.03	0.79
6	0.01	0.17	0.77	0.03	0.05	12	12	0.05	0.74
12	0.00	0.12	0.85	0.01	0.02	24	24	0.02	0.81
18	0.00	0.14	0.82	0.03	0.03	24	24	0.03	0.79
23	0.00	0.13	0.85	0.03	0.02	22	22	0.03	0.80

Astronomical Forecast

HR	SM (M)	RMSE (M)	CF	NOF	POF	MDPO (HR)	MDNO (HR)	WOF	NFUF
1	0.04	0.17	0.70	0.02	0.06	38	38	-	-
6	0.04	0.18	0.67	0.02	0.07	26	12	-	-
12	0.04	0.19	0.64	0.03	0.06	50	48	-	-
18	0.04	0.19	0.64	0.03	0.07	70	58	-	-
23	0.03	0.18	0.66	0.03	0.05	38	62	-	-

Principal Component Direction Current: Statistics are given by forecast age in hours for principal component direction currents at Bolivar Roads for the GBM in Table 8.4.17 and at Morgans Point for the HSCM in Table 8.4.18, respectively. The error is equal to model minus observation with the signal mean, SM, and RMS error, RMSE, given as informal statistics. A speed of 26 cm/s is used to determine the central frequency, CF, with 52 cm/s levels used to determine the negative outlier frequency, NOF, and positive outlier frequency, POF. The number of consecutive forecast hours times 24 (forecast interval in hours) associated with each hour during which the model exceeded the observation by more than 104 cm/s, MDPO, or was less than the observation by more than 104 cm/s, MDNO, are also given. The worst case outlier frequencies, WOF, associated with a reference level of 52 cm/s and the nowcast/forecast utility frequency using a 10 cm/s reference level, NFUF, are provided as well. Note at both current stations, the quality of the forecast does not degrade significantly over the 24 hour period. At Bolivar Roads, in the GBM, the model forecast is superior to the other two forecast types. Note this is not the case at Morgans Point, in the HSCM, due possibly to the omission of freshwater inflows as later discussed.

Table 8.4.17. Bolivar Roads Principal Component Direction Currents Forecast April 2000 -March 2001 Evaluation Statistics. Note the reference level is 26 cm/s.

GBM Forecast

HR	SM (M/S)	RMSE (M/S)	CF	NOF	POF	MDPO (HR)	MDNO (HR)	WOF	NFUF
1	0.06	0.19	0.89	0.00	0.02	0	0	0.00	0.97
6	0.15	0.26	0.72	0.00	0.04	0	0	0.03	0.90
12	0.08	0.24	0.80	0.00	0.03	24	0	0.02	0.87
18	0.01	0.19	0.87	0.01	0.00	0	0	0.00	0.94
23	-0.01	0.20	0.87	0.02	0.02	22	0	0.01	0.90

Astronomical plus Persistence Forecast

HR	SM (M/S)	RMSE (M/S)	CF	NOF	POF	MDPO (HR)	MDNO (HR)	WOF	NFUF
1	0.00	0.31	0.62	0.05	0.04	0	5	0.05	0.74
6	0.10	0.34	0.57	0.04	0.10	0	2	0.10	0.75
12	0.04	0.34	0.60	0.05	0.07	24	5	0.07	0.68
18	-0.03	0.32	0.60	0.05	0.03	24	5	0.03	0.71
23	-0.06	0.33	0.64	0.07	0.03	0	22	0.07	0.76

Astronomical Forecast

HR	SM (M/S)	RMSE (M/S)	CF	NOF	POF	MDPO (HR)	MDNO (HR)	WOF	NFUF
1	0.03	0.33	0.64	0.02	0.08	19	0	-	-
6	0.02	0.30	0.71	0.03	0.04	12	0	-	-
12	0.08	0.30	0.66	0.03	0.07	0	0	-	-
18	0.02	0.28	0.68	0.03	0.03	0	0	-	-
23	0.05	0.31	0.68	0.01	0.08	22	0	-	-

Table 8.4.18 Morgans Point Principal Component Direction Current Forecast April 2000 -March 2001 Evaluation Statistics. Note the reference level is 26 cm/s.

HSCM Forecast

HR	SM (M/S)	RMSE (M/S)	CF	NOF	POF	MDPO (HR)	MDNO (HR)	WOF	NFUF
1	0.13	0.24	0.72	0.00	0.02	0	0	0.02	0.35
6	0.15	0.25	0.74	0.00	0.04	0	0	0.04	0.25
12	0.12	0.25	0.72	0.01	0.02	0	0	0.02	0.38
18	0.16	0.27	0.63	0.00	0.03	0	0	0.03	0.29
23	0.09	0.22	0.78	0.00	0.01	0	0	0.01	0.52

Astronomical plus Persistence Forecast

HR	SM (M/S)	RMSE (M/S)	CF	NOF	POF	MDPO (HR)	MDNO (HR)	WOF	NFUF
1	0.06	0.25	0.71	0.01	0.03	0	0	0.04	0.49
6	0.07	0.24	0.79	0.00	0.03	0	0	0.03	0.64
12	0.05	0.25	0.69	0.01	0.02	0	0	0.03	0.62
18	0.09	0.27	0.66	0.01	0.02	0	0	0.02	0.55
23	0.02	0.24	0.71	0.00	0.02	0	0	0.02	0.57

Astronomical Forecast

HR	SM (M/S)	RMSE (M/S)	CF	NOF	POF	MDPO (HR)	MDNO (HR)	WOF	NFUF
1	0.10	0.18	0.87	0.00	0.01	0	0	-	-
6	0.08	0.17	0.89	0.00	0.01	0	0	-	-
12	0.08	0.14	0.90	0.00	0.01	0	0	-	-
18	0.08	0.16	0.93	0.00	0.01	2	0	-	-
23	0.10	0.19	0.87	0.00	0.01	0	0	-	-

8.4.3.4 Water Level Event Evaluation

The experimental nowcast/forecast system is evaluated with respect to both high and low water level events over the formal evaluation period April 2000 through March 2001. A high water level event is considered to occur when observed, nowcast, forecast, or astronomical tide predicted water levels exceed MHHW by a certain level. A low water level event is considered to occur when observed, nowcast, forecast, or astronomical tide predicted water levels fall below MLLW by a certain level. A 10 cm level was used for both high and low water events. A success is defined as the joint occurrence of observed and nowcast or forecast or astronomical tide predicted water level events; e.g., the model or astronomical tide predicts the event. A failure is defined as the occurrence of an observed water level event which is not present in the nowcast or forecast or astronomical tide predicted water levels. A false alarm is the opposite condition of a failure.

Of concern is the ability of the nowcast and forecasts to reproduce the low water events associated with northerly winds following winter cold frontal passages. The nowcast results are contrasted with the observations and the astronomical tide predictions at Morgans Point in Figure 8.4.8. We note the ability of the nowcasts to reproduce the low water events on 20 and 21 January. The forecast results at Morgans Point are given in Figure 8.4.9. The adjusted forecast, y, is determined as y =mx + b, where x is the unadjusted forecast in the top panel, and m=1.05 with b=-7.5 cm. We note the excellent ability of the adjusted forecast in predicting the 20-21 January low water event.

In addition, we seek the number of months in the nowcast, forecast, and tide prediction time series in which the number of successes is greater than the sum of the number of failures and false alarms. We divide by 12, the total number of months in the evaluation period, to obtain a overall skill score as given in Table 8.4.19. Note with respect to this measure as well, the nowcast, forecast, and adjusted forecasts are all superior to the tide prediction.

8.4.3.5 Physical Interpretation of the Statistical Evaluation

With respect to water levels, the greatest discrepancy appears to be in the timing of zero crossing, high, and low waters. Based on a 29-day harmonic analysis of the April 1996 calibration (Schmalz, 2001), the GBM agrees in phase with the tides by order 6 minutes through the Entrance, by mid Bay at Eagle Point the phase lag is order 60 minutes, and at Morgans Point the phase lag is reduced to order 1 minute. The amplitude response is damped by order 10% throughout the Bay. RMS errors based on the 24 constituent amplitude and phase differences are estimated to be order 3 cm or less than 10% of the mean diurnal tide range throughout the Bay. A similar response is observed in the HSCM up to Morgans Point. Above Morgans Point through the Port of Houston the HSCM response is slightly improved over GBM results with RMS errors 3 to 4 cm. The bottom roughness coefficient z_0= 0.2 cm used in the April 1996 calibration has been reduced to z_0= 0.15 cm during the one year evaluation period to reduce the damping of the tidal response. Based on the RMS errors of the tidal response only a 3 cm improvement can be made for a perfect tidal model set. Since the overall RMS errors are order 10 cm on the nowcast and range to 12.6 cm at Morgans Point on the forecast, the model forcing errors (wind, atmospheric pressure, river inflow, and subtidal open boundary) must be considered. At present five PORTS stations (see Figure 8.4.1) are used to describe the wind and sea level pressure fields

Table 8.4.19. Nowcast/Forecast System Water Level Event Skill Measure Results: April 2000 - March 2001. Note the values are obtained by dividing the number of months in which the number of successes is greater than the sum of the number of failures and false alarms by 12.

Station: Analysis	Nowcast	Tide Prediction	Forecast	Adjusted Forecast
Galveston Pleasure Pier:				
Total	9/12	2/12	9/12	9/12
High	7/12	2/12	5/12	4/12
Low	6/12	3/12	5/12	6/12
Morgans Point:				
Total	9/12	2/12	8/12	9/12
High	7/12	0/12	4/12	5/12
Low	4/12	2/12	4/12	6/12

during the nowcast. No station is located over the Trinity Bay region and this may result in some errors for some nowcasts. During the forecast, the NWS/AVN model wind and sea level pressure results on a one degree by one degree resolution are used resulting in a nearly spatially constant wind and pressure field over the Bay, which may not be representative in some cases. River inflows are specified during the nowcast based on the latest USGS observation. The flows are assumed constant over the 24 hour nowcast. During the forecast, NWS/WGRFC forecast flows are used based on a 6 hour interval for the Trinity and San Jacinto Rivers. A persistence of the USGS nowcast flow on the Buffalo Bayou is used on the forecast. While these procedures are appropriate during the dry season during thunderstorm activity and wet weather periods, the inclusion of the Houston metroplex runoff flows may improve results during both nowcast and forecast. This would necessitate the development of a rainfall/runoff model for the City of Houston. On the nowcast, no additional effort was made to refine the nontidal forcings along the GBM open boundary due to lack of observations. Note on the forecast again only the Galveston Pleasure Pier nontidal signal is used.

With respect to the principal component prediction depth currents, significant discrepancies occur in the flood, ebb, and slack timings at both Bolivar Roads and Morgans Point and are further indicated in the RMS error at Morgans Point. Based on a 29-day harmonic analysis of the April 1996 calibration (Schmalz, 2001), the GBM model 24 constituent amplitude weighted gains are order 90% through the Bay entrance, drop to 70% at Redfish Bar, and fall to 40% at Morgans Point. Phase errors range from 25-50 minute lags in the entrance to 15 to 45 minutes leads at Redfish Bar, and Morgans Point, respectively. In the HSCM, the current phase differences are less than 20 minutes through the Entrance but are advanced up to 30 to 40 minutes at Redfish Bar and Morgans Point, respectively. Based on the tidal analysis and towed ADCP analysis, it appears that the currents and topography exhibit higher spatial variability than the grid cell resolution in both models. The fine resolution HSCM cell sizes are order the width dimension of the Houston Ship Channel (100m) and to represent dynamics within the channel a factor of ten increase in resolution (10 grid cells over the channel width) would be required. This is presently outside the realm of computational time requirements for a useful nowcast/forecast cycle. RMS errors estimated from the tidal constituent amplitude and phase differences in the GBM are 15 cm/s throughout the Bay and in the HSCM range from 16 to 30 cm/s in the Entrance and are reduced to 9 cm/s at Morgans Point. The Morgans Point nontidal current effect is larger than at Bolivar Roads since the total RMS errors are order 20 cm/s at Bolivar Roads with mean diurnal range of 200 cm/s and are the same order at Morgans Point with a mean dirurnal range of 100 cm/s. This tends to suggest that the freshwater inflows may be underestimated at Morgans Point. This is also confirmed by noting that the ebb flow errors are larger than the flood flow errors at Morgans Point as shown in Table 8.4.12.

With respect to salinity, surface measurements were not available over the last six months of the evaluation period due to biofouling problems. No bottom measurements have ever been available, which means that the salt wedge excursion cannot be measured. Precipitation and evaporation are not included in the models and a zero flux condition is assumed. Surface PORTS measurements are used to adjust the initial salinity field for each nowcast if available but are not sufficiently dense to determine the horizontal salinity gradient. ADCP/CTD Houston Ship Channel survey measurements made in September 1999, indicated order 8 PSU stratification within the channel, which could not be maintained by the sigma coordinate representation employed in the GBM and HSCM (Schmalz, 2000a). It is felt that 5 sigma levels represents an upper bound for the shallow (order two meter depth) Galveston Bay. Diffusion along sigma surfaces suggests that a generalized vertical coordinate, higher order vertical interpolation or a higher order horizontal differencing scheme for the baroclinic pressure gradient may be needed.

With respect to temperature, some surface and bottom measurements at the ADCP locations had been available, but have not been made during the last six-months of the evaluation period due to biofouling problems. The initial temperature fields are revised based on PORTS measurements if available. The SST field is developed and persisted throughout the nowcast and forecast period. Alternatively, the initial field is assumed to satisfy a no flux surface condition throughout the nowcast/forecast period. The use of satellite derived SST analysis may be necessary to define the Bay to shelf horizontal temperature gradients.

8.4.4 Summary and Future Directions

An experimental nowcast/forecast system has been developed and evaluated for Galveston Bay and the Houston Ship Channel. The experimental system has been evaluated over the one year period April 2000 through March 2001 using the NOS (1999) formal procedures. In general, the water surface elevation nowcast and forecast results meet or exceed the acceptance criteria, except for the timings of high and low waters. For principal component direction currents, the acceptance criteria are generally met, except for the timings of the zero crossings (slack before ebb and slack before flood), peak ebb, and peak flood currents.

Since computations are performed on two geometrically dissimilar grids, grid convergence indices are not able to be reliably computed. Recently, Schmalz (2007) has considered approaches for performing grid convergence and numerical uncertainty estimates for individual nowcast/forecast cycles using the methods discussed in Chapter 2.

Measurement issues include the CTD sensor biofouling and ADCP cabling problems. The biofouling has prevented (S,T) measurements to be made at a number of stations. The ADCP at the Redfish Bar station was removed and continued cable snagging both at Morgans Point and Bolivar Roads has persisted. For an operational nowcast/forecast system these two issues will need to be addressed. One possibility is to develop systems analysis techniques to fill in (recover) missing station data. Missing data for forcing (salinity, temperature, water level at Galveston Pleasure Pier, wind and atmospheric pressure) and validation data (currents and Bay water levels) will need to be recovered. Statistical methods or data recovery methods developed in Section 8.3 are candidates for consideration. To lower the quantity of missing data, the use of horizontal ADCPs mounted on channel markers and improved antifouling sealants for the CTD sensor package will be pursued.

From a modeling perspective, horizontal ADCPs for measuring velocity cross-sections at different locations along the Houston Ship Channel (HSC) are needed to provide additional validation data for the gravitational circulation. Monthly or event triggered CTD surveys along the HSC are also needed to characterize the movement of the salt wedge for further model validation. During major events, alternate forcings may need to be incorporated. Higher resolution wind fields would be used over the Bay during tropical storms and hurricane events in conjunction with the development of an overland flooding capability. In addition, hourly flow forecasts would be used for the Trinity and San Jacinto Rivers to more accurately depict flood period hydrographs. For major rainfall events, a separate rainfall-runoff model for the City of Houston would be used to define the major inflows along the Port of Houston.

The development of the Galveston Bay Houston Ship Channel nowcast/forecast system is an iterative measurement-modeling process. As new measurements become available, the adequacy of the present numerical approach is reassessed and revised accordingly. Methods to perform the assessment will be upgraded to include the model response frequency, reliability index, and performance measures presented in Section 8.3. An experimental nowcast/forecast system provides interim products until the acceptance criteria (NOS, 1999) are met and a formal operational system is established. Even then, additional improvements will continue to be sought. The role of hindcasting is critical to the continued development of the system. As trial nowcast/forecast cycles diverge from observations, the troublesome cycles are repeated in hindcast mode using alternative approaches. In addition, it is envisioned that a hindcast on demand capability will also be included within the system, by which over requested time periods hydrodynamic variables will be simulated in hindcast mode using previous nowcast inputs and compared with observations.

With respect to forecasts, the use of artificial neural network techniques similar to those presented in Section 8.2 (e.g. mutiple input (forcing functions)/multiple station output (water level and current errors)), will be pursued to attempt to adjust model forecasts.

8.4.5 Acknowledgments

Dr. Bruce B. Parker (retired), Chief of the Coast Survey Development Laboratory, conceived of this project and provided leadership and critical resources. Dr. Frank Aikman provided the day to day leadership and encouragement to take on the writing of this chapter. Discussions with Dr. Kurt W. Hess on modeling shallow bay systems were of great assistance. Mr. Philip H. Richardson was instrumental in the development of the wind and atmospheric pressure field interpolation procedures and the water level event analysis programs.

8.4.6 References

Appell, G. F., T. N. Mero, T. D. Bethem, and G. W. French, 1994: The development of a real-time port information system, *IEEE Journal of Oceanic Engineering*,

19(2), 149-157.

Barnes, S. L., 1973: Mesoscale objective map analysis using weighted time-series observations, *NOAA Technical Memorandum ERL NSSL-62*, National Severe Storms Laboratory, Norman, OK.

Bethem, T. D., and H. R. Frey, 1991: Operational physical oceanographic real-time data dissemination. *Proceedings, IEEE Oceans 91*, 865 – 867

Black, K., 1987: A numerical sediment transport model for application to natural estuaries, harbours and rivers, In *Applied Numerical Modelling: Applications to Marine Systems*, ed. (John Noye), North-Holland, Amsterdam.

Blumberg, A. F., and G. L. Mellor, 1987: A description of a three-dimensional coastal ocean circulation model. *Three-Dimensional Coastal Ocean Models*, (ed. Heaps), American Geophysical Union, Washington, DC, 1 - 16.

Hess, K. W., 1994: Tampa Bay Oceanography Project: Development and application of the numerical circulation model. NOAA, National Ocean Service, Office of Ocean and Earth Sciences, *NOAA Technical Report NOS OES 005*, Silver Spring, MD.

Kalnay, E., M. Kanamitsu, and W.E. Baker, 1990: Global numerical weather prediction at the National Meteorological Center, *Bulletin of the American Meteorological Society*, 71, 1410-1428.

Ives, D. C. and R. M. Zacharias, 1987: Conformal mapping and orthogonal grid generation, *AIAA/SAE/ASME/ASEE 23rd Joint Propulsion Conference*, Paper No. 87-2057.

Lin, S-J., W. C. Chao, Y. C. Sud, and G. K. Walker, 1994: A class of the van-Leer transport schemes and its application to the moisture transport in a general circulation model, *Monthly Weather Review*, 122, 1575-1593.

Martin, P. J., 1985: Simulation of the mixed layer at OWS November and Papa with several models, *Journal of Geophysical Research*, 90, C1, 903-916.

Mellor, G. L., 1996: Princeton Ocean Model Users Guide, *Program in Oceanic and Atmospheric Sciences*, Princeton University.

National Geophysical Data Center, 1988: NOS Hydrographic Data Base-Expanded, Digital Bathymetric Data for U.S. Coastal Waters, *Data Announcement 87-MGG-12*, NOAA/NGDC, Boulder, CO.

NOS, 1999: NOS procedures for developing and implementing operational nowcast and forecast systems for PORTS, NOAA Technical Report NOS CO-OPS 0020, Silver Spring, Maryland.

Orlando, S. P., L. P. Rozas, G. H. Ward, and C. J. Klein, 1993: Salinity characterization of Gulf of Mexico estuaries, *NOAA Tech Memorandum, Office of Ocean Resources Conservation and Assessment*, Silver Spring, MD.

Schmalz, R. A., 1994: Application and documentation of the Long Island Sound three-dimensional circulation model, Long Island Sound Oceanography Project, Summary Report, Volume 1, NOAA, National Ocean Service, Office of Ocean and Earth Sciences, *NOAA Technical Report NOS OES 003*, Silver Spring, MD.

Schmalz, R. A., 1996: National Ocean Service Partnership: DGPS-supported hydrosurvey, water level measurement, and modeling of Galveston Bay: development and application of the numerical circulation model. NOAA, National Ocean Service, Office of Ocean and Earth Sciences, *NOAA Technical Report NOS OES 012*, Silver Spring, MD.

Schmalz, R. A., 1998a: Development of a nowcast/forecast system for Galveston Bay, *Proceedings of the 5th International Estuarine and Coastal Modeling Conference*, ASCE, 441-455.

Schmalz, R. A., 1998b: Design of a nowcast/forecast system for Galveston Bay, *Proceedings of the 2nd Conference on Coastal Atmospheric and Oceanic Prediction and Process*, AMS, 15-22.

Schmalz, R. A., 1998c: Initial evaluation of a nowcast/forecast system for Galveston Bay during September 1997, *Proceedings of the Ocean Community Conference*, MTS, 426-430.

Schmalz, R. A., 2000a: High-resolution Houston Ship Channel ADCP and CTD survey: model-data intercomparisons, *NOAA Technical Report NOS CS 6*, Silver Spring, MD.

Schmalz, R. A., 2000b: Demonstration of a nowcast/forecast system for Galveston Bay, *Proceedings of the 6th Estuarine and Coastal Modeling Conference*, ASCE, New Orleans, LA, 868-883.

Schmalz, R. A., 2001: Three-dimensional hydrodynamic model developments for a Galveston Bay nowcast/forecast system, *NOAA Technical Report NOS CS 9*, Silver Spring, MD.

Schmalz, R. A., 2007: Validation of the NOS Galveston Bay nowcast/forecast system hydrodynamic models: Review and New Perspectives, *Proceedings of the World Environmental and Water Congress*, ASCE/EWRI, Tampa, FL, CD-ROM.

Schureman, P., 1958: Manual of harmonic analysis and prediction of tides. *U.S. Department of Commerce, Coast and Geodetic Survey, Special Publication No. 98 [revised 1940 edition, reprinted 1988]*, Rockville, MD.

Sheinin, Dmitry, 1999: Emails 5 and 6 July, sheinin@gcm.mit.edu., Massachusetts Institute of Technology, Cambridge, Massachusetts.

Temple, R. F., D. L. Harrington, J. A. Martin, 1977: Monthly temperature and salinity measurements of continental shelf waters of the northwestern Gulf of Mexico, 1963-1965, *NOAA Technical Report NMFS SSRF-707*, Rockville, MD.

Wilken, J. L., 1988: A computer program for generating two-dimensional orthogonal curvilinear coordinate grids, *Woods Hole Oceanographic Institution (unpublished manuscript)*, Woods Hole, MA.

Williams, R. G., H. R. Frey, and T. D. Bethem, 1990: Houston Ship Channel/Galveston Bay current prediction quality assurance miniproject, *NOAA Technical Memorandum NOS OMA 53*, Rockville, Maryland.

Willmott C. J., S.G. Ackleson, R.E. Davis, J.J. Feddema, K. M. Klink, D.R. Legates, J. O'Donnell, and C. M. Rowe, 1985: Statistics for the evaluation and comparison of models, *Journal of Geophysical Research*, 90, 8995-9005

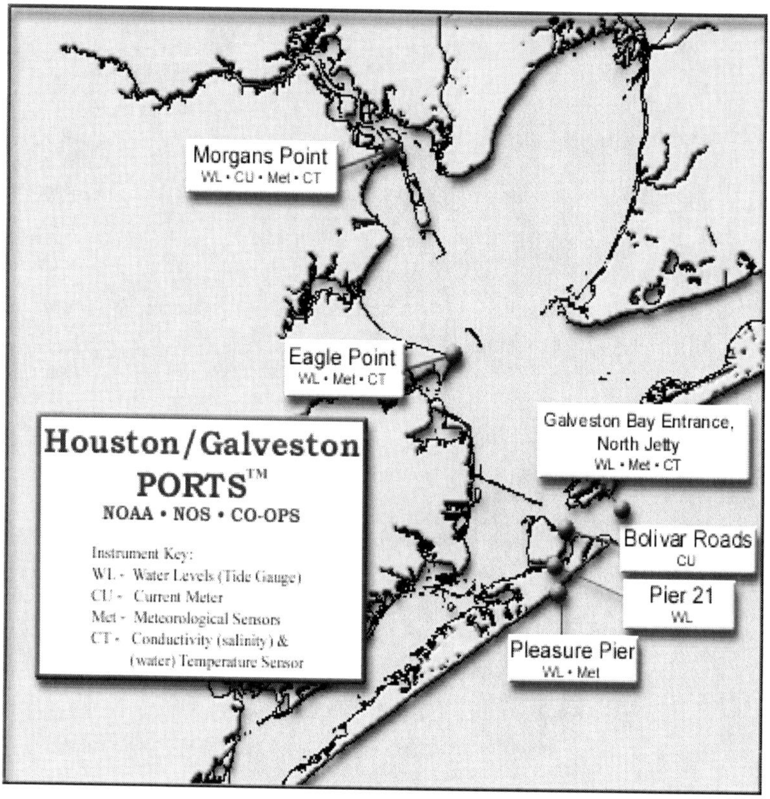

Figure 8.4.1 Galveston Bay Base Map. (PORTS station locations are indicated, where W = water level gauge, M= meteorological station, and C= current meter.)

Houston/Galveston PORTS Text Screen

```
      Houston/Galveston PORTS, NOAA/NOS at  2:00 pm CST  February 13, 2003
--------------- TIDES ------------------------------ CURRENTS ----------------
Morgans Point        0.0 ft.        : Morgans Point       0.3 kts.(F), 002°T
Eagle Point          0.7 ft.,Rising : Bolivar Roads       1.6 kts.(F), 285°T
Pier 21              1.2 ft.,Rising :
North Jetty          1.6 ft.        : (F)lood,(S)lack,(E)bb,towards °True
Pleasure Pier        1.8 ft.,Falling:-------------------Salinity---S.G.--W.Temp
                                    : Morgans Point     ***************   55°F
                                    : Eagle Point       ***************   58°F
                                    : North Jetty                         54°F
                                    : Pleasure Pier                       58°F
--- METEOROLOGICAL ----- Wind Speed/Dir ------------- Air Pressure -- Air Temp
Morgans Point      12 knots from ESE, gusts to  16    1020 mb,Falling    62°F
Eagle Point        10 knots from ESE, gusts to  10    1020 mb,Falling    59°F
North Jetty        13 knots from ESE, gusts to  14    1019 mb,Falling    59°F
Pleasure Pier      10 knots from E  , gusts to  11    1020 mb,Falling    55°F
------------------------------------------------------------------------------
 **** - Data not displayed as a result of quality control monitoring. For more
 information, go to http://co-ops.nos.noaa.gov/corms_status.html, or call CORMS
 at 301-713-2540.
```

Figure 8.4.2 Houston/Galveston PORTS Screen Data Format

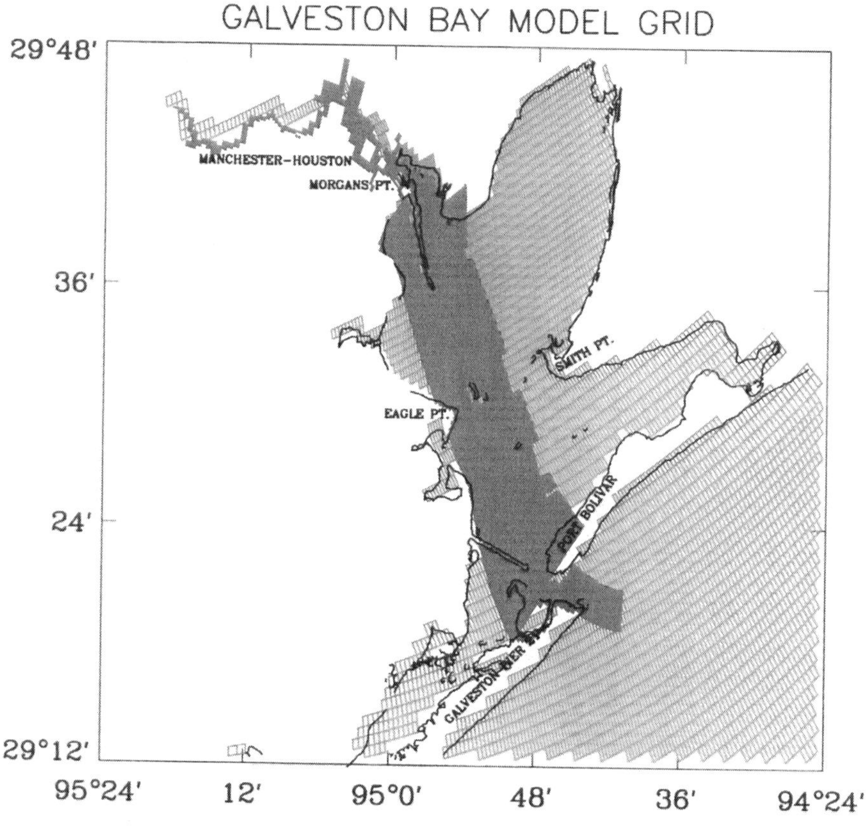

Figure 8.4.3 Galveston Bay Model Grid with Houston Ship Channel Model Grid in green.

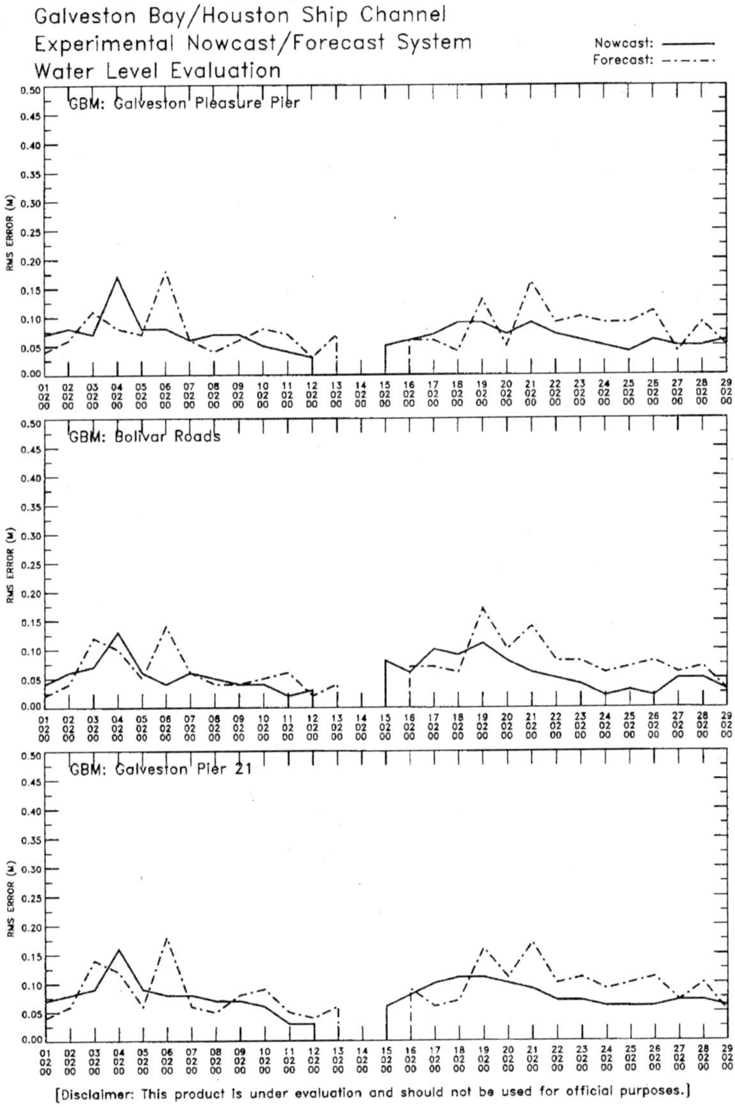

Figure 8.4.4 Gulf Coast and Lower Galveston Bay Water Level Nowcast (solid)/Forecast (dashed) RMS Errors.

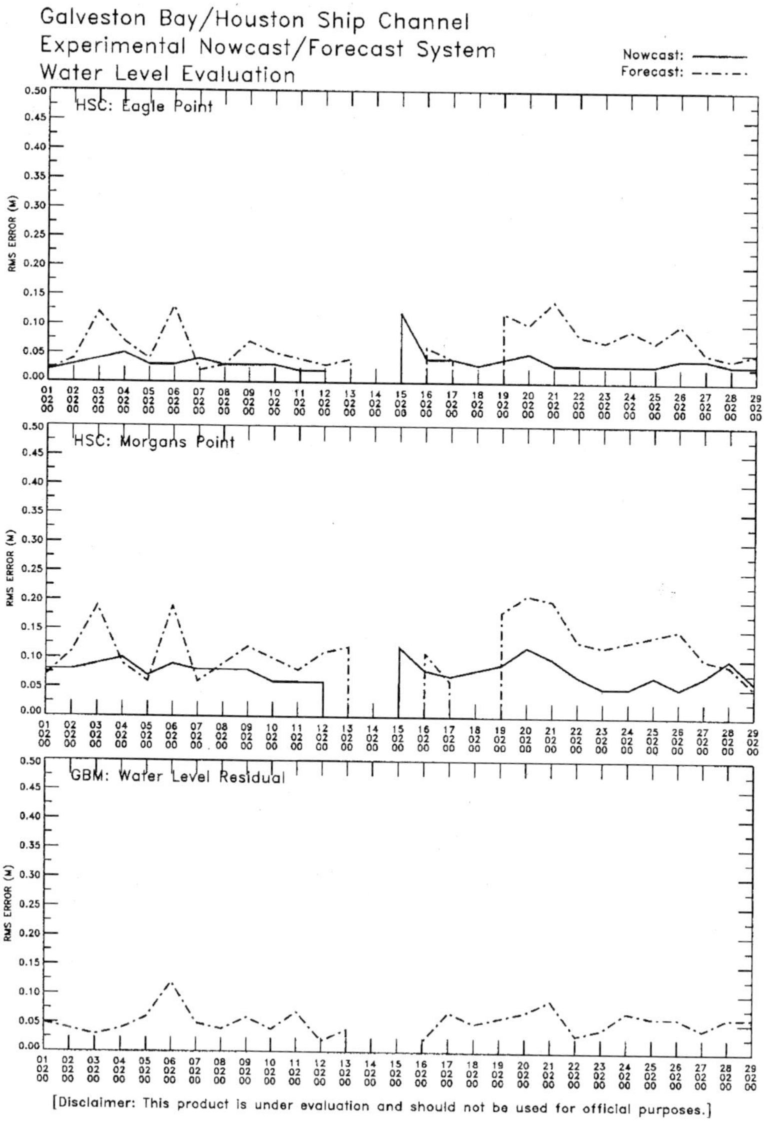

Figure 8.4.5 Upper Galveston Bay Water Level Nowcast (solid)/Forecast (dashed) RMS Errors.

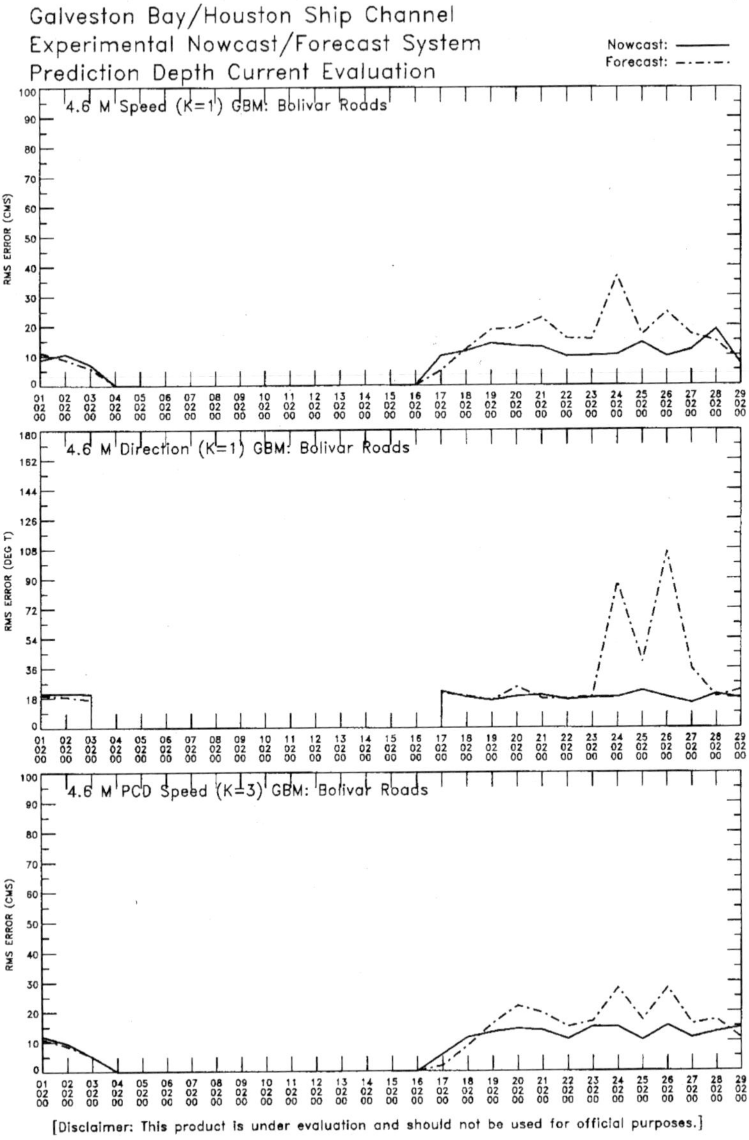

Figure 8.4.6 Bolivar Roads Prediction Depth Current Nowcast (solid)/Forecast (dashed) RMS Errors.

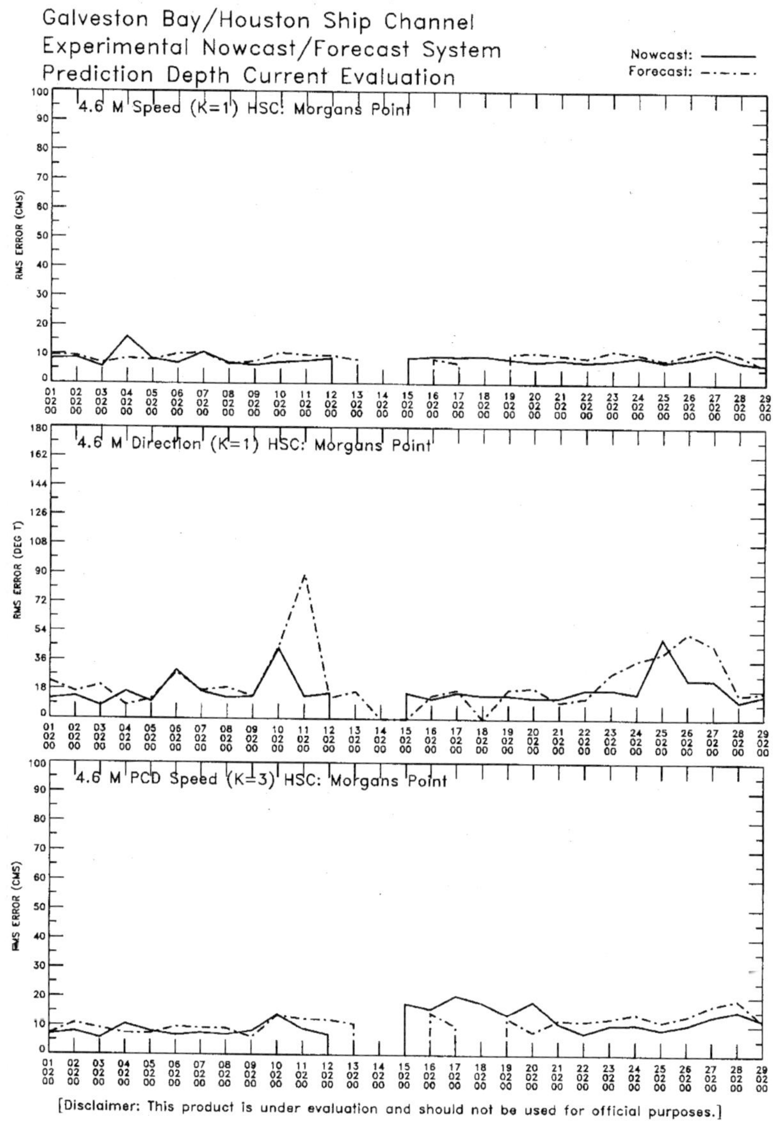

Figure 8.4.7 Morgans Point Prediction Depth Current Nowcast (solid)/Forecast (dashed) RMS Errors.

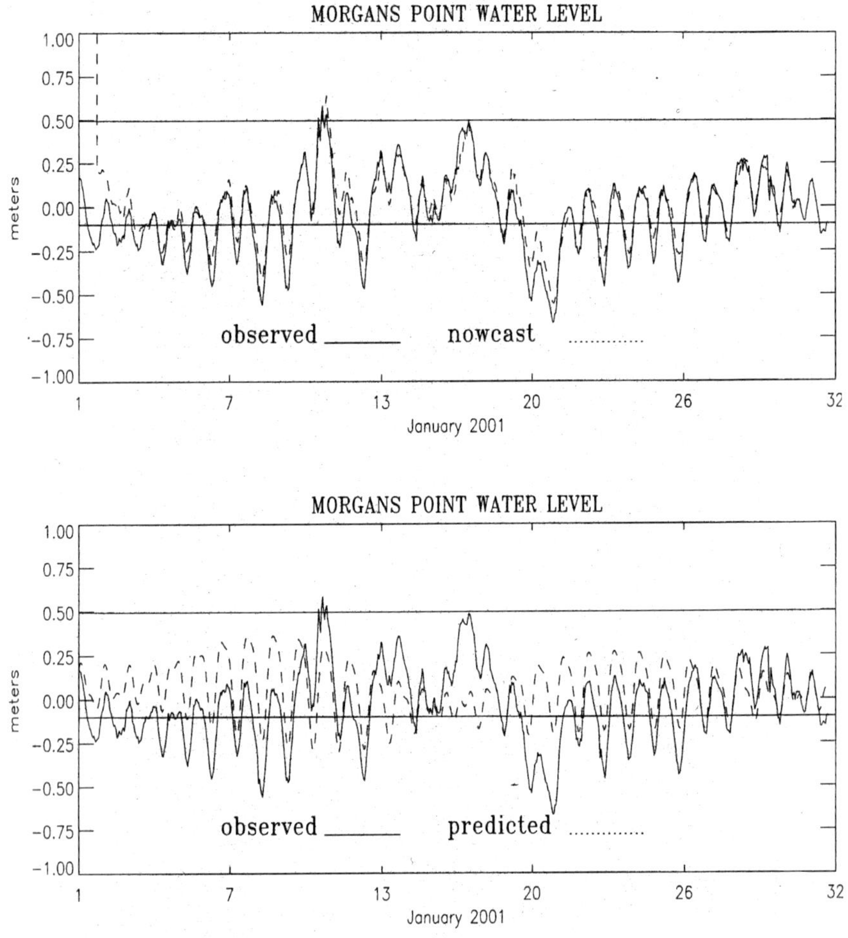

Figure 8.4.8 Morgans Point Nowcast Results: January 2001 Note observed water level results are compared with the norecast time series in the top panel. In the lower panel, the observed water levels are compared with the tronomical tide predictions. The high and low water event critical levels are shown as solid horizontal lines. Note the January 1 nowcast was missed due to computer system problems.

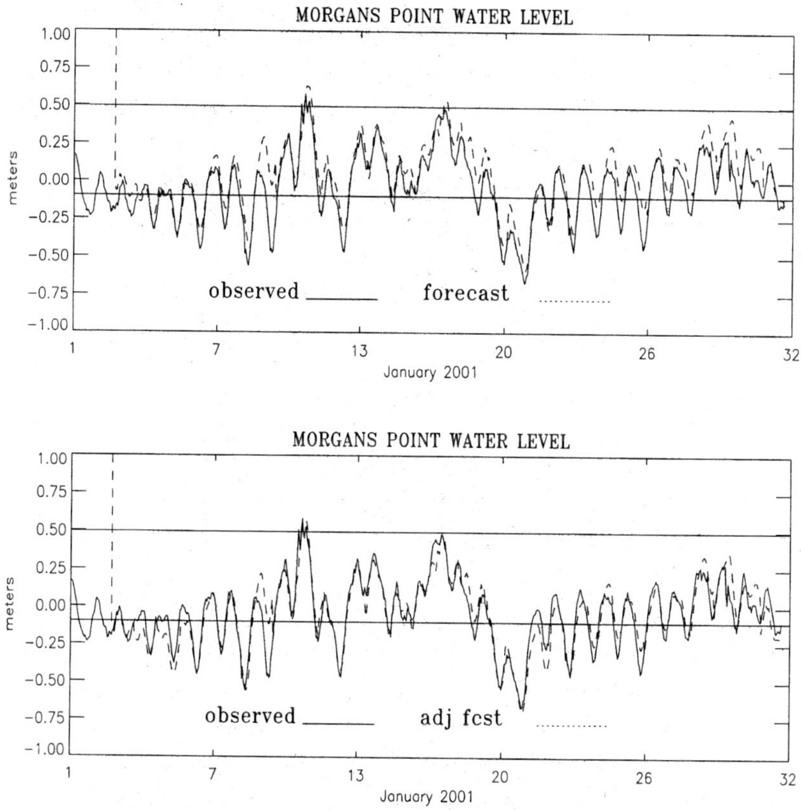

Figure 8.4.9 Morgan's Point Forecast Results: January 2001 Note observed water level results are compared with the forecast time series in the top panel. In the lower panel, the observed water levels are compared with the adjusted forecast results with bias equal –7.5 cm and gain equal 1.05. The high and low water event critical levels are shown as solid horizontal lines. Note the January 1 and 2 forecasts were missed due to computer system problems.

CHAPTER 9

FINDINGS, CONCLUSIONS AND RECOMMENDATIONS

Sam S.Y. Wang

9.1 INTRODUCTION

This is the Report of the ASCE EWRI Task Committee on Three Dimensional Free Surface Flow Model Verification and Validation Monograph. This chapter summarizes the findings, conclusions and recommendations of all members of the Task Committee which have been supported by the majority of the Committee.

9.2 THE NEED OF MODEL VERIFICATION AND VALIDATION

Rapid advancements in numerical modeling methodology and computer technology in recent decades have led wide-spread applications of computational modeling to scientific research, engineering design, project planning and management decisions, environmental impact assessment, among others. More and more highly sophisticated free surface models have been developed and utilized to conduct research and design, and to support the project decision by optimizing multiple-objective function under multiple constraints. Some of these projects are for large scale water infrastructural systems with high costs and long-term effects. Professionals in the field have become more concerned about the model's quality and reliability. Professional societies and hydraulic research institutions have supported a number of efforts trying to develop some means for measuring the computational model's quality. The Three-Dimensional Free-Surface Flow Model Verification and Validation Task Committee was established by the ASCE and its EWRI to meet this need.

The free-surface flows are governed by non-linear partial differential equations, which are the mathematical representations of the physical principles together with the required empirical functions. To study a system governed by this set of nonlinear equations in a highly irregular domain, one has to transform the mathematical representation into its numerical equivalent or numerical model in a discretized domain and obtain the numerical solutions of field variables with the assistance of highly efficient computer. During this lengthy and complicated mathematical derivation, implementation of discretization and sophisticated numerical solution, and coding process, there are plenty of possibilities to commit errors. Some typical examples are:

- During the formulation of idealized mathematical models, a physical system of continuous medium is usually assumed that it can be isolated by an imaginary control surface. This allows the system in the control volume to be governed by mathematical model and the effects of the surrounding media on the system to be included by application of boundary conditions. Obviously, the isolation assumption may cause inaccuracy in the solution;

- When a numerical model is developed through a complicated series of mathematical derivations, human mistakes can, and usually are, inadvertently committed;

- By adopting some acceptable mathematical manipulations to simplify the equations of numerical model for easy solutions, it may cause numerical diffusion, dispersion, spurious oscillations, etc,; even though no mathematical principles or rules have been violated;

- Sometimes, the implementation of alternative numerical scheme(s) for enhancing the numerical solution's stability, rate of convergence of iterations, computing efficiency, etc., may lead the solution to be inconsistent with the original governing equation or to converge to a different equation;

- An adoption of an inappropriate closure law or empirical function may limit the model's capability or applicability in predicting certain physical process;

- Even if the numerical model (code) is free of all mistakes, one may still wish to know what the bounds of calculation errors accumulated are during the simulation.

Obviously, there might be other errors and/or mistakes made unintentionally during the lengthy and sometimes involved mathematical derivations to discretize and formulate the numerical model, to implement a numerical solution scheme, and to code a computer program. These unintentional bugs are hard to find. To meet the demand of the clients, the model developers have applied various verification and validation methods. These include: from the test of symmetry for making sure the model results are indeed symmetrical when simulating a symmetrical problem, to mass conservation and Gaussian cone translation/rotation, to grid refinement convergence, to the comparison with known analytic solution, and to the validation by obtaining agreements with the field data directly. The ones cited here are only a few typical examples.

The method adopted most often has been the validation by comparing the model results with the field measurements directly. Because once good agreement is achieved, one can claim that the model is validated. Lately, more and more professionals have found that the agreements were achieved sometimes by "fine-tuning" the model parameters exclusively. In some cases, it was done locally around each data measuring location and repeatedly until the almost perfect match was obtained for one location at a time. This approach may be referred to perhaps as a demonstration of the skill of parametric fine-tuning. It can be accepted as calibration of model parameters, provided it is performed properly. It is unacceptable as a model validation. More instances about the abuse of the fine-tuning method have been found recently, which reveals that by adjusting the value of certain model parameters, modelers can even alter the simulation results to their liking. For example, by altering the river bed friction factor, say Manning's n-value, the streamline distribution in a river bendway could be changed. Examples like this are very damaging to the confidence of the value of numerical simulation models; and thus, the method of model validation by fine-tuning the model parameter should be carefully examined. This is another reason the professional societies and professionals in the field have decided that there is a need to establish a comprehensive and systematic procedure for performing rigorous numerical model verification and validation.

9.3 THE METHODOLOGY OF FLOW MODEL VERIFICATION AND VALIDATION

A systematic verification and validation procedure has been developed by the Task Committee. It was presented in Chapter 7 with a practical example. Only the key information is briefly summarized here.

The Systematic Procedure for Verification and Validation of 3D Free-Surface Flow Models recommended by the Task Committee includes three major aspects: Mathematical (Code) Verification, Physical Process Validation, and Application Site Validation. The Committee recommends also for the model testers to conduct a Calculation Verification within the second and the third step. When a numerical model has been developed for the study of a problem, the modeler is advised to test the model by at least the first two steps. The model user needs to perform the last step of this procedure, the Application Site Test before carrying out an application project. Of course, the model developer could also be the user.

As the first step, the model is tested by a set of selected test cases based on analytic solutions, which are either obtained analytically or manufactured (or prescribed) using an inverse method. Some of the sample test cases have been presented in Chapters 3 and 4. Testers are strongly encouraged to develop their own test case to serve their special needs. If significant disagreement (other than some minor numerical error due to approximation, truncation, round-off, etc.) between the numerical and analytical solutions is found, one can be sure that there must be some errors or mistakes in the numerical model during the mathematical derivations, numerical solution, or program coding, etc. The above conclusion is due to the fact that both the numerical model results and the analytic solutions have been obtained from the same set of governing differential equations, with the same boundary and initial conditions, having the same size and geometry of the solution domain, and using the same values of the physical parameters. Other than the approximation, truncation and round-off errors, the solutions should agree with each other.

Unlike the physical process and field validations, the Mathematical (Code) Verification has no way to conceal any mathematical mistakes, because the model parameters have been prescribed, so that no fine tuning is needed or allowed. Therefore, once disagreement between the numerical and analytic solutions is found, the modeler has to correct the model's (code's) defects and/or improve the model's capability to achieve the agreement. The Mathematical (Code) Verification can serve another purpose that is to determine the order of convergence or accuracy by a grid refinement analysis. This quantitative assessment can also provide the user with the information to select a model of higher accuracy among the models available. Therefore, this first step is very important, rather than trivial, as some critics have claimed.

After a numerical model is proven mathematically correct, the next question to answer is whether it can predict the correct basic physics or basic physical processes essential to a physical system to be studied. By performing a series of Physical Process Validation tests, and comparing the numerical model solutions of the flow fields with the measurements from an experiment conducted in a laboratory under the same geometric and physical conditions, one can evaluate whether the numerical model has the capability of predicting or reproducing the basic physical processes measured or observed in the physical experiments. If reasonable agreements are obtained from a set of selected test cases covering all essential physical processes of the problem to be studied, the model has passed the Physical Process Validation. Otherwise, the numerical model's governing differential equations have to be corrected or modified. Some modifications for considerations are: empirical closure laws (such as turbulence closure), terms omitted in the differential equations, etc.

After having passed these two major test steps, modelers may claim that the model is certified to be correct mathematically and capable of simulating a list of basic physical processes which have been validated. To validate a general purpose model, a more comprehensive list of physical process validations needs to be carried out.

In order to apply a numerical model to the study of a real-life flow field with reasonable accuracy, the model needs to go through one more validation step, the Application Site (or Field) Validation. To conduct this validation step properly, one needs to: (1) collect sufficient sets of field data at well-designed time and locations with an appropriate distribution of data collection points; (2) use a few sets of the data for calibrating the values of the site-specific parameters of the numerical model so that the effects of the unique geophysical, hydrological and hydraulic characteristics and conditions of the study site have been included in the model parameters; and (3) conduct validation tests of the calibrated model based on the field data set having not been used during calibration. If a reasonable agreement is obtained by comparing the numerical solutions with the measured field data sets, then the model is said to have passed the Application Site Validation.

After a numerical model has successfully passed this systematic three-step verification and validation procedure, it is ready to be applied only to the investigation of the particular site-specific real-life field problem.

One needs to, however, recognize that a model with site-specific parameters being calibrated using the field measurements at a particular site and a particular time, may not be automatically applicable to a different site without sufficient justification. Even at the same site, it may not be applied at a different time, especially under a drastically different forcing, such as an extreme event, or at a significantly lapse of time, say months or years later. In these situations, one may need to re-calibrate and re-validate the model before application. Therefore, the Application Site (or Field) Validation needs to be conducted on a case-to-case basis, rather than once and for all.

9.4 THE TEST CASES PROVIDED

Twenty two test cases have been developed by the Task Committee Members. They have been presented in Chapters 3, 4, 5 and 6 in four categories: Mathematical (Code) Verification by Analytic Solutions, or Manufactured (or Prescribed) Solutions; Physical Process Validation by laboratory experiments; and Application Site Validation by field data. Each test case serves a unique purpose. Some of them are designed to test the mathematical correctness and/or numerical accuracy, such as the numerical model solution's consistency, order of convergence, error magnitude estimate via the Grid Convergence Index, etc. Some others are intended to determine the model's capabilities in satisfying the physical principles; reproducing the basic physical process; predicting the field phenomena of the real-life problem realistically with acceptable accuracy; etc.

Each case has been designed to be easy to set up and execute, and is provided with either the analytic solutions or physical measurements for making comparison with the numerical results. Even though efforts have been made to have these cases cover a wide spectrum of three-dimensional free surface flows, they can not cover all possible cases needed by all users. Therefore, model testers are strongly advised to design additional test cases of their own or select them from literature to serve their specific needs. As emphasized in Chapter 7, a more complete and systematic verification and validation test needs a series of test cases rather than just one test case. Model testers need to have a well-thought out plan to select a series of cases sufficient to test all capabilities and the accuracy of the numerical model to be used for an investigation.

9.5 RECOMMENDATIONS

The Task Committee on 3D Free Surface Flow Model Verification and Validation, after having worked on and discussed these issues extensively for more than 10 years, has decided to submit the following recommendations, which include the following.

(1) The ASCE and EWRI consider raising the quality standard of flow simulation model by encouraging model developers and users to conduct a rigorous and comprehensive series of verification and validation tests of all numerical models before applying the model to the investigations of real-world problems;

(2) The ASCE encourages its members to adopt the three-step verification and validation procedure, which include: Mathematical (Code) Verification based on either linear or nonlinear analytic solutions, Physical Process Validation (including formal calculation verification) using measurements of laboratory experiments, and Application Site Validation including calculation verification based on the field data collected at the study site;

(3) Before the release of a numerical model, it has to pass the Mathematical (Code) Verification test to insure that it is mathematically correct and has an acceptable order of convergence, and a selected series of Physical Process Validation tests to insure that it has the capability of reproducing a list of important basic physical processes having been tested;

(4) Before a user is to use a model for a site-specific application study, the model has to be tested by an Application Site Validation, which should satisfy three conditions: (i) sufficient amount of high quality data must be collected, (ii) an appropriate portion of the field data is only used for calibrating the site-specific model parameters, and (iii) the calibrated numerical model must be validated using the field data, which has not been used to conduct the calibration;

(5) The EWRI and ASCE Journals should lead the free-surface flow modeling community to improve standards by encouraging the use of the three-step Verification and Validation Procedure including a quantitative Calculation Verification within Physical Process Validations and, where feasible, in Application Site Validations.

(6) Most importantly, the traditional "**calibration only**" **approach should not be accepted** by professional societies and professionals in the fields as the model validation. This policy should be adopted by EWRI and ASCE journals, just as its sister societies, ASME, AIAA, etc. have done several years ago. If ASCE-EWRI accepts the Task Committee's recommended three-step Verification and Validation procedure, it shall be a leader in this field.

(7) With the rapid ongoing advances of numerical modeling methodology and in physical measurement technology, the model verification and validation methods have to be advanced rapidly as well. Therefore, there is an urgent need to continue the development of new techniques and upgrades of the existing model Verification and Validation methods. This need may be served by either establishing a permanent task or technical committee specifically for this purpose or by expanding the role of Computational Hydraulic Committee to assume this important responsibility to insure that the quality of numerical models is maintained; and

(8) Due to strong interest and active research in the field, EWRI may want to consider the publication of a series of monographs or a professional journal in the field of Computational Model Verification and Validation.

APPENDIX A

INPUT DATA FOR TEST CASES

Editors: Yan Ding and Sam S.Y. Wang

A.1 INTRODUCTION

Appendix A gives description of input data for test cases included in Chapter 5 and Chapter 6, which are downloadable from the website http://www.ncche.olemiss.edu/publishing/.

Each data set begins with a README file, which gives essential instructions on how to properly utilize the data provided. In most cases, especially for those in Chapter 5, user needs to generate a mesh for the test selected. If a numerical grid is given, such as in some cases in Chapter 6, user may want to use it to save time and effort.

A.2 INPUT DATA SETS FOR CHAPTER 5

A.2.1 Free Overfall Flow Test Case

The data files and their instruction files are summarized in Table A.2.1. Note that the file types are categorized as: Data=data file; Image=image file; Text=text file including instructions.

For the instructions of the data files, users are suggested to read the README files carefully.

Table A.2.1 Data files in the dataset of free overfall flow test case

Filename	Type	Note
OVERFALL.XLS	Data	Measured velocity, pressure, and shear stress
README.TXT	Text	Instruction of file OVERFALL.XLS

A.2.2 Delft U-Shaped Channel Flow Test Case

The data files and their instruction files are summarized in Table A.2.2. For the instructions of the data files, users are suggested to read file README.doc carefully.

Table A.2.2 Data files in the dataset of Delft U-shaped channel flow test case

Filename	Type	Note
DISPLAY_COS.XLS	Data	Longitudinal velocity profiles in the water depth, see Figure 5.4.2 for the locations of the sections, and Figure 5.4.3 for illustrations of some sections
DISPLAY_SIN.XLS	Data	Transverse velocity profiles in the water depth, see Figure 5.4.2 for the locations of the sections, and Figure 5.4.4 for illustrations of some sections
DISPLAY_FIG13_F.XLS	Data	Water surface elevations shown in Figure 5.4.7
README.DOC	Text	Instructions of the validation data in the test case

A.2.3 Riprap Test Facility

This test case contains three datasets with three different discharges, in each the observed water surface elevations and velocities in a number of cross sections are provided. The data files and their instruction files are summarized in Table A.2.3, in which 'xxx', a three digital number stands for the filename. Users need to be aware that the unit in the data is English. For the instructions of the data files, users are suggested to read file README2.doc carefully.

Table A.2.3 Data files in the dataset of Riprap test case

Filename	Type	Note
xxxM.49	Data	Measured velocities and water surface elevations under the condition of 49.5 cfs (1 cubic feet per second=0.02832m^3/s) discharge
xxxM.101	Data	Measured velocities and water surface elevation in which discharge is 101 cfs
xxxM.150	Data	Measured velocities and water surface elevation in which discharge is 150 cfs
README2.DOC	Text	Instructions of the data in the case

A.2.4 Flow in a Channel with a Spur Dike Test Case

The data files and their instruction files are summarized in Table A.2.4. For the instructions of the data files, users are suggested to read file README.TXT.

Table A.2.4 Data files in the dataset of spur dike test case

Filename	Type	Note
EXPERIMENT DATA00.XLS	data	Experimental data about case setup, velocity profiles in water depth direction and transverse direction, shear stress, etc.
README.TXT	Text	Instructions of the data

A.2.5 Flow around a Submerged Trapezoidal Spur Dike Test Case

The data files and their instruction files are summarized in Table A.2.5. For the instructions of the data files, users are suggested to read file README-DIKE.doc carefully.

Table A.2.5 Data files in the dataset of submerged spur dike test case

Filename	Type	Note
3D_MEASURE_FLAT.TXT	data	Measured data about velocity, fluctuation velocity, turbulent stress, turbulence kinetic energy, temperature, and discharge.
FIGURE 3.XLS	Data	Data for locations of the spur dike
T0MAGDRNOSTRND.XLS	data	Values of normalized bed shear stress
README-DIKE.DOC	Text	Instructions of the data in the case

A.2.6 Flows around Groyne and in Harbor

The data files and their instruction files are summarized in Table A.2.6. For the instructions of the data files, users are suggested to read file README.DAT carefully.

Table A.2.6 Data files in the dataset of flows around groyne and in harbor test case

Filename	Type	Note
ST_xxxx.DAT	Data	Measured data in station xxxx, where xxxx indicates a 4-digital number
README.DAT	Text	Instructions of the data

A.3 INPUT DATA SETS FOR CHAPTER 6

A.3.1 Chesapeake Bay Test Case

The data files and their instruction files are summarized in Table A.3.1. Note that the file types are categorized as: Data=data file; Image=image file; Text=text file including instructions. For the instructions of the data files, users are suggested to read their README files carefully.

Table A.3.1 Data files in the dataset of Chesapeake Bay test case

Filename	Type	Note
CHESBAY.DEP	Data	Cell-centered water depth
READ_ME.DEP	Text	Instructions of the file CHESBAY.DEP
CHESBAY.GRD	Data	Grid coordinates file

READ_ME.GRD	Text	Instructions of the file CHESBAY.GRD
CHESBAY.PS	image	See Figure 6.4.4
PRIV_Q1.PS	image	See Figure 6.4.2
PRIV_Q2.PS	image	See Figure 6.4.3
PSAL036.PS	Image	See Figure 6.4.10
PSAL040.PS	image	See Figure 6.4.7
PSAL065.PS	Image	September 1983 Salinity, Station 65
PSAL121.PS	image	September 1983 Salinity, Station 121
PTEM036.PS	image	See Figure 6.4.11
PTEM040.PS	image	See Figure 6.4.8
PTEM065.PS	Image	September 1983 Temperature, Station 65
PTEM121.PS	image	September 1983 Temperature, Station 121
PTIDE1.PS	image	See Figure 6.2.5, September 1983 Tidal Heights, Chesapeake Bay Bridge Tunnel
PTIDE2.PS	Image	September 1983 Tidal Heights, Hampton Roads, VA & Lewisetta, Potomac, VA
PTIDE3.PS	image	September 1983 Tidal Heights, Colonial Beach, VA & Solomon's Island, MD
PTIDE4.PS	image	September 1983 Tidal Heights, Annapolis, MD & Havre de Grace, MD
PUV036.PS	Image	September 1983 U & V Velocity, Station 36
PUV040.PS	Image	September 1983 U & V Velocity, Station 40
PUV065.PS	Image	September 1983 U & V Velocity, Station 65
PUV121.PS	Image	September 1983 U & V Velocity, Station 121
PWIND.PS	Image	See Figure 6.4.9
READ_PS.ME	Text	Instructions of image files
SAL_OBS.DAT	Data	Salinity data at the stations in the bay for model validation
READ_OBS.SAL	Text	Instruction of file SAL_OBS.DAT
SEP83.RIVR	Data	River inflows
READ_ME.RIVR	Text	Instruction of SEP83.RIVR
SEP83.RIVRT	Data	Water temperature at the river inflows
READ_ME.RIVRT	Text	Instruction of SEP83.RIVRT
SEP83.SA15V	Data	Initial salinity on 1 September 1983
READ_ME.SA15V	Text	Instructions of SEP83.SA15V
SEP83.TE15V	Data	Initial temperature on 1 September 1983
READ_ME.TE15V	Text	Instruction of SEP83.TE15V
SEP83.TEK	Data	Data for Surface heat exchange
READ_ME.TEK	Text	Instruction of SEP83.TEK
SEP83.TIDE	Data	Water surface elevation at the mouth of the bay
READ_ME.TIDE	Text	Instruction of SEP83.TIDE
SEP83.TSATE	Data	Salinity and temperature at the mouth of the bay
READ_ME.TSATE	Text	Instruction of SEP83.TSATE
SEP83.WIND	Data	Wind field data
READ_ME.WIND	Text	Instruction of SEP83.WIND

STATM.PS	Image	Figure 6.4.1, Chesapeake Bay map showing Moored Stations
STATWT.PS	Image	Figure 6.4.6
TEM_OBS.DAT	Data	Temperature data at the stations in the bay for model validation
READ_OBS.TEM	Text	Instruction of file TEM_OBS.DAT
TID_OBS.DAT	Data	Water elevations at the stations in the bay (see Figure 6.4.1)
READ_OBS.TID	Text	Instruction for reading data in file TID_OBS.DAT
VEL036.DAT	Data	Water velocity data at Station 36 on Figure 6.4.6
READ_036.VEL	Text	Instruction of file VEL036.DAT
VEL040.DAT	Data	Water velocity data at Station 40 on Figure 6.4.6
READ_040.VEL	Text	Instruction of file VEL040.DAT
VEL065.DAT	Data	Water velocity data at Station 65 on Figure 6.4.6
READ_065.VEL	Text	Instruction of file VEL065.DAT
VEL121.DAT	Data	Water velocity data at Station 121 on Figure 6.4.6
READ_121.VEL	Text	Instruction of file VEL121.DAT

A.3.2 San Francisco Bay Test Case

The data files and their instruction files are summarized in Table A.3.2. For the instructions of the data files, users are suggested to read file README.txt carefully.

Table A.3.2 Data files for the San Francisco Bay test case

Filename	Type	Note
H	Data	500 meter square grid of San Francisco Bay bathymetry
wse.txt	Data	Water surface elevation boundary conditions
salic.txt	Data	Salinity initial conditions
sal_bc.txt	Data	Salinity boundary conditions
flow_bc.txt	Data	Flow boundary conditions
gg9801.vel	Data	ADCP data for January 1998 at the Golden Gate
gg9801.fil	Data	Low-pass filtered ADCP data for January 1998 at the Golden Gate
rich5.vel	Data	ADCP data for April to July 1998 near the Richmond-San Rafael Bridge
rich5.fil	Data	low-pass filtered ADCP data for April to July 1998 near the Richmond-San Rafael Bridge
README.txt	Text	Instructions on files
alameda_wl.csv	.csv	Water level validation data from Alameda
dumbarton_wl.csv	.csv	Water level validation data from the Dumbarton Bridge
mallard_wl.csv	.csv	Water level validation data from Mallard Island
mare_wl.csv	.csv	Water level validation data from Mare Island

martinez_wl.csv	.csv	Water level validation data from Martinez
psp_wl.csv	.csv	Water level validation data from Point San Pablo
portchicago_wl.csv	.csv	Water level validation data from Port Chicago
Presidio_wl.csv	.csv	Water level validation data from Presidio
redwoodcity_wl.csv	.csv	Water level validation data from Redwood City
richmond_wl.csv	.csv	Water level validation data from Richmond
mallard_sal.csv	.csv	Salinity validation data from Mallard Island
martinez_sal_top.csv	.csv	Salinity validation data from Mallard Island (upper probe only)
martinez_sal_bot.csv	.csv	Salinity validation data from Mallard Island (bottom probe only)
pier24_sal.csv	.csv	Salinity validation data from Pier 24 (Bay Bridge) (upper and bottom probes)
psp_sal.csv	.csv	Salinity validation data from Point San Pablo (upper and bottom probes)
presidio_sal.csv	.csv	Salinity validation data from Presidio
sanmateobr_sal.csv	.csv	Salinity validation data from San Mateo Bridge (upper and bottom probes)
selby_sal.csv	.csv	Salinity validation data from Selby (Wickland Oil Terminal) (upper and lower probes)

A.3.3 Apalachicola Bay Test Case

The data files and their instruction files are summarized in Table A.3.3. For the instructions of the data files, users are suggested to read file README.TXT.

Table A.3.3 Data files in the dataset of Apalachicola Bay test case

Filename	Type	Note
Initials.txt	Data	Initial salinity.
Depthgridcenter.txt	Data	Bathymetry data given at grid center
Oceane_ele.txt	Data	Water levels at open ocean boundary
Ocean_s.txt	Data	Salinity and temperature at open ocean boundaries
River_dis.txt	Data	Freshwater inputs from Apalachicola River and distributaries
Wind.txt	Data	Wind speeds and directions.
Els397_validation.txt	Data	Water level in the bay for model validation at station s397
Els399_validation.txt	Data	Water level in the bay for model validation at station s399
S_validation.txt	Data	Salinity data at stations in the bay for model validation
V_validation.txt	Data	Velocity at station s400 for model validation
Readme.txt	Text	Instructions of the data files in the case

A.3.4 Melfort Bay Test Case

This test case, in fact, contains two complete data sets for the application site validation. One was observed in 1990. Another was 1999. The number 90 and 99 used in data file names designate the year when the data set was measured. The data files and their instruction files are summarized in Table A.3.4. For the instructions of the data files, users are suggested to read the README file.

Table A.3.4 Data files in the dataset of Melfort Bay test case

Filename	Type	Note
MELDORF90.GRD	Data	Bathymetric data observed in 1990
MELDORF99.GRD	Data	Bathymetric data observed in 1999
TIDES90.AST	Data	Harmonic constants of tidal constituents
BLAUORT90.WL	Data	Tidal elevation at station Blauort in 1990
TERTIUS90.WL	Data	Tidal elevation at station Teritus in 1990
TRISCHEN90.WL	Data	Tidal elevation at station Trischen in 1990
BLAUORT99.WL	Data	Tidal elevation at station Blauort in 1999
TERTIUS99.WL	Data	Tidal elevation at station Teritus in 1999
TRISCHEN99.WL	Data	Tidal elevation at station Trischen in 1999
BUESUM90.METRO	Data	Meteorological data in 1990
BUE99.METRO	Data	Meteorological data in 1999
FLACKSTROM90.WL	Data	Tidal elevation at Flackstrom in 1990
STEERTLOCH90.WL	Data	Tidal elevation at Steertloch in 1990
FLACKSTROM99.WL	Data	Tidal elevation at Flackstrom in 1999
STEERTLOCH99.WL	Data	Tidal elevation at Steertloch in 1999
BUESUM99.WL	Data	Tidal elevation at Buesum in 1999
P3Axxx.DAT	Data	Velocity transect measurenments
READ_ADCP.TXT	Text	Instructions for velocity data files

A.3.5 Tokyo Bay Test Case

This data set is described on the basis of a triangular finite elemental mesh. A data driver code, README.F, is to facilitate the user to read and extract data in the set. The data files and their instruction files are summarized in Table A.3.5. For the instructions of the data files, users are suggested to read file README.TXT.

Table A.3.5 Data files in the dataset of Tokyo Bay test case

Filename	Type	Note
README.F	Text	A FORTRAN77 code is to read the data files provided in the database. Users may utilize this program as a driver to output the part of the data or all of the data. Using the program, you can

		handle the complex data set of a finite element mesh, and confirm the meanings of the data. A list of variables and their descriptions are provided in the code.
MESH.DAT	Data	The data set of multiple-leveled finite element mesh includes the coordinates in horizontal plane and the nodal points of triangular element connections;
WIND_VEL.DAT	Data	A wind velocity data file consisting of horizontal vectors of wind velocities in surface mesh
LEVEL.DAT	Data	The total nodal number in each level mesh, and the total number of triangular elements in each level
NOD_CONK.DAT	Data	The nodal relationship (connection) in vertical direction in the level meshes, including the downstairs nodal points from a nodal point on a level
ELE_CONK.DAT	Data	the elemental relationship (connection) in vertical direction in the level meshes, including the downstairs elements from an element on a level
NOD_RIVER.DAT	Data	The nodal points of rivers;
NOD_OPNBC.DAT	Data	a data set for specifying the open boundary incident wave boundary)
TABLE1.XLS	Data	The data file shown in Table 1 in the paper
TABLE2.XLS	Data	The data file shown in Table 2 in the paper, i.e., harmonic constants of northward and eastward velocity components (Upper Layer)
TABLE3.XLS	Data	The data file shown in Table 3 in the paper, i.e., harmonic constants of northward and eastward velocity components (Middle Layer)
TABLE4.XLS	Data	The data file shown in Table 4 in the paper, i.e., harmonic constants of northward and eastward velocity components (Lower Layer)
TABLE5.XLS	Data	The data file shown in Table 5 in the paper, i.e., measured steady flow at 10 stations in Tokyo Bay
TABLE6.XLS	Data	The data file shown in Table 6 in the paper, i.e., physical constants in the model
TABLE78.XLS	Data	The data file shown in Table 7 and Table 8 in the paper, i.e., tidal constituents of incident wave and river inflow velocities into Tokyo Bay
NOD_WVGAUG.XLS	Data	includes the nodal points related to the eight wave gauges retrieved from the surface mesh
NOD_VELOBS.XLS	Data	includes the nodal points related to the ten velocity observation stations retrieved from the

		mesh shown in the file mesh.dat.
SURF_MESH.DAT	Data	The data set of surface mesh including the water depth and wind velocities defined in the surface mesh
OUTLET_WL.DAT	Data	includes the time series of water elevations in the open boundary of Tokyo Bay. In the file, the left column is the time (unit: second), the right column is the corresponding water elevation.
README.TXT	Text	Instructions of the data files

A.3.6 Victoria Bendway Test Case

This data set was generated for validating an existing hydrodynamic model, CCHE3D. If a user has some knowledge about the CCHE3D or CCHE2D model, one may immediately copy the data to start the simulation. A FORTRAN code, READ_GEO.F, can be used for handling the bathymetric data in GEO.DAT. The data files and their instruction files are summarized in Table A.3.6. For the instructions of the data files, users are suggested to read the README file.

Table A.3.6 Data files in the dataset of Victoria Bendway test case

Filename	Type	Note
READ_GEO.F	Text	FORTRAN code to read bathymetry, boundary roughness
READ_MEA_DATA.F	Text	FORTRAN code to read velocity data
FLOWB.DAT	Data	Discharge and surface elevations
GEO.DAT	Data	Bathymetry, roughness, and boundary
MEASURE.DAT	Data	Velocity data
NEWMESH.GEO.DAT	Data	Mesh in the main channel, its transverse section are normal to the main flow. This mesh is for projecting the helical, not for computation.
BND.DAT	Data	Distribution of unit discharge and flow angle (degree) at inlet section
README	Text	Instructions for the data files

Appendix B

FORCING TERM FORMULATIONS AND FORTRAN CODES FOR USING THE METHOD OF MANUFACTURED SOLUTIONS

Contributor: Yafei Jia

B. 1 INTRODUCTION

To apply the Method of Manufactured Solution to conducting the Mathematical Verification of a non-linear model, the derivations for obtaining the Forcing or Source Terms on the right hand side of the modified governing equations could be quite tedious. For the convenience of users, the contributor of the test case has the forcing terms of all governing differential equations derived and the source code needed to compute the values of these terms for each equation at each given point in the solution domain are provided below. These forcing values should be added to each one of the algebraic equations of the numerical model before execution, so that the numerical solutions of your modified boundary value problem can be compared to the manufactured solution for the purpose of mathematical verification. Detailed introductions are in Chapter 4, Section 4.3 and 4.4.

B. 2 FORCING TERM FORMULATIONS

Elementary differentiations to non-linear functions of the manufactured solutions were performed. First and second order partial derivatives in terms of all three coordinates (x, y and z) were obtained and the analytical expressions of these derivatives are listed in the following. The forcing function or term for each equation is formed by assembling the these function values and derivatives (such as u, $\partial u/\partial x$,...) *exactly* as the differential equation to be verified.

B.2.1 The time dependent manufactured solution: *Function I*

The manufactured solution for the free surface flows

$$u = \sin y \cos^2 \frac{x}{2} \cos \frac{\pi}{2} (1 - \frac{z}{h}) \sin t$$

$$v = -\sin x \cos^2 \frac{y}{2} \cos \frac{\pi}{2} (1 - \frac{z}{h}) \sin t$$

$$w = -\lambda \cos \frac{x}{2} \cos \frac{y}{2} \cos \frac{\pi}{2} (1 - \frac{z}{h}) \sin t$$

$$h = \lambda \cos \frac{x}{2} \cos \frac{y}{2} \cos t + h_0$$

$$p = C_p \cos(x) \cos(y) \cos(\frac{\pi}{2} \frac{z}{h})$$

Boundary conditions:

At z=0, $u_0 = 0, v_0 = 0, w_0 = 0$.

At surrounding boundaries, $x = -\pi$, $x = \pi$, and $y = -\pi$, $y = \pi$, all vector components are zero:

$$u_{x=-\pi} = 0, v_{x=-\pi} = 0, w_{x=-\pi} = 0,$$
$$u_{x=\pi} = 0, v_{x=\pi} = 0, w_{x=\pi} = 0.$$

At top boundary $z=h$, one has velocity distribution on the top surface

$$u_h = \sin y \cos^2 \frac{x}{2} \sin t$$

$$v_h = -\sin x \cos^2 \frac{y}{2} \sin t$$

$$w_h = -\lambda \cos \frac{x}{2} \cos \frac{y}{2} \sin t$$

The free surface kinematic equation

$$\frac{\partial h}{\partial t} + u_h \frac{\partial h}{\partial x} + v_h \frac{\partial h}{\partial y} - w_h = 0$$

is satisfied,

$$\frac{\partial h}{\partial x} = -\frac{\lambda}{2} \sin \frac{x}{2} \cos \frac{y}{2} \cos t$$
$$\frac{\partial h}{\partial y} = -\frac{\lambda}{2} \cos \frac{x}{2} \sin \frac{y}{2} \cos t$$

$$C = \cos[\frac{\pi}{2} (1 - \frac{z}{h})]$$

$$S=\sin[\frac{\pi}{2}(1-\frac{z}{h})]$$

$$\frac{\partial C}{\partial x}=-\frac{\pi z}{2h^2}\cdot S\cdot\frac{\partial h}{\partial x}$$
$$\frac{\partial C}{\partial y}=-\frac{\pi z}{2h^2}\cdot S\cdot\frac{\partial h}{\partial y}$$
$$\frac{\partial C}{\partial z}=\frac{\pi}{2h}\cdot S$$

$$\frac{\partial S}{\partial x}=\frac{\pi z}{h^2}\cdot C\cdot\frac{\partial h}{\partial x}$$
$$\frac{\partial S}{\partial z}=-\frac{\pi}{2h}\cdot C$$

$$\frac{\partial^2 C}{\partial x^2}=\frac{\pi z}{h^3}(\frac{\partial h}{\partial x})^2\cdot S+\frac{\pi z}{2h^2}\cdot S\cdot\frac{\partial^2 h}{\partial x^2}-\frac{\pi z}{2h^2}\frac{\partial S}{\partial x}\frac{\partial h}{\partial x}$$
$$\frac{\partial^2 C}{\partial x\partial y}=-\frac{\lambda\pi z}{2h^3}\frac{\partial h}{\partial y}\sin\frac{x}{2}\cos\frac{y}{2}\cdot S\cdot\cos t-\frac{\lambda\pi z}{8h^2}\sin\frac{x}{2}\sin\frac{y}{2}\cdot S\cdot\cos t+\frac{\lambda\pi z}{4h^2}\sin\frac{x}{2}\cos\frac{y}{2}\frac{\partial S}{\partial y}\cos t$$
$$\frac{\partial^2 C}{\partial x\partial z}=\frac{\lambda\pi}{4h^2}\sin\frac{x}{2}\cos\frac{y}{2}\cdot S\cdot\cos t+\frac{\lambda\pi z}{4h^2}\sin\frac{x}{2}\cos\frac{y}{2}\frac{\partial S}{\partial z}\cos t$$
$$\frac{\partial^2 C}{\partial y^2}=-\frac{\lambda\pi z}{2h^3}\frac{\partial h}{\partial y}\cos\frac{x}{2}\sin\frac{y}{2}\cdot S\cdot\cos t+\frac{\lambda\pi z}{8h^2}\cos\frac{x}{2}\cos\frac{y}{2}\cdot S\cdot\cos t+\frac{\lambda\pi z}{4h^2}\cos\frac{x}{2}\sin\frac{y}{2}\frac{\partial S}{\partial y}\cos t$$
$$\frac{\partial^2 C}{\partial y\partial z}=\frac{\lambda\pi}{4h^2}\cos\frac{x}{2}\sin\frac{y}{2}\cdot S\cdot\cos t+\frac{\lambda\pi z}{4h^2}\cos\frac{x}{2}\sin\frac{y}{2}\frac{\partial S}{\partial z}\cos t$$

$$\frac{\partial^2 C}{\partial z^2}=\frac{\pi}{2h}\frac{\partial S}{\partial z}$$

$$\frac{\partial u}{\partial x}=-\frac{1}{2}\sin x\sin y\cdot C\cdot\sin t+\sin y\cos^2\frac{x}{2}\frac{\partial C}{\partial x}\sin t$$
$$\frac{\partial u}{\partial y}=\cos^2\frac{x}{2}\cos y\cdot C\cdot\sin t+\sin y\cos^2\frac{x}{2}\frac{\partial C}{\partial y}\sin t$$
$$\frac{\partial u}{\partial z}=\cos^2\frac{x}{2}\sin y\frac{\partial C}{\partial z}\sin t$$

$$\frac{\partial^2 u}{\partial x^2}=-\frac{1}{2}\cos x\sin y\cdot C\cdot\sin t-\sin x\sin y\frac{\partial C}{\partial x}\sin t+\sin y\cos^2\frac{x}{2}\frac{\partial^2 C}{\partial x^2}\sin t$$

$$\frac{\partial^2 u}{\partial x\partial y} = -\frac{1}{2}\sin x\cos y\cdot C\cdot\sin t - \frac{1}{2}\sin x\sin y\frac{\partial C}{\partial y}\sin t$$

$$+\cos^2\frac{x}{2}\cos y\frac{\partial C}{\partial x}\sin t + \sin y\cos^2\frac{x}{2}\frac{\partial^2 C}{\partial x\partial y}\sin t$$

$$\frac{\partial^2 u}{\partial x\partial z} = -\frac{1}{2}\sin x\sin y\frac{\partial C}{\partial z}\sin t + \sin y\cos^2\frac{x}{2}\frac{\partial^2 C}{\partial x\partial z}\sin t$$

$$\frac{\partial^2 u}{\partial y^2} = -\cos^2\frac{x}{2}\sin y\cdot C\cdot\sin t + 2\cos y\cos^2\frac{x}{2}\frac{\partial C}{\partial y}\sin t + \sin y\cos^2\frac{x}{2}\frac{\partial^2 C}{\partial y^2}\sin t$$

$$\frac{\partial^2 u}{\partial y\partial z} = \cos y\cos^2\frac{x}{2}\frac{\partial C}{\partial z}\sin t + \sin y\cos^2\frac{x}{2}\frac{\partial^2 C}{\partial y\partial z}\sin t$$

$$\frac{\partial^2 u}{\partial z^2} = \cos^2\frac{x}{2}\sin y\frac{\partial^2 C}{\partial z^2}\sin t$$

$$\frac{\partial v}{\partial x} = -\cos x\cos^2\frac{y}{2}\cdot C\cdot\sin t - \sin x\cos^2\frac{y}{2}\frac{\partial C}{\partial x}\sin t$$

$$\frac{\partial v}{\partial y} = \frac{1}{2}\sin x\sin y\cdot C\cdot\sin t - \sin x\cos^2\frac{y}{2}\frac{\partial C}{\partial y}\sin t$$

$$\frac{\partial v}{\partial z} = -\cos^2\frac{y}{2}\sin x\frac{\partial C}{\partial z}\sin t$$

$$\frac{\partial^2 x}{\partial x^2} = \cos^2\frac{y}{2}\sin x\cdot C\cdot\sin t - 2\cos x\cos^2\frac{y}{2}\frac{\partial C}{\partial x}\sin t - \sin x\cos^2\frac{y}{2}\frac{\partial^2 C}{\partial x^2}\sin t$$

$$\frac{\partial^2 v}{\partial x\partial y} = \frac{1}{2}\sin y\cos x\cdot C\cdot\sin t + \frac{1}{2}\sin x\sin y\frac{\partial C}{\partial x}\sin t$$

$$-\cos^2\frac{y}{2}\cos x\frac{\partial C}{\partial y}\sin t - \sin x\cos^2\frac{y}{2}\frac{\partial^2 C}{\partial x\partial y}\sin t$$

$$\frac{\partial^2 v}{\partial x\partial z} = -\cos x\cos^2\frac{y}{2}\frac{\partial C}{\partial z}\sin t - \sin x\cos^2\frac{y}{2}\frac{\partial^2 C}{\partial x\partial z}\sin t$$

$$\frac{\partial^2 v}{\partial y^2} = \frac{1}{2}\sin x\cos y\cdot C\cdot\sin t + \sin x\sin y\frac{\partial C}{\partial y}\sin t - \sin x\cos^2\frac{y}{2}\frac{\partial^2 C}{\partial y^2}\sin t$$

$$\frac{\partial^2 v}{\partial y\partial z} = \frac{1}{2}\sin x\sin y\frac{\partial C}{\partial z}\sin t - \sin x\cos^2\frac{y}{2}\frac{\partial^2 C}{\partial y\partial z}\sin t$$

$$\frac{\partial^2 v}{\partial z^2} = -\sin x\cos^2\frac{y}{2}\frac{\partial^2 C}{\partial z^2}\sin t$$

$$\frac{\partial w}{\partial x} = \frac{\lambda}{2}\sin\frac{x}{2}\cos\frac{y}{2}\cdot C\cdot\sin t - \lambda\cos\frac{x}{2}\cos\frac{y}{2}\frac{\partial C}{\partial x}\sin t$$

$$\frac{\partial w}{\partial y}=\frac{\lambda}{2}\cos\frac{x}{2}\sin\frac{y}{2}\cdot C\cdot\sin t-\lambda\cos\frac{x}{2}\cos\frac{y}{2}\frac{\partial C}{\partial y}\sin t$$

$$\frac{\partial w}{\partial z}=-\lambda\cos\frac{x}{2}\cos\frac{y}{2}\frac{\partial C}{\partial z}\sin t$$

$$\frac{\partial^2 w}{\partial x^2}=\frac{\lambda}{4}\cos\frac{x}{2}\cos\frac{y}{2}\cdot C\cdot\sin t+\lambda\sin\frac{x}{2}\cos\frac{y}{2}\frac{\partial C}{\partial x}\sin t-\lambda\cos\frac{x}{2}\cos\frac{y}{2}\frac{\partial^2 C}{\partial x^2}\sin t$$

$$\frac{\partial^2 w}{\partial x\partial y}=-\frac{\lambda}{4}\sin\frac{x}{2}\sin\frac{y}{2}\cdot C\cdot\sin t+\frac{\lambda}{2}\sin\frac{x}{2}\cos\frac{y}{2}\frac{\partial C}{\partial y}\sin t$$

$$+\frac{\lambda}{2}\cos\frac{x}{2}\sin\frac{y}{2}\frac{\partial C}{\partial x}-\lambda\cos\frac{x}{2}\cos\frac{y}{2}\frac{\partial^2 C}{\partial x\partial y}\sin t$$

$$\frac{\partial^2 w}{\partial x\partial z}=\frac{\lambda}{2}\sin\frac{x}{2}\cos\frac{y}{2}\frac{\partial C}{\partial z}\sin t-\lambda\cos\frac{x}{2}\cos\frac{y}{2}\frac{\partial^2 C}{\partial x\partial z}\sin t$$

$$\frac{\partial^2 w}{\partial y^2}=\frac{\lambda}{4}\cos\frac{x}{2}\cos\frac{y}{2}\cdot C\cdot\sin t+\lambda\cos\frac{x}{2}\sin\frac{y}{2}\frac{\partial C}{\partial y}\sin t-\lambda\cos\frac{x}{2}\cos\frac{y}{2}\frac{\partial^2 C}{\partial y^2}\sin t$$

$$\frac{\partial^2 w}{\partial y\partial z}=\frac{\lambda}{2}\cos\frac{x}{2}\sin\frac{y}{2}\frac{\partial C}{\partial z}\sin t-\lambda\cos\frac{x}{2}\cos\frac{y}{2}\frac{\partial^2 C}{\partial y\partial z}\sin t$$

$$\frac{\partial^2 w}{\partial z^2}=-\lambda\cos\frac{x}{2}\cos\frac{y}{2}\frac{\partial^2 C}{\partial z^2}\sin t$$

$$\frac{\partial p}{\partial x}=-\sin x\cos y\cos(\frac{\pi}{2}\frac{z}{h})+\cos x\cos y\sin(\frac{\pi}{2}\frac{z}{h})\frac{\pi z}{2h^2}\frac{\partial h}{\partial x}$$

$$\frac{\partial p}{\partial y}=-\cos x\sin y\cos(\frac{\pi}{2}\frac{z}{h})+\cos x\cos y\sin(\frac{\pi}{2}\frac{z}{h})\frac{\pi z}{2h^2}\frac{\partial h}{\partial y}$$

$$\frac{\partial p}{\partial z}=-\cos x\cos y\sin(\frac{\pi}{2}\frac{z}{h})\frac{\pi}{2h}$$

$$\frac{\partial^2 p}{\partial x^2}=-\cos x\cos y\cos(\frac{\pi}{2}\frac{z}{h})-2\sin x\cos y\sin(\frac{\pi}{2}\frac{z}{h})\frac{\pi z}{2h^2}\frac{\partial h}{\partial x}$$

$$+\cos x\cos y[-\cos(\frac{\pi}{2}\frac{z}{h})(\frac{\pi z}{2h^2})^2(\frac{\partial h}{\partial x})^2-\sin(\frac{\pi}{2}\frac{z}{h})\frac{\pi z}{h^3}(\frac{\partial h}{\partial x})^2+\sin(\frac{\pi}{2}\frac{z}{h})\frac{\pi z}{2h^2}\frac{\partial^2 h}{\partial x^2}]$$

$$\frac{\partial^2 p}{\partial y^2}=-\cos x\cos y\cos(\frac{\pi}{2}\frac{z}{h})-2\cos x\sin y\sin(\frac{\pi}{2}\frac{z}{h})\frac{\pi z}{2h^2}\frac{\partial h}{\partial y}$$

$$+\cos x\cos y[-\cos(\frac{\pi}{2}\frac{z}{h})(\frac{\pi z}{2h^2})^2(\frac{\partial h}{\partial y})^2-\sin(\frac{\pi}{2}\frac{z}{h})\frac{\pi z}{h^3}(\frac{\partial h}{\partial y})^2+\sin(\frac{\pi}{2}\frac{z}{h})\frac{\pi z}{2h^2}\frac{\partial^2 h}{\partial y^2}]$$

$$\frac{\partial^2 p}{\partial z^2} = -\cos x \cos y \cos(\frac{\pi}{2}\frac{z}{h})(\frac{\pi}{2h})^2$$

Source term development:

Source terms for the momentum equations

$$F_u = \frac{\partial u}{\partial t} + u\frac{\partial u}{\partial x} + v\frac{\partial u}{\partial y} + w\frac{\partial u}{\partial z} - \nu_t(\frac{\partial^2 u}{\partial x^2} + \frac{\partial^2 u}{\partial y^2} + \frac{\partial^2 u}{\partial z^2}) - \nu_t(\frac{\partial^2 u}{\partial x^2} + \frac{\partial^2 v}{\partial y \partial x} + \frac{\partial^2 w}{\partial z \partial x})$$

$$F_v = \frac{\partial v}{\partial t} + u\frac{\partial v}{\partial x} + v\frac{\partial v}{\partial y} + w\frac{\partial v}{\partial z} - \nu_t(\frac{\partial^2 v}{\partial x^2} + \frac{\partial^2 v}{\partial y^2} + \frac{\partial^2 v}{\partial z^2}) - \nu_t(\frac{\partial^2 u}{\partial x \partial y} + \frac{\partial^2 v}{\partial y^2} + \frac{\partial^2 w}{\partial z \partial y})$$

$$F_w = \frac{\partial w}{\partial t} + u\frac{\partial w}{\partial x} + v\frac{\partial w}{\partial y} + w\frac{\partial w}{\partial z} - \nu_t(\frac{\partial^2 w}{\partial x^2} + \frac{\partial^2 w}{\partial y^2} + \frac{\partial^2 w}{\partial z^2}) - \nu_t(\frac{\partial^2 u}{\partial x \partial z} + \frac{\partial^2 v}{\partial y \partial z} + \frac{\partial^2 w}{\partial z^2})$$

$$F_h = 0$$

$$F_p = \frac{\partial^2 p}{\partial x^2} + \frac{\partial^2 p}{\partial y^2} + \frac{\partial^2 p}{\partial z^2}$$

B.2.2 The manufactured solution: *Function II*

$$\frac{\partial u}{\partial x} = \cos(x)\cos(y)\sin(\frac{\pi}{2}\frac{z}{h}) + \sin(x)\sin(y)[\cos(2\pi\frac{z}{h}) - 1]$$
$$+\sin(x)\cos(y)(-\frac{\pi}{2}\frac{z}{h^2})\cos(\frac{\pi}{2}\frac{z}{h})\frac{\partial h}{\partial x} - \cos(x)\sin(y)(2\pi\frac{z}{h^2})\sin(2\pi\frac{z}{h})\frac{\partial h}{\partial x}$$

$$\frac{\partial u}{\partial y} = -\sin(x)\sin(y)\sin(\frac{\pi}{2}\frac{z}{h}) - \cos(x)\cos(y)[\cos(2\pi\frac{z}{h}) - 1]$$
$$+\sin(x)\cos(y)(-\frac{\pi}{2}\frac{z}{h^2})\cos(\frac{\pi}{2}\frac{z}{h})\frac{\partial h}{\partial y} - \cos(x)\sin(y)(2\pi\frac{z}{h^2})\sin(2\pi\frac{z}{h})\frac{\partial h}{\partial y}$$

$$\frac{\partial u}{\partial z} = \frac{\pi}{2h}\sin(x)\cos(y)\cos(\frac{\pi}{2}\frac{z}{h}) + \frac{2\pi}{h}\cos(x)\sin(y)\sin(2\pi\frac{z}{h})$$

$$\frac{\partial^2 u}{\partial x^2} = -u$$

$$+2\cos(x)\cos(y)(-\frac{\pi}{2}\frac{z}{h^2})\cos(\frac{\pi}{2}\frac{z}{h})\frac{\partial h}{\partial x} + 2\sin(x)\sin(y)(2\pi\frac{z}{h^2})\sin(2\pi\frac{z}{h})\frac{\partial h}{\partial x}$$

$$+\sin(x)\cos(y)[(-\frac{\pi}{2}\frac{z}{h^2})(\frac{\pi}{2}\frac{z}{h^2})\sin(\frac{\pi}{2}\frac{z}{h}) + (\pi\frac{z}{h^3})\cos(\frac{\pi}{2}\frac{z}{h})](\frac{\partial h}{\partial x})^2$$

$$+\cos(x)\sin(y)[(2\pi\frac{z}{h^2})^2\cos(2\pi\frac{z}{h}) + (4\pi\frac{z}{h^3})\sin(2\pi\frac{z}{h})](\frac{\partial h}{\partial x})^2$$

$$-[\sin(x)\cos(y)(\frac{\pi}{2}\frac{z}{h^2})\cos(\frac{\pi}{2}\frac{z}{h}) + \cos(x)\sin(y)(2\pi\frac{z}{h^2})\sin(2\pi\frac{z}{h})]\frac{\partial^2 h}{\partial x^2}$$

$$\frac{\partial^2 u}{\partial y^2} = -u$$

$$-2\sin(x)\sin(y)(-\frac{\pi}{2}\frac{z}{h^2})\cos(\frac{\pi}{2}\frac{z}{h})\frac{\partial h}{\partial y} - 2\cos(x)\cos(y)(2\pi\frac{z}{h^2})\sin(2\pi\frac{z}{h})\frac{\partial h}{\partial y}$$

$$-[\sin(x)\cos(y)(\frac{\pi}{2}\frac{z}{h^2})\cos(\frac{\pi}{2}\frac{z}{h}) - \cos(x)\sin(y)(2\pi\frac{z}{h^2})\sin(2\pi\frac{z}{h})]\frac{\partial^2 h}{\partial y^2}$$

$$+\sin(x)\cos(y)[(-\frac{\pi}{2}\frac{z}{h^2})(\frac{\pi}{2}\frac{z}{h^2})\sin(\frac{\pi}{2}\frac{z}{h}) + (\pi\frac{z}{h^3})\cos(\frac{\pi}{2}\frac{z}{h})](\frac{\partial h}{\partial y})^2$$

$$+\cos(x)\sin(y)[(2\pi\frac{z}{h^2})^2\cos(2\pi\frac{z}{h}) + (4\pi\frac{z}{h^3})\sin(2\pi\frac{z}{h})](\frac{\partial h}{\partial y})^2$$

$$\frac{\partial^2 u}{\partial z^2} = -(\frac{\pi}{2h})^2\sin(x)\cos(y)\sin(\frac{\pi}{2}\frac{z}{h}) + (\frac{2\pi}{h})^2\cos(x)\sin(y)\cos(2\pi\frac{z}{h})$$

$$\frac{\partial^2 u}{\partial x \partial y} = -\cos(x)\sin(y)\sin(\frac{\pi}{2}\frac{z}{h}) + \sin(x)\cos(y)[\cos(2\pi\frac{z}{h}) - 1]$$

$$+[\cos(x)\cos(y)\frac{\partial h}{\partial y} - \sin(x)\sin(y)\frac{\partial h}{\partial x}](-\frac{\pi}{2}\frac{z}{h^2})\cos(\frac{\pi}{2}\frac{z}{h})$$

$$+[\sin(x)\sin(y)\frac{\partial h}{\partial y} - \cos(x)\cos(y)\frac{\partial h}{\partial x}](2\pi\frac{z}{h^2})\sin(2\pi\frac{z}{h})$$

$$+[\sin(x)\cos(y)(-\frac{\pi}{2}\frac{z}{h^2})\cos(\frac{\pi}{2}\frac{z}{h}) - \cos(x)\sin(y)(2\pi\frac{z}{h^2})\sin(2\pi\frac{z}{h})]\frac{\partial^2 h}{\partial x \partial y}$$

$$+\sin(x)\cos(y)[-(\frac{\pi}{2}\frac{z}{h^2})^2\sin(\frac{\pi}{2}\frac{z}{h}) + (\pi\frac{z}{h^3})\cos(\frac{\pi}{2}\frac{z}{h})]\frac{\partial h}{\partial x}\frac{\partial h}{\partial y}$$

$$+\cos(x)\sin(y)[(2\pi\frac{z}{h^2})^2\cos(2\pi\frac{z}{h}) + (4\pi\frac{z}{h^3})\sin(2\pi\frac{z}{h})]\frac{\partial h}{\partial x}\frac{\partial h}{\partial y}$$

$$\frac{\partial^2 u}{\partial x \partial z} = \cos(x)\cos(y)(\frac{\pi}{2h})\cos(\frac{\pi}{2}\frac{z}{h}) - \sin(x)\sin(y)(\frac{2\pi}{h})\sin(2\pi\frac{z}{h})$$
$$+\sin(x)\cos(y)[(\frac{\pi^2}{4}\frac{z}{h^3})\sin(\frac{\pi}{2}\frac{z}{h}) - \frac{\pi}{2h^2}\cos(\frac{\pi}{2}\frac{z}{h})]\frac{\partial h}{\partial x}$$
$$-\cos(x)\sin(y)[(4\pi^2\frac{z}{h^3})\cos(2\pi\frac{z}{h}) + \frac{2\pi}{h^2}\sin(2\pi\frac{z}{h})]\frac{\partial h}{\partial x}$$

$$\frac{\partial^2 u}{\partial y \partial z} = -\sin(x)\sin(y)(\frac{\pi}{2h})\cos(\frac{\pi}{2}\frac{z}{h}) + \cos(x)\cos(y)(\frac{2\pi}{h})\sin(2\pi\frac{z}{h})$$
$$+\sin(x)\cos(y)[(\frac{\pi^2}{4}\frac{z}{h^3})\sin(\frac{\pi}{2}\frac{z}{h}) - \frac{\pi}{2h^2}\cos(\frac{\pi}{2}\frac{z}{h})]\frac{\partial h}{\partial y}$$
$$-\cos(x)\sin(y)[(4\pi^2\frac{z}{h^3})\cos(2\pi\frac{z}{h}) + \frac{2\pi}{h^2}\sin(2\pi\frac{z}{h})]\frac{\partial h}{\partial y}$$

Derivatives for v velocity:

$$\frac{\partial v}{\partial x} = \sin(x)\sin(y)\sin(\frac{\pi}{2}\frac{z}{h}) + \cos(x)\cos(y)[(\cos(2\pi\frac{z}{h}) - 1]$$
$$+\cos(x)\sin(y)(\frac{\pi}{2}\frac{z}{h^2})\cos(\frac{\pi}{2}\frac{z}{h})\frac{\partial h}{\partial x} + \sin(x)\cos(y)(2\pi\frac{z}{h^2})\sin(2\pi\frac{z}{h})\frac{\partial h}{\partial x}$$
$$\frac{\partial v}{\partial y} = -\cos(x)\cos(y)\sin(\frac{\pi}{2}\frac{z}{h}) - \sin(x)\sin(y)[\cos(2\pi\frac{z}{h}) - 1]$$
$$+\cos(x)\sin(y)(\frac{\pi}{2}\frac{z}{h^2})\cos(\frac{\pi}{2}\frac{z}{h})\frac{\partial h}{\partial y} + \sin(x)\cos(y)(2\pi\frac{z}{h^2})\sin(2\pi\frac{z}{h})\frac{\partial h}{\partial y}$$
$$\frac{\partial v}{\partial z} = -\frac{\pi}{2h}\cos(x)\sin(y)\cos(\frac{\pi}{2}\frac{z}{h}) - \sin(x)\cos(y)(\frac{2\pi}{h})\sin(2\pi\frac{z}{h})$$

$$\frac{\partial^2 v}{\partial x^2} = -v$$
$$+2\sin(x)\sin(y)[(-\frac{\pi}{2}\frac{z}{h^2})\cos(\frac{\pi}{2}\frac{z}{h})\frac{\partial h}{\partial x} + 2\cos(x)\cos(y)(2\pi\frac{z}{h^2})\sin(2\pi\frac{z}{h})\frac{\partial h}{\partial x}$$
$$-\cos(x)\sin(y)[-(\frac{\pi}{2}\frac{z}{h^2})^2\sin(\frac{\pi}{2}\frac{z}{h}) + (\pi\frac{z}{h^3})\cos(\frac{\pi}{2}\frac{z}{h})](\frac{\partial h}{\partial x})^2$$
$$-\sin(x)\cos(y)[(2\pi\frac{z}{h^2})^2\cos(2\pi\frac{z}{h}) + (4\pi\frac{z}{h^3})\sin(2\pi\frac{z}{h})](\frac{\partial h}{\partial x})^2$$
$$+[\cos(x)\sin(y)(\frac{\pi}{2}\frac{z}{h^2})\cos(\frac{\pi}{2}\frac{z}{h}) + \sin(x)\cos(y)(2\pi\frac{z}{h^2})\sin(2\pi\frac{z}{h})]\frac{\partial^2 h}{\partial x^2}$$

$$\frac{\partial^2 v}{\partial y^2} = -v$$

$$- 2\cos(x)\cos(y)(-\frac{\pi}{2}\frac{z}{h^2})\cos(\frac{\pi}{2}\frac{z}{h})\frac{\partial h}{\partial y} - 2\sin(x)\sin(y)(2\pi\frac{z}{h^2})\sin(2\pi\frac{z}{h})\frac{\partial h}{\partial y}$$

$$+ [\cos(x)\sin(y)(\frac{\pi}{2}\frac{z}{h^2})\cos(\frac{\pi}{2}\frac{z}{h}) + \sin(x)\cos(y)(2\pi\frac{z}{h^2})\sin(2\pi\frac{z}{h})]\frac{\partial^2 h}{\partial y^2}$$

$$+ \cos(x)\sin(y)[(\frac{\pi}{2}\frac{z}{h^2})^2\sin(\frac{\pi}{2}\frac{z}{h}) - (\pi\frac{z}{h^3})\cos(\frac{\pi}{2}\frac{z}{h})](\frac{\partial h}{\partial y})^2$$

$$- \sin(x)\cos(y)[(2\pi\frac{z}{h^2})^2\cos(2\pi\frac{z}{h}) + (4\pi\frac{z}{h^3})\sin(2\pi\frac{z}{h})](\frac{\partial h}{\partial y})^2$$

$$\frac{\partial^2 v}{\partial z^2} = (\frac{\pi}{2h})^2\cos(x)\sin(y)\sin(\frac{\pi}{2}\frac{z}{h}) - \sin(x)\cos(y)(\frac{2\pi}{h})^2\cos(2\pi\frac{z}{h})$$

$$\frac{\partial^2 v}{\partial x\partial y} = \sin(x)\cos(y)\sin(\frac{\pi}{2}\frac{z}{h}) - \cos(x)\sin(y)[\cos(2\pi\frac{z}{h}) - 1]$$

$$+ \sin(x)\sin(y)(-\frac{\pi}{2}\frac{z}{h^2})\cos(\frac{\pi}{2}\frac{z}{h})\frac{\partial h}{\partial y} + \cos(x)\cos(y)(2\pi\frac{z}{h^2})\sin(2\pi\frac{z}{h})\frac{\partial h}{\partial y}$$

$$+ \cos(x)\cos(y)(\frac{\pi}{2}\frac{z}{h^2})\cos(\frac{\pi}{2}\frac{z}{h})\frac{\partial h}{\partial x} - \sin(x)\sin(y)(2\pi\frac{z}{h^2})\sin(2\pi\frac{z}{h})\frac{\partial h}{\partial x}$$

$$- \cos(x)\sin(y)[-(\frac{\pi}{2}\frac{z}{h^2})^2\sin(\frac{\pi}{2}\frac{z}{h}) + (\pi\frac{z}{h^3})\cos(\frac{\pi}{2}\frac{z}{h})]\frac{\partial h}{\partial x}\frac{\partial h}{\partial y}$$

$$+ \sin(x)\cos(y)[-(2\pi\frac{z}{h^2})^2\cos(2\pi\frac{z}{h}) - (4\pi\frac{z}{h^3})\sin(2\pi\frac{z}{h})]\frac{\partial h}{\partial x}\frac{\partial h}{\partial y}$$

$$+ [\cos(x)\sin(y)(\frac{\pi}{2}\frac{z}{h^2})\cos(\frac{\pi}{2}\frac{z}{h}) + \sin(x)\cos(y)(2\pi\frac{z}{h^2})\sin(2\pi\frac{z}{h})]\frac{\partial^2 h}{\partial x\partial y}$$

$$\frac{\partial^2 v}{\partial x\partial z} = \sin(x)\sin(y)(\frac{\pi}{2h})\cos(\frac{\pi}{2}\frac{z}{h}) - \cos(x)\cos(y)(\frac{2\pi}{h})\sin(2\pi\frac{z}{h})$$

$$+ \cos(x)\sin(y)[(-\frac{\pi^2}{4}\frac{z}{h^3})\sin(\frac{\pi}{2}\frac{z}{h}) + (\frac{\pi}{2h^2})\cos(\frac{\pi}{2}\frac{z}{h})]\frac{\partial h}{\partial x}$$

$$+ \sin(x)\cos(y)[(4\pi^2\frac{z}{h^3})\cos(2\pi\frac{z}{h}) + (\frac{2\pi}{h^2})\sin(2\pi\frac{z}{h})]\frac{\partial h}{\partial x}$$

$$\frac{\partial^2 v}{\partial y\partial z} = -\cos(x)\cos(y)(\frac{\pi}{2h})\cos(\frac{\pi}{2}\frac{z}{h}) + \sin(x)\sin(y)(\frac{2\pi}{h})\sin(2\pi\frac{z}{h})$$

$$+ \cos(x)\sin(y)[-(\frac{\pi^2}{4}\frac{z}{h^3})\sin(\frac{\pi}{2}\frac{z}{h}) + (\frac{\pi}{2h^2})\cos(\frac{\pi}{2}\frac{z}{h})]\frac{\partial h}{\partial y}$$

$$+ \sin(x)\cos(y)[(4\pi^2\frac{z}{h^3})\cos(2\pi\frac{z}{h}) + (\frac{2\pi}{h^2})\sin(2\pi\frac{z}{h})]\frac{\partial h}{\partial y}$$

Derivatives of w velocity:

$$\frac{\partial w}{\partial x}=-\frac{A}{2\pi}\sin(2x)[\sin(2\pi\frac{z}{h})-2\pi\frac{z}{h}\cos(2\pi\frac{z}{h})+2\pi]$$
$$+\frac{A}{2\pi}[\sin^2(x)\cos^2(y)-\cos^2(x)\sin^2(y)](4\pi^2\frac{z^2}{h^3})\sin(2\pi\frac{z}{h})\frac{\partial h}{\partial x}$$

$$\frac{\partial w}{\partial y}=\frac{A}{2\pi}\sin(2y)[\sin(2\pi\frac{z}{h})-2\pi\frac{z}{h}\cos(2\pi\frac{z}{h})+2\pi]$$
$$+\frac{A}{2\pi}[\sin^2(x)\cos^2(y)-\cos^2(x)\sin^2(y)](4\pi^2\frac{z^2}{h^3})\sin(2\pi\frac{z}{h})\frac{\partial h}{\partial y}$$

$$\frac{\partial w}{\partial z}=-2A\pi\frac{z}{h^2}[\sin(x)^2\cos(y)^2-\cos^2(x)\sin^2(y)]\sin(2\pi\frac{z}{h})\quad)$$

$$\frac{\partial^2 w}{\partial x^2}=-\frac{A}{\pi}\cos(2x)[\sin(2\pi\frac{z}{h})-2\pi\frac{z}{h}\cos(2\pi\frac{z}{h})+2\pi]$$
$$+\frac{A}{\pi}\sin(2x)(4\pi^2\frac{z^2}{h^3})\sin(2\pi\frac{z}{h})\frac{\partial h}{\partial x}$$
$$+\frac{A}{2\pi}[\sin^2(x)\cos^2(y)-\cos^2(x)\sin^2(y)][(-8\pi^3\frac{z^3}{h^5})\cos(2\pi\frac{z}{h})-(12\pi^2\frac{z^2}{h^4})\sin(2\pi\frac{z}{h})](\frac{\partial h}{\partial x})^2$$
$$+\frac{A}{2\pi}[\sin^2(x)\cos^2(y)-\cos^2(x)\sin^2(y)](4\pi^2\frac{z^2}{h^3})\sin(2\pi\frac{z}{h})\frac{\partial^2 h}{\partial x^2}$$

$$\frac{\partial^2 w}{\partial y^2}=\frac{A}{\pi}\cos(2y)[\sin(2\pi\frac{z}{h})-2\pi\frac{z}{h}\cos(2\pi\frac{z}{h})+2\pi]$$
$$-\frac{A}{\pi}\sin(2y)(4\pi^2\frac{z^2}{h^3})\sin(2\pi\frac{z}{h})\frac{\partial h}{\partial y}$$
$$+\frac{A}{2\pi}[\sin^2(x)\cos^2(y)-\cos^2(x)\sin^2(y)][(-8\pi^3\frac{z^3}{h^5})\cos(2\pi\frac{z}{h})-(12\pi^2\frac{z^2}{h^4})\sin(2\pi\frac{z}{h})](\frac{\partial h}{\partial y})^2$$
$$+\frac{A}{2\pi}[\sin^2(x)\cos^2(y)-\cos^2(x)\sin^2(y)](4\pi^2\frac{z^2}{h^3})\sin(2\pi\frac{z}{h})\frac{\partial^2 h}{\partial y^2}$$

$$\frac{\partial^2 w}{\partial z^2}=-A[\sin(x)^2\cos(y)^2-\cos^2(x)\sin^2(y)][4\pi^2\frac{z}{h^3}\cos(2\pi\frac{z}{h})+\frac{2\pi}{h^2}\sin(2\pi\frac{z}{h})]$$

$$\frac{\partial^2 w}{\partial x\partial y}=\frac{A}{2\pi}(4\pi^2\frac{z^2}{h^3})\sin(2\pi\frac{z}{h})[\sin(2x)\frac{\partial h}{\partial y}-\sin(2y)\frac{\partial h}{\partial x}]$$
$$+\frac{A}{2\pi}[\sin^2(x)\cos^2(y)-\cos^2(x)\sin^2(y)][(-8\pi^3\frac{z^3}{h^5})\cos(2\pi\frac{z}{h})-(12\pi^2\frac{z^2}{h^4})\sin(2\pi\frac{z}{h})]\frac{\partial h}{\partial x}\frac{\partial h}{\partial y}$$
$$+\frac{A}{2\pi}[\sin^2(x)\cos^2(y)-\cos^2(x)\sin^2(y)](4\pi^2\frac{z^2}{h^3})\sin(2\pi\frac{z}{h})\frac{\partial^2 h}{\partial x\partial y}$$

$$\frac{\partial^2 w}{\partial x \partial z} = -\frac{A}{2\pi}\sin(2x)(4\pi^2\frac{z}{h^2})\sin(2\pi\frac{z}{h})$$

$$+\frac{A}{2\pi}[\sin^2(x)\cos^2(y) - \cos^2(x)\sin^2(y)][(8\pi^2\frac{z}{h^3})\sin(2\pi\frac{z}{h}) + (8\pi^3\frac{z^2}{h^4})\cos(2\pi\frac{z}{h})]\frac{\partial h}{\partial x}$$

$$\frac{\partial^2 w}{\partial y \partial z} = \frac{A}{2\pi}\sin(2y)(4\pi^2\frac{z}{h^2})\sin(2\pi\frac{z}{h})$$

$$+\frac{A}{2\pi}[\sin^2(x)\cos^2(y) - \cos^2(x)\sin^2(y)][(8\pi^2\frac{z}{h^3})\sin(2\pi\frac{z}{h}) + (8\pi^3\frac{z^2}{h^4})\cos(2\pi\frac{z}{h})]\frac{\partial h}{\partial y}$$

Source term development:

Source terms for the momentum equations

$$F_u = u\frac{\partial u}{\partial x} + v\frac{\partial u}{\partial y} + w\frac{\partial u}{\partial z} - \nu_t(\frac{\partial^2 u}{\partial x^2} + \frac{\partial^2 u}{\partial y^2} + \frac{\partial^2 u}{\partial z^2}) - \nu_t(\frac{\partial^2 u}{\partial x^2} + \frac{\partial^2 v}{\partial y \partial x} + \frac{\partial^2 w}{\partial z \partial x})$$

$$F_v = u\frac{\partial v}{\partial x} + v\frac{\partial v}{\partial y} + w\frac{\partial v}{\partial z} - \nu_t(\frac{\partial^2 v}{\partial x^2} + \frac{\partial^2 v}{\partial y^2} + \frac{\partial^2 v}{\partial z^2}) - \nu_t(\frac{\partial^2 u}{\partial x \partial y} + \frac{\partial^2 v}{\partial y^2} + \frac{\partial^2 w}{\partial z \partial y})$$

$$F_w = u\frac{\partial w}{\partial x} + v\frac{\partial w}{\partial y} + w\frac{\partial w}{\partial z} - \nu_t(\frac{\partial^2 w}{\partial x^2} + \frac{\partial^2 w}{\partial y^2} + \frac{\partial^2 w}{\partial z^2}) - \nu_t(\frac{\partial^2 u}{\partial x \partial z} + \frac{\partial^2 v}{\partial y \partial z} + \frac{\partial^2 w}{\partial z^2})$$

$$F_h = 0$$

B.3 FORTRAN CODES

Fortran Codes for calculating the forcing terms needed for conducting the mathematical verification of non-linear models using Method of Manufactured Solution can be downloaded from the website http://www.ncche.olemiss.edu/publishing/ASCE_VV/Appendix_B/ .

Index